Mechanochemical Organic Synthesis

Davor Margetić
Vjekoslav Štrukil
Ruđer Bošković Institute

ELSEVIER

AMSTERDAM • BOSTON • HEIDELBERG • LONDON • NEW YORK • OXFORD
PARIS • SAN DIEGO • SAN FRANCISCO • SINGAPORE • SYDNEY • TOKYO

Elsevier
Radarweg 29, PO Box 211, 1000 AE Amsterdam, Netherlands
The Boulevard, Langford Lane, Kidlington, Oxford OX5 1GB, UK
50 Hampshire Street, 5th Floor, Cambridge, MA 02139, USA

British Library Cataloguing-in-Publication Data
A catalogue record for this book is available from the British Library

Library of Congress Cataloging-in-Publication Data
A catalog record for this book is available from the Library of Congress

ISBN: 978-0-12-802184-2

Publisher: Cathleen Sether
Acquisition Editor: Katey Birtcher
Editorial Project Manager: Jill Cetel
Production Project Manager: Anitha Sivaraj
Designer: Maria Ines Cruz

Typeset by TNQ Books and Journals

Contents

Preface

Responsible management of natural resources and the diminution of environmental impact of chemical industry in many ways rely on organic chemists, and development of novel ecofriendly organic methods and processes is gaining in importance. During 2010s mechanosynthesis by automated ball milling has become more widely applied among researchers. Since the potential of the method for greener synthesis has been recognized, more accounts appear and the number of research papers published in scholarly journals is steadily increasing. Thus, there was a need to collect the information scattered in the literature and organize an up-to-date overview.

In this book organic chemists could find the advantages and shortcomings of the automated ball-milling technique and its applicability. In addition to benefits of excluding solvents from reaction, by this way increasing the concentration and kinetics, as well as better reproducibility than mortar and pestle manual grinding, exciting opportunities lay before organic chemists to discover new products and reaction pathways which differ from conventional synthesis, thus complementing the conventional reaction conditions. Synthetic chemists will surely find it appealing to carry out reactions in a new and unique way without solvent effects, and in most protocols without special precaution against moisture. We strongly believe that this field of chemistry will be further developed and find its application to variety of organic transformations.

Application of environment friendly methods in organic synthesis is an ongoing program in our research group focusing on application of extremely high pressures and microwave-assisted synthesis, and this book is a part of our activity in the area of mechanochemistry.

This book is intended for a broad audience, for the most part to practical organic chemists working on development of new "greener" techniques for organic synthesis, describing the advantages, possibilities, and downturns of the mechanosynthetic method. Studies of reaction mechanisms in solid state have importance to physical organic chemists, whereas industrial chemists interested in green technologies could also find a wealth of information. Academic educators and advanced organic chemistry students will find relevant chapter devoted to undergraduate organic chemistry laboratory, which was initiated by our involvement in postgraduate-level course entitled "Special techniques in organic synthesis."

The first chapter gives a review on practical aspects of mechanochemical organic synthesis, which is followed by an in-depth literature overview of examples divided into chapters, according to the reaction types; however, in some cases additional examples could be found in the other chapters, particularly for cycloadditions and multicomponent or cascade reactions. These chapters are followed by chapters devoted to applications of ball milling to fullerene and supramolecular chemistry. Final chapter gives examples of ball-milling reactions which could be integrated in undergraduate organic chemistry courses. Throughout the book, standard reaction workup involving product extraction with solvent and purification is not specified, but rather examples given which differ from this standard procedure by complete avoidance of the use of organic solvents. Wherever data were available, a comparison with conventional synthesis is given. A great care has been taken to provide enough experimental details, so that synthetic procedures can be applied on the new reactions without the need to consult the original research papers.

Technical help of members of Laboratory for Physical Organic Chemistry is acknowledged. Elsevier editorial staff is thanked for the opportunity to publish this work and their continuous assistance during the project. We would like to thank Mr. Milan Trenc for the book cover illustration. Croatian Science Foundation is thanked for financial support (grant no. 9310). We are also very grateful to colleagues who introduced mechanochemistry to us, Professor Koichi Komatsu (Kyoto, D.M.) and Professor Tomislav Friščić (McGill, V.Š.). Great thanks go to our families, especially to our children Dominik, Karlo Ken, and Marina for their patience and understanding during the preparation of this manuscript.

Davor Margetić
Vjekoslav Štrukil
Ruđer Bošković Institute
Zagreb, December 2015

List of Abbreviations

API Active pharmaceutical ingredients
ATR-IR spectroscopy Attenuated total reflectance infrared spectroscopy
Bn Benzyl
Boc tert-Butyloxycarbonyl
BQ Benzoquinone
BTC 1,3,5-Benzenetricarboxylate
n-Bu n-Butyl
CAN Cerium ammonium nitrate
CDC Cross-dehydrogenative coupling
CDI Carbonyl diimidazole
CPD Cyclopentadiene
CP-MAS Cross-polarization magic angle spinning
DABCO 1,4-Diazabicyclo[2.2.2]octane
DBU 1,8-Diazabicyclo[5.4.0]undec-7-ene
DCB 1,3-Dichlorobenzene
DCC N,N′-Dicyclohexylcarbodiimide
DDQ 2,3-Dichloro-5,6-dicyano-p-benzoquinone
DIC N,N′-Diisopropylcarbodiimide
DIEA Diisopropylethylamine
DKR Dynamic kinetic resolution
DMAD Dimethylacetylene dicarboxylate
DMAP 4-(Dimethylamino)pyridine
DMF N,N-Dimethylformamide
DMSO Dimethylsulfoxide
DMT 4,4′-Dimethoxytrityl
DSC N,N′-Disuccinimidyl carbonate
EDC·HCl N-Ethyl-N′-(3-dimethylaminopropyl)carbodiimide hydrochloride
EDTA Ethylenediaminetetraacetic acid
Et Ethyl
EtOAc Ethyl acetate
Fmoc 9-Fluorenylmethyloxycarbonyl
FRP Functional resin particles
Gn Graphene
GO Graphene oxide
n-Hex n-Hexyl
HMDS Hexamethyldisilazane
HPLC High-performance liquid chromatography
HRMS High-resolution mass spectrometry
HSVM High-speed vibration mill
IBA 2-Iodosobenzoic acid
IBX 2-Iodoxybenzoic acid
ICP-MS Inductively coupled plasma mass spectrometry
ILAG Ion- and liquid-assisted grinding
INA Isonicotinic acid

IR spectrocopy Infrared spectroscopy
KIE Kinetic isotope effect
LAG Liquid-assisted grinding
MA Maleic anhydride
MALDI Matrix-assisted laser desorption/ionization
MAPP Maleated poly(propylene)
MAS Magic angle spinning
mCPBA m-Chloroperbenzoic acid
Me Methyl
MeCN Acetonitrile
MI Maleic imide
MOF Metal organic framework
MSDS Material safety data sheet
MW Microwave
MWCNT Multiwalled carbon nanotube
NBS N-Bromosuccinimide
NCS N-Chlorosuccinimide
NHS ester N-Hydroxysuccinimidyl ester
NIS N-Iodosuccinimide
NMR Nuclear magnetic resonance
PCBs Polychlorinated biphenyls
PCP Pentachlorophenol
PEG Polyethylene glycol
Ph Phenyl
PM Planetary ball mill
PMMA Poly(methyl)methacrylate
POLAG Polymer-assisted grinding
iPr Isopropyl
PTC Phase transfer catalyst
PTFE Polytetrafluoroethylene
PXRD analysis Powder X-ray diffraction analysis
RCM Ring closing metathesis
rpm Rounds per minute
SDG Solvent-drop grinding
SEM Scanning electron microscopy
ssNMR spectrocopy Solid-state nuclear magnetic resonance spectroscopy
SWNH Single walled nanohorn
SWNT Single walled nanotube
TBAB Tetrabutylammonium bromide
TBAI Tetrabutylammonium iodide
TBDMS t-Butyldimethylsilyl
TBDPS t-Butyldiphenylsilyl
TLC Thin layer chromatography
TMS Trimethylsilyl
TOF Turnover frequencies
Tr Triphenylmethyl (trityl)
TsOH p-Toluenesulfonic acid
UNCA Urethane-protected α-amino acid N-carboxyanhydride
Z Benzyloxycarbonyl

CHAPTER

1 Practical Considerations in Mechanochemical Organic Synthesis

CHAPTER OUTLINE

1.1 A HISTORICAL PERSPECTIVE

From the beginnings of mankind, mechanical treatment and processing of grains and seeds emerged as the first engineering technology in food preparation [1]. Later on, treating raw materials like minerals and ores in the same way, allowed the production of finely powdered paints and medicines. Prototypical mortars differing in material, shape, size, and decorations, found at many archaeological sites throughout the world, testify to early developments of tools that were intended to make use of mechanical force exerted by a hand. In those primitive grindstones, a stone ball was devised as a substitute to what later would become a pestle. Following the advancement of technology, simple grindstones eventually evolved into a variety of stylized mortars. Beautiful pieces named *molcajete* made of basalt stone, typical for pre-Hispanic Mesoamerican cultures, represent one such addition. Dating back to several thousand years, the Aztec and Maya people extensively used them for crushing and grinding spices and for preparation of salsas and guacamole. Even today, molcajete is a must-have kitchen utensil in traditional Mexican cuisine. Another example is a *metate* or mealing stone, traditionally used

Mechanochemical Organic Synthesis. http://dx.doi.org/10.1016/B978-0-12-802184-2.00001-7

by Mesoamerican cultures to grind lime-treated maize and to prepare food. However, mealing stones are not tied with Mexico only since variations are found all over the world (Fig. 1.1).

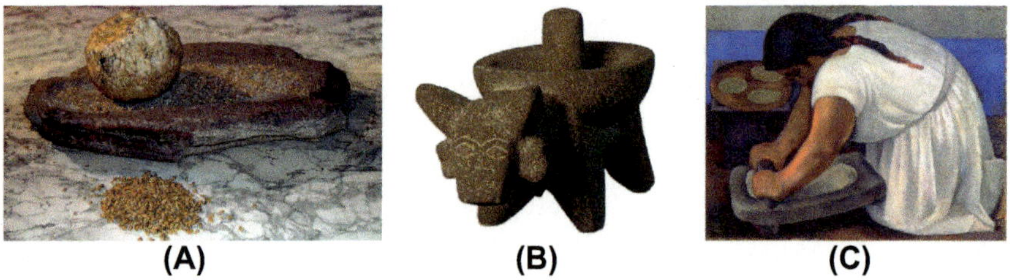

FIGURE 1.1

(A) A primitive grindstone, (B) Molcajete, a traditional Mexican tool for crushing and grinding spices, (C) Diego Rivera's *La Molendera* (The Woman Grinder, 1924) showing traditional use of metate in making tortillas.

In his booklet *De Lapidibus* (On Stones), Theophrastus of Eresus described what is believed to be the first surviving testimony on relationship between grinding, as a means to introduce a mechanical force, and a chemical change as a consequence thereof [2]. It was a reduction of native cinnabar to the liquid mercury metal in the presence of vinegar, carried out in a copper mortar with a copper pestle:

$$HgS(s) + Cu(s) \rightarrow Hg(l) + CuS(s)$$

In this, according to today's standards, mechanochemical reaction, vinegar was used to eliminate side-reactions that often take place during milling in air. Interestingly, it would take another 2000 years for scientists to rediscover the effects of added liquid on the course of mechanochemically promoted reactions in what has come to be known as solvent-drop grinding (SDG), kneading, or liquid-assisted grinding (LAG) [3].

By entering the middle ages, much of the knowledge collected during the ancient times was lost or forgotten. However, mortar and pestle continued to be the primary tool for mechanical treatment of substances, and an alchemist laboratory could not be imagined without this grinding equipment, as nicely illustrated in Jan Van der Straet's painting *The Alchemist's Studio* (Fig. 1.2). While the medieval awakened interest in unraveling the mysteries of nature and new revolutionary ideas of renaissance have greatly contributed to the development of all fields of science, a systematic approach in studying mechanochemical reactions was left out until the second half of the 19th century.

Although Michael Faraday performed reduction of silver chloride with zinc, copper, tin, and iron by manual grinding demonstrating that chemical changes induced by means of mechanical agitation was a common knowledge [4], it was only with Walthère Spring and Matthew Carey Lea stepping up on the stage when mechanochemical phenomena started to be investigated in a systematic fashion [5]. While Spring, led by an aspiration to grasp the formation of minerals inside the earth's crust, focused his research on the effect of high pressure on phase

FIGURE 1.2

Jan Van der Straet's "*The Alchemist's Studio*" (1571).

transformations and chemical reactions, M. C. Lea explored the behavior of silver, gold, mercury, and platinum halides (Cl, Br, and I) under the conditions of static pressure and shearing forces during manual agitation. He found that the potential of large static pressure to bring about a chemical reaction is much less pronounced compared to weak shearing forces exerted on the system as a result of manual grinding. Notably, the most important and cited result of Lea's research on mechanochemistry, which earned him the title of "father of mechanochemistry," is the observation that mechanical grinding leads to effects different from those induced by heat in thermochemical reactions. The two examples that illustrate this were mechanochemical decomposition of silver and mercury chlorides to silver, liquid mercury, and chlorine gas, as opposed to melting (AgCl) and sublimation ($HgCl_2$) without decomposition upon exposure to heat:

$$2AgCl(s) \rightarrow 2Ag(s) + Cl_2(g)$$

$$HgCl_2(s) \rightarrow Hg(l) + Cl_2(g)$$

As far as the mechanochemical synthesis involving organic molecules goes, the earliest documented example dates back in 1893 in a paper published by Ling and

Baker [6]. In their contribution, an equimolar mixture of metadichloroquinol **1** and metadichloroquinone **2** was ground in a mortar to yield tetrachloroquinhydrone co-crystal **3** (Scheme 1.1A). It took nearly 100 years for chemists to again engage in mechanochemically promoted organic reactions. The work published by Paul in the early 1980s dealt with syntheses of quinhydrone charge–transfer complexes in the solid state by means of grinding with a mortar and pestle [7]. Specifically, mechanochemical co-crystallization of benzoquinone **4** with several hydroquinones, for example, naphthalene-1,4-diol **5** and 2-methylhydroquinone **7** resulted in charge–transfer molecular complexes **6** and **8** (Scheme 1.1B and C) that are difficult to synthesize in solution due to redox self-isomerization.

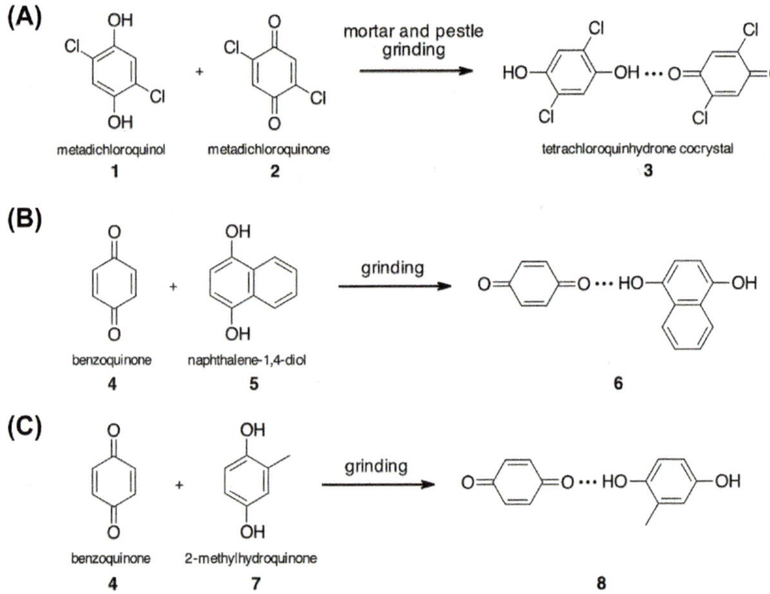

SCHEME 1.1

Solid-state synthesis of charge–transfer complexes.

Toda et al. succeeded in preparing several co-crystals of different compounds that normally undergo redox isomerization, but did not display charge-transfer interaction by agitation in a mechanical test tube shaker [8]. During the late 1980s, Etter's group embarked on a research project aimed at understanding hydrogen bonding preferences and the viability of solution evaporation and solid-state grinding as synthetic methods to prepare co-crystals [9]. Their research has shown that solvent-free (or neat grinding) approach for promoting co-crystallization is a simple and rapid alternative to solution-based methods that enables preparation of a wide range of hydrogen-bonded co-crystals. The work by Etter, and also Caira, published in 1995 marked the beginning of an extensive application of manual neat grinding and ball milling as preparative methods for the synthesis of co-crystals of active pharmaceutical ingredients (API), and offered new opportunities for the solid-state chemistry in pharmaceutical science [10].

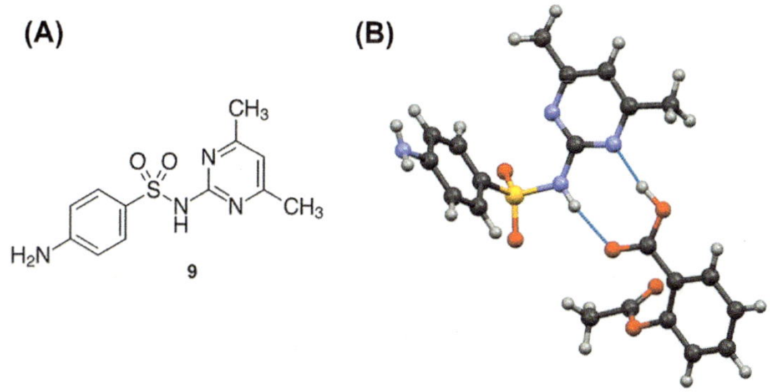

SCHEME 1.2

(A) Structure of the drug sulfadimidine **9** and (B) hydrogen-bonded molecular complex between sulfadimidine and acetylsalicylic acid (aspirin) in the co-crystal.

Caira demonstrated that mechanochemical approach to crystallize the drug sulfadimidine **9** with a series of carboxylic acid co-crystal formers, including benzoic, anthranilic, salicylic, and acetylsalicylic acids (aspirin), resulted in an efficient synthesis of the corresponding API co-crystals which were shown to be identical to those prepared by solution crystallization (Scheme 1.2). Besides providing co-crystals that have already been known from solution chemistry, Kuroda et al. demonstrated that mechanochemical approach by neat grinding can afford crystal forms that could not be obtained by other preparative methods, such as growing crystals from solution or melt. In their work, a mixture of racemic *bis*-β-naphthol **10** and benzoquinone **4** was ground with a mortar and pestle to yield **11**, a two-component co-crystal of form I with 1:1.5 stoichiometry [11]. Solution crystallization resulted in a 1:1 form II while crystallization from melt provided form III. While the structure of form III is still unknown, single crystal X-ray analysis of form I revealed its crystal structure composed of finite hydrogen-bonded assemblies unlike form II where naphthol and benzoquinone molecules are linked through hydrogen bonds into infinite tapes. The progress of the mechanochemical synthesis of form I can easily be monitored by visual inspection since the charge-transfer from the donor molecule to the acceptor molecule results in an intensive color change (Fig. 1.3).

Despite success of solvent-free mechanochemical grinding approach to making various forms of co-crystals, and even ones that are unobtainable by conventional solution-based methods, there are cases where neat grinding fails to afford a co-crystal product. A groundbreaking discovery that has revolutionized mechanochemistry came in 2002 with the work of Jones et al. [12]. Investigation of the potential of grinding using a ball mill to synthesize 1,3,5-cyclohexane tricarboxylic acid (**12**) co-crystals with hexamethyltetramine (**13**), 4,4′-bipyridine (**14**) and 4,7-phenanthroline (**15**) led the authors to the realization that the addition of a small amount of solvent (typically one to two drops) to the reactant mixture significantly enhances the kinetics of co-crystal formation (Scheme 1.3). While the 1:1 co-crystal between **12** and **13** is readily formed upon neat grinding for 20 min, the corresponding (**12**)(**14**) and (**12**)(**15**)$_2$ co-crystals form only partially after 1 h of neat grinding as evidenced by powder X-ray diffraction (PXRD) analysis.

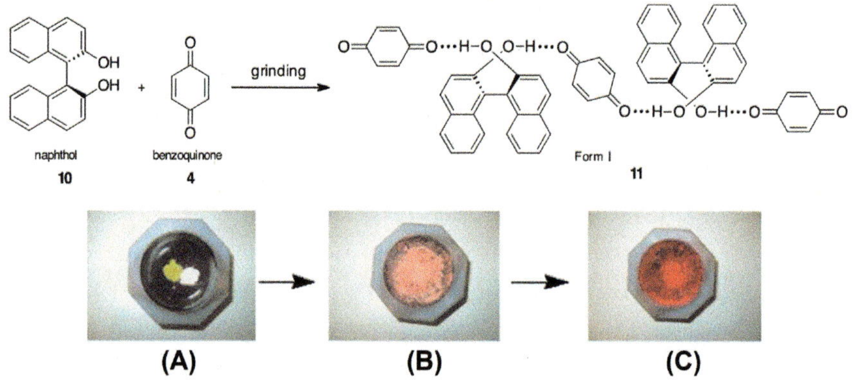

FIGURE 1.3

Color change of the co-ground crystals of racemic *bis*-β-naphthol **10** (white) and benzoquinone **4** (yellow); (A) before grinding, (B) during the grinding, and (C) the final product.

Reproduced from Kuroda R, Imai Y, Tajima N. Generation of a co-crystal phase with novel coloristic properties via *solid state grinding procedures. Chem Commun 2002;37:2848–9 with permission from the Royal Society of Chemistry.*

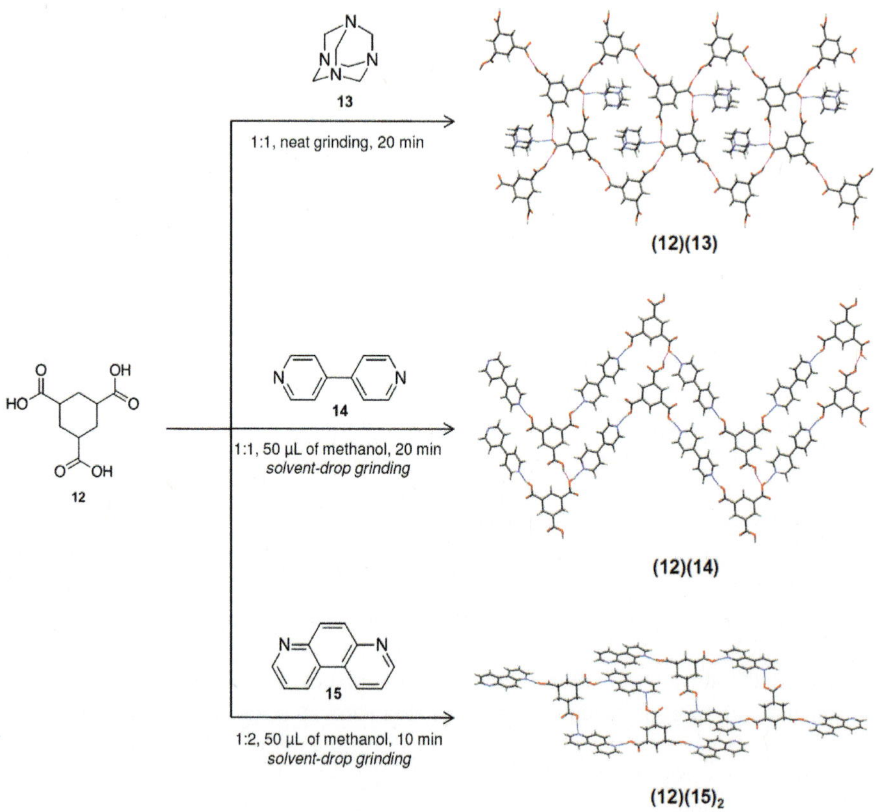

SCHEME 1.3

Co-crystallization of 1,3,5-cyclohexane tricarboxylic acid via solvent-drop grinding.

However, when 50 μL (2 drops) of methanol was added to the starting mixture, both **(12)(14)** and **(12)(15)**$_2$ co-crystals quantitatively (based on PXRD) formed after only 20 and 10 min of milling, respectively. The same accelerating effect was observed with other solvents like water, acetonitrile, and ethyl acetate. Interestingly, cyclohexane was found not to have the effect on the co-crystallization rate, with no reaction taking place. The authors offered an explanation of the observed effect and linked it to the solubility of components in a particular solvent. As a rule of thumb, they propose that the choice of the solvent should be based on at least partial solubility of the original components.

In a follow-up paper published in 2002, Trask et al. explored the utility of SDG to selectively prepare the two polymorphs of an API co-crystal [13] consisting of caffeine (**16**) and glutaric acid (**17**). Neat grinding or SDG with nonpolar solvents like n-hexane, cyclohexane, and heptane afforded predominantly monoclinic form I, structure of which is characterized by overlaying 2-D sheets containing hydrogen-bonded ribbons. In such an arrangement, a nonpolar cleavage plane (200) between the stacks of ribbons along the *b* axis could be identified.

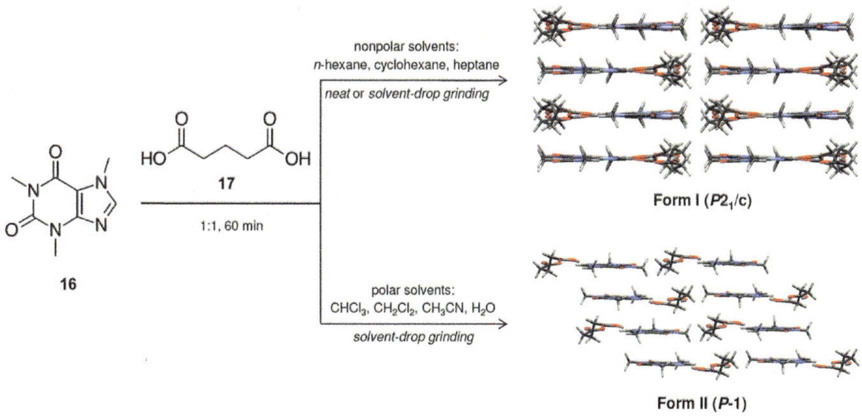

SCHEME 1.4

Polymorph control by solvent-drop grinding during co-crystallization of caffeine with glutaric acid.

Contrary to the tertiary structure of form I, in the case of triclinic form II, produced exclusively by SDG using polar solvents such as chloroform, dichloromethane, acetonitrile, and water, the 2-D sheets with hydrogen-bonded ribbons are staggered so that the nonpolar cleavage plane is not observed. To account for the observed polymorph selectivity, the authors propose that nonpolar solvents could stabilize this plane during grinding-induced cleavage (Scheme 1.4).

In the following years, the interest in grinding as a reliable method to prepare co-crystals continued to grow, resulting in an accumulation of experimental observations on the effect of solvent in SDG. Apart from influencing the kinetics, the reactions conducted in the presence of a few drops of solvent were found to proceed in quantitative yields affording highly crystalline products as opposed to neat grinding where partial amorphization takes place, as evidenced by PXRD and terahertz time-domain spectroscopy analyses [14]. Studies have also demonstrated the superiority of SDG in the control over polymorphic outcome of co-crystallization [13] as well as the control of stoichiometric composition in products by simply adjusting the ratio of starting components [15]. Eventually, the term "solvent-drop grinding" or SDG was replaced with "liquid-assisted grinding" or LAG to more accurately describe to role of the added liquid during such grinding experiments [16]. Although SDG implies the connection between the accelerating effect and solubility of the components being ground, later studies on model co-crystal systems have shown that the success of co-crystallization by adding a small catalytic amount of a solvent/liquid is not entirely related to solubility effects, but rather that it affects the kinetics, that is, the rate of co-crystallization process [17]. By introducing the term "liquid-assisted grinding," a misinterpretation on the role of the added liquid tends to be avoided. Although the exact mechanism of LAG is still unknown, the proposed term targets the general role of a solvent in mechanochemically aided reactions and encompasses liquid-phase effects that stretch beyond solubility.

The next step in utilizing this novel mechanochemical reactivity was to advance from the construction of supramolecular assemblies via hydrogen or halogen bonds to covalent bond formation. In organic synthesis, majority of reactions carried out in the solid state have been performed by neat (solvent-fee) grinding [18]. However, the introduction of LAG as a means to enhance the kinetics of grinding processes provided synthetic chemists with a powerful tool to access organic molecules. In 2011, Aakeröy et al. demonstrated the use of SDG for an efficient functionalization of cavitands [19]. In their work, the last step in decorating the cavitand framework **18** was the transformation of the aldehyde group to an oxime in **19** by a condensation reaction with excess of hydroxylammonium chloride/sodium hydroxide (6 equiv.), carried out with a mortar and pestle. A mixture of ethanol and dichloromethane was used as the solvent for SDG. The advantages of SDG in this case were recognized in terms of short reaction time (5 min), high yield, clean reaction without by-product formation, and reduction in the amount of solvent normally used in the solution synthesis (Scheme 1.5).

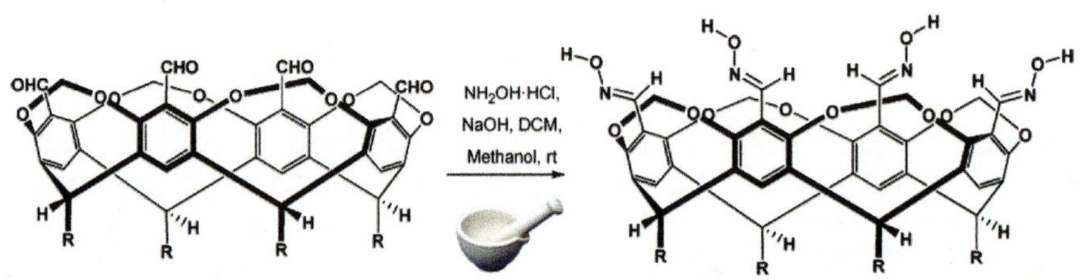

18 **19**

SCHEME 1.5

Functionalization of cavitand structure by solvent-drop grinding.

Adapted with permission from Aakeröy CB, Chopade PD. Oxime decorated cavitands functionalized through solvent-assisted grinding. Org Lett 2011;13:1–3. copyright 2011 American Chemical Society.

In another paper, the mechanochemical methods of neat grinding and LAG were applied for a quantitative synthesis of mono- and bis-thiourea derivatives, typically used as organocatalysts and anion sensors, by a click-type reaction between aliphatic and aromatic diamines and aromatic isothiocyanates [20]. Different mechanochemical reaction conditions such as milling time, ball size, and neat grinding or LAG were optimized in the synthesis of mono-thiourea **25** and bis-thiourea **26** derived from *o*-phenylenediamine **24** (Scheme 1.6). The stoichiometry of the product was simply controlled by the amine:isothiocyanate ratio, that is, mono-thiourea **25** resulted from the 1:1 reaction, while bis-products were obtained in 1:2 reactions, respectively. LAG was shown to be an effective way of achieving quantitative conversions in short reaction times, for example, in the synthesis of the binaphthyl chiral catalyst (*R*)-**23** where LAG using methanol enabled almost complete conversion of the intermediate mono-thiourea to the bis-thiourea product. Apart from providing the amine–isothiocyanate coupling products in quantitative yields, LAG reactions also led to crystalline materials in contrast to neat grinding experiments where a high degree of amorphization was observed, based on broad features in the PXRD patterns. The authors also demonstrated that the mechanochemical syntheses of mono- and bis-thioureas could be scaled up to 1 g without noticeable loss of product yield. Mechanochemical strategy employed in this work, in comparison with the conventional solution synthesis, has proven superior with regard to control of product stoichiometry in organic reactions as well, similar to previously recognized stoichiometric control during mechanosynthesis of hydrogen- and halogen-bonded co-crystals [15] and coordination polymers [21].

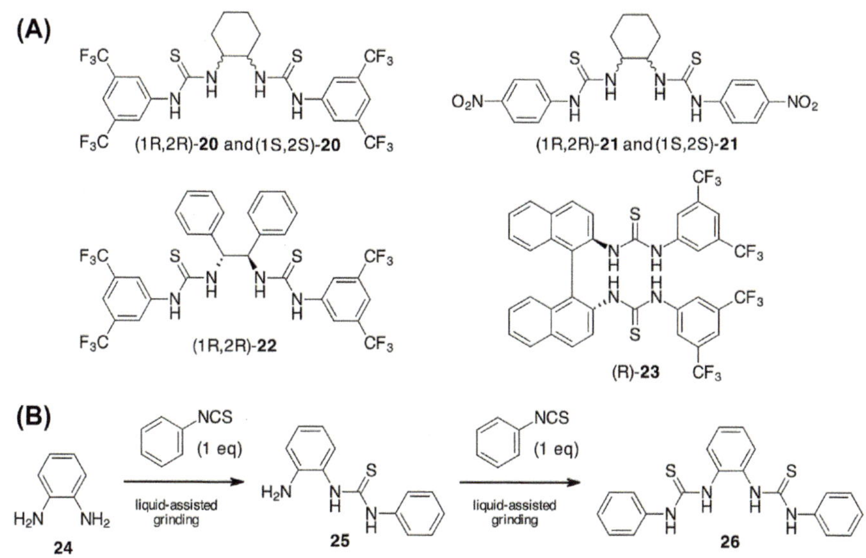

SCHEME 1.6

(A) Bis-thioureas prepared by mechanochemical click-type reaction between amines and aromatic isothiocyanates. (B) Stoichiometric control in LAG mechanosynthesis of sterically congested mono- and bis-thioureas from *o*-phenylenediamine.

Besides LAG, two more mechanochemical methods that influence the reactivity in the solid state have been discovered. These are the so-called "ion- and liquid-assisted grinding" or ILAG and "polymer-assisted grinding" or POLAG. The discovery of ILAG came with an extensive research on the application of mechanochemistry in the field of metal–organic framework (MOF) synthesis. In 2010, Friščić et al. demonstrated that the addition of a small catalytic quantity of an inorganic salt had a profound effect on the rate of MOF synthesis, as well as on controlling the polymorphic outcome of such a synthesis [22]. In this work, the authors conducted a thorough investigation of several parameters that influenced MOF synthesis utilizing powder X-ray diffraction and infrared and solid-state NMR spectroscopy (see Section 1.4 for more information). As the model compound, a pillared MOF $[Zn_2(ta)_2(dabco)]$ (**27**) that crystallizes in either a tetragonal (**27a**) or a hexagonal (**27b**) polymorphic form, was selected. Relying on the previous report on the mechanochemical transformation of zinc oxide to a pillared MOF, neat grinding a 1:1:0.5 mixture of ZnO, terephthalic acid (Hta) and 1,4-diazabicyclo[2.2.2]octane (dabco) for 60 min afforded only the dabco-Hta salt. When the experiment was repeated under LAG conditions using DMF as the grinding liquid, partial reaction resulting in the formation of the hexagonal form **27b** was observed (Fig. 1.4A–C). Interestingly, the solvothermal reaction in DMF, using zinc nitrate as the source of zinc ions, produces exclusively the tetragonal form **27a**. This observation led the authors to attempt the ILAG mechanosynthesis by adding a small amount of sodium nitrate to the reaction mixture. Indeed, the PXRD analysis after only 20 min revealed complete disappearance of ZnO reflections yielding a mixture of **27a** and **27b** polymorphs. Other nitrate salts (KNO_3, $RbNO_3$, $CsNO_3$, and NH_4NO_3) also induced the formation of **27a**, which was the only product after prolonged grinding. In contrast to nitrates, sulfate salts (Na_2SO_4, K_2SO_4, and $(NH_4)_2SO_4$) selectively yielded the hexagonal form **27b** after 30 min. An anion-templating effect was proposed as an explanation of the observed reactivity. Further solid-state direct polarization (DP) magic-angle spinning (MAS) ^{15}N and ^{23}Na NMR studies revealed salt inclusion in the structure of MOF, evidenced by two distinct signals (free and MOF-included salt) in the corresponding spectra. While both cation and anion undergo inclusion, sulfate anions have been found to possess high templating efficiency with low inclusion levels (Fig. 1.4D–E).

The most recent contribution to enhanced mechanochemical reactivity was published by Jones et al. in 2015 [23]. While co-crystal synthesis by mechanochemical approach has proven superior to other methods, the inability of neat grinding to afford co-crystal products in some cases, as well as the apparent disadvantage of LAG in terms of unintentional formation of solvates, motivated the authors to further explore the possibilities of increasing the rate of mechanochemical co-crystallization by a polymeric macromolecular catalyst. Three model co-crystal systems were chosen to test the performance of polyethylene glycol (PEG) catalysts under ball-milling conditions. PEGs with different molecular weight from 200 to 10,000 were used (three liquids and three crystalline solids). Investigation of the co-crystallization of caffeine and citric acid showed that the efficiency of polymer-assisted reaction was diminished at very low (1 and 5 wt%) and very high (30 and 60 wt%) polymer content, suggesting that there is an optimum loading of the polymer in these reactions. Although all samples contained some starting material, prolonged milling with

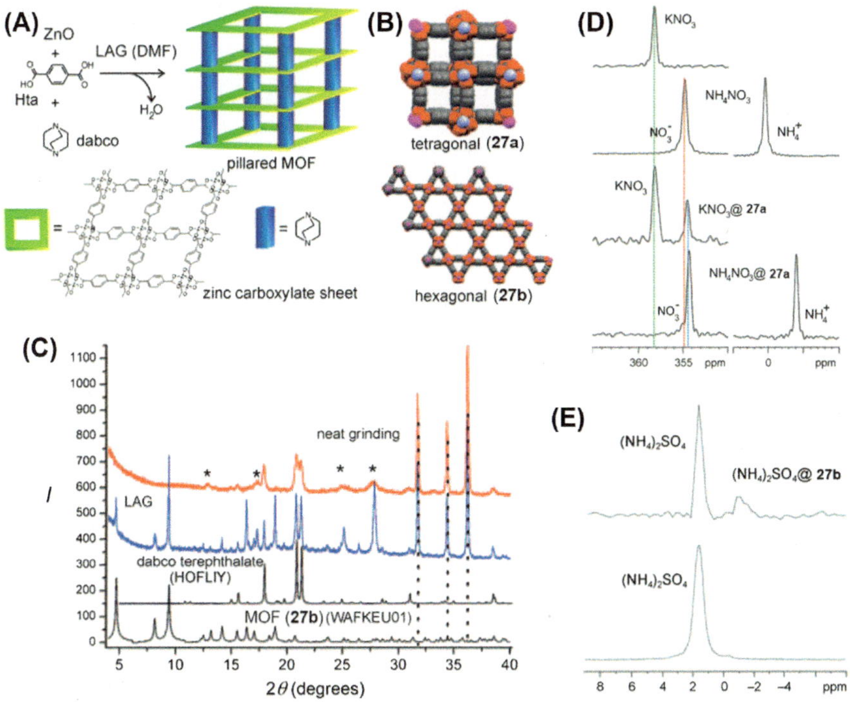

FIGURE 1.4

(A) Synthesis of MOF **27**, (B) tetragonal (**27a**) and hexagonal (**27b**) polymorphs, (C) PXRD patterns (top to bottom) of 60-min neat and LAG reactions and simulated patterns of dabco-Hta salt and **27b** form, (D) DP-MAS ^{15}N ssNMR spectra of (top to bottom): pure $K^{15}NO_3$, pure $^{15}NH_4{}^{15}NO_3$, **27a** prepared using $K^{15}NO_3$, and **27a** prepared using $^{15}NH_4{}^{15}NO_3$, and (E) DP-MAS ^{15}N ssNMR spectra of a sample of **27b** prepared by ILAG with $(^{15}NH_4)_2SO_4$ (top, w=0.7%) and pure $(^{15}NH_4)_2SO_4$ (bottom).

10 wt% of PEG led to a significant reduction of the residual reactants in the mixture. Also, it was established that the physical state of the polymer (liquid vs. solid) as well as the polymer chain length (PEG 200 to PEG 10,000) had no effect on the co-crystallization rate under milling conditions. However, the kinetic effects on the solid-state co-crystallization resulting from the addition of the polymeric catalyst were obvious and were rationalized through assisting the co-crystal nucleation and growth. For that reason, the authors have proposed to refer to this novel mechanochemical reactivity as POLAG (Fig. 1.5).

The other two co-crystal systems showed similar behavior in the presence of PEG catalysts. While in the phenazine–mesaconic acid co-crystallization 5 wt% of PEG was found as the optimal loading, an interesting observation during POLAG

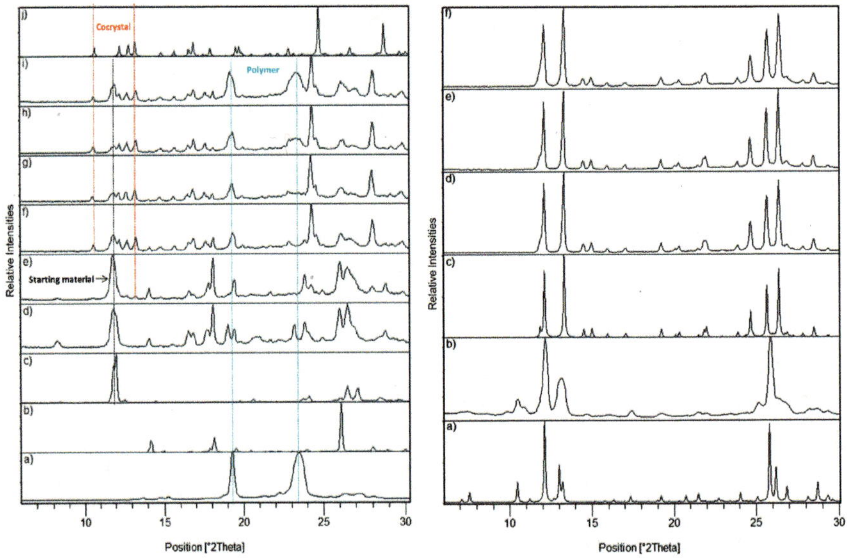

FIGURE 1.5

Left: PXRD analysis of the effect of PEG loading on the co-crystallization of caffeine with citric acid. From bottom to top: (A) pure PEG 10,000, (B) pure citric acid, (C) pure caffeine, (D) caffeine/citric acid physical mixture neatly ground for 20 min, (E–I) the mixture ground for 20 min with the addition of 1, 5, 10, 30, and 60 wt% of PEG 10,000, respectively, (J) simulated pattern of caffeine–citric acid co-crystal. Right: PXRD patterns of: (A) calculated caffeine–anthranilic acid, form I, (B) caffeine–anthranilic acid, form I prepared by neat grinding, (C) calculated caffeine–anthranilic acid, form II, caffeine–anthranilic acid prepared by (D) LAG with acetonitrile, and POLAG using 1 wt% of (E) PEG 200 and (F) PEG 10,000.

synthesis of caffeine–anthranilic acid co-crystal was the control of the polymorphic outcome of the reaction. Neat grinding led to the thermodynamically stable form I and POLAG with 1 wt% of PEG 200 or 10,000 gave the metastable form II, otherwise produced only by LAG using acetonitrile.

1.2 MODERN LABORATORY INSTRUMENTATION FOR MECHANOSYNTHESIS

In the last 15 years mechanochemistry has become increasingly popular among synthetic chemists as an environmentally friendly and highly efficient means to conduct chemical reactions, especially in the fields of crystal engineering and organic synthesis.

A large proportion of published papers still relies on the traditional way of performing mechanochemical reactions by using mortar and pestle. While this piece of equipment is regarded almost as a trademark of mechanochemistry, it suffers from several disadvantages. The energy input in these reactions is a result of manual work and depending on the person's strength, it is likely that the energy introduced into the system is not constant. This usually becomes a problem if longer grinding times are required which consequently affects proper mixing and energy transfer leading to irreproducible results. Also, typical laboratory mortars have a limited processing capacity and are not intended for large-scale synthesis. Finally, there is a safety problem as the person conducting an experiment becomes exposed to numerous hazards. However, application of automated ball mills circumvents these issues and enables a precise control over the energy input by adjusting milling frequency and time, whereas milling accessories (jars and balls) influence the grinding process by selecting the milling material, number, and size of balls and other technological parameters.

The development of commercial ball mills started in the second half of the 19th century when the first tumbler mill revolutionized cement clinker processing. Since the energy of the impacts in these tumbler mills was low, further advancement in designing more powerful milling devices led to inventions of stirred-ball attritors in the 1920s and vibration ball mills in the 1930s. In 1961 Fritsch company released the first model of a planetary ball mill, while at the same time SPEX SamplePrep launched their most famous SPEX 8000 mixer mill [5]. The majority of organic mechanochemical reactions presented in this book have been typically carried out on a laboratory scale in either a planetary ball mill or a mixer mill (also referred to as a shaker mill) that are robust, energetic, and easy to use. For that reason, the following sections will focus only on these two types of ball mills providing a brief overview of their main characteristics [24].

1.2.1 PLANETARY BALL MILLS

In a technical sense, planetary ball mills developed from tumbler mills. The idea behind the design of the first planetary mill was to replace the gravitational force as the source of impact in tumbler mills with the centrifugal force by vertically positioning the milling vessel. The vessels (Fig. 1.6A and B) are arranged on a rotating support disc (v_2) while spinning in the opposite direction about their own axes (v_1) at the same time. Such movement of the milling vessels is reminiscent of the planetary motion around the central star, hence the name of this type of ball mill. The content of milling jars (balls and the material being ground) in the planetary ball mill is accelerated by the centrifugal force and slides along the inside walls of the jar, what is often referred to as the friction mode. Because the velocities v_1 and v_2 have the opposite directions, the centrifugal forces exerted on the jar content will also alternate, causing the balls and the ground material to lift off from the jar walls (the so-called release point). At that moment, the content leaves the centrifugal trajectory and travels freely through the jar until it collides with the opposite wall—this is called the impact mode. The friction and impact modes therefore constitute the main mechanisms by which materials are processed inside the planetary ball mill. With respect to technical characteristics, a typical planetary ball mill has one, two,

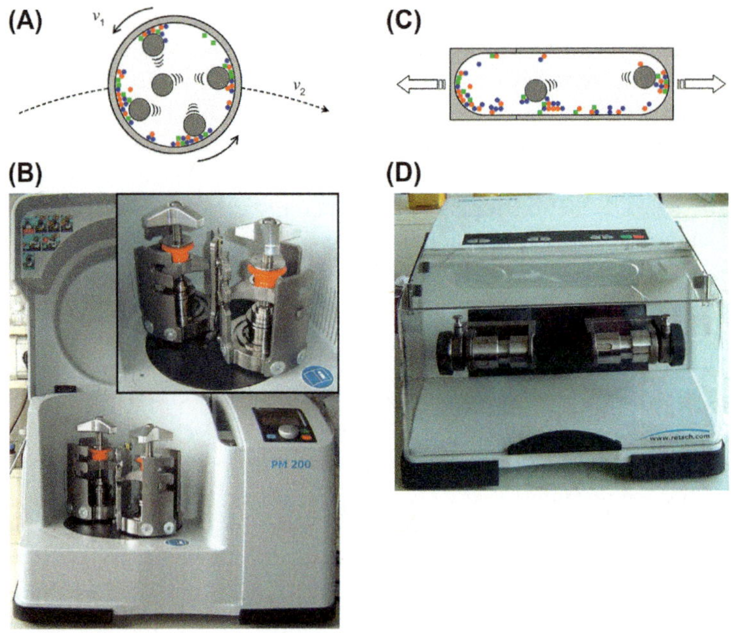

FIGURE 1.6

(A) Schematic representation of the working principle of a planetary ball mill (top view); (B) Retsch PM200 planetary ball mill (*inset*: two milling stations with jars attached); (C) schematic representation of the working principle of a mixer mill; and (D) Retsch MM400 mixer mill with clamped grinding jars.

or four milling stations where milling jars are mounted. The rotation speed (v_2) is easily controlled and can be set between 30 and 650 rpm with milling times up to 99 min, depending on the model. The capacity of the jars usually ranges from 12 to 500 mL and the available milling media can be chosen from stainless and hardened steel, sintered corundum, tungsten carbide, silicon nitride, zirconium oxide, agate, or plastic polyamide. As optional accessories, the manufacturers offer temperature and pressure measuring systems that are attached to the milling jars.

1.2.2 MIXER (SHAKER) BALL MILLS

Another type of commonly used ball mills for laboratory research is the mixer or shaker mill. In contrast to planetary ball mill where the grinding jars are spun in a vertical position, in mixer mills the jars are tightly clamped horizontally and swung back and forth with oscillating frequency up to 30 Hz in commercially available instruments. The horizontal oscillations accelerate the jar content and cause the balls to impact against the opposite end of the jar (Fig. 1.6C and D). With the amplitude of the clamp motion in about 2–5 cm range and its operating frequency up to 1800 cycles per minute, the balls can reach high velocities (of the order of several m s^{-1}), depending on the model. This leads to large forces exerted on the ground material, which classifies mixer mills as high-energy alternative to planetary mills. As in the case of planetary ball milling, both friction and impact modes provide energy into the system, with the impact mode being the dominant one. The maximum operating frequency and the milling time in these commercial mills are usually 30 Hz and 99 min, respectively. Grinding jars are available in different sizes (1.5–50 mL) and materials (hardened steel, stainless steel, tungsten carbide, agate, zirconium oxide, and Teflon).

1.2.3 CUSTOM-MADE BALL MILLS

Although investigations by Ondruschka and Stolle have shown that the ball mill type does not influence the yield or selectivity in organic reactions (assuming the same reaction time and scale) [25], there are cases where energy input in planetary or mixer mills is simply not sufficient to induce chemical changes (eg, fullerene chemistry). Apart from commercially available ball mills, the specific requirements imposed by the nature of studied systems that are not met by the current manufacturers in the market, force researchers to look for alternative custom-made solutions. The two examples of such home-built mills are high-speed vibration mills located at Kyoto University, Japan and Central Queensland University, Australia (Fig. 1.7).

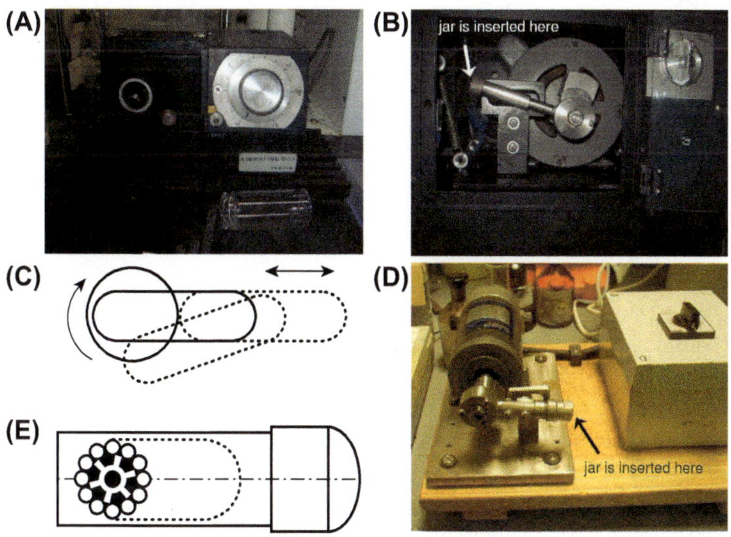

FIGURE 1.7

(A) A home-built high-speed vibration mill and (B) the interior of the mill. The grinding jar is inserted into the tilted metal tubing in front. *(By courtesy of Prof. Yasujiro Murata, Kyoto University.)* (C) In HSVM the jar oscillates in different planes as opposed to mixer mills. (D) A different version of the 60 Hz-HSVM at Central Queensland University, and (E) the grinding jar designed for this instrument.

They differ from the commercial instruments in the specific design that allows milling at much higher frequencies (up to 60 Hz), thus providing substantially more mechanical energy into the system. In the case of vibration mills, the jar containing the milling ball and the sample is shaken along its long axis horizontally, but unlike standard mixer mills, more degrees of freedom are allowed which is reflected in a slight fluctuation of the motion along the short jar axis (Fig. 1.7C). Both mills are installed on a timber support to reduce the vibrations.

1.2.4 MILLING PARAMETERS

Several milling parameters can influence the course and the outcome of an organic mechanochemical reaction, and should therefore be taken into account when optimizing milling processes. These usually include milling media and number and size of balls, jar filling degree, operating frequency, milling time, and addition of grinding auxiliaries. In a nice demonstration, Stolle et al. investigated the effect of a number of milling parameters on the oxidative cleavage of β-pinene (**28**) to produce nopinone (**29**) under ball-milling conditions [26] (Scheme 1.7).

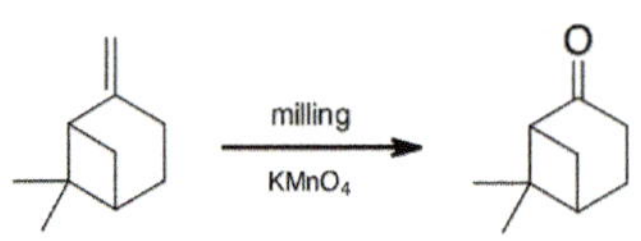

β-pinene
28

nopinone
29

SCHEME 1.7

Mechanochemical oxidative cleavage of β-pinene.

The reaction was first performed in a planetary ball mill using potassium permanganate ($KMnO_4$) as the oxidant, and without optimization of reaction parameters a decent yield of 56% was obtained after 20 min. By changing variables such as rotation frequency (ν_{rot}), milling time (t), number and ball size, milling material, grinding auxiliary, oxidant, and different stoichiometric ratios of the reactants, the mechanochemical oxidative cleavage was optimized. The reactions were carried out in zirconium oxide milling jars (45 mL) charged with six 15 mm balls made of the same material, with the maximum frequency set to 800 rpm (or 13.3 Hz). Aluminum oxide (acidic α-Al_2O_3 for chromatography) and quartz sand (SiO_2) were tested as grinding auxiliaries. The application of the grinding auxiliary in this case is necessary because liquid organic components need to be adsorbed. Otherwise, a paste-like mixture forms where the efficiency of inelastic collisions is significantly reduced. Although several solid auxiliaries were screened, Al_2O_3 and SiO_2 performed the best. Interestingly, a combination of $KMnO_4$ and montmorillonite as the solid support resulted in quantitative conversion of β-pinene, but without formation of nopinone. Due to acidic character of montmorillonite, β-pinene underwent isomerization and dimerization reactions yielding a mixture of several products, which was established by GC–MS analysis.

Optimization of $KMnO_4$ quantity and rotation frequency showed that the highest yield and selectivities were achieved with β-pinene/$KMnO_4$ ratio of 0.33 at 800 rpm. Selectivity was slightly better with Al_2O_3 (95% vs. 91%) while quartz support gave higher yield (89% vs. 79%). Manual grinding in a mortar resulted in poor yields and selectivities. Further optimization revealed that increasing rotation frequency and milling time generally led to higher yields, independent of the grinding auxiliary. However, in the case of SiO_2, the yield was more affected by increasing values of ν_{rot} at a given milling time than with Al_2O_3 support.

Next, the authors investigated how the number of balls used for milling influenced the conversion of β-pinene and selectivity. In the case of aluminum oxide auxiliary, the conversion dramatically rose from 40% to 80% when three balls were used instead of two. With quartz sand, this increase was even more pronounced from 20% to 70%. Addition of more balls did not significantly influence the conversion and values around 90% could be achieved with seven balls in a 45 mL grinding jar. In terms of selectivity, slightly better results were obtained with four to seven balls using Al_2O_3 support. With the total amount of reactants kept constant, the increase in the number of balls was equivalent to the increase of their total mass that reflected in higher levels of energy input during milling. Lower selectivities observed with fewer number of balls indicated that the activation energy for the formation of by-products was lower than for the oxidative cleavage leading to nopinone. Except for the number of balls, their size represented as the ball diameter can be varied. As a rule of thumb, a large number of smaller balls is preferred over a small number of large balls (providing the total mass of balls is kept constant).

Another important parameter when considering mechanochemical reactions is the type of material from which the grinding jars and balls are made. Physical properties such as density, hardness, or resistance to abrasion and chemical stability may play a key role in determining the outcome of a mechanochemical reaction [27]. Typical materials for milling accessories are stainless or hardened steel, zirconium oxide, tungsten carbide, sintered corundum, agate, and Teflon. Based on their densities and hardness, tungsten carbide provides very high energy input into the milling system ($\rho = 14.8\,g\,cm^{-3}$, hardness 1200) followed by steel ($\rho = 7.8–7.9\,g\,cm^{-3}$, hardness 550–750). Moderate levels of energy input are usually realized by using zirconium oxide jars and balls ($\rho = 5.9\,g\,cm^{-3}$, hardness 1200) while corundum, although being very hard material (hardness 1750), enables low-energy conditions due to its relatively low density ($\rho = 3.9\,g\,cm^{-3}$). Finally, agate ($\rho = 2.7\,g\,cm^{-3}$, hardness 100) and Teflon ($\rho = 2.1\,g\,cm^{-3}$) are classified as materials with very low energy input. Whereas stainless steel is not the best option if cross-contamination with iron is to be avoided (see Section 1.3), materials like tungsten carbide and zirconium oxide are better choice due to much lower abrasion. They are also chemically resistant which makes them ideal for milling corrosive substances, frequently encountered in organic synthesis. Although milling material is generally an important variable, in the case of oxidative cleavage of β-pinene it was demonstrated that changing material from zirconium oxide to hardened steel, sintered corundum, or agate did not have a significant effect on the yield of nopinone. Even with tools made of hardened steel, which is known to react in the presence of oxidizing agents, results were similar as with very low energy input agate jars and balls.

1.3 CONTAMINATION FROM WEAR IN ORGANIC MECHANOSYNTHESIS

While contamination from wear under high-energy ball milling has long been recognized as a technical problem in inorganic chemistry, where these processes can lead to substantial effects on the composition of the material being milled [28], the reports on metal leaching in mechanochemical organic reactions remain scarce. During mechanochemical synthesis of *N*-(3-nitrophenyl)-*N′*-phenylurea (**30**) by a coupling reaction between 3-nitroaniline and phenyl isocyanate, it was observed that neat grinding, apart from being inefficient, also led to high contamination of the sample with iron as evidenced by broad features in the ^{1}H NMR spectrum (Fig. 1.8) [29]. LAG reaction afforded the urea product quantitatively demonstrating its superiority over neat grinding, but the contamination was still present. Solubilization of the product in methanol by gentle heating resulted in precipitation of iron particles upon cooling. Following filtration over a short plug of celite, metal-free urea **30** was obtained with splitting patterns now clearly visible in the corresponding ^{1}H NMR spectrum (Fig. 1.8). Fülöp also encountered technical problems arising from paramagnetic iron particles abraded from the stainless steel milling equipment, which led to extreme broadening of the lines in the NMR spectra [30]. To eliminate such contamination, further mechanochemical syntheses of 2-spiroquinazolinones were carried out by using two 15 mm zirconium oxide balls, leading to better resolved NMR spectra. Baek et al. employed high-energy ball milling for graphene functionalization, and as a result of such intensive treatment under harsh milling conditions, significant metal contamination was observed. Instead of changing the milling material, the nature of the graphene product allowed a simple acidic wash that effectively removed traces of metal particles incorporated into the final product [31].

Boldyrev et al. described another approach to minimizing metal contamination from wear [32]. In an interesting application of ball milling for derivatization of a complex natural product lappaconitine **31**, two derivatives exhibiting lower toxicity were synthesized. Since the products investigated in this study display biological activity, it was necessary to reduce the metal content arising from wear to minimum by performing lining of the vial and the balls with an appropriate reagent. Oxidation of lappaconitine **31** to lappoconitine-N^{20}-oxide **32** was achieved with *m*-chloroperoxybenzoic acid (*m*-CPBA) by milling in an agate 60 mL vial charged with four 10 mm agate balls using a Spex 8000 vibratory ball mill. Prior to addition of the reactants, the authors performed a 10-min pre-grinding with hydrogen peroxide–urea complex with the purpose of lining the vial walls and balls. To address the problem of inefficient mixing in a conglutinated mass, 40–50% wt of aluminum oxide was added which prevented the formation of a paste-like reaction mixture. Such intervention increased the yield to 64%. In another example, deethylation reaction at the same reaction center (nitrogen N^{20}) to afford N^{20}-deethyllappaconitine **33** was carried out by *N*-bromosuccinimide (NBS) dealkylation. In this case, NBS was also employed for lining of steel grinding vial and 3 mm steel balls, which preceded the deethylation reaction (Scheme 1.8).

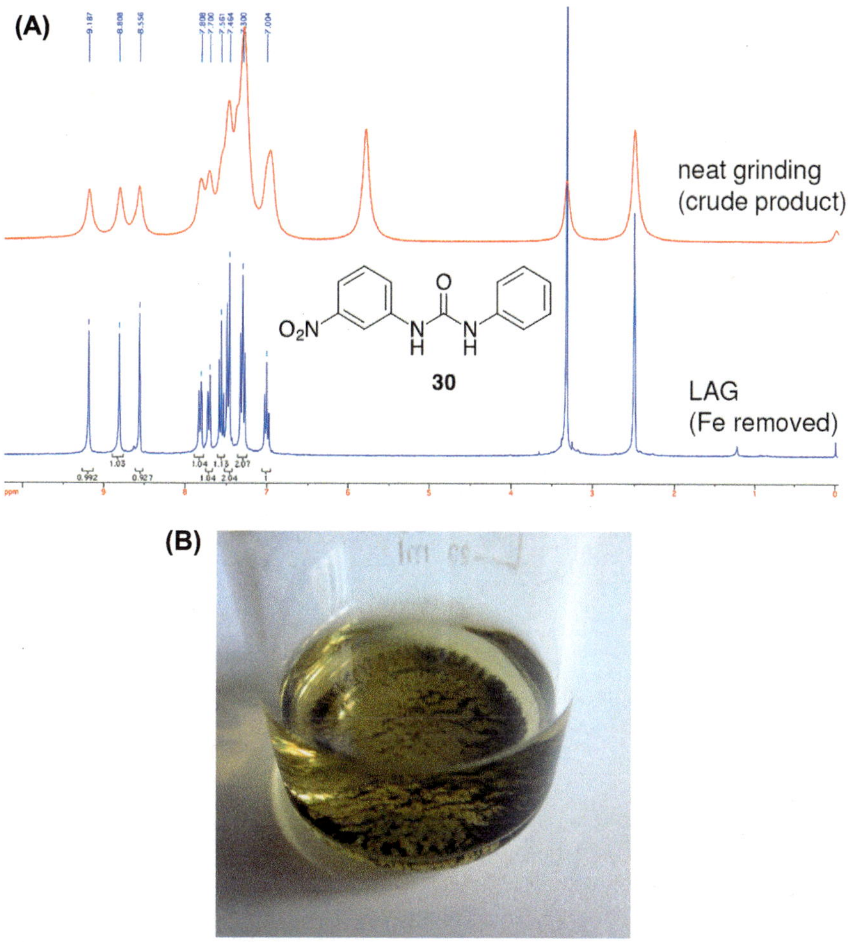

FIGURE 1.8

The effect of metal impurities from wear on the appearance of ^{1}H NMR spectrum of *N*-(3-nitrophenyl)-*N′*-phenylurea (**30**). (A) Crude product from neat grinding (red) and iron-free LAG product. (B) The precipitated iron particles after solubilization of the urea product from LAG reaction.

By courtesy of the authors, unpublished results.

These examples illustrate the occurrence of metal contamination during mechanochemical treatment of organic molecules without affecting the reactivity and yields, however wear in organic mechanosynthesis can also lead to incompatibility in a chemical sense, which especially holds for metal-catalyzed reactions. In a recently published demonstration of olefin cross-metathesis reaction conducted under ball-milling conditions [33], the authors reported on an observation that the model metathesis reaction of styrene to produce stilbene proceeded with diminished

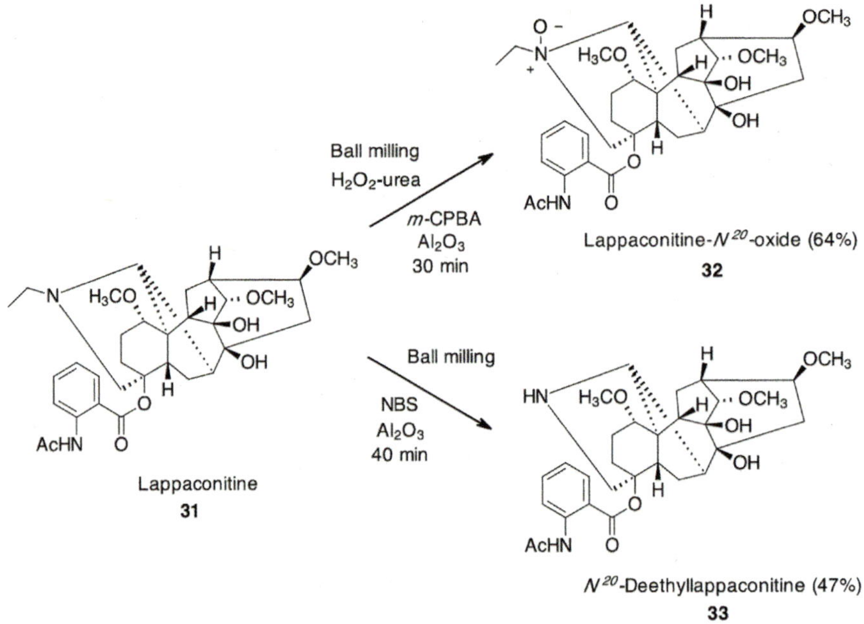

SCHEME 1.8

Functionalization of lappaconitine alkaloid **31** in a Spex 8000 vibratory ball mill. Hydrogen peroxide–urea complex and NBS were used for lining of milling jars and balls to reduce the contamination from wear.

yields and high irreproducibility when using stainless steel grinding jars and balls. After 30 min milling with 1 mol% of second-generation Grubbs catalyst, yields of stilbene ranged from none to 92%. Also, adding the required amount of the catalyst in two to four portions instead of all at once turned out to have a significant impact on improving the yield. Low yields and irreproducibility could partly arise from poor mixing due to formation of a thick shell around the grinding ball (Fig. 1.9). Switching to in-house made Teflon vessels, charged with an abrasive inert solid auxiliary (eg, sodium chloride), enabled reproducible formation of stilbene in 89–96% isolated yields. These results strongly suggested that the stainless steel jars were not compatible with ruthenium catalysts screened in the study, presumably due to enhanced catalyst reduction by impact of steel media against the steel vessel. Interestingly, while olefin metathesis suffers from wear contamination, there are cases where abrasion of milling media is advantageous in terms of facilitating a chemical reaction. Most notable examples are copper-catalyzed Sonogashira and alkyne–azide coupling reactions conducted in copper vials with copper balls, reported by Mack et al. [34,35].

The first attempt to systematically investigate contamination from iron during organic mechanosynthesis was reported in 2015 by Metro et al. [36]. The effect of wear from grinding jars and balls on the metal contamination levels was investigated in the synthesis of *N*-benzylbenzamide, prepared by amide coupling

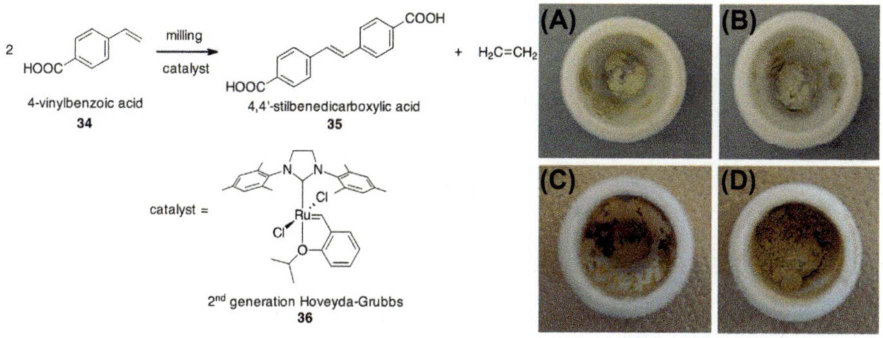

FIGURE 1.9

Left: Mechanochemical olefin cross-metathesis reaction of 4-vinylbenzoic acid (**34**) catalyzed by second-generation Hoveyda–Grubbs catalyst **36**. Right: The appearance of the reaction mixture during synthesis in a Teflon grinding jar. (A) Aggregation of the contents of a cross-metathesis reaction involving 4-vinylbenzoic acid (catalyst loading 2 mol%) about a stainless steel ball after 15 min milling; (B) continued aggregation after 2 h milling; (C) partial dispersion of the same mixture after 5 h milling is accompanied by a reaction and a change in color; and (D) the same metathesis reaction conducted for 2 h neat grinding using NaCl as a milling auxiliary.

Adapted with permission from Do J-L, Mottillo C, Tan D, Štrukil V, Friščić T. Mechanochemical ruthenium-catalyzed olefin metathesis. J Am Chem Soc 2015;137:2476–9. Copyright 2015 American Chemical Society.

reaction between benzoic acid and benzylamine hydrochloride mediated by *N,N'*-carbonyldiimidazole (CDI). A typical workup consisted of a brief 5-min ball milling the crude reaction mixture with distilled water followed by filtration and drying of the amide precipitate. To evaluate the effect of milling material on the reaction efficiency, stainless steel, zirconium oxide, agate, or Teflon grinding jars and balls were used throughout these experiments, and the metal contamination was determined by inductively coupled plasma mass spectrometry (ICP-MS) or gravimetric analysis. A standard synthetic procedure, based on milling reactants in a 12 mL stainless steel jar with 50 stainless steel balls (5 mm diameter) using a planetary ball mill at 500 rpm for a total of 20 min (including 5-min aqueous workup), resulted in contamination levels of 366 ppm of iron and 169 ppm of chromium, the two main components of stainless steel. Solubilization of the product in ethyl acetate, followed by filtration over a cotton plug or 0.2 μm PTFE filter reduced these values < 30 ppm. Next, the reaction was carried out in a 10 mL stainless steel jar with two 10 mm stainless steel balls using a vibratory ball mill operating at 30 Hz. In this case, the metal contamination was much higher, with 1400 ppm of Fe and 1000 ppm of Cr. Repeating the same reaction in a brand new grinding jar led to even higher metal content in the amide product, reaching values of 7400 ppm of Fe and 1400 ppm of Cr. The same trend was observed when the reaction was scaled down to fit the 5 mL jar charged with two 5 mm balls.

Switching from high-density stainless steel (ρ=7.8 g cm^{-3}) to lighter material such as zirconium oxide (ρ=5.9 g cm^{-3}) indeed reduced the contamination levels of iron (151 ppm), but significantly increased zirconium content (12,200 ppm). The lowest contamination with 48 ppm of Fe and <100 ppm of Si was achieved with 50 mL agate jars in combination with four 12 mm agate balls using the planetary ball mill. Changing the jar and ball material to Teflon resulted in 2400 ppm of Fe, indicated by gravimetric analysis. If a single 18 mm rubber ball was used with Teflon jar, the iron content dropped to below 1000 ppm which was the detection limit of the gravimetric method. Scaling-up the reaction to 22.5 mmol in steel milling equipment using fifty 10 mm balls led to significant wear and Fe content of 42,000 ppm according to gravimetric analysis, while ten 18 mm rubber balls reduced this level to 3400 ppm. Investigation of the effect of milling time on contamination from wear, studied on benzyl hydrocinnamide synthesis, showed that increasing the milling time was accompanied by higher metal content in the final product. These results demonstrate how different milling material can have a substantial effect on the level of metal contamination in mechanochemical organic synthesis and that proper choice of milling equipment, as well as optimization of the milling time, should be compatible with the intended use of the products.

1.4 ANALYSIS AND MONITORING OF MECHANOCHEMICAL REACTIONS

1.4.1 ANALYTICAL METHODS IN ORGANIC SOLID-STATE CHEMISTRY

Although standard analytical methods which rely on dissolution of the analyte in the appropriate solvent have been used for the analysis of solid-state mechanochemical reactions (eg, NMR and HPLC, Ref. [20]), the facts that many analytes arising from the solid-state reactions display poor solubility and that solution environment required for such procedures may induce further changes in the system impose the necessity for exclusively solid-state analytical methods. In this respect, powder X-ray diffraction (PXRD) is the first choice analysis in solid-state chemistry. While this analytical technique is traditionally related to inorganic chemistry, organic materials can also be amenable to PXRD analysis [37]. The products of mechanochemical reactions are typically fine powders, and PXRD allows for a fast qualitative analysis of reaction mixtures and products by comparing the PXRD patterns of the starting materials and the products. If the crystal structures of the reaction products have already been determined, a simple comparison of the measured PXRD pattern with the simulated one from the single-crystal diffraction data (deposited in the Cambridge Structural Database, CSD) provides a quick access to information about the completeness of the reaction and the polymorphic purity of the products. An illustrative example of application of PXRD analysis in organic mechanosynthesis has been published by Friščić et al. where manual grinding and ball milling were utilized for an environment-friendly synthesis of a wide range of simple aliphatic and aromatic thioureas with the purpose of establishing a "solvent-free research laboratory" [38]. As such, the analytical methods used to screen the mechanochemical products also relied on the overall reduction in bulk solvent, so the PXRD analysis, among others, has been chosen as the primary tool to characterize the milling products. In the case of *N*-(4-nitrophenyl)-*N*′-morpholinothiourea (**37**), PXRD analysis of the products obtained from solution synthesis carried out in dichloromethane at two different concentrations showed that two polymorphs were obtained. When the reaction between morpholine and 4-nitrophenyl isothiocyanate was done at low concentration, a monoclinic form I precipitated out. On the other hand, performing the reaction at higher concentration resulted in the precipitation of a new, orthorhombic polymorph denoted as form II. After screening for solvents to grow single crystals of these two polymorphs suitable for the structural X-ray diffraction analysis, it was found that form I easily crystallized from a range of solvents, whereas ethyl acetate was the appropriate solvent for crystallization of the form II. By comparing the simulated PXRD patterns of the two polymorphs with the measured one obtained from the mechanochemically synthesized product, it was revealed that LAG, using acetonitrile as the grinding liquid, selectively afforded only the second polymorph, form II. Interestingly, in an attempt to crystallize the product from acetonitrile, single crystals of the monoclinic form I were isolated. In the crystal structure of form I, individual molecules of *N*-(4-nitrophenyl)-*N*′-morpholinothiourea form an infinite chain via N—H···O hydrogen bonds between the N—H group of one molecule and the oxygen atom from the morpholine substituent of the neighboring molecule. This unexpected bonding pattern was replaced in the form II with the typical N—H···S hydrogen bonds encountered in other thiourea derivatives. The asymmetric unit in the crystal structure of form II contains four crystallographically independent molecules ($Z' = 4$). A pair of two crystallograpically distinct molecules is held together by N—H···S bonds and assembled into a chain in an alternating fashion, giving rise to two crystallographically independent hydrogen-bonded chains (Fig. 1.10).

In this extensive study of thiourea synthesis by coupling isothiocyanates with amines under mechanochemical conditions of neat and LAG, a total of 49 different thiourea derivatives were prepared. Since the supramolecular chemistry of simple symmetrical and nonsymmetrical diarylthioureas had not previously been studied in detail, the information from the PXRD patterns of the milling products in this work also allowed for a systematic analysis of structural features of diarylthioureas. The three main self-assembly motifs were identified. A characteristic supramolecular synthon based on bifurcated N—H···S hydrogen bonds was found in most cases, whereby the individual thiourea molecules were assembled into chains aligned in a head-to-head (parallel molecular dipoles) or head-to-tail (antiparallel molecular dipoles) manner. In the case of sterically hindered thioureas, such as *N*-(2,6-dimethylphenyl)-*N*′-phenylthiourea, the third motif arising from discrete centrosymmetric dimers based on $R_2^2(8)$ supramolecular synthon was found. A detailed look at the measured PXRD patterns for all diarylthioureas revealed that solid-state structures of these compounds could be classified into two structural families, denoted as family I and family II (Fig. 1.11).

In the family I, the chains of hydrogen-bonded thiourea molecules form parallel stacks in a supramolecular architecture that results in a characteristic and intensive (200) reflection at the Bragg angle range 5–7 degrees in the PXRD pattern. This

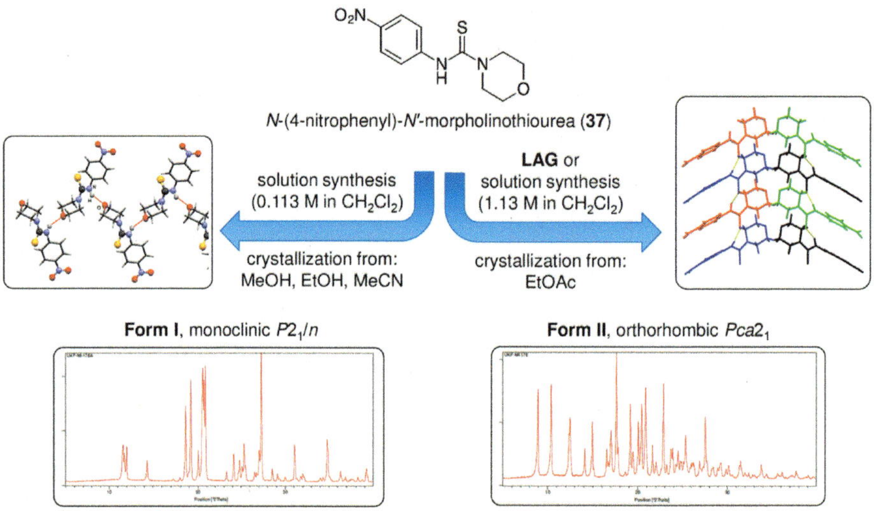

FIGURE 1.10

Application of powder X-ray diffraction for the analysis of polymorphic purity of *N*-(4-nitrophenyl)-*N*′-morpholinothiourea (**37**) samples prepared by conventional solution synthesis in dichloromethane and mechanochemical liquid-assisted grinding approach. At low concentration, polycrystalline powder of form I precipitates out, while of form II precipitates from more concentrated solution and mechanochemical synthesis.

reflection corresponds to the width of the supramolecular stack, but also to the lattice separation between the sulfur atoms in neighboring stacks, thus accounting for its high intensity. The characteristic feature in the PXRD patterns of compounds belonging to the family II is the (110) reflection at the Bragg angle range 8–10 degrees. Similarly to family I, this reflection originates from the diffraction on the (110) plane that contains strongly scattering sulfur atoms. In the family II, the hydrogen-bonded thiourea chains are stacked in a herringbone pattern with an angle between neighboring stacks of 44 degrees (Fig. 1.11). The analysis of PXRD patterns of thioureas bearing ortho-substituted phenyl groups showed that the crystal structures of these derivatives cannot be simply divided into structural families, presumably due to formation of discrete $R_2^2(8)$ hydrogen-bonded dimers.

Unlike typical inorganic compounds, organics are considered as softer materials and thus during harsh ball milling can lose crystallinity and become amorphous, resulting in poor diffraction and overall reduction in the quality of data obtained. Unless these changes provide some mechanistic information (eg, amorphous intermediates), the utility of PXRD in such cases would be limited. However, assuming there are no issues with the diffraction, PXRD is also becoming a tool that enables structure determination of organic molecules without the need to grow single crystals from bulk solvent [39]. This is particularly useful in cases where poor solubility prevents recrystallization and crystal growth or when solution-based crystallization

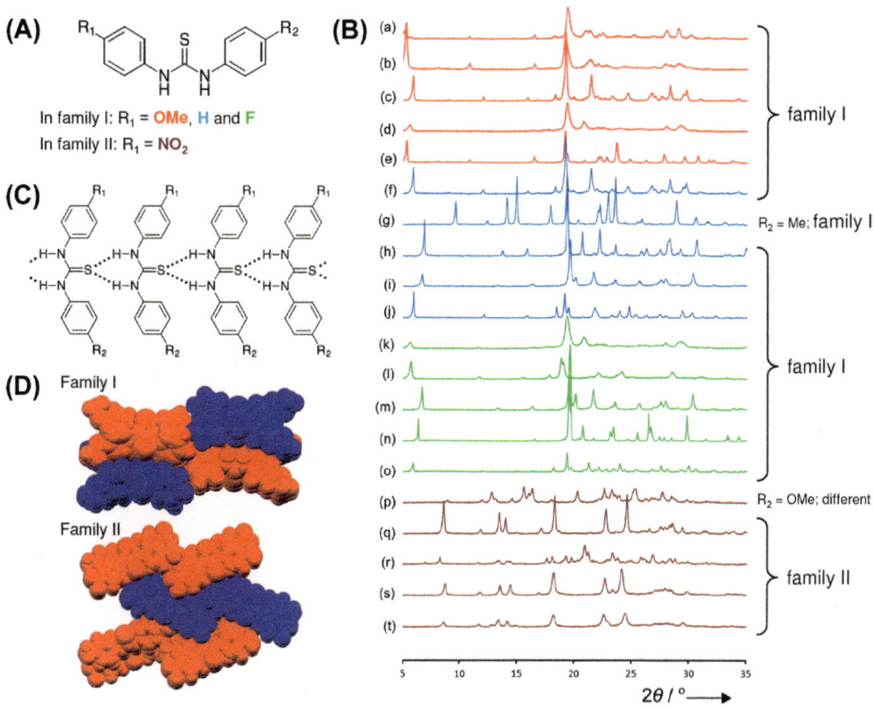

FIGURE 1.11

(A) The schematic representation of *p*-substituted diarylthioureas. (B) PXRD analysis of a series of aromatic thioureas. (C) A characteristic self-assembly motif of diarylthioureas in the solid state by the bifurcated N—H···S hydrogen bonds (head-to-head or parallel molecular dipoles arrangement is shown). (D) Arrangement of molecules in structural families I and II viewed along the N—H···S hydrogen-bonded chains.

Adapted from Štrukil V, Igrc MD, Fábián L, Eckert-Maksić M, Childs SL, Reid DG, et al. A model for a solvent-free synthetic organic research laboratory: click-mechanosynthesis and structural characterization of thioureas without bulk solvents. Green Chem 2012;14:2462–73 with permission from the Royal Society of Chemistry.

yields polymorphs different from mechanochemical solid-state synthesis. The first successful structure determination from PXRD data of a mechanochemically synthesized product was reported by Harris et al. in 2003 [40]. It was the structure of co-crystal obtained by solvent-free manual grinding a mixture of benzoquinone, *bis*-β-naphthol, and anthracene in 2:2:1 ratio. Similarly to previously reported benzoquinone-*bis*-β-naphthol 1:1.5 co-crystal [11], the PXRD analysis showed that the mechanochemical product is different from the one isolated from solution, rendering structure determination from powder diffraction data as the only alternative.

Structure determination from PXRD data has been regarded as an invaluable tool to elucidate the structures of active pharmaceutical ingredient (API) co-crystals [41]. Trask et al. described the mechanochemical synthesis of several caffeine co-crystal with simple aliphatic carboxylic acid co-crystal formers [42]. While co-crystallization of caffeine (**16**) with acetic acid **38** in 1:1 M ratio to yield (**16**)(**38**) was feasible only by grinding, the 1:1 (**16**)(**39**) co-crystal with trifluoroacetic acid **39**, also prepared using the solid-state reaction, was later found to crystallize from a solution upon seeding with the mechanochemical product. In these two cases, the respective crystal structures of API co-crystals were determined from the PXRD data, shown in Fig. 1.12A and B. Emmerling et al. reported in 2015 on the mechanochemical synthesis, mechanism of formation and structure determination from powder diffraction data for theobromine-oxalic acid 2:1 co-crystal [43]. Neat grinding of a 2:1 mixture of theobromine **40** and oxalic acid dihydrate **41**$\cdot 2H_2O$ yielded quantitatively a new co-crystal form $(\mathbf{40})_2(\mathbf{41})$. Indexing procedure revealed a $P2_1/c$ monoclinic unit cell, followed by Rietveld refinement that provided a plausible crystal structure of the 2:1 API co-crystal.

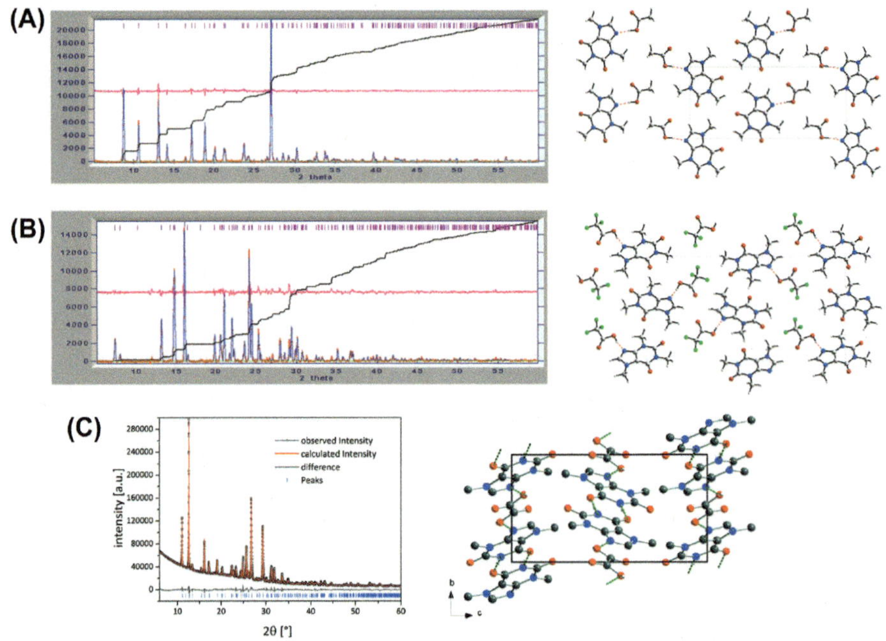

FIGURE 1.12

Crystal structure determination of API co-crystals directly from powder X-ray diffraction data: (A) Rigid-body Rietveld refinement of the crystal structure of (**16**)(**38**) co-crystal (left) and packing diagram viewed down [010] direction and (B) rigid-body Rietveld refinement of the crystal structure of (**16**)(**39**) co-crystal (left) and packing diagram viewed down [001] direction. *(Adapted with permission from Trask AV, van de Streek J, Motherwell WDS, Jones W. Achieving polymorphic and stoichiometric diversity in cocrystal formation: importance of solid-state grinding, powder X-ray structure determination, and seeding. Cryst Growth Des 2005;5:2233–41. Copyright 2005 American Chemical Society.)* (C) Rietveld refinement of the crystal structure of 2:1 theobromine-oxalic acid co-crystal $(\mathbf{40})_2(\mathbf{41})$ (left) and packing of molecules in the unit cell (right) along *a*-axis. *(Published by The Royal Society of Chemistry.)*

In a 2015 paper, the molecular and crystal structures of three aromatic thiocarbamoyl benzotriazoles were successfully determined for the first time directly from PXRD data [44]. These compounds deemed impossible to isolate as they readily decompose in solution environment and were consequently regarded only as reactive intermediates. However, the LAG mechanochemical approach enabled their quantitative synthesis from the respective substituted anilines and bis(benzotriazolyl)methanethione (**42**) in 1:1 M ratio and yielded the desired products as highly crystalline powders which were subsequently subjected to PXRD structural analysis. Based on the high-quality powder diffraction data collected on a synchrotron source, an indexing procedure in the first step gave monoclinic unit cells containing four molecules for all three compounds. Structure determination using TOPAS software [45] followed by Rietveld refinement, showed one crystallographically independent molecule per asymmetric unit. The studied thiocarbamoyl benzotriazole derivatives (4-methoxyphenyl **43**, 4-chlorophenyl **44** and 2-aminophenyl **45**) are nonplanar molecules with the angle between the benzotriazole and phenyl groups in range 65–86 degrees. However, the thiocarbamoyl group and the phenyl substituents are almost coplanar in all cases. The presence of N—H group allows these molecules to self-assemble into centrosymmetric dimers, either by forming long hydrogen bonds toward the benzotriazole nitrogen atom (as in 4-methoxyphenyl **43** and 4-chlorophenyl derivatives **44**), or the *ortho*-positioned amino group in 2-aminophenyl derivative **45** (Fig. 1.13).

Another analytical method, which is not affected by crystalline-to-amorphous state transition as PXRD, is infrared spectroscopy (IR). A KBr pellet technique or more conveniently attenuated total reflectance (ATR) spectroscopy provides a quick route to qualitative and quantitative analysis of samples from mechanochemical reactions. In this sense, ATR is a preferred method as it is simple and faster compared to KBr pellet, while at the same time eliminating the need for a solid support. Additionally, making a pellet requires additional grinding of the analyte in a mortar with the solid support, which in some cases interacts with the analyte and chemically modifies it. The most prominent example of such behavior was described by Braga et al. who noticed the transformation of the organometallic zwitterion $[Co^{III}(\eta^5\text{-}C_5H_4COOH)(\eta^5\text{-}C_5H_4COO)]$ (**46**) upon grinding with potassium bromide during preparation of KBr pellets for IR analysis [46].

The authors observed large differences in the carbonyl group stretching vibration frequencies when the spectra were taken in Nujol in contrast to spectra recorded using KBr pellets. An attempt to crystallize the organometallic compound **46** from KBr solution afforded single crystals of material which was structurally characterized to reveal the formation of a hybrid organometallic-inorganic salt with the composition $[Co^{III}(\eta^5\text{-}C_5H_4COOH)(\eta^5\text{-}C_5H_4COO)]_2\cdot K^+Br^-$, $\mathbf{46}_2\cdot K^+Br^-$. This supramolecular complex consists of potassium cations encapsulated within a cage formed by four zwitterions (Fig. 1.14). When a 2:1 mixture of **46** and KBr was manually ground, the PXRD analysis of the product obtained confirmed the identity of the mechanochemically prepared complex to that isolated by solution crystallization. Also, the C=O stretching frequencies in the IR spectrum of the ground product recorded in Nujol matched the frequencies observed in KBr pellets, further corroborating the structure

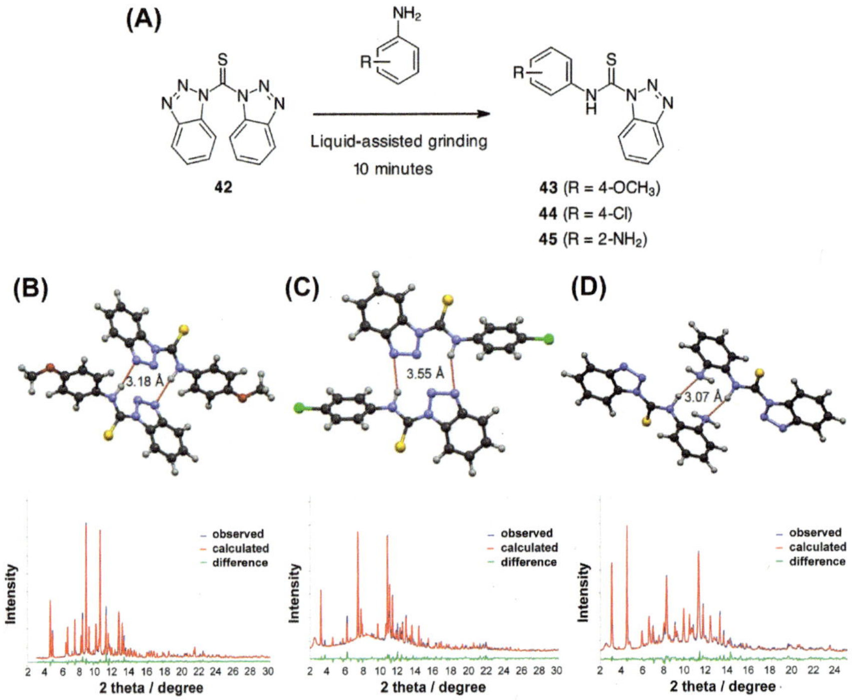

FIGURE 1.13

(A) Mechanosynthesis of thiocarbamoyl benzotriazoles **43–45**. The molecular structures and self-assembly of aromatic thiocarbamoyl benzotriazoles **43–45** in the solid-state and Rietveld refinement of the corresponding crystal structures for (B) 4-methoyxphenyl, (C) 4-chlorophenyl, and (D) 2-aminophenyl derivatives. The powder X-ray diffraction data were collected using a synchrotron source.

of the hybrid salt. The authors also carried out analogous grinding experiments with potassium and ammonium hexafluorophosphate salts and established the generality of inorganic cation encapsulation with **46** during mechanochemical synthesis.

Solid-state nuclear magnetic resonance (ssNMR) spectroscopy has emerged over the years as a powerful analytical method in solid-state chemistry, especially with the advancements in techniques that allow the acquisition of high-resolution spectra [47]. In the broadest sense, ssNMR is mostly applied in characterization of crystalline materials as a means to support PXRD structural analyses by providing information on the number of molecules in the asymmetric unit or the symmetry of the occupied positions within the unit cell. Another major field of application is the structural characterization of amorphous and disordered solids where standard X-ray diffraction-based techniques fail to give detailed structural information. When discussing ssNMR in the context of API polymorphism and synthesis of co-crystals,

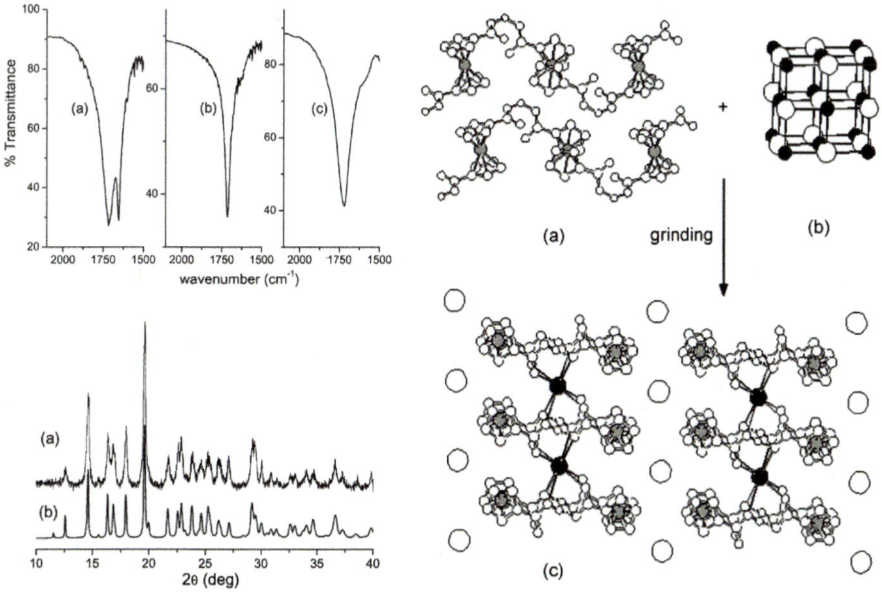

FIGURE 1.14

Upper left: Comparison of IR spectra of (A) **46** in Nujol, (B) **46** in KBr, and (C) **46**$_2$ K^+Br^- in Nujol. Right: The supramolecular rearrangement leading from crystalline **46** (A) and KBr (B) to crystalline **46**$_2$ K^+Br^- (C). ● = K^+ ions, ○ = Br^- ions. Bottom left: Comparison between the measured (A) and calculated (B) X-ray diffraction patterns of **46**$_2$ K^+Br^-.

Adapted from Braga D, Maini L, Polito M, Grepioni F. Unexpected solid-solid reaction upon preparation of KBr pellets and its exploitation in supramolecular cation complexation. Chem Commun 2002;37:2302–2303 with permission from the Royal Society of Chemistry.

pharmaceutical industry has recognized its potential to assist in the discovery and characterization of novel forms, which is an important issue from the intellectual property rights and drug licensing perspective. In contrast to solution NMR where interactions that affect the nuclear spins are averaged due to rapid and random movement of molecules in a solution, a typical feature of ssNMR spectrum is the line broadening. Resulting from several effects such as anisotropic chemical shielding, dipole–dipole coupling (between nuclear dipoles) and quadrupole coupling (for nuclei with spin I greater than 1/2), the line broadening is successfully suppressed by the development of experimental techniques such as cross-polarization magic angle spinning or CP-MAS. On the practical side, ssNMR has many advantages as an analytical method; it is nondestructive, requires a relatively small amount of a powdered sample for analysis, has nuclear-site specificity, and displays sensitivity to hydrogen bonding and local conformational changes. A number of one- and two-dimensional experiments are routinely available to research chemists and allow the analysis of different nuclei and internuclear connectivity [48].

While numerous papers on the characterization of mechanochemical products by utilizing ssNMR have been reported, including MOFs [49], organometallics [50], and co-crystals [51], the intention of this chapter is to provide the reader with representative examples where ssNMR techniques have been applied in the realm of mechanochemical organic synthesis. In 2002, Balema et al. disclosed a mechanochemical synthesis of several phosphonium salts [52], and later on the mechanosynthesis of phosphorus ylides and the first example of solvent-free Wittig reaction under milling conditions [53]. In both cases, the progress of the reactions and the identity of the final products were analyzed by means of CP-MAS ^{31}P ssNMR spectroscopy. The authors reported on mechanochemical approach to the synthesis of phosphonium salts by milling triphenylphosphine (**47**) and 2-bromoacetophenone in an equimolar ratio in a hardened-steel vial with steel balls. After ball-milling for 1 h, CP-MAS ^{31}P ssNMR analysis showed a complete conversion of the starting material ($\delta = -5$ ppm for Ph_3P) to phenacyltriphenylphosphonium bromide salt **48** as evidenced by the sharp signal at $\delta = 23$ ppm and comparison with the commercial sample of the salt. Following the same procedure, phosphonium salt **49**, derived from 2-bromo-2-phenylacetophenone, was prepared in quantitative yield. This reactant was previously found the yield a mixture of *C*- and *O*-phosphorylated products during solution synthesis. Interestingly, the mechanochemical solid-state approach afforded exclusively *C*-phosphorylated compound whose identity was confirmed by comparing the CP-MAS ^{31}P ssNMR spectra of the milling product with the recrystallized pure sample of **49** (Fig. 1.15).

Further development of this strategy resulted in mechanochemical synthesis of stabilized, semistabilized, and nonstabilized phosphorus ylides starting from phosphonium salts and potassium carbonate [53]. Whereas the stabilized ylides were isolated in pure form, the semistabilized and nonstabilized analogues were trapped by a solid carbonyl compound. In that way, the authors succeeded in performing the first Wittig reaction under solid-state ball-milling conditions. Two stable phosphorus ylides derived from benzoylmethylene and carbethoxymethylene phosphonium bromides were synthesized in near-quantitative yields after 3–4 h of solvent-free ball milling. Again, the CP-MAS ^{31}P ssNMR spectroscopy revealed that the quantitative formation of benzoylmethylene ylide **51** from **48** took 3 h as evidenced by a single sharp signal at 17 ppm. The analysis of the reaction mixture after 1 h indicated that the reaction was still in progress since the signal of the starting phosphonium bromide **48** at 23 ppm could also be detected. In the case of nonstabilized ylide **52**, 2-naphthaldehyde was used to trap the reactive ylide in the solvent-free Wittig reaction yielding 2-vinylnaphthalene and triphenylphosphine oxide as the products. The CP-MAS ^{31}P ssNMR analysis of the reaction mixture after 10 h still showed the presence of unreacted methyltriphosphonium bromide **50** along with triphenylphosphine oxide, and it took further 10 h of milling to reach completion.

In the paper, discussed above in terms of a systematic PXRD analysis of aromatic thiourea derivatives, the authors have also described structural characterization of five randomly selected thioureas directly from the powder diffraction data collected on a laboratory diffractometer [38]. Whereas the structure determination procedure readily gave reasonable crystal structures for three derivatives, the ambiguities with two other structures (*N*-(4-fluorophenyl)-*N'*-isopropylthiourea (**53**) and *N*-(4-chlorophenyl)-*N'*-(4-methoxyphenyl) thiourea (**54**)) could not have been

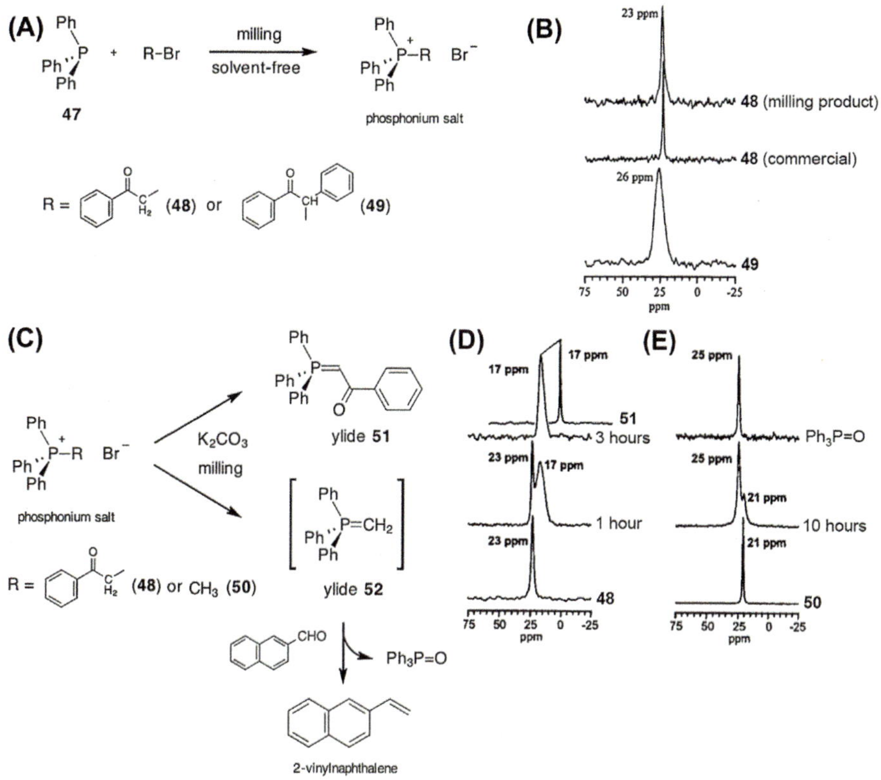

FIGURE 1.15

(A) Solvent-free mechanochemical synthesis and (B) the CP-MAS ^{31}P ssNMR spectra of phosphonium salts **48** and **49**. *(Adapted from Balema VP, Wiench JW, Pruski M, Pecharsky VK. Solvent-free mechanochemical synthesis of phosphonium salts. Chem Commun 2002;37:724–25 with permission from the Royal Society of Chemistry.)* (C) Mechanosynthesis of stabilized ylide **51** and trapping of nonstabilized ylide **52** by 2-naphthaldehyde in the solvent-free Wittig reaction; (D) the CP-MAS ^{31}P ssNMR analysis of the formation of ylide **51**; and (E) triphenylphospine oxide as the byproduct of the Wittig reaction. *(Adapted with permission from Balema VP, Wiench JW, Pruski M, Pecharsky VK. Mechanically induced solid-state generation of phosphorus ylides and the solvent-free wittig reaction. J Am Chem Soc 2002;124:6244–5. Copyright 2002 American Chemical Society.)*

resolved based only on PXRD data. In these cases, the solid-state CP-MAS ^{13}C NMR spectroscopy was employed to provide additional information that was necessary to successfully complete the structure determination (Fig. 1.16). Although the preliminary crystal structure of isopropyl derivative **53** was reasonable in terms of molecular geometry and supramolecular interactions, the initial solution consistently indicated the presence of two crystallographically independent molecules in the asymmetric

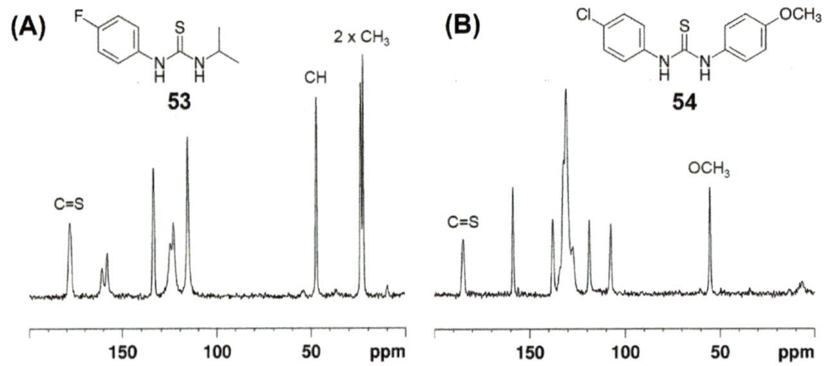

FIGURE 1.16

The CP-MAS ^{13}C ssNMR spectra of (a) *N*-(4-fluorophenyl)-*N′*-isopropylthiourea (**53**) and (b) *N*-(4-chlorophenyl)-*N′*-(4-methoxyphenyl)thiourea (**54**). In both cases, no peak splitting is observed, indicating the presence of one thiourea molecule in the respective asymmetric units ($Z' = 1$).

Adapted from Štrukil V, Igrc MD, Fábián L, Eckert-Maksić M, Childs SL, Reid DG, et al. A model for a solvent-free synthetic organic research laboratory: click-mechanosynthesis and structural characterization of thioureas without bulk solvents. Green Chem 2012;14:2462–73 with permission from the Royal Society of Chemistry.

unit ($Z' = 2$). To verify this, a CP-MAS ^{13}C ssNMR spectrum of **53** was recorded, which in the aliphatic region showed three sharp signals corresponding to the three carbon atoms of the isopropyl group. Also, the characteristic thiourea carbon atom signal emerged as a slightly broad singlet, in agreement with the presence of two quadrupolar nitrogen nuclei. Since no splitting of the signals was observed in the spectrum, this experiment suggested the presence of only one crystallographically independent **53** molecule in the asymmetric unit ($Z' = 1$) and the preliminary structure obtained from PXRD data was finally discarded.

In the case of thiourea **54**, the indexing of the measured PXRD pattern suggested isostructurality with the previously reported crystal structure of the symmetrical bis(4-methoxyphenyl) derivative. This observation, in combination with the reduction in molecular symmetry, implied that the space groups were to be chosen so as to provide a disordered structure with the full retention of space group symmetry, or a fully ordered structure where the space group symmetry was reduced. To resolve this dilemma, CP-MAS ^{13}C ssNMR spectroscopy was used and, as in the previous case, confirmed that only one **54** molecule is present in the asymmetric unit ($Z' = 1$) based on the intensive and sharp singlet of the methoxy carbon atom. This finding narrowed down the choice of potential space groups to $P2_12_12$, $P2_12_12_1$, and $Pna2_1$. By performing a careful Rietveld refinement, the $P2_12_12$ space group was identified as the most suitable one. Single crystal structure determination, carried out to verify the correctness of the $P2_12_12$ structure, revealed a unit cell identical to the one derived from the PXRD data. However, the space group was determined as *Pbcm*, identical to the reported structure of the bis(4-methoxyphenyl) derivative. Such highly symmetrical space group requires the **54** molecule to be disordered over a mirror plane, what was in contrast with the information from the CP-MAS ^{13}C ssNMR spectrum. The authors offered a tentative explanation of the observed inconsistency between single crystal X-ray diffraction and ssNMR measurements as a result of single crystals being composed of microscale domains of different orientations, which lead to a pseudo-centrosymmetric average structure observed by using single crystal X-ray diffraction. In contrast, individual domains would display such organization that affords sharp NMR signals in the solid-state spectra. That the non-centrosymmetric $P2_12_12$ structure of **54** derived from PXRD and ^{13}C ssNMR data was the correct one, was verified by comparison of the powder pattern simulated from the single crystal data and the measured one followed by Rietveld refinement.

As a part of the spectroscopic and structural characterization of aromatic thiocarbamoyl benzotriazoles as a class of compounds particularly sensitive to solution environment, the application of ssNMR enabled their structures to be confirmed without resorting to solvent-based analytical methods [44]. For example, the CP-MAS ^{13}C ssNMR spectroscopy aided the structure determination from PXRD data for 4-methoxyphenyl derivative **43** (Fig. 1.17). In the ^{13}C spectrum, a sharp singlet corresponding to methoxy group carbon atom at 57.7 ppm as well as the thiocarbamoyl carbon signal at 174.9 ppm indicated the presence of one thiocarbamoyl benzotriazole molecule in the asymmetric unit ($Z'=1$), consistent with the PXRD structure. On the other hand, the solid-state ^{13}C NMR spectrum of 4-aminophenyl derivative **55** displayed peak splitting for the thiocarbamoyl carbon atom, suggesting $Z'=2$. The 4-methoxyphenyl derivative **43** and bis(thiocarbamoyl benzotriazole) derived from 1,4-phenylenedimane **56** were also characterized by CP-MAS ^{15}N ssNMR. In these spectra, signals arising from four magnetically nonequivalent nitrogen atoms were identified, in full agreement with the proposed structures for aromatic thiocarbamoyl benzotriazoles.

Although more often applied in inorganic chemistry, particle characterization by scanning electron microscopy (SEM) has found application in the study of MOFs and co-crystals prepared by mechanochemical synthesis. James et al., in 2006 reported the first rapid and quantitative mechanosynthesis of a microporous MOF $[Cu(INA)_2]$ (INA = isonicotinic acid) by solvent-free milling of copper acetate with INA for 10 min [54]. SEM technique was used to study the particle morphology of MOF samples that were ground for 1 and 5 min prior to aging for one week. SEM analysis revealed that samples obtained by 1-min milling consisted of larger crystals with more clearly defined faces and edges, in contrast to 5-min samples (Fig. 1.18). These different morphologies suggest that grinding time may be important in determining the bulk properties of MOFs prepared by the mechanochemical solvent-free method.

In a different study by the same group, SEM was utilized to investigate the differences between neat grinding and LAG synthesis of another MOF system $Cu_3(BTC)_2$ (BTC = 1,3,5-benzenetricarboxylate), commonly known as HKUST-1 [55]. In accordance with a previous study, it was found that LAG provides better defined crystallites with a typical size in the range 50–100 nm. In an attempt to activate the product

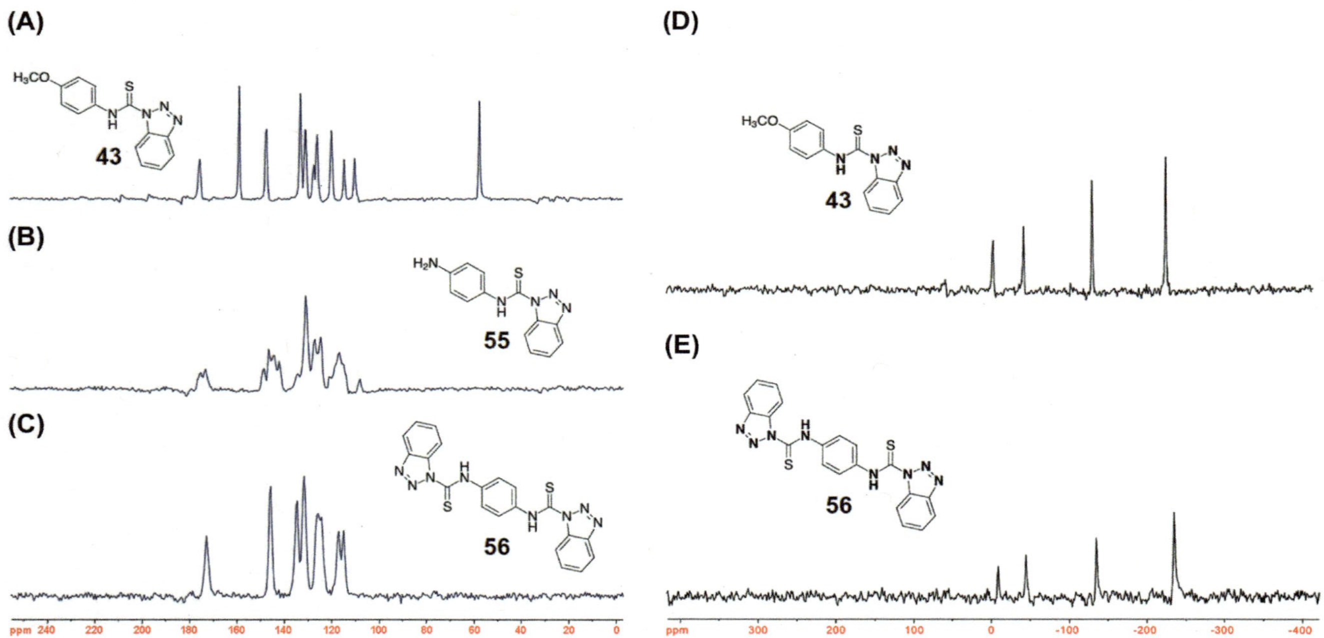

FIGURE 1.17

The CP-MAS ^{13}C ssNMR spectra of (A) *N*-(4-methoxyphenyl)thiocarbamoyl benzotriazole **43**, (B) *N*-(4-aminophenyl)thiocarbamoyl benzotriazole **55**, and (C) *N,N′*-bis(benzotriazole-1-thiocarbamoyl)-1,4-phenylenediamine **56**. The CP-MAS ^{15}N ssNMR spectra of (D) *N*-(4-methoxyphenyl) thiocarbamoyl benzotriazole **43** and (E) *N,N′*-bis(benzotriazole-1-thiocarbamoyl)-1,4-phenylenediamine **56**. Notably, the carbon atom signals in the spectrum of 4-aminophenyl derivative are split, suggesting two crystallographically independent molecules in the asymmetric unit ($Z' = 2$).

FIGURE 1.18

SEM images of $[Cu(INA)_2]$ samples ground for 1 and 5 min after 1 week aging.

Reproduced from Pichon A, Lazuen-Garay A, James SL. Solvent-free synthesis of a microporous metal-organic framework. CrystEngComm 2006;8:211–4 with permission from the Royal Society of Chemistry.

by simple heating, the authors encountered problems (measured BET surface area of $278\,m^2\,g^{-1}$) which was not the case with samples obtained by solution synthesis (BET $900–2200\,m^2\,g^{-1}$). However, if washing with ethanol is applied prior to activation by heating, a significant increase in BET surface area is observed ($965–1364\,m^2\,g^{-1}$). A possible explanation is that the pores within the mechanochemical product might be blocked by traces of unreacted material or nonvolatile by-products. This was supported by SEM images of 15-min neat grinding samples that reveal the removal of some surface particles by washing. Interestingly, SEM analysis also showed that smoother mesoscale texture was obtained after grinding for 15 min than after 5 min, consistent with the observed hysteresis in N_2 sorption isotherms for samples ground for 5 min. Such hysteretic isotherms suggest mesoporosity in the form of cracks and fissures in combination with microporosity (Fig. 1.19).

In 2015, Jones et al. employed SEM technique for the analysis of particle size after mechanochemical co-crystallization of caffeine with citric and anthranilic acids in the presence of different amounts of a PEG additive during POLAG [23]. It was found that the physical state of the polymer influenced the particle size of the caffeine-citric acid co-crystal in such a way that when 10 wt% of a liquid polymer (PEG 200, 300, and 400) was added, the particle size was larger than the particle size of the samples prepared with 10 wt% of solid polymers like PEG 3000, PEG 6000, and PEG 10,000. The shape of the co-crystal particles remained similar in all analyzed samples, irrespective of the polymer physical state (Fig. 1.20). SEM analysis performed on caffeine-anthranilic acid co-crystal samples revealed that LAG and POLAG using a liquid polymer (PEG 200) yielded powders with similar particle size, whereas in the case of POLAG with PEG 10,000 the powder particle size was smaller and more homogeneous.

1.4.2 MONITORING OF MECHANOCHEMICAL REACTIONS

The analytical techniques briefly discussed in the previous section are also employed to monitor how mechanochemical processes evolve with time. Such time-dependent

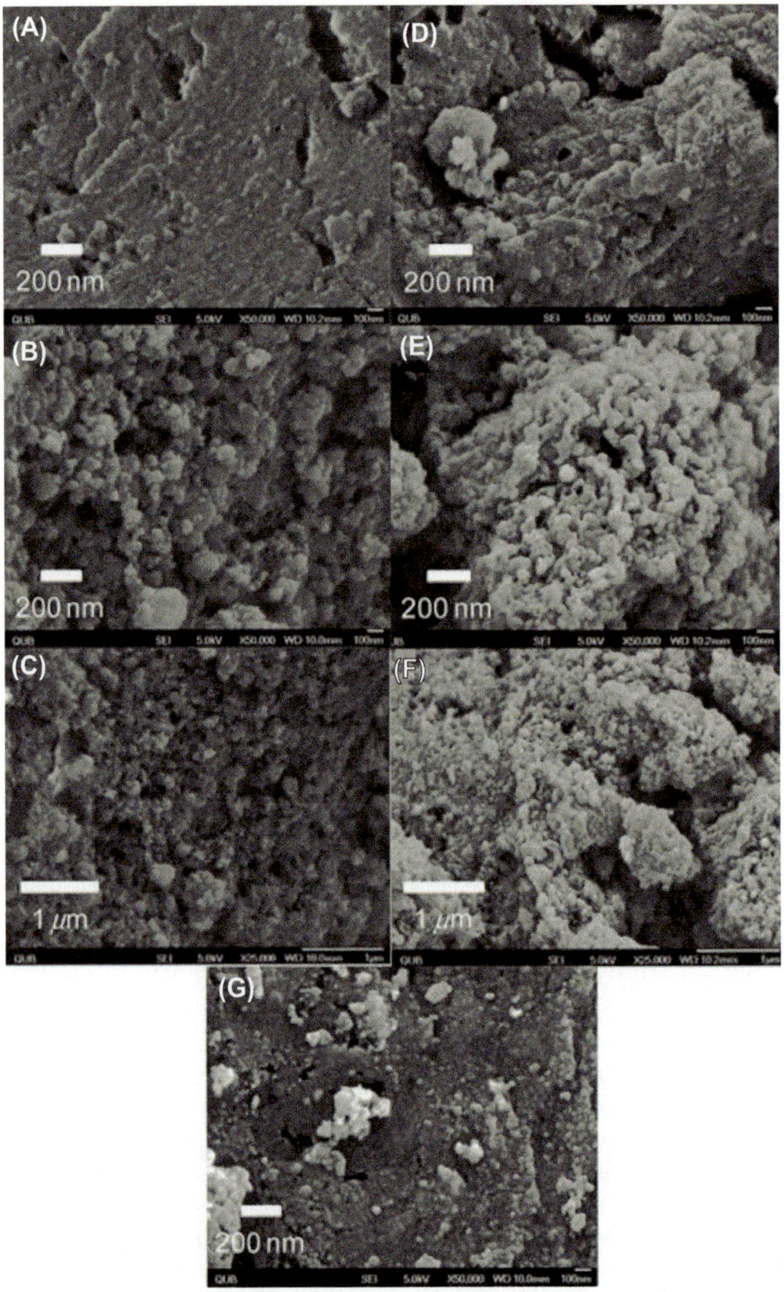

FIGURE 1.19

SEM images of $Cu_3(BTC)_2$ prepared by (A) 15 min neat grinding followed by washing with ethanol, (B) 15 min LAG followed by (C) washing with ethanol, (D) 5 min neat grinding followed by washing with ethanol, (E) 5 min LAG followed by (F) washing with ethanol, and (G) 15 min neat grinding without washing.

Reproduced from Yuan W, Lazuen Garay A, Pichon A, Clowes R, Wood CD, Cooper AI, et al. Study of the mechanochemical formation and resulting properties of an archetypal MOF:Cu3(BTC)2 (BTC = 1,3,5-benzenetricarboxylate). CrystEngComm 2010;12:4063–5 with permission from the Royal Society of Chemistry.

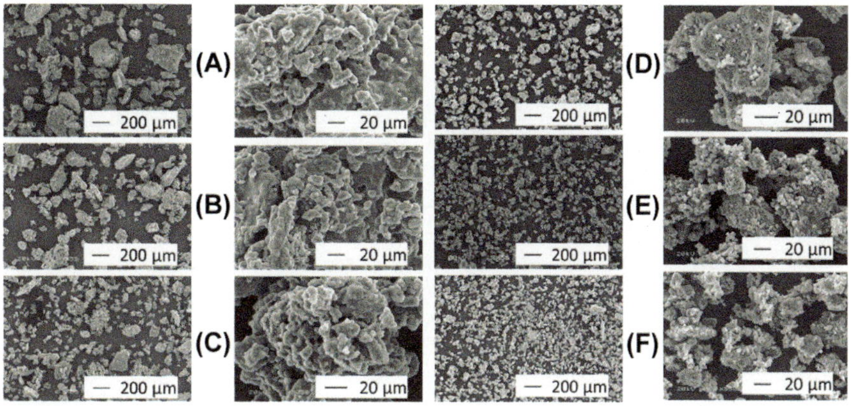

FIGURE 1.20

Left: SEM images of caffeine–citric acid co-crystal samples prepared using 10wt% of (A) PEG 200, (B) PEG 300, (C) PEG 400, (D) PEG 3000, (E) PEG 6000, and (F) PEG 10,000.

Copyright (2015) Wiley. Used with permission from Hasa D, Schneider Rauber G, Voinovich D, Jones W. Cocrystal formation through mechanochemistry: from neat and liquid-assisted grinding to polymer-assisted grinding. Angew Chem Int Ed 2015;54:7371–5.

measurements are important from the aspect of better understanding the underlying mechanisms that govern the observed solid-state reactivity and may allow for more efficient control and optimization of milling synthesis, especially for the scale-up purposes. However, a major obstacle in real-time in situ studies of mechanochemical reactions is imposed by the milling equipment that has not been designed to meet the specific requirements of such measurements. For that reason, up until only recently, the majority of analyses of the progress of mechanochemical reactions relied exclusively on ex situ measurements.

1.4.2.1 Ex Situ Monitoring

Ex situ analysis, commonly performed by deliberately interrupting the ongoing mechanochemical reaction, is typically carried out by using PXRD, infrared or Raman spectroscopy. In a study of mechanochemical synthesis of cadmium–cyanoguanidine (cnge) coordination polymers, Friščić et al. resorted to time-resolved PXRD analysis to monitor the progress of these reactions [21]. Ball-milling cadmium chloride and cnge for 30 min in an equimolar ratio under neat or LAG conditions with two 7 mm balls (each 1.4 g) or a single 10 mm ball (4.0 g) resulted in quantitative synthesis of a 1:1 complex Cd(cnge)Cl_2. Strikingly, in an attempt to prepare a lower dimensionality 1:2 complex $Cd(cnge)_2Cl_2$ from a 1:2 mixture of $CdCl_2$ and cnge, solvent-free milling with 7 mm balls yielded only the 1:1 complex and unreacted cnge while some 1:2 product formation was observed under LAG conditions. The quantitative mechanosynthesis of 1:2 complex required harsher conditions (neat or LAG) using the 10 mm grinding ball, based on an ex situ PXRD measurements (Fig. 1.21). Both of the products were fully characterized by IR and ssNMR spectroscopy, and their crystal structures were determined from the PXRD data collected on a laboratory diffractometer.

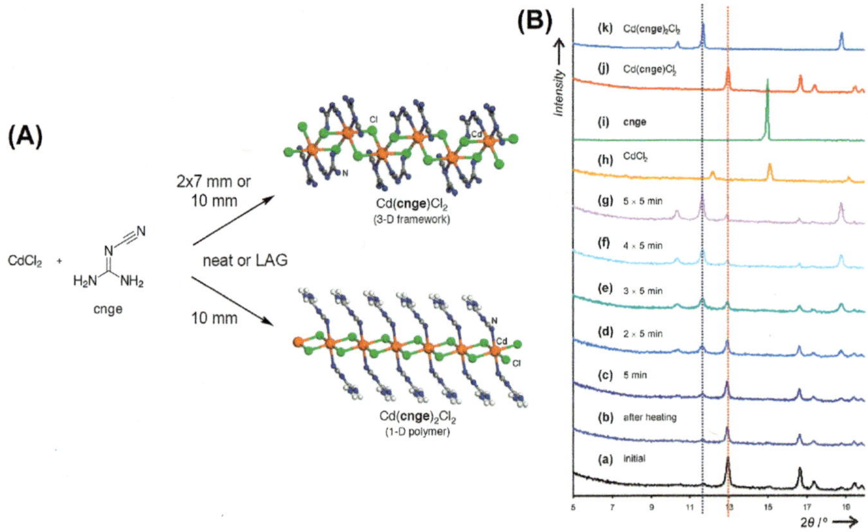

FIGURE 1.21

(A) Mechanosynthesis of cadmium–cyanoguanidine coordination polymers, (B) Ex situ powder X-ray diffraction analysis of the 1:2 reaction between $CdCl_2$ and cnge carried out by preheating at 85°C for 1 h followed by 5-min milling cycles. Notably, the 1:1 Cd(cnge)Cl_2 intermediate gradually transforms to 1:2 product Cd(cnge)$_2Cl_2$.

Reproduced from Štrukil V, Fábián L, Reid DG, Duer MJ, Jackson GJ, Eckert-Maksić M, et al. Towards an environmentally-friendly laboratory: dimensionality and reactivity in the mechanosynthesis of metal-organic compounds. Chem Commun 2010;46:9191–3 with permission from the Royal Society of Chemistry.

Conducting the 1:2 reaction with the 10 mm ball in six 5-min milling cycles, where the thermal effect was minimized, afforded only a mixture of the 1:1 and 1:2 complexes, in contrast to continuous milling for 10–30 min which gave exclusively the 1:2 product. Also, minimization of the mechanical effect in aging experiments at 85°C did not lead to transformation to the 1:2 product, suggesting that the temperature increase alone was not enough to drive the reaction to completion. However, when the reaction mixture was kept for 1 h at 85°C prior to each 5-min milling cycle, the ex situ PXRD analysis showed that the formation of a thermodynamically stable 1:2 complex Cd(cnge)$_2Cl_2$ proceeded via a 1:1 intermediate Cd(cnge)Cl_2 within 30 min. These results indicated that kinetic hindrances during the transformation of the 1:1 to 1:2 complex could be overcome by a joint effect of increased grinding force and average sample temperature, demonstrating that mechanical and thermal effects in the mechanosynthesis of Cd–cnge coordination polymers could not be separated.

The synthesis of *N*-(benzotriazole-1-thiocarbamoyl)benzimidazole-2-thione (**58**) by LAG milling *ortho*-phenylenediamine (**24**) with two equivalents of bis(benzotriazolyl) methanethione (**42**) for 2 h provides an example of an organic reaction monitoring using time-resolved ex situ PXRD analysis [44] (Fig. 1.22). While the corresponding 1:1 reaction quantitatively afforded the thiocarbamoyl benzotriazole **45**, in this case the desired

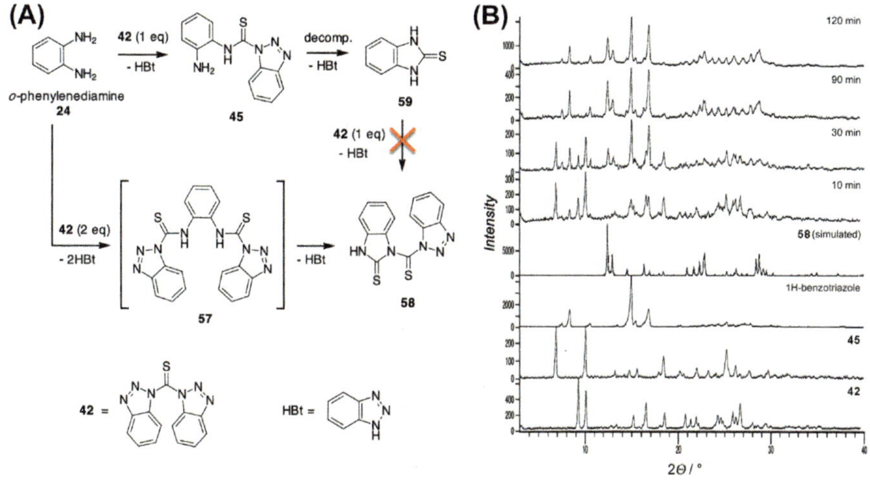

FIGURE 1.22

(A) Reaction mechanism for the mechanosynthesis of thiocarbamoyl benzimidazole-2-thione **58**. (B) Time-resolved powder X-ray diffraction analysis of the 1:2 mechanochemical reaction between *o*-phenylenediamine **24** and methanethione **42** using one 10-mm grinding ball. After 10 min, the major product is mono-thiocarbamoylated intermediate **45** which undergoes transformation to **58** within 2 h of milling.

bis(thiocarbamoyl) benzotriazole **57** could not be isolated. Instead, the benzimidazole thione derivative **58** and 1H-benzotriazole were the only products. However, separate milling experiments excluded the thiocarbamoylation of benzimidazole-2-thione (**59**) as the reaction path leading to **58**. Based on these observations and the ex situ PXRD, the authors proposed compound **57** as the intermediate, reactivity of which hindered its direct detection by PXRD. In the last step, **57** undergoes an intramolecular cyclization to yield **58** and benzotriazole by-product.

A nice demonstration of the application of ex situ attenuated total reflectance IR spectroscopy (ATR-IR) was published by Stolle et al. in 2015 [56]. The hydroamination reaction between *o*-aminophenol (**60**) and dimethyl acetylenedicarboxylate (**61**) was found to rapidly proceed solvent-free in a mortar or in a planetary ball mill to quantitatively yield the benzoxazine product **62**, according to ATR-IR analysis. Since simple mixing of the reactants also provided the product, it was recognized that a fast and direct analysis such as ATR-IR, which excludes any pretreatment of the sample, would be required in this case. Characteristic absorption bands for the product lactone carbonyl stretching vibration at 1762 cm^{-1}, the reactant ester carbonyl vibration at 1720 cm^{-1}, along with the aromatic ring bending at 750 cm^{-1} were used to monitor the progress of the hydroamination reaction. Full conversion was indicated by the ratio (A_v) of transmission minima from the lactone stretching and ring bending being >0.46 and the absence of the ester vibration at 1720 cm^{-1} (Fig. 1.23A).

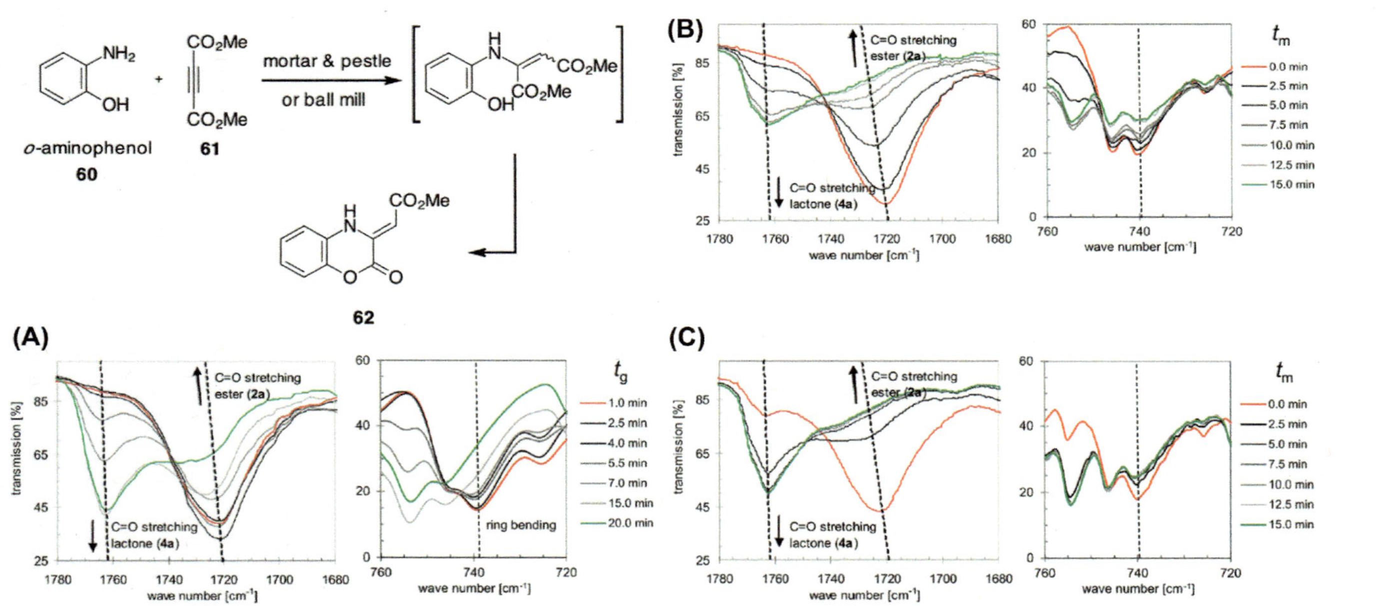

FIGURE 1.23

Monitoring of benzoxazine **62** synthesis in a mortar with pestle using ATR-IR spectroscopy. (A) Continuous grinding at different t_g, (B) 1-min grinding followed by ATR measurement at different t_m, and (C) 5-min grinding followed by ATR measurement at different t_m.

Due to possible pressure-induced reaction during ATR-IR measurement, the reaction time t_r was separated into three components: grinding time (t_g), sample transfer time (t_{st}, typically <40 s), and measurement time (t_m). Fig. 1.23B shows changes in the intensities during ATR-IR measurement after the initial 1-min grinding in a mortar, while Fig. 1.23C illustrates the changes after the initial 5-min grinding. Quantification of analogous experiments under short (t_g<4 min) mixer and vibrational ball-milling conditions and simple mixing with a spatula showed that the reaction continued during ATR spectra collection for approx. t_m=7.5 min until equilibrium with A_v values of c. 0.20–0.25 were reached. On the other hand, treating the reactants in a mortar with pestle for 1–5 min resulted in a steady increase of A_v values to c. 0.40 within 15 min of the ATR measurement. Longer grinding time ($t_g \geq 7$ min) led to low reaction rates with respect to the measurement time t_m, indicating full conversion prior to the analysis. The authors concluded that for the investigation of the progress of solvent-free reactions the sample preparation and measurement times should be kept at minimum, while the additional solvent-based treatment during workup and preparation should be avoided.

An alternative to IR spectroscopic measurements is to employ Raman spectroscopy for a quantitative analysis. The first example of such ex situ measurements on mechanochemically promoted reactions is provided in a paper by James et al. [57]. In this work, a mechanochemical synthesis of ZIF-6 from zinc oxide and imidazole (HIm) was selected as the model reaction. ZIF-6 is a zeolitic-type coordination polymer with the composition $Zn(Im)_2{\cdot}nH_2O$, one of several forms that arise from this combination of starting material depending on the reaction conditions. In order to collect samples for the analyses, ball milling was interrupted periodically over a period of 90 min and fresh samples were taken out from the grinding jar. Additional measurements showed that the reaction did not continue detectably if the milling process was stopped before completion. By monitoring the disappearance of the absorption band at 1322 cm^{-1} and the emergence of the band at 970 cm^{-1}, the extent of the reaction in terms of reacted imidazole α_{HIm} was calculated.

Integration of the kinetic data collected at different milling frequencies from the $(1-\alpha_{HIm})$ versus time plot, gave straight lines in the second order $1/(1-\alpha_{HIm})$ versus time plot showing that the rate of the reaction is a function of milling frequency (Fig. 1.24). In such a model, the instantaneous rate is proportional to the amounts of each reactant remaining, rate $= k(1-\alpha_{HIm})(1-\alpha_{ZnO})$, what is a solid-state equivalent of a simple solution or gas phase second order reaction. It was somewhat surprising to find that the kinetics of a milling reaction would display such behavior, and that the overall rate of the reaction is simply determined by the frequency of reactive contacts between reactants. The authors have also noticed an interesting feature of the studied reaction: the rates at different frequencies are virtually insensitive to temperature changes during milling, especially at higher frequencies, where frictional heating contributes to a rise in the average sample temperature. Based on these results, the authors proposed a fluid-like model for the description of mechanochemical reactions.

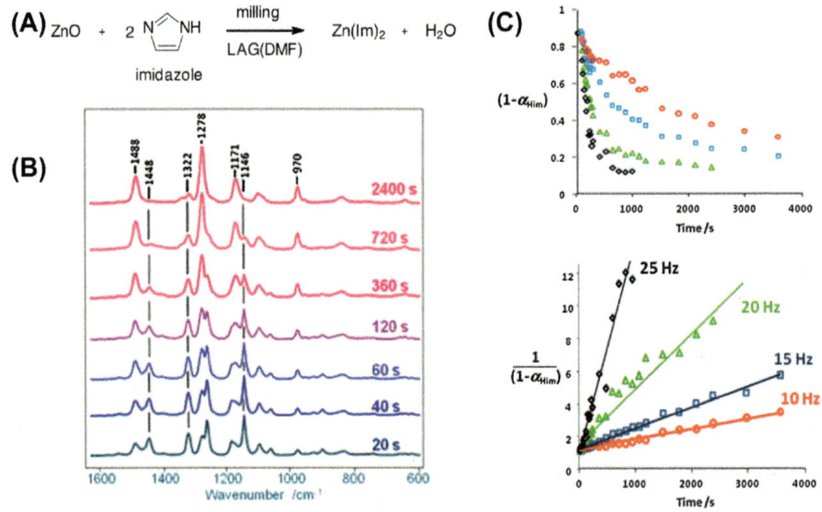

FIGURE 1.24

(A) Mechanosynthesis of ZIF-6 studied by ex situ Raman spectroscopy. (B) Time-dependent Raman spectra with characteristic bands used to monitor the progress of the model reaction. (C) Kinetic profiles at different milling frequencies (top) and the integrated second-order plot (bottom).

Adapted from Ma X, Yuan W, Bell SEJ, James SL. Better understanding of mechanochemical reactions: Raman monitoring reveals surprisingly simple 'pseudo-fluid' model for a ball milling reaction. Chem Commun 2014;50:1585–7 with permission from the Royal Society of Chemistry.

1.4.2.2 In Situ Monitoring

The above examples of ex situ analysis of mechanochemical reactions provide important information on their course and even allow kinetic data to be extracted. But as pointed out by Stolle using ATR-IR spectroscopy, the mechanochemically-induced reactions may continue despite the mechanical effect by interrupting/stopping the grinding or when the milling process has been suspended [56]. Such phenomena may lead to inconsistencies between the ex situ measurements and the actual state of the reaction taking place in a sealed grinding jar. Obviously, in order to alleviate these artifacts of stepwise analysis, in situ analysis would be required. Accomplishing this, however, had not been an easy task and it was not until recently that such in situ measurements were made possible.

The breakthrough in real-time in situ monitoring of mechanochemical reactions came in 2013 when Friščić et al. employed, for the first time, high-energy synchrotron X-rays to study the time-dependent diffraction of samples undergoing a solid-state reaction under ball-milling conditions [58]. For this purpose, two essential modifications to the existing milling equipment had to be made. Firstly, a standard laboratory Retsch MM200 ball mill was modified so as to allow for direct X-ray diffraction measurements using a synchrotron beam. Secondly, the grinding jars used in most of the experiments were manufactured from transparent poly(methyl)methacrylate (PMMA, also known as Plexiglas or Perspex) which has an amorphous structure and low absorption of X-rays (Fig. 1.25A). Steel and aluminum jars were also used, especially for neat grinding where plastic jars rapidly deteriorated [59]. After passing through the PMMA jar, the diffracted beam hits a two-dimensional detector that enables high-quality diffraction patterns to be collected with resolution in seconds.

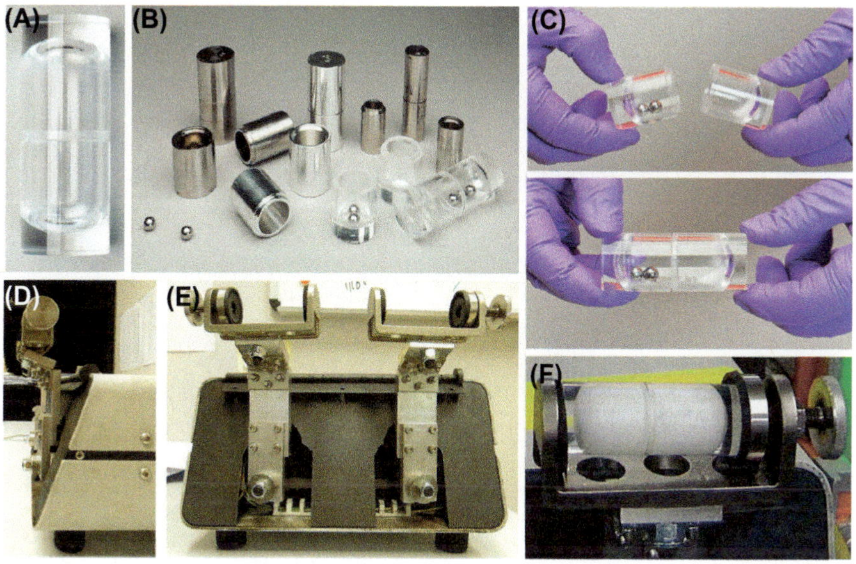

FIGURE 1.25

(A) A 14-mL PMMA milling jar used for in situ PXRD monitoring; (B) steel, aluminum, or PMMA jars of different sizes; (C) handling of the two halves of the PMMA milling jar before snapping them together (the left side contains the milling liquid and milling balls and the right side contains the reactants); (D) side view and (E) front view of the modified Retsch MM200 mill with the milling stations lifted above the mill to allow for the X-ray beam to pass above the mill and through the jars, and (F) secured PMMA jar on the milling station.

Reprinted by permission from Macmillan Publishers Ltd: Halasz I, Kimber SAJ, Beldon PJ, Belenguer AM, Adams F, Honkimäki V, et al. In situ and real-time monitoring of mechanochemical milling reactions using synchrotron X-ray diffraction. Nat Protoc 2013;8:1718–29, copyright 2013.

The technique was first tested on the mechanochemical synthesis of zeolitic imidazolate frameworks (ZIFs) by milling zinc oxide with 2-methyl-(HMeIm) or 2-ethylimidazole (HEtIm) under LAG and ILAG conditions. In situ monitoring revealed that the LAG synthesis of ZIF-8 from ZnO and HMeIm followed the nucleation and growth mechanism that was identified by a sigmoidal curve in the diffracted intensity vs. time plot. The measurements also showed that ILAG synthesis of ZIF-8 using ammonium nitrate proceeded at a much faster rate with the

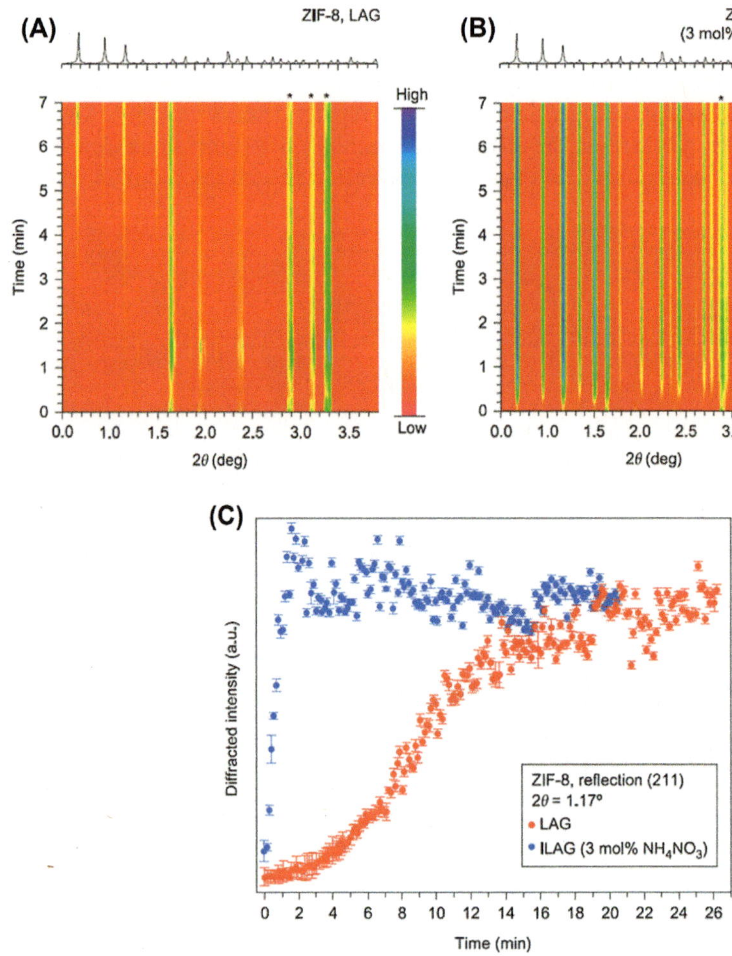

FIGURE 1.26

Time-resolved diffractograms for the (A) LAG and (B) ILAG synthesis of ZIF-8. (C) Time-dependent variation of the intensity of the (211) reflection of the ZIF-8 product.

Reprinted by permission from Macmillan Publishers Ltd: Friščić T, Halasz I, Beldon PJ, Belenguer AM, Adams F, Kimber SAJ, et al. Real-time and in situ *monitoring of mechanochemical milling reactions. Nat Chem 2013;5:66–73, Copyright 2013.*

maximum conversion reached in only 4 min (Fig. 1.26). Interestingly, in the case of ZIF derived from ZnO and HEtIm, the interconversion of several framework topologies in a stepwise reaction sequence was readily observed. The analysis of the diffracted intensities for all crystalline phases involved enabled a detailed insight into the mechanism of mechanochemical solid-state $Zn(EtIm)_2$ formation. This in situ analytical technique also provided information on the effect of different amounts of grinding liquid used in LAG, represented by η [17a], on the stability of each intermediate framework. It was established that decreasing amounts of liquid phase (low η values) shortened lifetimes of the intermediate with analcime topology, demonstrating that the role of liquid phase in LAG is not restricted only to providing a medium that facilitates solid-state diffusion and mass transfer.

In the next step, more precise in situ measurements of mechanochemical ZIF synthesis were conducted by using microcrystalline silicon as an internal scattering standard [60]. By employing silicon, the problems related to quantification of the amorphous phase during milling and variations in the amount of diffracting sample were successfully eliminated. Mechanochemical reaction between ZnO and imidazole was selected as a model reaction to monitor the kinetics of ZIF zni-$Zn(Im)_2$ formation. The authors found that the rate at which ZnO reflections disappeared best fit the first-order kinetics, providing additional evidence to the pseudo-fluid model of mechanochemical reactions put forward by James et al. [57]. Also, this study showed that the amorphous content in LAG and ILAG syntheses was much higher than previously established by ex situ analysis. Further in situ PXRD investigation of acid-induced ZIF-8 amorphization led the authors to discovery of a novel, previously unrecognized topology of imidazolate frameworks named katsenite, *kat*-$Zn(MeIm)_2$ [61]. Katsenite phase was observed upon crystallization of the amorphous phase, however, its occurrence was not reproducible suggesting the importance of nucleation during mechanochemical reactions. Although unstable at elevated temperatures and in solution environment, its stability under solvent-free conditions at room temperature allowed full characterization by structure determination from PXRD data, solid-state NMR, ATR-IR, SEM, thermogravimetric analysis, and porosity measurements (Fig. 1.27).

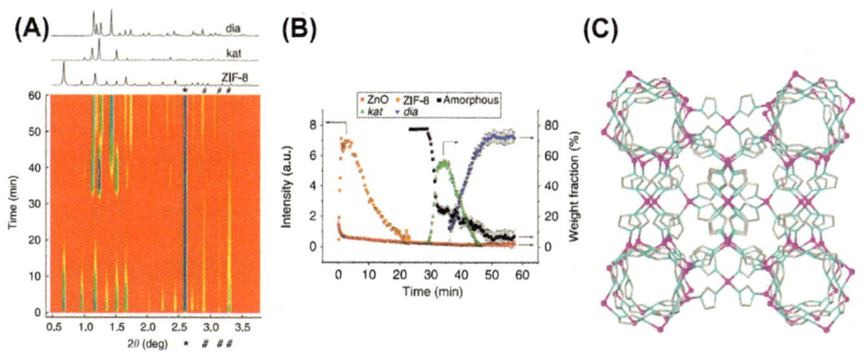

FIGURE 1.27

(A) Time-resolved diffractogram for the LAG reaction of zinc oxide and 2-methylimidazole with aqueous acetic acid as the grinding liquid during mechanochemical ZIF-8 → amorphous → *kat*-$Zn(MeIm)_2$ → *dia*-$Zn(MeIm)_2$ transformation (*kat*, katsenite phase; *dia*, diamondoid phase). (B) Plot for the evolution of each phase, and (C) structure of *kat*-$Zn(MeIm)_2$ viewed along the crystallographic *c*-axis.

Reprinted by permission from Macmillan Publishers Ltd: Katsenis AD, Puškarić A, Štrukil V, Mottillo C, Julien PA, Užarević K, et al. In situ X-ray diffraction monitoring of a mechanochemical reaction reveals a unique topology metal-organic framework. Nat Commun 2015;6:6662.

The real-time in situ synchrotron PXRD technique found its next application in the mechanistic study of mechanochemically promoted co-crystallization of carbamazepine and nicotinamide as representatives of active pharmaceutical ingredients (API) [62]. In the case of carbamazepine:saccharin co-crystal, in situ analysis confirmed previous studies by cryomilling that neat grinding led to amorphization while LAG synthesis rapidly produced the co-crystal. The other investigated system involved two-step neat grinding synthesis of nicotinamide:suberic acid 2:1 co-crystal. The ex situ PXRD analysis had revealed a stepwise mechanism with the 1:1 co-crystal as the intermediate. Whereas LAG synthesis was too fast for ex situ approach, in situ monitoring enabled the discovery of an unknown phase that preceded the rapid formation of the 1:1 intermediate (Fig. 1.28). However, the authors were not successful at full characterization of the new phase due to its pronounced instability.

While in situ synchrotron PXRD measurements are proven as an invaluable tool for monitoring the course of mechanochemical synthesis of MOFs and API

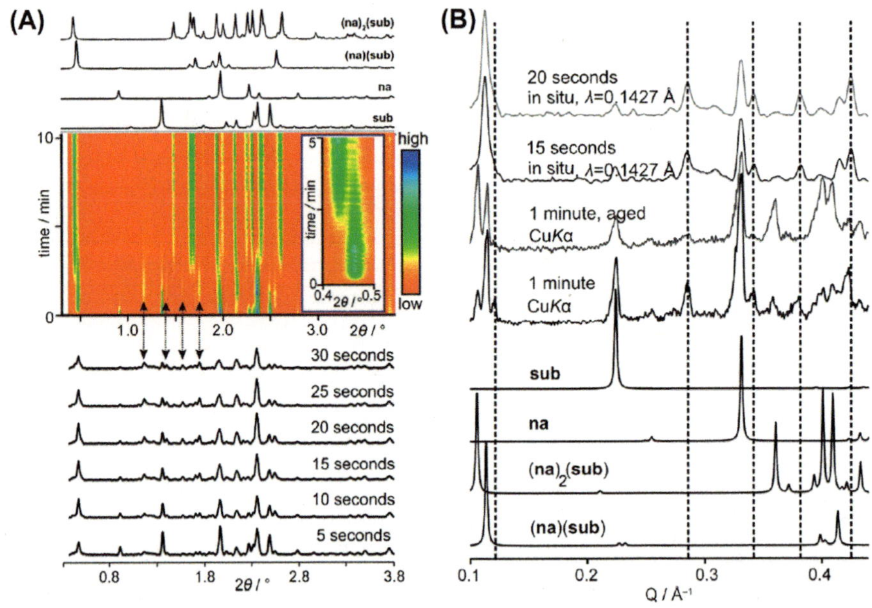

FIGURE 1.28

(A) Time-resolved diffractogram for the LAG synthesis of nicotinamide (na):suberic acid (sub) 2:1 co-crystal demonstrating the formation of the 1:1 intermediate co-crystal. The *arrows* indicate the unknown phase reflections. (B) The comparison of the in situ synchrotron patterns (after 15 and 20s) with the laboratory PXRD patterns of a sample after 1 min of milling and aging. *Dotted lines* indicate the reflections corresponding to the unknown phase.

Copyright (2013) Wiley. Used with permission from Halasz I, Puškarić A, Kimber SAJ, Beldon PJ, Belenguer AM, Adams F, et al. Real-time in situ powder X-ray diffraction monitoring of mechanochemical synthesis of pharmaceutical cocrystals. Angew Chem Int Ed 2013;52:11538–41.

co-crystals, there are a few limitations that are incompatible with its routine, day-to-day application. First of all, the need for a synchrotron source, accompanied by substantial financial costs, makes it relatively inaccessible to a broader range of researchers. Secondly, from the aspect of organic chemistry, the "softness" of organic crystals often leads to significant amorphization due to high-energy milling, a situation where in situ PXRD measurements can provide only limited information about the molecular-level structural changes during such mechanochemical synthesis. These limitations can be successfully alleviated by employing molecular spectroscopy analytical methods such as infrared and Raman spectroscopy.

With the development of readily available laser sources incorporated into portable and easy-to-handle spectrometers, Raman spectroscopy has found many applications in various types of analyses. One such application was introduced in 2014 as a real-time in situ methodology for monitoring the course of chemical reactions under milling conditions, which is fully complementary to the existing PXRD technique [63]. The detection of the time-resolved Raman scattering signal was made possible by using translucent plastic grinding jars. The experimental setup, shown in Fig. 1.29A, consists of a Raman probe on an adjustable stand placed underneath

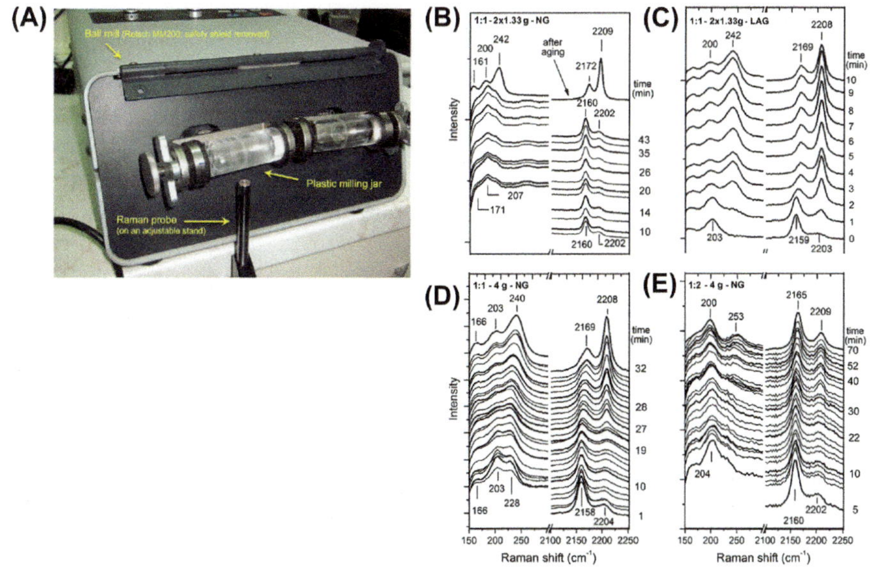

FIGURE 1.29

(A) An experimental setup for real-time in situ Raman spectroscopy monitoring of mechanochemical reactions. Time-resolved Raman spectra for $CdCl_2$–cnge neat grinding (B) and LAG (C) 1:1 reaction using two 7 mm balls. (D) and (E) show the spectral changes during neat grinding of 1:1 and 1:2 $CdCl_2$–cnge mixtures using a single 10 mm ball, respectively.

the plastic milling jar that is horizontally attached to one milling station of Retsch MM200 mill. The probe can be moved up and down so as to allow fine adjustment of the laser focus to be 1–2 mm inside the jar. The new methodology was first validated by monitoring the progress of mechanochemical synthesis of cadmium coordination polymers and nicotinamide-suberic acid co-crystals. These model reactions were previously studied by either ex situ (Cd–cnge polymers) or in situ PXRD analysis (nicotinamide:suberic acid co-crystals), and Raman spectroscopic measurements were assumed to provide complementary information on the mechanistic aspects of mechanochemical reactions induced in a stepwise (ex situ analysis) or continuous (in situ analysis) manner.

Through monitoring the changes in the spectral region around 2200 cm^{-1}, characteristic for cyano group stretching vibration, and around 200 cm^{-1} typical for coordinated cadmium ions, in situ Raman spectroscopy revealed interesting features of mechanochemical complexation of cadmium chloride with cyanoguanidine (Fig. 1.29B–E). In contrast to ex situ study, it was found that neat grinding of 1:1 or 1:2 $CdCl_2$-cnge mixtures with two 7 mm balls did not result in the formation of 1:1 complex Cd(cnge)Cl_2. However, when these samples were taken out of the plastic or steel jar and analyzed after being exposed to air for several minutes to several days, quantitative formation of the 1:1 complex was confirmed. Switching from two small balls to a heavier one (10 mm) however, quantitatively afforded the 1:1 complex in 30 min, as suggested by the emergence of bands at 2169 and 2208 cm^{-1}. In the case of 1:2 reaction, a direct formation of the corresponding 1:2 complex Cd$(cnge)_2Cl_2$, was observed, contrasting the ex situ analysis where 1:1 complex was proposed as the intermediate. Surprisingly, when a slightly smaller ball (2.9 g) was used, the time-resolved spectra revealed the formation of 1:1 intermediate after 27 min followed by its complete disappearance to afford 1:2 complex after 40 min of milling. Finally, LAG reactions using methanol as the grinding liquid were investigated. While neat grinding with 7 mm balls was shown to yield only a reactant mixture, LAG facilitated the 1:1 reaction and quantitatively afforded Cd(cnge)Cl_2 in just 4 min. With a 1:2 stoichiometric ratio, LAG first provided the 1:1 intermediate, followed by quantitative transformation to the 1:2 product. These results, obtained by in situ Raman measurements, clearly demonstrate the sensitivity of mechanochemical reactions to environmental conditions and sample manipulation during ex situ analysis.

In following studies, the focus of real-time in situ Raman spectroscopic measurements has been shifted from coordination polymers and co-crystals toward mechanochemical organic and organometallic reactions. Monitoring of solid-state reaction between equimolar quantities of 4′-(*N*,*N*-dimethylamino)-4-nitroazobenzene and palladium acetate with glacial acetic acid as an additive revealed facile formation of the expected dimeric monopalladated complex via aromatic C—H bond activation [64]. When this complex was allowed to further react with two equivalents of palladium acetate, Raman measurements clearly indicated the formation of dicyclopalladated complex, which can only be accessed by milling. In another study, the authors investigated the mechanochemical reaction of 4-nitrobenzoyl azide with *p*-phenylenediamine and followed the course of the reaction using in situ Raman spectroscopy by monitoring the changes in carbonyl stretching vibration upon nucleophilic substitution [65]. Interestingly, the effect of the grinding liquid in LAG experiments could be correlated with the basicity of the liquid expressed in terms of Gutmann's donor number. The reaction was found to proceed for c. 16 min when acetonitrile was used in LAG, in contrast to LAG with ethanol (8 min) or DMF which took only 4 min. The base-catalyzed nature of the studied nucleophilic substitution reaction was additionally supported by ILAG using basic sodium acetate that accelerated the reaction (6.5 min) in comparison to pure LAG with ethanol.

Another exciting contribution of in situ Raman spectroscopic monitoring was the discovery and identification of aromatic thiocarbamoyl benzotriazoles as bench-stable synthetic equivalents of isothiocyanates [44]. In an attempt to utilize bis(benzotriazolyl)methanethione (**42**) as a reagent to promote solid-state synthesis of thiourea **65** from *p*-bromoaniline (**63**), in situ monitoring revealed an intermediate species with characteristic Raman shifts positioned at 544, 1041, 1159, 1248, and 1511 cm^{-1}, appearing about 2 min into milling and disappearing concomitantly with the appearance of thiourea product (Fig. 1.30). Since thiocarbamoyl benzotriazoles

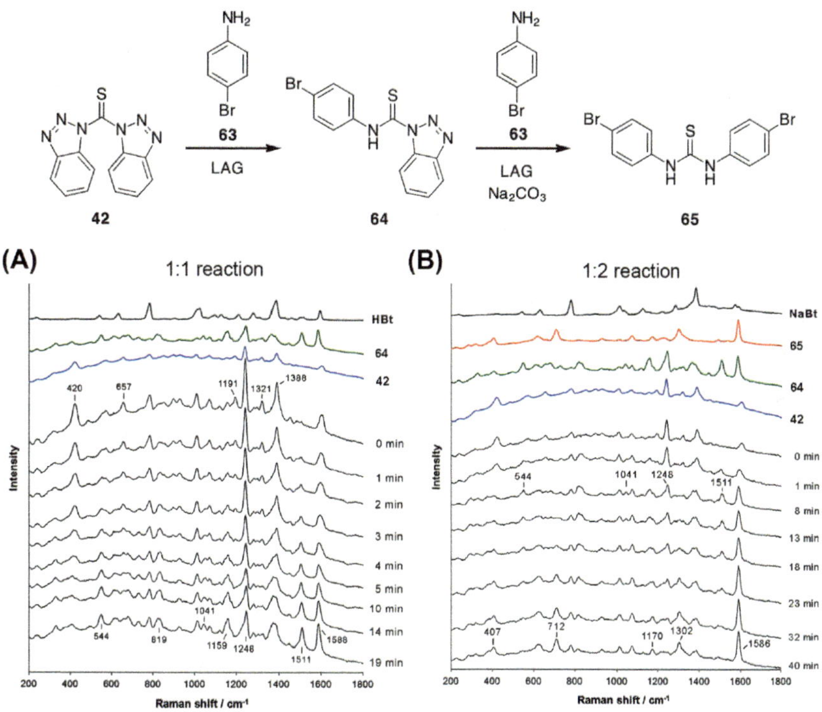

FIGURE 1.30

Time-resolved Raman spectra for (A) the LAG synthesis of thiocarbamoyl benzotriazole **64** and (B) the LAG synthesis of thiourea **65** using sodium carbonate as the base (HBt = 1H-benzotriazole).

had been proposed as reactive intermediates during thiourea solution synthesis from **42**, Raman spectroscopy was again used to follow the course of the solid-state LAG reaction between **42** and *p*-bromoaniline (**63**) in a 1:1 stoichiometry. It was established that the 1:1 reaction quantitatively yielded the product **64** which corresponded to the observed intermediate in the 1:2 reaction. As it is described in the previous section, the identity of thiocarbamoyl benzotriazoles was confirmed by solid-state analytical methods (see Section 1.4.1).

The latest advancement in real-time in situ monitoring of mechanochemical reactions was achieved by the Emmerling group [66]. By combining the synchrotron PXRD measurements with Raman spectroscopy, a tandem methodology for simultaneous collection of in situ PXRD and Raman data was devised. Such an approach allowed for an unprecedented study of formation of model 3D MOFs (eg, ZIF-8 and HKUST-1), layered structures, and organic molecular compounds. Since in situ PXRD and Raman spectroscopy have their own specific benefits, the combination of the two enables a detailed study of milling processes on the level of both molecular and crystalline structures, providing in this way reliable data for mechanistic studies of mechanochemical reactions.

REFERENCES

[1] Lynch AJ, Rowland CA. The history of grinding, society of mining. Littleton (CO): Metallurgy and Exploration, Inc.; 2005.

[2] Baláž P. Mechanochemistry in nanoscience and minerals engineering. (Berlin): Springer Verlag; 2008.

[3] Braga D, Maini L, Grepioni F. Mechanochemical preparation of co-crystals. Chem Soc Rev 2013;42:7638–48.

[4] Faraday M. On the decomposition of chloride of silver, by hydrogen, and by zinc. Q J Sci Lit Arts 1820;8:374–6.

[5] Takacs L. The historical development of mechanochemistry. Chem Soc Rev 2013; 42:7649–59.

[6] Ling AR, Baker JL. Halogen derivatives of quinone. Part III. Derivatives of quinhydrone. J Chem Soc Trans 1893;63:1314–27.

[7] Patil AO, Curtin DY, Paul IC. Solid-state formation of quinhydrones from their components. Use of solid-solid reactions to prepare compounds not accessible from solution. J Am Chem Soc 1984;106:348–53.

[8] Toda F, Tanaka K, Sekikawa A. Host-guest complex formation by a solid-solid reaction. J Chem Soc Chem Commun 1987:279–80.

[9] Etter MC, Frankenbach GM. Hydrogen-bond directed cocrystallization as a tool for designing acentric organic solids. Chem Mater 1989;1:10–2.

[10] [a] Etter MC, Reutzel SM, Choo CG. Self-organization of adenine and thymine in the solid state. J Am Chem Soc 1993;115:4411–2.
[b] Caira MR, Nassimbeni LR, Wildervanck AF. Selective formation of hydrogen bonded cocrystals between a sulfonamide and aromatic carboxylic acids in the solid state. J Chem Soc Perkin Trans 1995;2:2213–6.

[11] Kuroda R, Imai Y, Tajima N. Generation of a co-crystal phase with novel coloristic properties *via* solid state grinding procedures. Chem Commun 2002;37:2848–9.

[12] Shan N, Toda F, Jones W. Mechanochemistry and co-crystal formation: effect of solvent on reaction kinetics. Chem Commun 2002;37:2372–3.

[13] Trask AV, Motherwell WDS, Jones W. Solvent-drop grinding: green polymorph control of cocrystallisation. Chem Commun 2004;39:890–1.

[14] Nguyen KL, Friščić T, Day GM, Gladden LF, Jones W. Terahertz time-domain spectroscopy and the quantitative monitoring of mechanochemical cocrystal formation. Nat Mater 2007;6:206–9.

[15] [a] Karki S, Friščić T, Jones W. Control and interconversion of cocrystal stoichiometry in grinding: stepwise mechanism for the formation of a hydrogen-bonded cocrystal. CrystEngComm 2009;11:470–81.
[b] Cinčić D, Friščić T, Jones W. A stepwise mechanism for the mechanochemical synthesis of halogen-bonded cocrystal architectures. J Am Chem Soc 2008;130:7524–5.

[16] Friščić T, Trask AV, Jones W, Motherwell WDS. Screening for inclusion compounds and systematic construction of three-component solids by liquid-assisted grinding. Angew Chem Int Ed 2006;45:7546–50.

[17] [a] Friščić T, Childs SL, Rizvi SAA, Jones W. The role of solvent in mechanochemical and sonochemical cocrystal formation: a solubility-based approach for predicting cocrystallisation outcome. CrystEngComm 2009;11:418–26.
[b] Friščić T, Jones W. Recent advances in understanding the mechanism of cocrystal formation via grinding. Cryst Growth Des 2009;9:1621–37.

[18] Tanaka K, Toda F. Solvent-free organic synthesis. Chem Rev 2000;100:1025–74.

[19] Aakeröy CB, Chopade PD. Oxime decorated cavitands functionalized through solvent-assisted grinding. Org Lett 2011;13:1–3.

[20] Štrukil V, Igrc MD, Eckert-Maksić M, Friščić T. Click mechanochemistry: quantitative synthesis of "ready to use" chiral organocatalysts by efficient two-fold thiourea coupling to vicinal diamines. Chem Eur J 2012;18:8464–73.

[21] Štrukil V, Fábián L, Reid DG, Duer MJ, Jackson GJ, Eckert-Maksić M, Friščić T. Towards an environmentally-friendly laboratory: dimensionality and reactivity in the mechanosynthesis of metal-organic compounds. Chem Commun 2010;46:9191–3.

[22] Friščić T, Reid DG, Halasz I, Stein RS, Dinnebier RE, Duer MJ. Ion- and liquid-assisted grinding: improved mechanochemical synthesis of metal-organic frameworks reveals salt inclusion and anion templating. Angew Chem Int Ed 2010;49:712–5.

[23] Hasa D, Schneider Rauber G, Voinovich D, Jones W. Cocrystal formation through mechanochemistry: from neat and liquid-assisted grinding to polymer-assisted grinding. Angew Chem Int Ed 2015;54:7371–5.

[24] Suryanarayana C. Mechanical alloying and milling. Prog Mater Sci 2001;46:1–184.

[25] Schneider F, Szuppa T, Stolle A, Ondruschka B, Hopf H. Energetic assessment of the Suzuki–Miyaura reaction: a curtate life cycle assessment as an easily understandable and applicable tool for reaction optimization. Green Chem 2009;11:1894–9.

[26] Szuppa T, Stolle A, Ondruschka B, Hopfe W. An alternative solvent-free synthesis of nopinone under ball-milling conditions: investigation of reaction parameters. ChemSusChem 2010;3:1181–91.

[27] Thorwirth R, Bernhardt F, Stolle A, Ondruschka B, Asghari J. Switchable selectivity during oxidation of anilines in a ball mill. Chem Eur J 2010;16:13236–42.

[28] Štefanić G, Krehula S, Štefanić I. The high impact of a milling atmosphere on steel contamination. Chem Commun 2013;49:9245–7.

[29] Portada T, Margetić D, Štrukil V. [unpublished results].

[30] Miklós F, Hum V, Fülöp F. Eco-friendly syntheses of 2,2-disubstituted- and 2-spiroquinazolinones. ARKIVOC 2014;(vi):25–37.

[31] Seo J-M, Jeon I-Y, Baek J-B. Mechanochemically driven solid-state Diels–Alder reaction of graphite into graphene nanoplatelets. Chem Sci 2013;4:4273–7.

[32] Pankrushina N, Nikitina I, Chernjak E, Myz C, Shakhtshneider T, Boldyrev V. Solvent-free mechanochemical modification of lappaconitine and piroxicam. Mater Manuf Process 2008;23:561–5.

[33] Do J-L, Mottillo C, Tan D, Štrukil V, Friščić T. Mechanochemical ruthenium-catalyzed olefin metathesis. J Am Chem Soc 2015;137:2476–9.

[34] Fulmer DA, Shearouse WC, Medonza ST, Mack J. Solvent-free Sonogashira coupling reaction *via* high speed ball milling. Green Chem 2009;11:1821–5.

[35] Cook TL, Walker JA, Mack J. Scratching the catalytic surface of mechanochemistry: a multi-component CuAAC reaction using a copper reaction vial. Green Chem 2013;15:617–9.

[36] Métro T-X, Bonnamour J, Reidon T, Duprez A, Sarpoulet J, Martinez J, Lamaty F. Comprehensive study of the organic-solvent-free CDI-mediated acylation of various nucleophiles by mechanochemistry. Chem Eur J 2015;21:12787–96.

[37] Karki S, Fábián L, Friščić T, Jones W. Powder X-ray diffraction as an emerging method to structurally characterize organic solids. Org Lett 2007;9:3133–6.

[38] Štrukil V, Igrc MD, Fábián L, Eckert-Maksić M, Childs SL, Reid DG, Duer MJ, Halasz I, Mottillo C, Friščić T. A model for a solvent-free synthetic organic research laboratory: click-mechanosynthesis and structural characterization of thioureas without bulk solvents. Green Chem 2012;14:2462–73.

[39] Harris KDM. Powder diffraction crystallography of molecular solids. Top Curr Chem 2012;315:133–78.

[40] Cheung EY, Kitchin SJ, Harris KDM, Imai Y, Tajima N, Kuroda R. Direct structure determination of a multicomponent molecular crystal prepared by a solid-state grinding procedure. J Am Chem Soc 2003;125:14658–9.

[41] Brüning J, Schmidt MU. The determination of crystal structures of active pharmaceutical ingredients from X-ray powder diffraction data: a brief, practical introduction, with fexofenadine hydrochloride as example. J Pharm Pharmacol 2015;67:773–81.

[42] Trask AV, van de Streek J, Motherwell WDS, Jones W. Achieving polymorphic and stoichiometric diversity in cocrystal formation: importance of solid-state grinding, powder X-ray structure determination, and seeding. Cryst Growth Des 2005;5:2233–41.

[43] Fischer F, Scholz G, Batzdorf L, Wilke M, Emmerling F. Synthesis, structure determination, and formation of a theobromine:oxalic acid 2:1 cocrystal. CrystEngComm 2015;17:824–9.

[44] Štrukil V, Gracin D, Magdysyuk OV, Dinnebier RE, Friščić T. Trapping reactive intermediates by mechanochemistry: elusive aryl N-thiocarbamoylbenzotriazoles as bench-stable reagents. Angew Chem Int Ed 2015;54:8440–3.

[45] Coelho AA. TOPAS. Version 4.1. Coelho Software. Brisbane (Australia); 2007.

[46] Braga D, Maini L, Polito M, Grepioni F. Unexpected solid-solid reaction upon preparation of KBr pellets and its exploitation in supramolecular cation complexation. Chem Commun 2002;37:2302–3.

[47] Potrzebowski MJ. What high-resolution solid-state NMR spectroscopy can offer to organic chemists. Eur J Org Chem 2003;7:1367–76.

[48] Vogt FG, Clawson JS, Strohmeier M, Edwards AJ, Pham TN, Watson SA. Solid-state NMR analysis of organic cocrystals and complexes. Cryst Growth Des 2009;9:921–37.

[49] Ashbrook SE, Dawson DM, Seymour VR. Recent developments in solid-state NMR spectroscopy of crystalline microporous materials. Phys Chem Chem Phys 2014;16:8223–42.

[50] Hernández JG, Macdonald NAJ, Mottillo C, Butler IS, Friščić T. A mechanochemical strategy for oxidative addition: remarkable yields and stereoselectivity in the halogenation of organometallic Re(I) complexes. Green Chem 2014;16:1087–92.

[51] Fernandes JA, Sardo M, Mafra L, Choquesillo-Lazarte D, Masciocchi N. X-ray and NMR crystallography studies of novel theophylline cocrystals prepared by liquid assisted grinding. Cryst Growth Des 2015;15:3674–83.

[52] Balema VP, Wiench JW, Pruski M, Pecharsky VK. Solvent-free mechanochemical synthesis of phosphonium salts. Chem Commun 2002;37:724–5.

[53] Balema VP, Wiench JW, Pruski M, Pecharsky VK. Mechanically induced solid-state generation of phosphorus ylides and the solvent-free wittig reaction. J Am Chem Soc 2002;124:6244–5.

[54] Pichon A, Lazuen-Garay A, James SL. Solvent-free synthesis of a microporous metal-organic framework. CrystEngComm 2006;8:211–4.

[55] Yuan W, Lazuen Garay A, Pichon A, Clowes R, Wood CD, Cooper AI, James SL. Study of the mechanochemical formation and resulting properties of an archetypal MOF: $Cu_3(BTC)_2$ (BTC = 1,3,5-benzenetricarboxylate). CrystEngComm 2010;12:4063–5.

[56] Weiße M, Zille M, Jacob K, Schmidt R, Stolle A. Hydroamination reactions of alkynes with *ortho*-substituted anilines in ball mills: synthesis of benzannulated N-heterocycles by a cascade reaction. Chem Eur J 2015;21:6511–22.

[57] Ma X, Yuan W, Bell SEJ, James SL. Better understanding of mechanochemical reactions: Raman monitoring reveals surprisingly simple 'pseudo-fluid' model for a ball milling reaction. Chem Commun 2014;50:1585–7.

[58] Friščić T, Halasz I, Beldon PJ, Belenguer AM, Adams F, Kimber SAJ, Honkimäki V, Dinnebier RE. Real-time and *in situ* monitoring of mechanochemical milling reactions. Nat Chem 2013;5:66–73.

[59] Halasz I, Kimber SAJ, Beldon PJ, Belenguer AM, Adams F, Honkimäki V, Nightingale RC, Dinnebier RE, Friščić T. *In situ* and real-time monitoring of mechanochemical milling reactions using synchrotron X-ray diffraction. Nat Protoc 2013;8:1718–29.

[60] Halasz I, Friščić T, Kimber SAJ, Užarević K, Puškarić A, Mottillo C, Julien P, Štrukil V, Honkimäki V, Dinnebier RE. Quantitative *in situ* and real-time monitoring of mechanochemical reactions. Faraday Discuss 2014;170:203–21.

[61] Katsenis AD, Puškarić A, Štrukil V, Mottillo C, Julien PA, Užarević K, Pham M-H, Do T-O, Kimber SAJ, Lazić P, Magdysyuk O, Dinnebier RE, Halasz I, Friščić T. *In situ* X-ray diffraction monitoring of a mechanochemical reaction reveals a unique topology metal-organic framework. Nat Commun 2015;6:6662.

[62] Halasz I, Puškarić A, Kimber SAJ, Beldon PJ, Belenguer AM, Adams F, Honkimäki V, Dinnebier RE, Patel B, Jones W, Štrukil V, Friščić T. Real-time *in situ* powder X-ray diffraction monitoring of mechanochemical synthesis of pharmaceutical cocrystals. Angew Chem Int Ed 2013;52:11538–41.

[63] Gracin D, Štrukil V, Friščić T, Halasz I, Užarević K. Laboratory real-time and *in situ* monitoring of mechanochemical milling reactions by Raman spectroscopy. Angew Chem Int Ed 2014;53:6193–7.

[64] Juribašić M, Užarević K, Gracin D, Ćurić M. Mechanochemical C–H bond activation: rapid and regioselective double cyclopalladation monitored by *in situ* Raman spectroscopy. Chem Commun 2014;50:10287–90.

[65] Tireli M, Juribašić Kulcsar M, Cindro N, Gracin D, Biliškov N, Borovina M, Ćurić M, Halasz I, Užarević K. Mechanochemical reactions studied by *in situ* Raman spectroscopy: base catalysis in liquid-assisted grinding. Chem Commun 2015;51:8058–61.
[66] Batzdorf L, Fischer F, Wilke M, Wenzel K-J, Emmerling F. Direct *in situ* investigation of milling reactions using combined X-ray diffraction and Raman spectroscopy. Angew Chem Int Ed 2015;54:1799–802.

CHAPTER

Carbon–Carbon Bond-Forming Reactions

2

CHAPTER OUTLINE

2.1 HECK COUPLING

Mechanochemical Heck coupling reaction was initially developed by Frejd et al. [1]. Electron-rich unsubstituted and monosubstituted aryl halides **1** were coupled with protected amino acrylates **2** employing 5 mol% of palladium catalyst to afford *Z* products **3** in 25–76% yields (Scheme 2.1). Sodium chloride was used as grinding

Mechanochemical Organic Synthesis. http://dx.doi.org/10.1016/B978-0-12-802184-2.00002-9

aid, and it was found that the reaction performed best when a relatively small amount was added (5 mg/mg aryl halide). In contrast from reactions in dimethylformamide (DMF) solution, a number of electron-deficient aryl halides showed no reaction (4-iodonitrobenzene, 2-bromo-6-methylpyridine, 4-iodophenol, and chlorobenzene), while 1,3-diiodobenzene and 1,3,5-triidobenzene afforded a complex mixtures (Table 2.1). For comparison, the model reaction mixture of iodobenzene and methyl-2-[(*tert*-butoxycarbonyl)-amino] acrylate was heated to 80°C for 60 min with and without stirring in test tube, to afford 33% and 18% yields, respectively, which is much lower than obtained by mechanosynthesis. The reaction temperature was chosen to fit that generated by the ball mill when run at full speed for an hour. Furthermore, model experiments performed in a microwave oven obtained the yields lower than those of the corresponding ball-milling experiments.

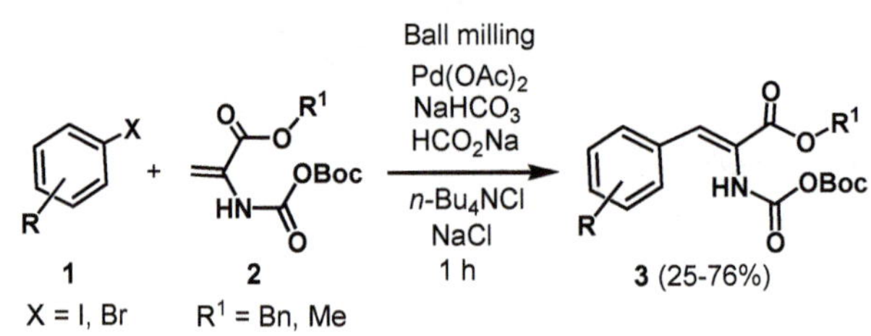

SCHEME 2.1

Heck coupling amino acrylates with aryl halides.

Table 2.1 Heck Coupling of Amino Acrylates With Aryl Halides[a]

Entry	Substrate	R^1	Yield (%)[b]
1	Iodobenzene	Bn	76
2	Iodobenzene	Me	77
3	Bromobenzene	Bn	57
4	Bromobenzene	Me	57
5	4-Iodoaniline	Me	88
6	4-Iodoanisole	Me	74
7	3-Iodobenzonitrile	Me	28
8	3-Bromofuran	Bn	13
9	2-Iodothiophene	Bn	25
10	2-Iodothiophene	Me	25

[a]*Fritsch Planetary Micro Mill Pulverisette 7, stainless steels containers, eight balls, Ar purged, 60 min, aminoacrylate: $NaHCO_3$:HCO_2Na:n-Bu_4NC:$Pd(OAc)_2$ 1.05:2.5:0.2:1.2:0.05, NaCl grinding auxiliary.*

[b]*Isolated yields.*

Reproduced from Tullberg E, Peters D, Frejd T. The Heck reaction under ball-milling conditions. J Organomet Chem 2004;689:3778–81. Copyright (2004), with permission from Elsevier.

These optimized Heck–Jeffery ball-milling conditions were applied in subsequent study to the synthesis of precursors of unnatural amino acids and indoles (Scheme 2.2) [2]. The reactions of aromatic iodides **4** with 2-*tert*-butoxycarbonylaminoacrylic acid methyl ester **5** yielded substituted coupling products **6** in modest to high yields within 1 h (Table 2.2). In these reactions poorer yields or no products were obtained for iodoanilines carrying the electron-withdrawing groups. Heck couplings with substrate carrying two amino groups in the standard reaction conditions gave the slightly poorer yield (35%) than under ball-milling conditions (entry 4, 44%). In addition, it was observed that some of the primary Heck coupling products **6** easily cyclized to give methyl-1*H*-indole-2-carboxylate derivatives **7** (Table 2.3). The other derivatives could be cyclized by heating in acetic anhydride.

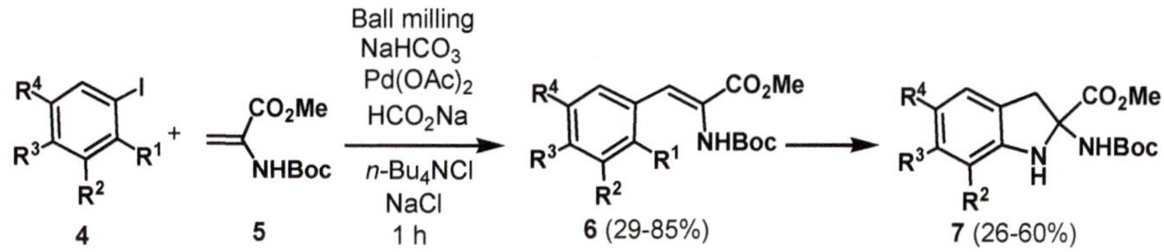

SCHEME 2.2

Heck coupling of amino acrylate with aryl iodides.

Table 2.2 Heck Coupling of Amino Acrylate With Aryl Iodides[a]

	Product				
Entry	**R^1**	**R^2**	**R^3**	**R^4**	**Yield (%)[b]**
1	NH_2			CN	29
2	NH_2			CO_2CH_3	31
3		NH_2			81
4		NH_2		NH_2	44
5	CH_3		NH_2		85
6		CH_3	NH_2		66
7	NH_2			F	–
8	NH_2		CF_3		43
9	OH				31
10			OH		64

[a]*Fritsch Planetary Micro Mill Pulverisette 7, stainless steels containers, eight balls, Ar purged, 60 min, aminoacrylate: $NaHCO_3$:HCO_2Na:n-Bu_4NC:$Pd(OAc)_2$ 1.05:2.5:0.2:1.2:0.05, NaCl grinding auxiliary.*

[b]*Isolated yields.*

Reproduced from Tullberg E, Schacher F, Peters D, Frejd T. Solvent-free Heck-Jeffery reactions under ball-milling conditions applied to the synthesis of unnatural amino acids precursors and indoles. Synthesis 2006:1183–9. With permission from Thieme Publishers.

Table 2.3 Yields of Methyl-1*H*-Indole-2-Carboxylate Derivatives[a]

Product					
Entry	**R^1**	**R^2**	**R^4**	**R^5**	**Yield (%)[b]**
1	NH_2		CN		–
2	NH_2		CO_2CH_3		–
3	OH				–
4	NH_2				60
5	NH_2	Cl	CH_3		46
6	NH_2			OCH_3	58
7	NH_2		Cl		26
8	NH_2		Br		35

[a]*Fritsch Planetary Micro Mill Pulverisette 7, stainless steels containers, eight balls, Ar purged, 60 min, aminoacrylate: $NaHCO_3$:HCO_2Na:n-Bu_4NC:$Pd(OAc)_2$ 1.05:2.5:0.2:1.2:0.05, NaCl grinding auxiliary.*

[b]*Isolated yields.*

Reproduced from Tullberg E, Schacher F, Peters D, Frejd T. Solvent-free Heck-Jeffery reactions under ball-milling conditions applied to the synthesis of unnatural amino acids precursors and indoles. Synthesis 2006:1183–9. With permission from Thieme Publishers.

Efficient mechanochemical protocol for synthesis of (*E*)-stilbenes by Heck reaction was reported by Su et al. (Scheme 2.3). [3]. Optimal reaction conditions in planetary mill for high product yields comprise of palladium catalyst (0.5 mol%) in conjunction with K_2CO_3 base, tetrabutylammonium bromide (TBAB; additive for stabilizing the metal colloids formed in situ, preventing their aggregation to larger-size particles), and silica gel as grinding auxiliary. Within 45 min to 1.5 h milling, (*E*)-stilbenes **10** were produced in high yields (Table 2.4). There was no significant difference in the yield of product between electron-deficient and electron-rich aryl halogenides. The coupling of 1-chloro-4-nitrobenzene and styrene performed by different methods showed that simple heating or grinding could not significantly promote the reaction. Authors propose that the combination of compressive stress, internal temperature, and efficient mixing ought to be the main factor which promoted the reaction by ball milling.

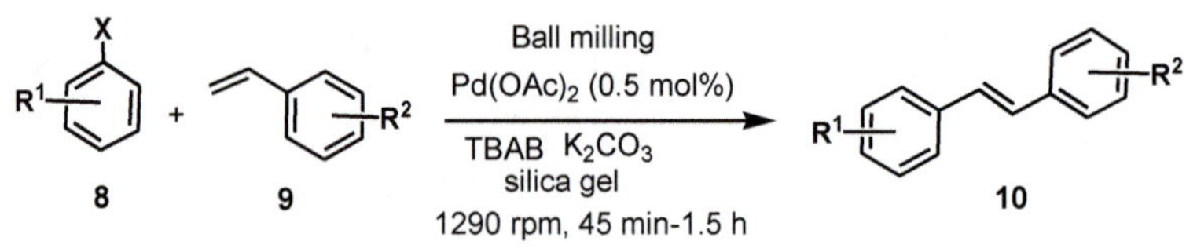

SCHEME 2.3

Synthesis of (*E*)-stilbenes by Heck reaction.

Table 2.4 Synthesis of (*E*)-Stilbenes by Heck Reaction[a]

Entry	X	R^1	R^2	Time	Yield (%)[b]
1	Br	4-EtOOC	H	45 min	90
2	Br	4-EtOOC	3,4-$(MeO)_2$	45 min	88
3	Br	H	3,5-$(MeO)_2$	45 min	90
4	Br	4-CH_3	H	45 min	87
5	Br	4-CH_3O	3,4-$(MeO)_2$	45 min	80
6	Br	3-F	H	45 min	89
7	Br	3,4,5-$(F)_3$	3,5-$(MeO)_2$	45 min	69
8	Cl	3-NO_2	H	45 min	85
9	Cl	4-NO_2	H	45 min	91
10	Cl	4-NO_2	4-Cl	45 min	81
11	Cl	3-Ac	H	1 h	76
12	Cl	4-Ac	4-Cl	1 h	77
13	Cl	4-CHO	H	1 h	67
14	Cl	4-MeO	H	1.5 h	0

[a]*AGO-2 planetary centrifugal mill, drum (105 mL), 8 mm balls (234 g), 1290 rpm, substrate: styrene:$Pd(OAc)_2$:TBAB:K_2CO_3 1:1.4:0.05:1:2.4, silica gel–milling auxiliary.*
[b]*Isolated yields.*
Reproduced with permission from Zhu X, Liu J, Chen T, Su W. Mechanically activated synthesis of (E)-stilbene derivatives by high-speed ball milling. Appl Organomet Chem 2012;26:145–7. Copyright (2012), Wiley.

For Mizoroki–Heck coupling of aryl halides with *tert*-butyl acrylate Lamaty et al. employed solid PEGs [poly(ethylene glycols)] polymers which were previously used as solvent and stabilizing/precipitating agents to separate the organic product from the metallic catalytic system (Scheme 2.4) [4]. During the course of the reaction PEG melted due to the slight heating of the ball mill, hence in order to increase the viscosity of the PEG shorter polymers having a lower melting point were employed and the yield increased by 20% by using PEG-2000-OH. In a typical run, a mixture of aryl iodide **11**, *tert*-butyl acrylate **12**, PEG, $Pd(OAc)_2$ (5 mol%), and K_2CO_3 as an inorganic base was grinded in a ball mill for 1 h at 30 Hz. While under traditional heating and microwave irradiation, PEG has been shown to reduce $Pd(OAc)_2$ to Pd(0), under ball-milling conditions, it was found necessary to add sodium formate as a reducing agent for the activation of $Pd(OAc)_2$. Due to the stabilizing effect of oxygen atoms in PEG, no phosphine ligand was necessary for the catalysis. In these optimized conditions, cinnamates were **13** produced in good to excellent yields (Table 2.5). However, reaction is limited in scope, since other acrylates and styrene gave very poor yields, and reactions with aryl chloride and bromide were unreactive.

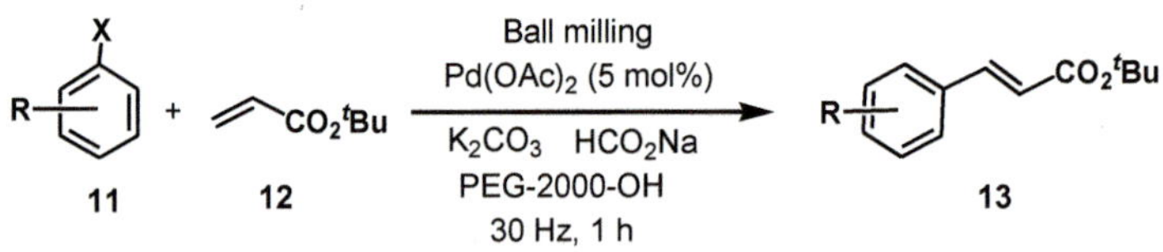

SCHEME 2.4

Cinnamate synthesis by Mizoroki–Heck reaction.

Table 2.5 Cinnamate Synthesis by Mizoroki–Heck Reaction[a]

Entry	R	X	Yield (%)[b]
1	H	I	100
2	4-MeO	I	73
3	4-NC	I	62
4	4-Br	I	68
5	4-OHC	I	100
6	4-NO_2	I	35
7	3-F	I	100
8	2-Me	I	29

[a]*Retsch MM200 ball mill, stainless steel reactor (10 mL), 2 × 7 mm balls, 30 Hz, 1 h, substrate: acrylate:$Pd(OAc)_2$:HCO_2Na:K_2CO_3 1:5:0.05:0.2:3, PEG-2000-OH.*

[b]*Isolated yields.*

Reproduced from Declerck V, Colacino E, Bantreil X, Martinez J, Lamaty F. Poly(ethylene glycol) as reaction medium for mild Mizoroki–Heck reaction in a ball-mill. Chem Commun 2012;48:11778–80. With permission from the Royal Society of Chemistry.

Application of rhodium catalyst in Heck reaction led to good substrate conversion and selectivity toward monoolefination in a directed aerobic C—H functionalization of acetanilides with olefines with catalytic amounts of rhodium and copper. The oxidative Heck reaction of arenes was carried by Bolm and coworkers by the aerobic coupling of acetanilides **14** with *n*-butyl acrylates **15** catalyzed with $[Cp^*RhCl_2]_2$ and $AgBF_4$ in the presence of $Cu(OAc)_2$ as redox mediator enabling the use of dioxygen as the terminal oxidant (Scheme 2.5) [5]. Oxygen was provided by simple flushing of the reactor vessel with O_2 before closing. Moderate to high yield was obtained after 15 h of milling, with selective formation of monosubstituted products **16** (Table 2.6). Styrene was also found to give the corresponding product in 32% yield. On the other hand, the increase of $[Cp^*RhCl_2]_2$–5 mol% and $Cu(OAc)_2$–10 mol%, in conjunction with $AgSbF_6$ (2 mol%) resulted in the loss of selectivity in reaction of unsubstituted acetanilide with *n*-butyl acrylate. Under these conditions, 62% conversion and a 93:7 mixture of the mono- and disubstituted products (**17** and **18**, respectively) was obtained (Scheme 2.5).

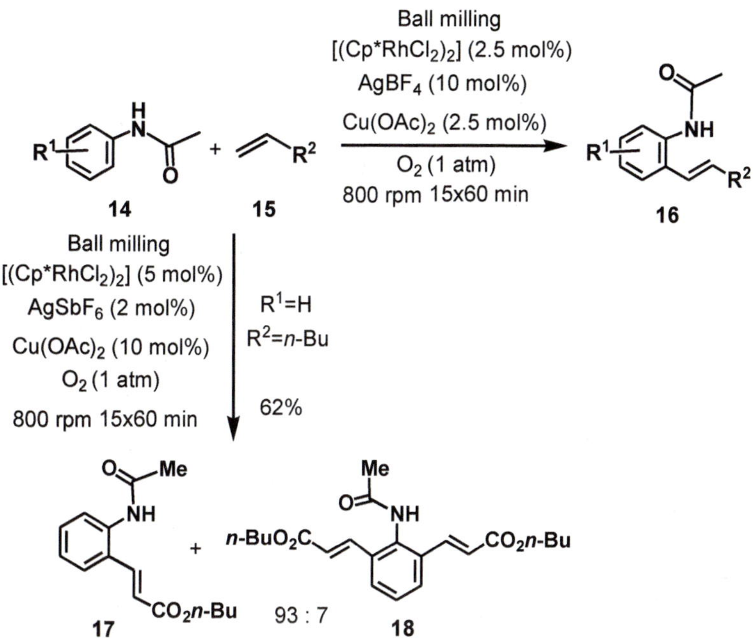

SCHEME 2.5

Oxidative Heck reaction of acetanilides with *n*-butyl acrylates.

Table 2.6 Oxidative Heck Reaction of Acetanilides With *n*-Butyl Acrylates[a]

Entry	R^1	R^2	Yield (%)[b]
1	H	CO_2nBu	71
2	H	CO_2nHex	57
3	H	CO_2Me	38
4	H	CO_2Bn	50
5	H	Ph	32
6	3-CH_3	CO_2nBu	78
7	3-Cl	CO_2nBu	37
8	4-F	CO_2nBu	55

[a]*Fritsch Planetary Micro Mill Pulverisette 7, ZrO_2 vessel (20mL), 16×5mm ZrO_2 balls, 800rpm, 15h, substrate:olefin:Rhcat:$AgBF_4$:$Cu(OAc)_2$ 1:1.1:0.025:0.1:0.025, O_2 (1 atm).*
[b]*Isolated.*

For this reaction, kinetic isotope effects (KIEs) were determined by competition reactions between deuterated and nondeuterated substrates **19** with olefin **20** to obtain insight into the reaction mechanism (Scheme 2.6). Under solvent-free conditions in a ball mill, a KIE value of 1.77 was obtained, suggesting a reaction mechanism in which the metal insertion into the C—H bond of the substrate is not turnover limiting. In solution (*tert*-amyl alcohol) at 120°C a similar KIE value of 1.48 was obtained, suggesting that reaction follows similar path in both conditions and the feasibility of determination of KIEs in ball-milling conditions.

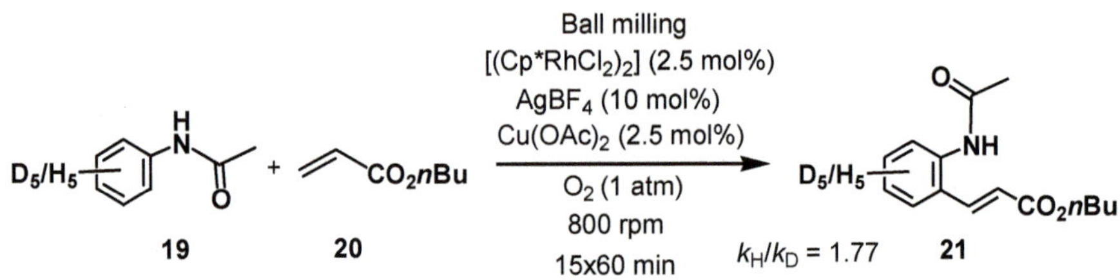

SCHEME 2.6

KIEs determination for mechanochemical oxidative Heck reaction.

2.2 SUZUKI COUPLING

Peters *et al.* have found out that potassium carbonate is the base of choice for Suzuki coupling of 2-bromonaphthalene and phenylboronic acid with palladium tetrakis(triphenylphosphine)-palladium(0) catalyst (5 mol%) in solvent-free mechanochemical conditions [6]. Homo-coupling of arylboronic acid to biphenyl was a significant side reaction, which could be avoided by addition of 2 equiv. of arylboronic acid. Sodium chloride was used as grinding auxiliary in Fritsch Planetary Micro Mill Pulverisette. With this improved protocol, yields of various biphenyls **24** increased up to 96% (selected results, Scheme 2.7). No attempts were made to keep air out of the reaction mixture demonstrating the simplicity of procedure. The normal scope of the Suzuki reaction is complemented by mechanosynthesis: aryl bromides that would be expected to be reactive under ordinary Suzuki conditions in solutions are generally poor substrates, whereas the electron-rich bromides which are not as good substrates in the solution reaction (3-bromoanisole and 3-bromoaniline) generally give high yields. From the results obtained by solid-state milling, a conclusion on the reactivity could be drawn: when the aryl bromide has a functionality that is expected to coordinate to palladium, the yields are very poor. Couplings with 2-bromobenzonitrile, 2-Br-benzaldehyde, 2-Br-3-NO_2-toluene, 4-Br-benzaldehyde, 5-Br-pyrimidine, 5-Br-indole, 3-Br-pyridine, 4-Br-benzonitrile, 2-Br-nitrobenzene, 4-I-nitrobenzene, 4′-Br-acetanilide, 4-Br-phenol, and 4-Br-pyridine gave yields less than 5%, or completely failed.

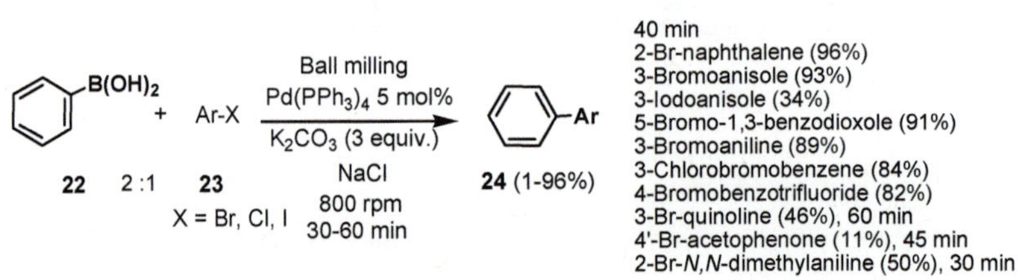

SCHEME 2.7

Suzuki coupling of phenylboronic acid with aryl halides.

Ligand-free palladium-catalyzed Suzuki reaction carried out in ball mill by Leadbeater and Klingensmith is an efficient variant of the reaction [7]. Optimization of conditions in 4-bromotoluene and phenylboronic acid reaction revealed that $Pd(OAc)_2$ catalyst is more active than $PdCl_2$ or $Pd_2(dba)_3$, and among bases tested (Na_2CO_3, NaOH, KF, and Et_3N), triethyl amine was the most satisfactory (Scheme 2.8). Sodium chloride was used as grinding auxiliary and interestingly, the yield did not drop when reaction time was reduced from 30 to 10 min. This methodology was found to be general and applicable to number of aryl halides **25** (Table 2.7). The reactions are rapid, the catalyst used is simple, and reactions were performed without the need to exclude air and moisture.

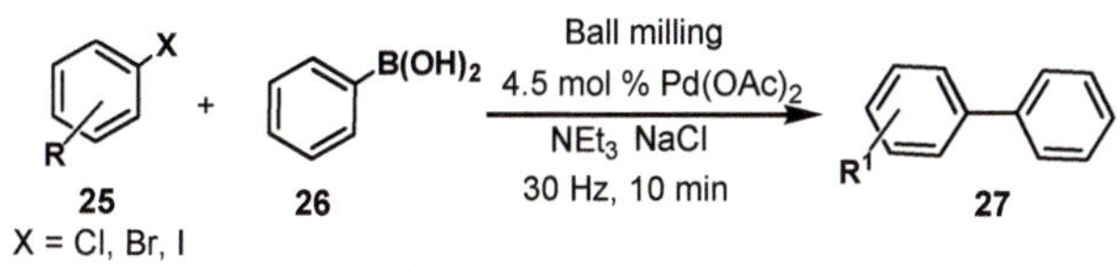

SCHEME 2.8

Suzuki coupling of phenylboronic acid with aryl halides.

Table 2.7 Suzuki Coupling of Phenylboronic Acid With Aryl Halides[a]

Entry	Aryl Halide	Yield (%)[b]
1	C_6H_5Br	96
2	4-NO_2-C_6H_4Br	97
3	4-COMe-C_6H_4Br	92
4	4-CH_3-C_6H_4Br	95
5	2-CH_3-C_6H_4Br	51
6	4-OCH_3-C_6H_4Br	79
7	3-OCH_3-C_6H_4Br	77
8	2-OCH_3-C_6H_4Br	33
9	4-CHO-C_6H_4Br	95
10	4-CO_2CH_3-C_6H_4Br	93
11	4-CH_3-C_6H_4I	20
12	4-CH_3-C_6H_4I	67[c,d]
13	4-CH_3-C_6H_4Cl	2[c]

[a]*Retsch MM200 ball mill, 30 Hz, stainless steel grinding jar, ArX: $PhB(OH)_2$:NEt_3:$Pd(OAc)_2$ 1:1:3:0.045, 10 min, NaCl grinding auxiliary.*
[b]*Isolated yields.*
[c]*30 min.*
[d]*3 equiv. $AgBF_4$.*
Reproduced from Klingensmith LM, Leadbeater NE. Ligand-free palladium catalysis of aryl coupling reactions facilitated by grinding. Tetrahedron Lett 2003;44:765–8.

Addition of silver salts as additives using 4-iodotoluene and phenylboronic acid as substrates increased yield from 20% to 67% (3 equiv. of $AgBF_4$). However, by using 3 equiv. of $AgNO_3$, together with formation of the Suzuki product, significant quantities of homo-coupled product deriving from the phenylboronic acid were isolated. In the absence of aryl halide, symmetrical biaryls **29** were obtained from boronic acids in moderate to good yields (Scheme 2.9, Table 2.8).

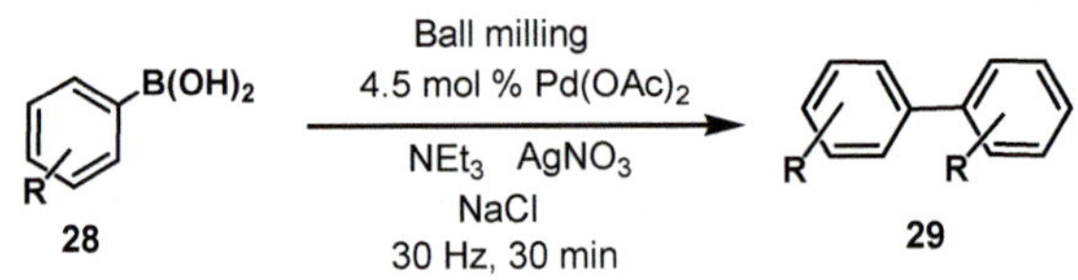

SCHEME 2.9

Boronic acid homo-coupling.

Table 2.8 Boronic Acid Homo-Coupling[a]

Entry	Aryl Halide	Yield (%)[b]
1	$C_6H_5B(OH)_2$	72
2	2-Naphthyl-$B(OH)_2$	82
3	4-CH_3-$C_6H_4B(OH)_2$	62
4	2-CH_3-$C_6H_4B(OH)_2$	41
5	4-CH_3-$C_6H_4B(OH)_2$	19

[a]*Retsch MM200 ball mill, 30 Hz, stainless steel grinding jar, $RB(OH)_2$:NEt_3:Pd$(OAc)_2$:$AgNO_3$ 1:3:0.045:3, 30 min, NaCl grinding auxiliary.*
[b]*Isolated yields.*
Reproduced from Reproduced from Klingensmith LM, Leadbeater NE. Ligand-free palladium catalysis of aryl coupling reactions facilitated by grinding. Tetrahedron Lett 2003;44:765–8. Copyright (2003), with permission from Elsevier.

Schneider and Ondruschka further investigated the C—C bond formation carried out by Suzuki protocol in solid-state mechanochemical conditions [8]. The standard coupling of phenylboronic acids with different aryl halides yielded 4-biphenyls **32** as products (Scheme 2.10). Reaction takes place with the help of solid-supported bases, whereas among different inorganic support systems tested, KF/Al_2O_3 was the most active. Other binary oxides, MgO, SiO_2, TiO_2, CeO_2, Fe_2O_3, and $TiSiO_4$ showed lower activity. The sufficient water content and high concentration of KF in home-made solid support was crucial for effective reaction, and preparation conditions and storage of KF-loaded aluminas influence the yield [9]. After the reaction, the crude product was extracted with water and ethyl acetate and analyzed by GC. Similar to the previous mechanochemical variants of Suzuki coupling, the most reactive halides used for Suzuki coupling were aryl bromides (Table 2.9). Variation of the boron-bearing component showed that substituents in *meta*-position dramatically decrease the yield of the coupled product (Table 2.10).

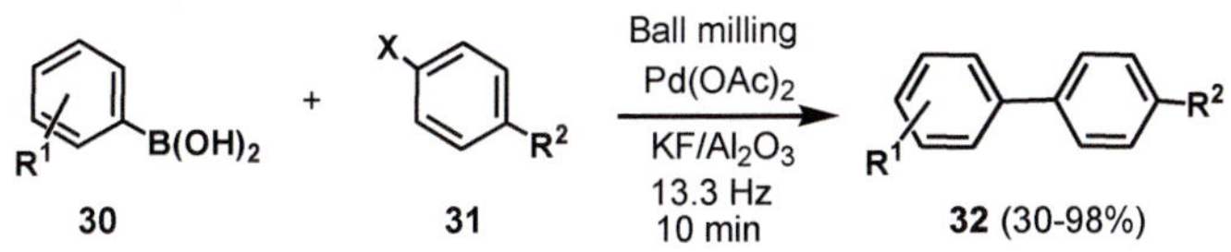

SCHEME 2.10

Suzuki coupling.

Table 2.9 Suzuki Reaction of Phenylboronic Acids With Different Aryl Halides[a]

Entry	Substrate	Yield (%)[b]
1	4-Cl-C_6H_4-$COCH_3$	0
2	4-Cl-C_6H_4-CH_3	0
3	4-Cl-C_6H_4-OCH_3	0
4	C_6H_5Cl	6
5	4-Br-C_6H_4-$COCH_3$	94
6	4-Br-C_6H_4-CH_3	70
7	4-Br-C_6H_4-OCH_3	93
8	C_6H_5Br	98
9	4-Br-C_6H_4-NO_2	84
10	3-Br-C_6H_4-CH_3	62
11	2-Br-C_6H_4-$COOCH_3$	95
12	2-Br-C_6H_2-1,3,5-$(CH_3)_3$	18
13	4-I-C_6H_4-$COCH_3$	97
14	4-I-C_6H_4-CH_3	37
15	4-I-C_6H_4-OCH_3	30
16	C_6H_4I	53

[a]*Fritsch Planetary Micro Mill model "Pulverisette 7," 800 rpm, 10 min, grinding beaker (75 mL), 6 × 15 mm balls, aryl halide:phenylboronic acid:Pd(OAc)$_2$ 1:1.25:0.03 ratio, KF-Al$_2$O$_3$ (32 wt% KF).*
[b]*Determined by GC.*

Reproduced with permission from Schneider F, Ondruschka B. Mechanochemical solid-state suzuki reactions using an in situ generated base. ChemSusChem 2008;1:622–5. Copyright (2008), Wiley.

Table 2.10 Suzuki Reaction of Bromoacetophenone With Different Phenylboronic Acids[a]

Entry	Substrate	Yield (%)[b]
1	$C_6H_5B(OH)_2$	94
2	4-CH_3-$C_6H_5B(OH)_2$	98
3	4-F-$C_6H_5B(OH)_2$	88
4	3-CHO_3-$C_6H_5B(OH)_2$	9

[a]*Fritsch Planetary Micro Mill model "Pulverisette 7," 800 rpm, 10 min, grinding beaker (75 mL), 6 × 15 mm balls, aryl halide:phenylboronic acid:Pd(OAc)$_2$ 1:1.25:0.03 ratio, KF-Al$_2$O$_3$ (32 wt% KF).*
[b]*Determined by GC.*

Reproduced with permission from Schneider F, Ondruschka B. Mechanochemical solid-state suzuki reactions using an in situ generated base. ChemSusChem 2008;1:622–5. Copyright (2008), Wiley.

The experiments of the influence of Pd concentration on the yield of 4-methylbiphenyl performed using 2.2 mol % Pd instead of 3.6 mol % as for the other experiments revealed that Pd loadings higher than 1 mol % led to satisfactory yields of coupling product. The application of lower Pd concentrations (<2 μmol) led to moderate conversions, while the reactions with ultralow amounts of Pd required either longer reaction times or co-grinding with higher-weight milling balls (ZrO_2, stainless steel and tungsten carbide).

Detailed investigation on influence of various parameters on the coupling results for phenylboronic acid/4-bromoacetophenone coupling revealed that zirconium oxide is the most effective material for beakers/milling balls [10]. Study also shows that optimal number and size of milling balls has to be empirically determined (six balls, 15 mm was found to be optimal). The best yield was obtained with the maximum rotation speed of 800 rpm, while increasing milling time (from 5, 10 to 15 min) the corresponding yield reaches its maxima for 200 and 400 rpm at $t=10$ min. Longer reaction times (15 min) support the decomposition of the product formed. The following order of the investigated parameters regarding their positive influence on the formation of product was established: rpm > milling time > size of milling balls > number of milling balls > grinding material.

In a separate study, Ondruschka et al. studied the energetics of Suzuki protocol in mechanochemical conditions and compared different energy modes [11]. The standard coupling of phenylboronic acid with different aryl halides yielded 4-biphenyls **35** in high yields (Scheme 2.11, Table 2.11). Variations in yields were found when different energy modes were applied, with the best yields obtained by combined mechanochemical/microwave procedure. In general, for Suzuki–Miyaura reaction, ball milling is proven to be a more effective tool for energy entry than microwave irradiation.

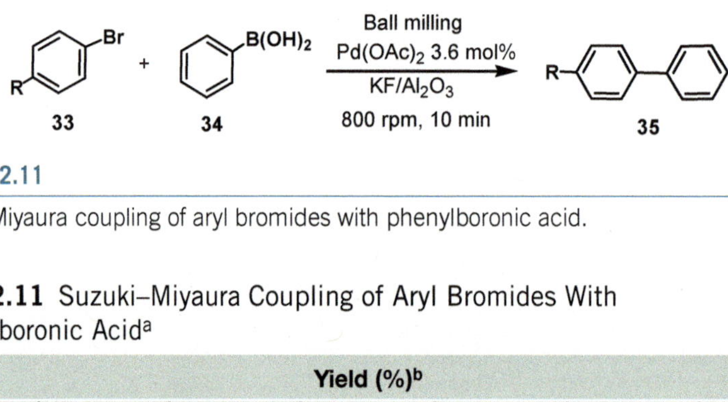

SCHEME 2.11

Suzuki–Miyaura coupling of aryl bromides with phenylboronic acid.

Table 2.11 Suzuki–Miyaura Coupling of Aryl Bromides With Phenylboronic Acid[a]

		Yield (%)[b]				
Entry	**R**	**BM1**	**BM2**	**MW1**	**MW2**	**COMB**
1	H	93	99	78	60	98
2	CH_3	74	63	65	54	82
3	OCH_3	93	56	80	49	96

Table 2.11 Suzuki–Miyaura Coupling of Aryl Bromides With Phenylboronic Acid[a]—cont'd

		Yield (%)[b]				
Entry	**R**	**BM1**	**BM2**	**MW1**	**MW2**	**COMB**
4	$COCH_3$	89	45	80	70	94
5	NO_2	78	83	75	66	80

[a]*BM1: Fritsch Pulverisette 7 Planetary Micro Mill, stainless steel beaker (45mL), 6×15mm agate balls, 800rpm, 10min, BM2: Fritsch MM301 mixer mill, stainless steel beaker (50mL), 6×30mm agate balls, 800rpm, 10min, MW1: MLS Praktika MW reactor, multimode P=300W, 15min, 150°C, MW2: CEM Discover MW reactor, monomode P=150W, 15min, 150°C, COMB: BM1+MW1; ArX:PhB(OH)$_2$:PdOAc$_2$ 1:1.2:0.036, KF-Al$_2$O$_3$ (32wt% KF).*

[b]*Determined by GC.*

Reproduced from Schneider F, Szuppa T, Stolle A, Ondruschka B, Hopf H. Energetic assessment of the Suzuki-Miyaura reaction: a curtate life cycle assessment as an easily understandable and applicable tool for reaction optimization. Green Chem 2009;11:1894–9. With permission from the Royal Society of Chemistry.

Cravotto described another variant of Pd-catalyzed Suzuki cross-coupling reaction in planetary ball mill in which cross-linked chitosan/Pd(II) catalyst was employed (Scheme 2.12) [12]. By milling of aryl halide, phenylboronic acid, potassium carbonate, and catalyst for 90 or 120min, a series of cross-coupled products **37** was obtained in high yields (Table 2.12). Upon completion, products were simply dissolved in ethyl acetate and supported catalyst, insoluble in the organic solvent, was separated by filtration and reused in a new reaction after regeneration. Homocoupling (Ullmann reaction) was the side reaction, and the amount of homo-coupled products **38** is considerably higher for the present CS-based catalyst system than for other Suzuki cross-coupling procedures in ball mills using $Pd(OAc)_2$ or $Pd(PPh_3)_4$ catalysts. The increased formation of homo-coupling products was evident in catalyst-recycling experiments. Noteworthy results were obtained with the less reactive aryl chlorides, and this is the first report in which aryl chlorides have been cross-coupled in a ball mill affording >10% yield. Obviously, the stronger C—Cl bond favors oxidative addition to the catalyst over the dehalogenation accompanied by homo-coupling seen in aryl iodides and bromides. When the catalyst was recycled three times, a loss in activity was observed, which was explained by the weight loss during catalyst recycling.

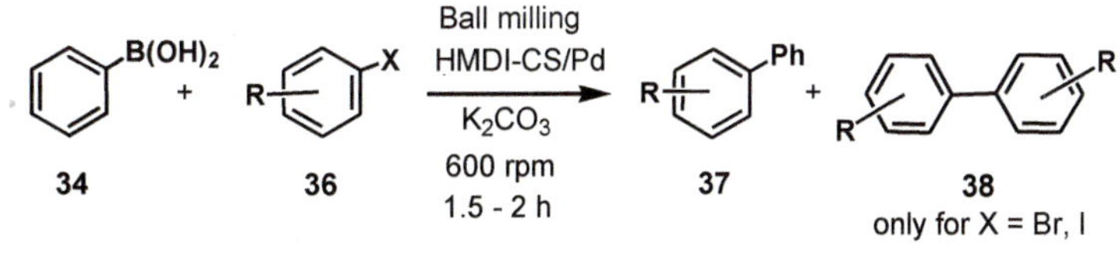

SCHEME 2.12

Suzuki cross-coupling reaction.

Table 2.12 Suzuki Reaction of Aryl Halides With Phenylboronic Acid[a]

Entry	Aryl Halide X	R	Time (min)	Conversion (%)[b]	Yield (%)	Homo-Coupling (%)
1	I	*p*-MeO	90	>99	74	25
2	I[c]	*p*-MeO	90	>99	64	35
3	I	*m*-OH	90	90	90	
4	Br	*p*-Ac	90	>99	84	15
5	Br	*p*-MeO	90	>99	72	22
6	Cl	H	120	>99	99	
7	Cl	*p*-NO_2	120	>99 (95)[c]	84 (75)[c]	
8	Cl	*p*-NO_2	120	82[d] (65)[e]	64[d] (50)[e]	
9	Cl	*p*-Ac	120	29	29	
10	Cl	*m*-Ac	120	40	36	
11	Cl	*o*-Ac	120	78	39	

[a]Retsch Planetary ball mill PM100, single milling stainless steel beaker (125 mL), milling balls (15 × 10 mm), 600 min^{-1}, ArX:$PhB(OH)_2$:K_2CO_3 1:1.1:2.
[b]Estimated by GC.
[c]Catalyst first reuse.
[d]Catalyst second reuse.
[e]Catalyst third reuse.
Reproduced from Cravotto G, Garella D, Tagliapietra S, Stolle A, Schüßler S, Leonhardt SES, et al. Suzuki cross-couplings of (hetero)aryl chlorides in the solid-state. N J Chem 2012;36:1304–7. With permission from the Royal Society of Chemistry.

The observed enhanced reactivity of aryl chlorides in mechanochemical conditions was employed in solid-state Suzuki cross-coupling of selected heteroaryl chlorides **39** and **40** with phenylboronic acid (Scheme 2.13), affording high yields of cross-coupling products **41** and **42**.

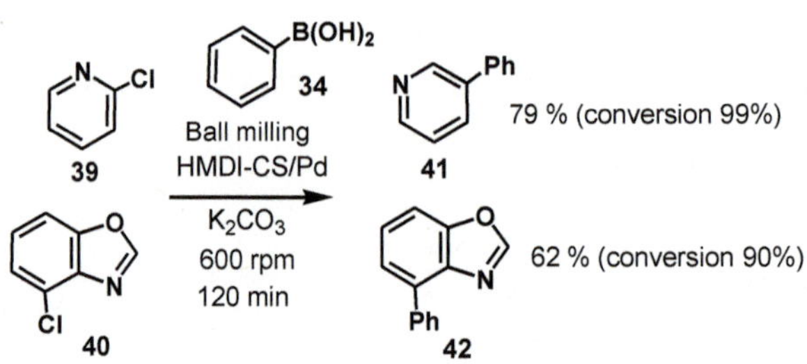

SCHEME 2.13

Suzuki cross-coupling reaction of heteroaryl chlorides.

2.3 SONOGASHIRA COUPLING

Sonogashira solvent-free reaction utilizing ball milling was reported by Mack et al. [13]. The C—C bond formation of an aryl halide and terminal alkyne is facilitated with the presence of palladium and copper catalysts (Scheme 2.14). A number of *p*-substituted acetylenes **45** was prepared by this reaction in 3–90% yield (selected results, Table 2.13), although time for reaction completion was rather long (17h). In some cases, significant amounts of disubstituted products were also obtained. The advantage of this mechanochemical procedure is the use of aerobic conditions without taking care about moisture in the air, while typically Sonogashira reactions are carried out under an inert atmosphere using dry solvents. Further improvement in comparison with other solvent-less methods, which only react with aryl iodides or strongly electron-withdrawing groups on the aryl ring, is that ball-milling conditions could be applied to wide range of substrates. The effects of the size and ball material on the results were examined, employing stainless steel, brass, bronze, aluminum, tungsten carbide, and aluminum oxide balls. The best yield in the shortest reaction time was obtained with tungsten carbide ball.

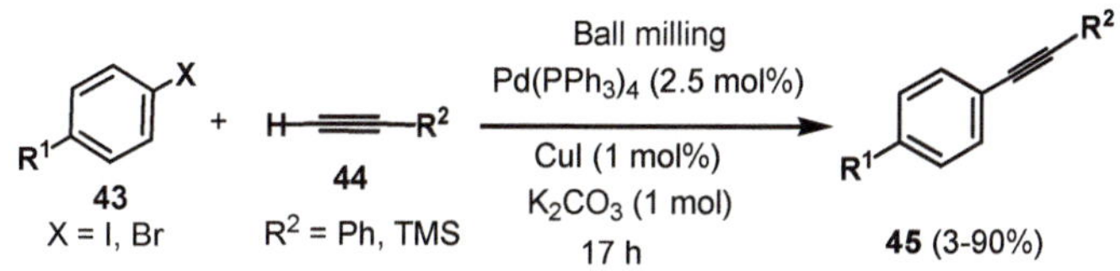

SCHEME 2.14

Sonogashira reaction.

Table 2.13 Sonogashira Reactions of *p*-Substituted Benzenes[a]

Entry	X	R	R′ = Ph Yield (%)	R′ = TMS Yield(%)
1	I	OMe	86	84
2	I	Me	84	40
3	I	H	95	93
4	I	Br	88	76
5	I	I	39[b]	30[c]
6	Br	OMe	Trace	Trace
7	Br	Me	33	26
8	Br	H	93	90
9	Br	Cl	7	4
10	Br	CHO	89	87
11	Br	NO_2	85	66
12	Br	Br	9[d]	5

[a]*8000M SpexCertiprep Mixer Mill, stainless steel vial (2.0×0.5inch), tungsten carbide ball (0.250 inch), ArX:acetylene:Pdcat:CuI:K_2CO_3 1:1.05:0.025:0.01:1.*
[b]*Isolated 63% of disubstituted product.*
[c]*Isolated 31% of disubstituted product.*
[d]*Isolated 48% disubstituted product.*
Reproduced from Fulmer DA, Shearouse WC, Medonza ST, Mack J. Solvent-free Sonogashira coupling reaction via high speed ball milling. Green Chem 2009;11:1821–5. With permission from the Royal Society of Chemistry.

Interesting is the development of concept of "copper-free" Sonogashira reaction in which CuI catalyst was not used (Scheme 2.15). In these reactions, coupling products **47** in various yields were still obtained (Table 2.14), probably due to the presence of trace amounts of copper in palladium reagent. When copper catalyst-free reaction was carried out with copper milling balls as a source of copper, yields increased to 31–88% (Table 2.15). Finally, copper vials in combination with copper balls as a source of copper afforded respectable yields in the range of 42–90% (Table 2.16).

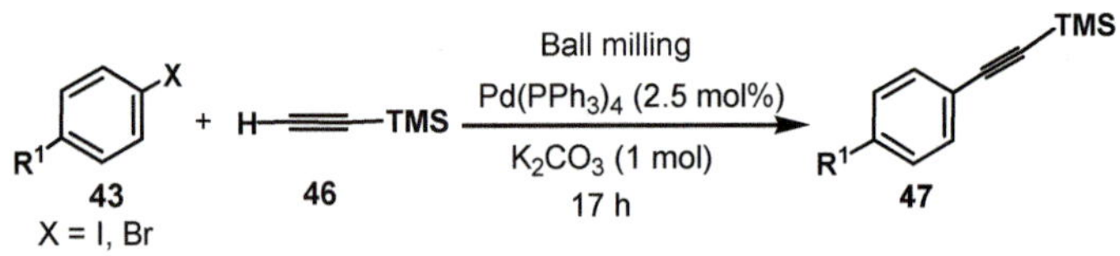

SCHEME 2.15

"Copper-free" Sonogashira reaction.

Table 2.14 "Copper-Free" Sonogashira Reactions of *p*-Substituted Benzenes[a]

Entry	X	R	"Copper-Free"[b] Yield (%)	Cu Ball[c] Yield (%)	Cu Ball + Cu Vial[d] Yield (%)
1	I	H	39	87	88
2	I	Me	17	46	83
3	I	Br	43	31	89
4	I	Cl	37	52	86
5	I	OMe	58	40	42
6	Br	CHO	44	85	84

[a]*8000M SpexCertiprep Mixer Mill.*
[b]*Stainless steel vial (2.0 × 0.5 inch), tungsten carbide ball (3/32 inch).*
[c]*Stainless steel vial (2.0 × 0.5 inch), copper ball (0.250 inch).*
[d]*Copper vial (2.0 × 0.5 inch), copper ball (3/32 inch), ArX:acetylene:Pdcat:K_2CO_3 1:1.05:0.025:1.*
Reproduced from Fulmer DA, Shearouse WC, Medonza ST, Mack J. Solvent-free Sonogashira coupling reaction via high speed ball milling. Green Chem 2009;11:1821–5. With permission from the Royal Society of Chemistry.

Table 2.15 Sonogashira Reactions of Aryl Iodides With Phenylacetylenes[a]

Entry	Aryl Iodide R[1]	Acetylene R[2]	Conversion (%)[b]	Selectivity to Product (%)
1	H	H	68	99
2	H	H	80 (ZrO_2)	99
3	*o*-Me	H	54	98
4	*p*-Me	H	47	>99
5	*p*-Me	H	74 (ZrO_2)	>99
6	*o*-MeO	H	51	98

Table 2.15 Sonogashira Reactions of Aryl Iodides With Phenylacetylenes[a]—cont'd

Entry	Aryl Iodide R^1	Acetylene R^2	Conversion (%)[b]	Selectivity to Product (%)
7	*p*-Ac	H	98	99
8	H	*p*-Me	67	97
9	H	*p*-Me	83 (ZrO_2)	98
10	*p*-Me	*p*-Me	55	98
11	*p*-Me	*p*-Me	77 (ZrO_2)	97
12	*p*-MeO	*p*-Me	53	98
13	*p*-MeO	*p*-MeO	61	>99
14	*p*-Ac	*p*-MeO	99	>99
15	*p*-MeO	*p*-F	53	96
16	*p*-Ac	*p*-F	93	99

[a]*Fritsch "Pulverisette" planetary ball mill, grinding beaker (45 mL, agate, ZrO_2), balls (6×15 mm; agate or ZrO_2), 800 rpm, 20 min, ArI:acetylene:$Pd(OAc)_2$:DABCO:K_2CO_3 2:2.5:0.05:2.5, SiO_2 grinding auxiliary.*
[b]*Estimated by GC.*
Reproduced from Thorwirth R, Stolle A, Ondruschka B. Fast copper-, ligand- and solvent-free Sonogashira coupling in a ball mill. Green Chem 2010;12:985–91. With permission from the Royal Society of Chemistry.

Table 2.16 Sonogashira Reactions of Aryl Halides With Phenylacetylene[a]

		Agate		ZrO_2	
Entry	Aryl Halide R	Conversion (%)[b]	Selectivity to Product (%)	Conversion (%)[b]	Selectivity to Product (%)
1	I	24	99	99	98
2		26	98	99	98
3		35	99	99	99
4	Br	4	98	71	98
5		1	98	61	99
6		3	99	52	98
7		34	99	97	98
8		26	98	99	97
9		94	99	99	98

[a]*Fritsch "Pulverisette" planetary ball mill, grinding beakers (45 mL, agate, ZrO_2), balls (6×15 mm), 800 rpm, 20 min, ArI:acetylene:$Pd(OAc)_2$:DABCO: 2:2.5:0.05:2.5, Al_2O_3 grinding auxiliary.*
[b]*Estimated by GC.*
Reproduced from Thorwirth R, Stolle A, Ondruschka B. Fast copper-, ligand- and solvent-free Sonogashira coupling in a ball mill. Green Chem 2010;12:985–91. With permission from the Royal Society of Chemistry.

Stolle and coworkers also investigated the synthetic utility of Sonogashira reaction in solvent-free mechanochemical conditions [14]. They found similar observation that coupling reaction could be carried out without the use of copper or additional ligands with $Pd(OAc)_2$ or $Pd(PPh_3)_4$ catalysts in the presence of 1,4-diazabicyclo[2.2.2]octane (DABCO) as base (Scheme 2.16). Using agate (a relatively lightweight material) for milling beakers and balls led to medium and high conversions of the aryl iodides. The use of more heavyweight-milling material (ZrO_2) resulted in significantly higher conversions for the same reaction conditions, caused by the higher-energy input of the milling balls. All reaction vessels were cleaned with aqua regia prior to use to avoid any contamination or memory effects. Application of $Pd(OAc)_2$ catalyst and SiO_2 as a grinding auxiliary was found to be the best for transformations of aryl iodides **48** to the respective Sonogashira coupling products **50** (Table 2.15). The variation of the acetylene reaction component seems to have little influence on the conversion of the aryl iodides, and reasonable yields were obtained, even for hydroxyl-bearing acetylenes. In addition, double Sonogashira reactions of *p*-iodoacetophenone with bis-ethynyl compounds were carried out in 63–75% yields.

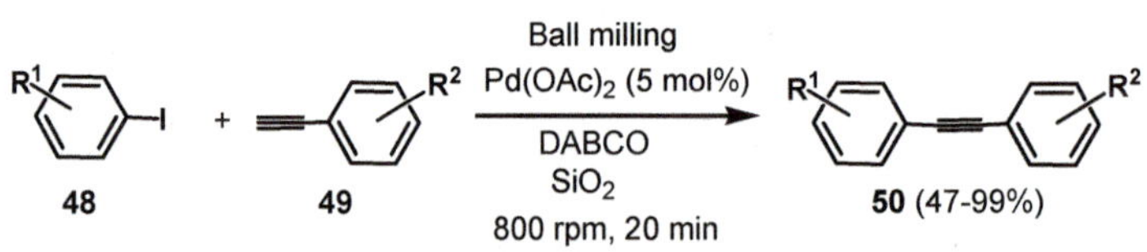

SCHEME 2.16

Sonogashira reaction.

The variation of amount of $Pd(OAc)_2$ in the range of 1–5 mol% showed linear dependence of the conversion for the reaction between *p*-iodoanisole and phenylacetylene. The turnover frequencies (TOF) for the coupling reactions of aryl iodides with phenylacetylene given in Table 2.17 range from 28 to 59 h^{-1}. Applied reaction conditions are advantageous due to short reaction times, in comparison with earlier study published by Mack (TOF: ~2 h^{-1}). A linear dependence of conversion and amount of DABCO is found and a 1.5-fold molar excess seems to be an upper limit for the reaction. Comparison with other common bases revealed that only quinuclidine showed slightly better conversion, but DABCO is more applicable, as cheaper and less-toxic reagent. Variation of grinding auxiliary (SiO_2 (fused quartz sand), SiO_2, CeO_2, TiO_2, γ-Al_2O_3 (neutral), α-Al_2O_3 (basic), and KF-Al_2O_3 (32 wt% KF)) indicated that both neutral and basic alumina significantly increase the conversion.

Table 2.17 Asymmetric Sonogashira Reaction *p*-Iodobromobenzene With Phenylacetylene[a]

Catalyst	Milling Material	Conversion (%)[b]	Selectivity to Product 55 (%)	Selectivity to Product 56 (%)	Selectivity to Product 57 (%)	Selectivity to Product 58 (%)
$Pd(OAc)_2$	Agate	87	81	10	0	0
$Pd(PPh_3)_4$	Agate	39	100	0	0	0
$Pd(OAc)_2$	ZrO_2	01	78	20	1	1
$Pd(PPh_3)_4$	ZrO_2	99	94	2	1	3

[a]*Fritsch "Pulverisette" planetary ball mill, grinding beakers (45 mL), balls (6 × 15 mm), 800 rpm, 20 min, PhI:acetylene:Pdcat:DABCO: 2:2.5:0.05:2.5, Al_2O_3 grinding auxiliary.*
[b]*Estimated by GC.*
Reproduced from Thorwirth R, Stolle A, Ondruschka B. Fast copper-, ligand- and solvent-free Sonogashira coupling in a ball mill. Green Chem 2010;12:985–91. With permission from the Royal Society of Chemistry.

Investigations on the influence of Pd catalyst and copper showed that the $Pd(PPh_3)_4$ in combination with Al_2O_3 grinding auxiliary facilitates the coupling reactions of aryl bromides **51** with phenylacetylenes **52** (Scheme 2.17, Table 2.16). The addition of CuI to $Pd(OAc)_2$ had no positive influence on the Sonogashira coupling under solvent-free reaction conditions. The conversion of coupling product decreased, which is caused by the increasing conversion to the homo-coupling product 1,4-diphenylbuta-1,3-diyne (Glaser reaction) [15]. In situ formation of $Pd(PPh_3)_4$ by addition of $Pd(OAc)_2$ and PPh_3 to the grinding jar had not occurred. This reaction system is advantageous to Mack's conditions with respect to the fact that neither additional copper milling balls nor containers made of copper were needed.

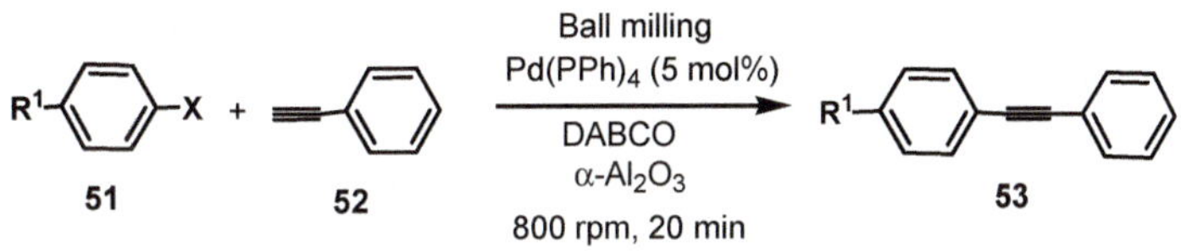

SCHEME 2.17

Sonogashira reaction of aryl halides with phenylacetylene, catalyst variation.

An asymmetric coupling of *p*-iodobromobenzene with phenylacetylene (Scheme 2.18, Table 2.17) showed that the selectivity for the product **55** as well as the conversion of *p*-iodobromobenzene were significantly higher for $Pd(PPh_3)_4$, than for

$Pd(OAc)_2$ with agate milling equipment. On the other hand, the double-coupled product 1,4-bis(phenylethynyl)benzene **56** is only formed with the use of $Pd(OAc)_2$ catalyst. Switching to higher weight–milling material (ZrO_2) afforded higher conversions, which was accompanied by a decreased selectivity for **55**. In these milling conditions, small amounts of deiodinated starting material **57** and the corresponding Sonogashira coupling product **58** were formed.

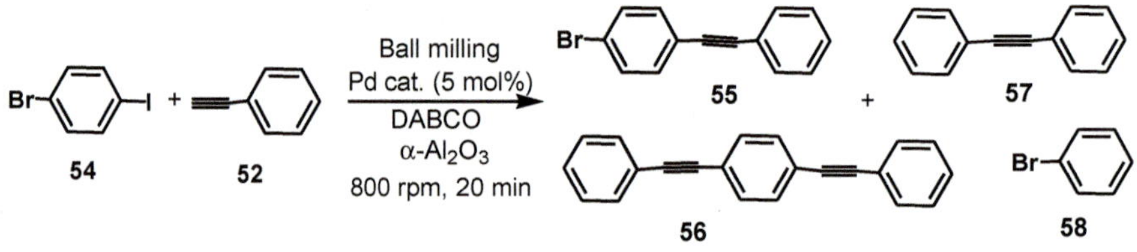

SCHEME 2.18

Asymmetric Sonogashira reaction.

2.4 GLASER COUPLING

Mechanochemical Glaser coupling (homo-coupling of terminal alkynes) was investigated in depth by Stolle et al. in the subsequent report [16]. Initial studies were carried with phenylacetylene (Scheme 2.19). The reactions were carried out in a ball mill by using zirconia (ZrO_2) as the material for the milling beakers and balls. Among the variety of milling auxiliaries employed, commercial KF-Al_2O_3 (KF content = 32 w%) in conjunction with DABCO was the most efficient. Other KF-enriched Al_2O_3 milling auxiliaries showed moderate to good activity. On the other hand, acidic, neutral, and basic Al_2O_3 itself, silica gel, quartz sand, KF-silica gel, and KF quartz sand could not promote coupling. Inorganic and organic bases screening revealed that DABCO is the most potent (Table 2.18). Obtained yield outperforms results obtained by other literature methods (microwave and stirring).

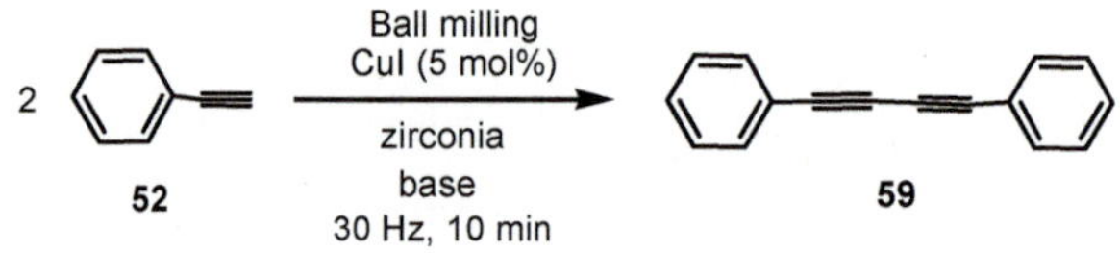

SCHEME 2.19

Mechanochemical Glaser coupling of phenylacetylene.

Table 2.18 Screening of Bases for Glaser Coupling of Phenylacetylene in Ball Mill[a]

Entry	Base	Yield (%)[b]
1		42
2	DABCO	96 (93)[c] [14]
3	Et_3N	51
4	Bu_4NBr	0
5	KOH	54
6	K_2CO_3	43

[a]*Retsch MM 301 vibration ball mill, ZrO_2 beaker (35 mL), ZrO_2 balls (12 × 10 mm) 30 Hz, 10 min, alkyne:base:CuI 1:1.25:0.05, KF-Al_2O_3 milling auxiliary.*
[b]*Estimated by GC.*
[c]*Isolated.*
Reproduced with permission from Schmidt R, Thorwirth R, Szuppa T, Stolle A, Ondruschka B, Hopf H. Fast, ligand- and solvent-free synthesis of 1,4-substituted Buta-1,3-diynes by Cu-catalyzed homocoupling of terminal alkynes in a ball mill. Chem Eur J 2011;17:8129–38. Copyright (2011), Wiley.

The necessity of oxygen presence for the catalytic cycle of copper Glaser coupling was demonstrated in experiments with CuI, in the presence of different oxidants which afforded higher yields. The addition of Oxone-, $(NH_4)_2S_2O_8$-, or H_2O_2-furnished product **59** in 64%, 72%, and 98% yield, respectively. To avoid addition of oxidants, reaction mixtures were air exposed after ball milling, which was beneficial in the case of copper, nickel, and cobalt catalysts, while other catalysts (palladium, ruthenium, and iron) showed much lower activity and influence of air to reaction yield. Copper catalysts were by far the most efficient: ball milling with CuI for 10 min followed by an immediate workup gave 1,4-diphenylbuta-1,3-diyne in 32%, while the exposure of reaction mixture to air for 30 min increased yield to >99%. Low yields obtained in the case of $PdCl_2$ catalyst was accompanied with the formation of aromatic cyclotrimerization products **61** and **62** (Scheme 2.20).

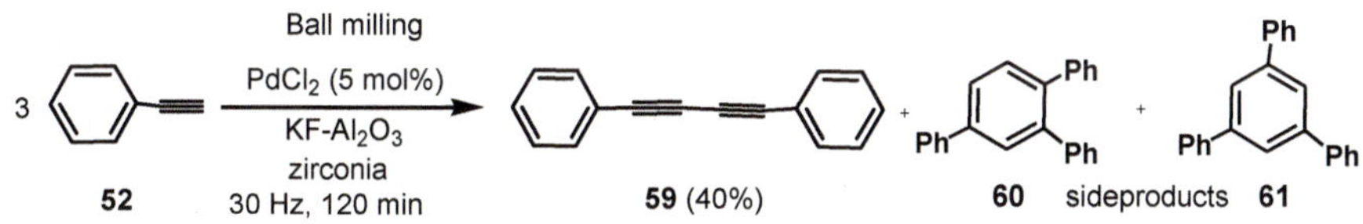

SCHEME 2.20

Cyclotrimerization side products in Glaser coupling of phenylacetylene.

A screening of the reaction scope applying optimal reaction conditions using different acetylene substrates **62** revealed that all substrates showed high selectivity in reaction conditions applied (Scheme 2.21, Table 2.19). For less-reactive alkynes, catalyst amount was doubled (10 mol %) and reaction time elongated. Solidifying the reagents by cooling down to the temperature of liquid nitrogen significantly lowered yield, or fully suppressed the reaction when constant cooling was applied. The lowered yield is in agreement with cryogenic milling results for the Knoevenagel condensation [43] and Tishchenko reaction [17]. Prolonged standing at room temperature resulted in unfreezing of the reactants, and the recovered molecular flexibility allowed the reaction to proceed (when beakers were cooled prior to milling, followed by 30 min milling, 88% yield was obtained).

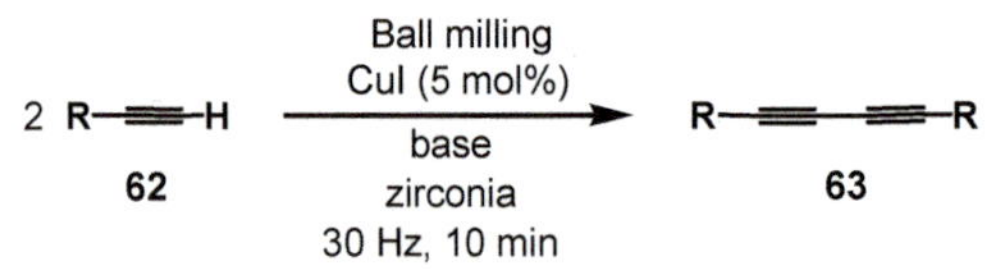

SCHEME 2.21

Glaser coupling of terminal alkynes.

Table 2.19 Glaser Coupling of Terminal Alkynes in Ball Mill[a]

		Workup A		Workup B	
Entry	**R**	**Yield (%)[b]**	**Selectivity (%)[b]**	**Yield (%)[b]**	**Selectivity (%)[b]**
1	C_6H_5	>99	>99	96	>99
2	C_6H_5[c]	3	98	13	99
3	C_6H_5[d]	–	–	1	99
4	o-Me-C_6H_4	88	97	99	>99[e]
5	p-Me-C_6H_4	88	95	77	>99
6	o-MeO-C_6H_4	72	89	92	>99[e]
7	m-MeO-C_6H_4	80	91	98	>99[e]
8	p-MeO-C_6H_4	68	90	70	>99
9	p-F-C_6H_4	90	98	93	>99
10	C_8H_{17}	92	93	96	>99[e]
11	$C_{10}H_{21}$	89	91	89	>99[e]
12	3-Pyridyl	88	92	0	–

[a]*Retsch MM 301 vibration ball mill, ZrO_2 beaker (35 mL), ZrO_2 balls (12 × 10 mm), 30 Hz, 10 min, alkyne: CuI 1:0.05, KF-Al_2O_3 milling auxiliary (γ-Al_2O_3, neutral, 60 w% KF); workup A: after milling the mixture was exposed to air for 30 min without further mixing; workup B: reaction was carried out with KF-Al_2O_3 (32 w%, commercial) and DABCO (2.5%) and subsequently worked up.*
[b]*Estimated by GC.*
[c]*Cryogenic milling: beakers were cooled with liquid nitrogen prior to milling.*
[d]*Cryogenic milling under constant cooling with liquid nitrogen.*
[e]*Catalyst (10 mol%), t = 20 min.*
Reproduced with permission from Schmidt R, Thorwirth R, Szuppa T, Stolle A, Ondruschka B, Hopf H. Fast, ligand- and solvent-free synthesis of 1,4-substituted Buta-1,3-diynes by Cu-catalyzed homocoupling of terminal alkynes in a ball mill. Chem Eur J 2011;17:8129–38.

Cross-coupling of two different alkynes provided a mixture of three products **66–68**, in which the homo-coupled products **66** and **68** were slightly preferred (Scheme 2.22), and thus in contrast to preferential formation of cross-coupled product in microwave conditions. Furthermore, the TMS substitution on the acetylene is readily removed, and homo-coupled product **66** obtained with conversion smaller than for phenylacetylene (Scheme 2.23). This result indicates that removal of TMS is the rate-determining step, whereas fluoride ion (KF) cleaves the TMS group in an analogous way as in solution. Glaser coupling with in situ TMS deprotection also works for *p*-CH_3- and *p*-CF_3-TMS-ethynylbenzenes, while the yields are considerably lower than for corresponding phenylacetylenes.

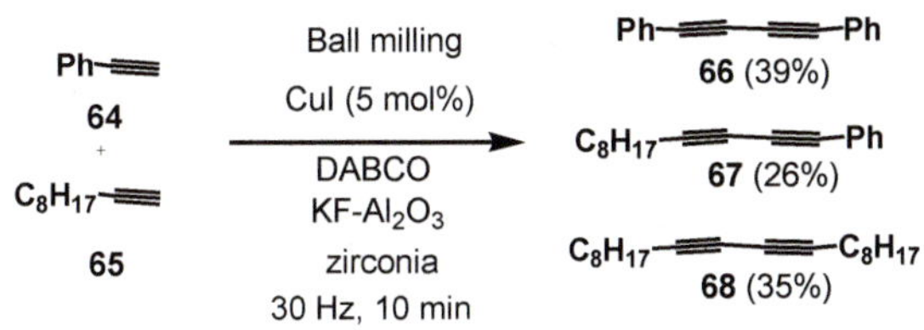

SCHEME 2.22

Cross-coupling Glaser coupling of terminal alkynes.

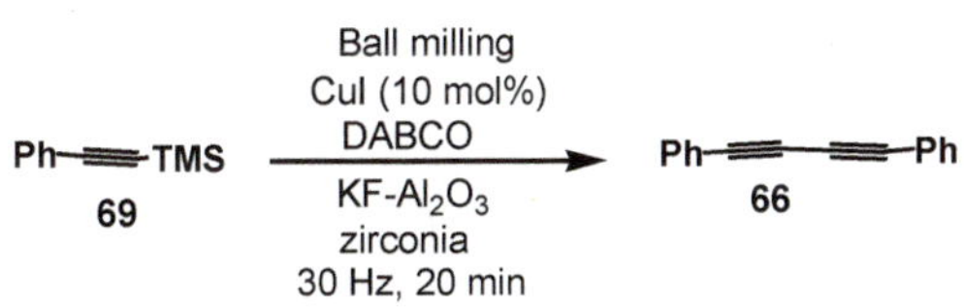

SCHEME 2.23

In situ TMS deprotection in Glaser coupling of terminal alkynes.

Comparison of methods and energy entry by means of the TOF for the different reaction protocols (ball milling, stirring, and microwave) indicate that ball milling is as powerful as microwave irradiation, giving TOFs of 115 and 98 h^{-1}, respectively [15].

Copper-, solvent-, and oxidant-free Glaser-type coupling reaction under mechanochemical conditions was developed by Mack and coworkers [18]. The use of tetrakis triphenylphosphine palladium catalyst gives high yields of head-to-head homo-coupling product **59**, while the use of bis(triphenylphosphine)palladium dichloride give high yields of the *trans*-enyne product **67** (Scheme 2.24, Table 2.20). Obtained yields are higher than the corresponding reactions carried out in solvent.

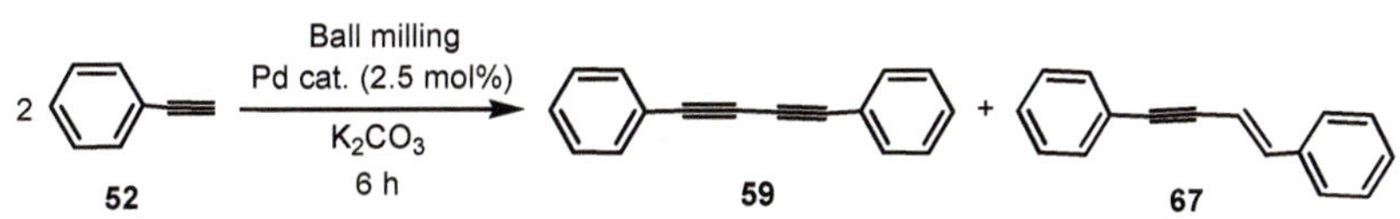

SCHEME 2.24

Glaser coupling of phenyl acetylene.

Table 2.20 Glaser Coupling Under Mechanical Milling[a]

Entry	Catalyst	Additive	Conditions	Product Ratio 59:67	Yield (%)[b]
1	$Pd(PPh_3)_4$		Stainless steel	100:0	74
2	$Pd(PPh_3)_4$		Teflon	100:0	73
3	$Pd(PPh_3)_4$	Br-ethylacetate	THF/DIEA	100:0	40[c]
4	$PdCl_2(PPh_3)_2$	PPh_3	Teflon	100:0	86
5	$PdCl_2(PPh_3)_2$	PPh_3	Stainless steel	100:0	90
6	$PdCl_2(PPh_3)_2$	CuI	THF/NEt_3	not given	75[c]
7	$PdCl_2(PPh_3)_2$		Stainless steel	14:86	89
8	$PdCl_2(PPh_3)_2$		Teflon	19:81	87
9	PS-$Pd(PPh_3)_4$		Stainless steel	12:88	93
10	PS-$Pd(PPh_3)_4$	PPh_3	Stainless steel	92:8	76
11	$Pd(TFA)_2$		Teflon	100:0	57
12	$Pd(TFA)_2$		Stainless steel	100:0	60
13	$Pd(OAc)_2$		Teflon	100:0	64
14	$Pd(OAc)_2$		Stainless steel	100:0	69
15	$Pd(OAc)_2$	Air	THF/DABCO	100:0	58[c]

[a]*8000M SpexCertiprep mixer mill, vial (2×0.5 inch), ball (0.125 inch), acetylene:Pd cat.:K_2CO_3 1:0.025:1.*
[b]*Isolated products.*
[c]*Literature data.*
Reproduced from Chen L, Betsegaw E, Lemma BE, Rich JS, Mack J. Freedom: a copper-free, oxidant-free and solventfree palladium catalysed homocoupling reaction. Green Chem 2014;16:1101–3. With permission from the Royal Society of Chemistry.

Polymer-supported $Pd(PPh_3)_4$ catalyst gives the *trans*-enyne as the major product, showing a different reactivity than the free catalyst. The addition of free triphenylphosphine or polymer-supported triphenylphosphine to the polymer-supported palladium catalyst gives the head-to-head homo-coupling product in high yields. The polymer-supported catalyst can only be recycled twice in the absence of addition triphenylphosphine and five times in the presence of excess triphenylphosphine, but in both cases the reactivity of the catalyst is greatly diminished. Ball milling carried out in a Teflon-milling vial showed that there is no need for copper or any additional metal. Furthermore, milling in Teflon in an inert atmosphere proceeded without change in yield or product ratio, suggesting that additional oxidant is not required under mechanochemical conditions. Optimized conditions were applied to various alkynes to obtain products **69** in high yields (Scheme 2.25, Table 2.21).

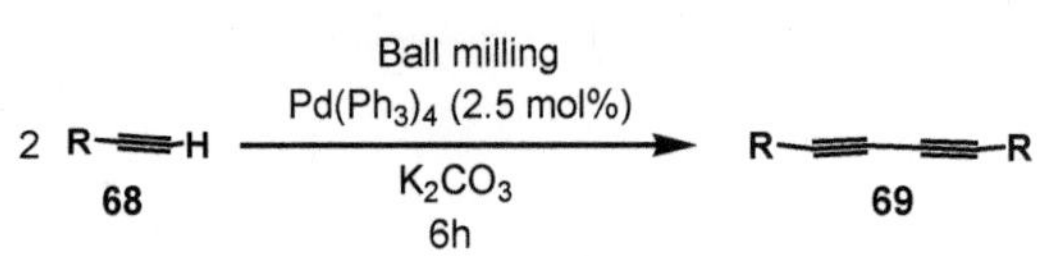

SCHEME 2.25

Glaser coupling of alkynes.

Table 2.21 Glaser Coupling Under Mechanical Milling[a]

Entry	R	Yield (%)[b]
1	C_6H_5	74
2	3-Pyridyl	71
3	4-Br-C_6H_4	82
4	*n*-Bu	62

[a]8000M SpexCertiprep mixer mill, vial (2×0.5 inch), ball (0.125 inch), acetylene:$Pd(Ph_3)_4$:K_2CO_3 1:0.025:1.
[b]Isolated products.
Reproduced from Chen L, Betsegaw E, Lemma BE, Rich JS, Mack J. Freedom: a copper-free, oxidant-free and solventfree palladium catalysed homocoupling reaction. Green Chem 2014;16:1101–3. With permission from the Royal Society of Chemistry.

2.5 MICHAEL REACTION

Wang and Komatsu have applied high-speed vibrational milling to solvent-free Michael reactions of chalkones and azachalkones **70** with ethylacetoacetate catalyzed by potassium carbonate (Scheme 2.26) [19]. Traditionally, this reaction is carried out in solution and catalyzed by strong bases. In these drastic conditions, side reactions often take place, which could be avoided in solvent-free conditions. In mechanochemical variant of this reaction, mixture of two diastereoisomeric products was formed by use of only 10% molar equiv. of K_2CO_3 (in home-made mill operating at 60Hz, in stainless steel capsule (9×26mm), with one 6mm ball) in very short time at room temperature, where the *anti*-isomer **71** dominates (from 72:28 to 93:7 ratio) and reactions proceed in high total yield (86–99%). The highest diastereoselectivity was obtained for cyano substrate (R=4-CN, X=CH). The advantages of mechanochemical procedure are milder reaction conditions, higher yields, shorter reaction time, and occurrence of no side reactions, which is obvious by comparison with reaction of chloro chalkone (R=4-Cl, X=CH) carried in ethanol. The better efficiency of the procedure was ascribed to the increased reaction rate resulting from ultimately high concentrations of reactants with no use of solvent. Furthermore, common side reactions are avoided owing to the use of catalytic amount of weak base K_2CO_3 and the solvent-free conditions, which led to high chemoselectivity.

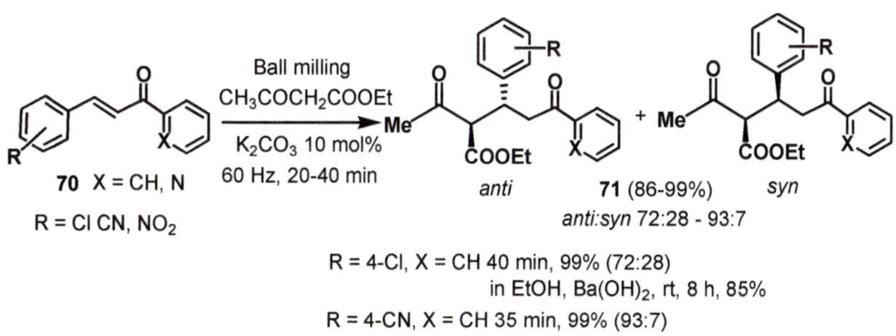

SCHEME 2.26

Michael reaction of (aza)chalkones and ethyl acetoacetate.

Replacement of ethyl acetoacetate with diethyl malonate was equally successful [20]. These Michael addition reactions generally gave products in remarkably high yields (76–99%) (Scheme 2.27, Table 2.22). By the application of different bases, chemoselectivity could be altered and besides Michael adducts **75**, products arising from second aldol condensation **76** and subsequent dehydration **77** were obtained, as in the reaction of 4-chlorochalkone and ethyl acetoacetate (Scheme 2.28, Table 2.23). The vibration frequency of the applied vibration mill had a significant effect on the Michael reaction (Scheme 2.28, Table 2.24). Higher vibration frequency considerably accelerated the reaction process: it was reasoned that the faster mill vibration transfers the higher mechanical energy, and therefore the local pressure, applied to the reaction system. When the mill vibrates at higher speed (or intensity of milling), the temperature of its capsule reaches a higher value, which may be advantageous for the diffusion process to promote homogenization and alloying of powders.

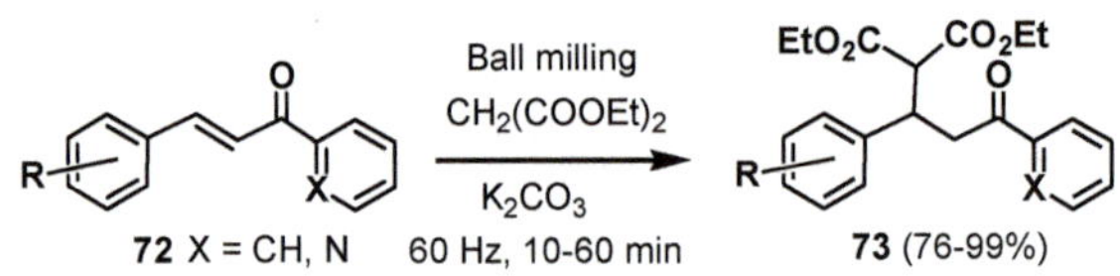

SCHEME 2.27

Michael reaction of chalcones and diethyl malonate.

Table 2.22 Michael Addition of Diethyl Malonate to Chalcones Under Ball Milling[a]

Entry	R	X	Reaction Time (min)[b]	Yield (%)[c]
1	H	CH	55	98
2	4-CH_3	CH	50	98
3	4-OCH_3	CH	60	91
4	4-NO_2	CH	35	99
5	3-NO_2	CH	10	99
6	4-CN	CH	15	99
7	4-Cl	CH	50	99
8	3,4-Cl_2	CH	30	99
9	3,4-OCH_2O-	CH	60	76
10	4-NO_2	N	25	99
11	3-NO_2	N	30	99
12	4-CN	N	30	99
13	4-Cl	N	60	92
14	3,4-Cl_2	N	30	99
16	3,4-OCH_2O-	N	25	98

[a]*Home-made high-speed vibrational ball mill, 3500 rpm, chalcone:ethyl acetoacetate: K_2CO_3 1:1:0.1 mol.*
[b]*Time needed for the reaction to complete, and the maximum reaction time is limited to 1 h.*
[c]*Product is essentially pure, only entries 3, 9, 13 were isolated by chromatography.*
Reproduced from Zhang Z, Dong Y-W, Wang G-W, Komatsu K. Highly efficient mechanochemical reactions of 1,3-dicarbonyl compounds with chalcones and azachalcones catalyzed by potassium carbonate. Synlett 2004:61–4. With permission from Thieme Publishers.

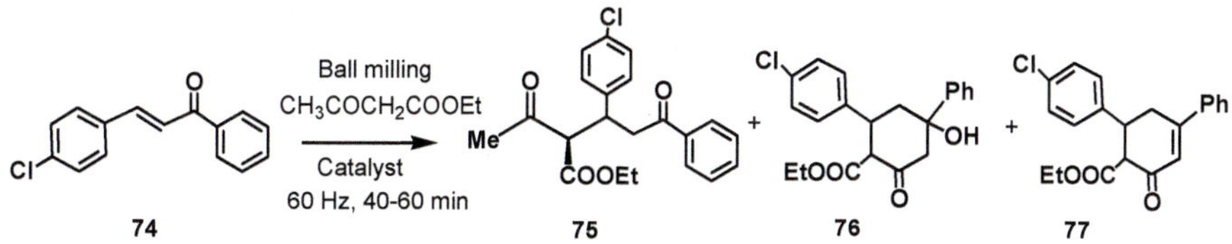

SCHEME 2.28

Michael reaction of 4-chlorochalcone and ethyl acetoacetate.

Table 2.23 Chemoselectivity of Michael Addition of 4-Chlorochalcone and Ethyl Acetoacetate[a]

Entry	Catalyst (equiv.)	Reaction Time (min)	Yield (%)[b]	Product Yield (%)[b]	Yield (%)[b]
			1	2	3
1	None	40	28		
2	K_2CO_3 (0.1)	40	99		
3	Na_2CO_3 (0.1)	60	47		
4	Na_2CO_3 (1.0)	60	68		
5	KF/Al_2O_3 (0.1)	60	28	17	
6	KF/Al_2O_3 (1.0)	60	46	24	
7	NaOH (0.1)	60	26	44	23
8	NaOH (1.0)	60	Trace	57	32

[a]*Home-made high-speed vibrational ball mill, 3500 rpm, chalcone:ethyl acetoacetate: K_2CO_3 1:1:0.1 mol.*

[b]*Isolated products.*

Reproduced from Zhang Z, Dong Y-W, Wang G-W, Komatsu K. Highly efficient mechanochemical reactions of 1,3-dicarbonyl compounds with chalcones and azachalcones catalyzed by potassium carbonate. Synlett 2004:61–4. With permission from Thieme Publishers.

Table 2.24 Michael Addition of 4-Chlorochalcone and Ethyl Acetoacetate Under Ball Milling at Different Vibration Frequencies[a]

Entry	Catalyst (equiv.)	Vibration Frequency (rpm)	Yield (%)[b]
1	None	3500	28
2	None	1800	Trace
3	None	1200	No reaction
4	K_2CO_3 (0.1)	3500	99
5	K_2CO_3 (0.1)	1800	47
6	K_2CO_3 (0.1)	1200	19

[a]*Home-made high-speed vibrational ball mill, 3500 rpm, the reactions at 1800 and 1200 rpm were run on a Retsch MM200 mixer mill, (aza)chalcone:ethyl acetoacetate: K_2CO_3 1:1:0.1 mol, 40 min.*

[b]*Isolated products.*

Reproduced from Zhang Z, Dong Y-W, Wang G-W, Komatsu K. Highly efficient mechanochemical reactions of 1,3-dicarbonyl compounds with chalcones and azachalcones catalyzed by potassium carbonate. Synlett 2004:61–4. With permission from Thieme Publishers.

To accomplish Michael reactions of nitrostyrenes **79** and 1,3-dicarbonyl substrates **78** in ball mill, Zhang et al. employed various bases [21]. In this variant of mechanochemical Michael reaction, addition of inorganic salts as catalysts was crucial to increase reaction yields. The best reaction conditions employ $CaCl_2$ as a catalyst jointly with Et_3N as the base, affording the desired products after 30 min (Scheme 2.29). Because $CaCl_2$ is very affordable and possesses low toxicity, it was selected as the best catalyst for this reaction (selected results, Table 2.25). Authors postulated that the catalytic cycle is initiated by the $CaCl_2$-aided deprotonation of the α-position of malonate to give calcium enolate. The activated species then undergoes nucleophilic addition to nitroalkene.

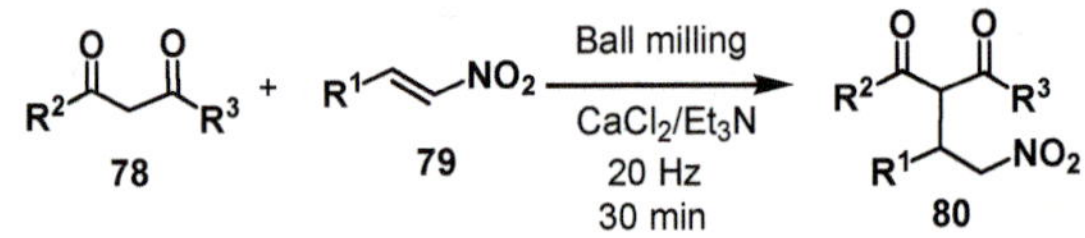

SCHEME 2.29

Michael reaction of nitrostyrenes.

Table 2.25 Michael Reaction of Nitrostyrenes[a]

Entry	R^1	R^2	R^3	Yield (%)[b]
1	C_6H_5	OEt	OEt	92
2	4-$NO_2C_6H_4$	OEt	OEt	77
3	3-$Cl_2C_6H_4$	OEt	OEt	51
4	2,4-$Cl_2C_6H_4$	OEt	OEt	80
5	2-Naphthyl	OEt	OEt	86
6	2-Furanyl	OEt	OEt	82
7	C_6H_5	OMe	OMe	87
8	C_6H_5	O^{i}Pr	O^{i}Pr	95
9	C_6H_5	Me	OEt	86
10	C_6H_5	Ph	OEt	79

[a]*Retsch MM400 mixer mill, stainless steel vessel (10 mL), 10 mm ball, 20 Hz, 30 min, substrate:styrene:Et_3N:catalyst 1:1.1:1:1 mol.*
[b]*Isolated products.*
Reproduced from Jia C, Chen D, Zhang C, Zhang Q, Cao B, Zhao Z. Mechanosynthesis of γ-nitro dicarbonyl compounds via CaCl2-catalyzed Michael addition. Tetrahedron 2013;69:7320–4.

2.5.1 ASYMMETRIC MICHAEL REACTION

Substantial efforts have been made by several groups to develop enantioselective Michael reaction in mechanochemical conditions. Michael addition of propanal (R^1=Me) **81** to *trans*-β-nitrostyrene (R^2=Ph) **82** was used as a model reaction for enantioselective pyrrolidine catalyst screening (Scheme 2.30). Šebesta and coworkers observed after 1 h of milling a complete conversion of reactants in the presence of catalyst **C1** (Table 2.26) [22]. The reaction rate increased markedly, compared with conventional magnetic stirring, and higher yields were obtained after 1 h of milling versus 96 h of stirring (entry 1). More important is the observation that the diastereoselectivity and the enantioselectivity of the reactions with various catalysts in the ball mill were higher than that of the conventional reaction (entries 1, 9, and 10) and the catalyst **C8** was the most efficient.

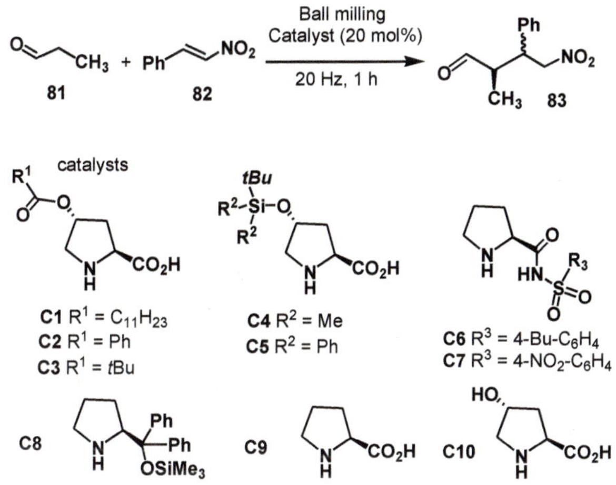

SCHEME 2.30

Enantioselective Michael reaction of propanal to *trans*-β-nitrostyrene.

Table 2.26 Enantioselective Michael Reaction of Propanal to *Trans*-β-Nitrostyrene[a]

Entry	Catalyst	Yield (%)[b]	dr (*syn:anti*)[c]	ee (*syn:anti*)[c]
1	C1	96 (55)[e]	93:7 (73:27)	84/86 (81/81)
2	C2	43	85:15	78/78
3	C3	38	74:26	68/67
4	C4	67	64:36	74/76
5	C5	37	52:48	78/80
6	C6	44	84:16	70/74
7	C7	82	88:12	70/76
8	C8	97	90:10	93/91
9	C9	96 (72)	71:29 (62:38)	58/60 (46/48)
10	C10	97 (52)	85:15 (73:27)	88/91 (87/86)

[a]*Planetary ball mill, two steel balls, 20 Hz, 1 h, aldehyde:nitroalkene:$NaHCO_3$:catalyst 1.2:1:0.2:0.2 mol.*
[b]*Isolated products.*
[c]*In parentheses: magnetic stirring, 96 h.*
Reproduced with permission from Veverková E, Poláčková V, Liptaková L, Kázmerová E, Mečiarová M, Toma Š, Šebesta R. Organocatalyst efficiency in the Michael additions of aldehydes to nitroalkenes in water and in a ball-mill. ChemCatChem 2012;4:1013–8. Copyright (2012), Wiley.

The most efficient catalysts in water and ball mill, *O*-lauroyl-derived catalyst **C1** and Hayashi–Jorgensen catalyst **C8** were used with a variety of aldehydes and nitroalkenes in aqueous solution, as well as in the ball mill (Scheme 2.31). The yields and diastereoselectivities were high in solution and in the ball mill (Table 2.27).

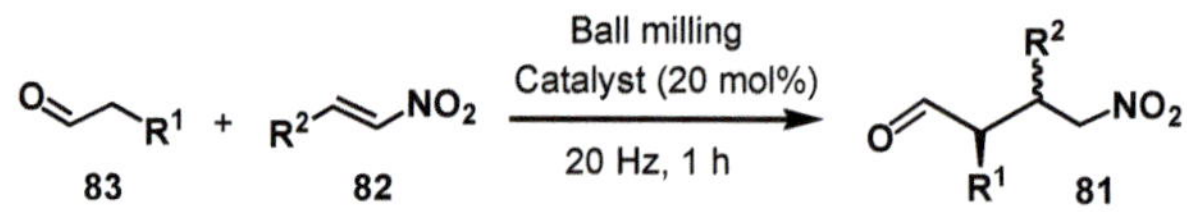

SCHEME 2.31

Enantioselective Michael reaction of nitroalkenes and aldehydes.

Table 2.27 Enantioselective Michael Reaction of Nitroalkenes and Aldehydes[a]

Entry	Product R[1]	R[2]	Catalyst/ Method	Yield (%)[b]	dr (*syn:anti*)[d]	ee (*syn:anti*)[e]
1	Me	Ph	C1/water[c]	94	91:9	97
2	*i*Pr	Ph	C1/water	91	93:7	92
3	Bn	Ph	C1/water	83	91:9	99
4	OBn	Ph	C1/water	56	53:47	95/83
5	Me	*i*Bu	C1/water	86	74:26	98/94
6	Me	COOEt	C1/water	78	64:36	n.d.
7	Me	Ph	C8/ball mill	97	90:10	93
8	*i*Pr	Ph	C8/ball mill	83	95:5	62
9	Bn	Ph	C8/ball mill	77	81:19	94
10	OBn	Ph	C8/ball mill	44	51:49	93/87
11	Me	*i*Bu	C8/ball mill	78	51:49	91/92
12	Me	COOEt	C8/ball mill	69	63:37	n.d.

ee, *enantiomeric excess;* dr, *diastereomeric ratio.*
[a]*Planetary ball mill, two steel balls, 20 Hz, 1 h, aldehyde:nitroalkene:NaHCO$_3$:catalyst 1.2:1:0.2:0.2 mol.*
[b]*Isolated products.*
[c]*Brine, 24 h, RT.*
[d]*NMR.*
[e]*HPLC.*
Reproduced with permission from Veverková E, Poláčková V, Liptaková L, Kázmerová E, Mečiarová M, Toma Š, Šebesta R. Organocatalyst efficiency in the michael additions of aldehydes to nitroalkenes in water and in a ball-mill. ChemCatChem 2012;4:1013–8. Copyright (2012), Wiley.

Xu et al. have carried out asymmetric Michael addition reaction in ball mill using squaramide catalyst (Scheme 2.32) [23]. Optimal reaction conditions in planetary ball mill feature milling of 1,3-diones **83** and aryl styrenes in the presence of the cinchona-derived squaramide catalyst **CA** (5 mol%). In these conditions, high yields and excellent level of enantioselectivity was achieved with variety of reactants (Table 2.28). Some less-active nucleophiles needed longer milling time, and slight erosion in yields and enantiomeric excesses was observed in these cases, perhaps as a result of increase in temperature during the long milling time. The results showed that the H-bonding-mediated enantioselective organocatalysis could be carried out under solvent-free conditions in a ball mill. The intense mixing of the reactants accelerates the reaction efficiently, but the diastereo- and enantioselectivities do not decrease, suggesting that H-bonding interactions, which are enhanced in the absence of solvents, are crucial for the course of the reaction.

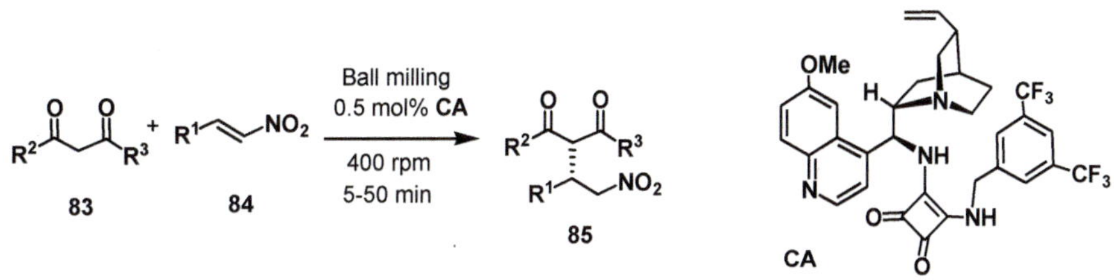

SCHEME 2.32

Asymmetric Michael reaction of 1,3-diones with nitrostryrenes.

Table 2.28 Asymmetric Michael Reaction of 1,3-Diones With Nitrostryrenes[a]

Entry	R^1	R^2	R^3	Time (min)	Yield (%)[b]	dr	ee (%)
1	C_6H_5	CH_3	CH_3	5	95		99
2	4-$CH_3C_6H_4$	CH_3	CH_3	5	93		96
3	4-FC_6H_4	CH_3	CH_3	10	88		99
4	2-BrC_6H_3	CH_3	CH_3	20	83		98
5	4-BrC_6H_4	CH_3	CH_3	10	93		94
6	2,4-$Cl_2C_6H_3$	CH_3	CH_3	30	72		98
7	Naphth-2-yl	CH_3	CH_3	10	86		91
8	Furan-2-yl	CH_3	CH_3	25	63		94
9	Thien-2-yl	CH_3	CH_3	30	92		92
10	C_6H_5	OCH_3	CH_3	10	90	1.6:1	99
11	C_6H_5	OC_2H_5	CH_3	5	95	1.5:1	99
12	C_6H_5	O^tBu	CH_3	25	86	1.4:1	91
13	C_6H_5	OCH_3	OCH_3	30	80		96
14	C_6H_5	OC_2H_5	C_6H_5	50	75	1.7:1	96

[a]*Fritsch Planetary Micro Mill Pulverisette 7, stainless steel vessel (45mL), 80×2mm balls, 400rpm, dione:styrene:CA 1:4:0.005mol.*

[b]*Isolated products.*

Reproduced from Wang Y-F, Chen R-X, Wang K, Zhang B-B, Li Z-B, Xu D-Q. Fast, solvent-free and hydrogen-bonding-mediated asymmetric Michael addition in a ball mill. Green Chem 2012;14:893–5. With permission from the Royal Society of Chemistry.

Mechanochemical effects induced by ball milling on thiourea-catalyzed asymmetric Michael addition reactions were investigated in the solvent-free coupling of α-nitrocyclohexanone **86** and aryl-substituted nitroalkenes **87** by Bolm et al. (Scheme 2.33). [24]. Chiral thiourea catalysts were chosen due to hydrogen bonding which plays a major role in organocatalytic process and the expectation that this reaction could benefit from applying the mechanochemical technique by the high local reagent concentration, the lack of solvent interference to hydrogen bonding, and the efficient reagent mixing in the solvent-free media. Screening of the number of chiral thiourea catalysts and optimization of reaction parameters in ball mill revealed that these assumptions were correct and that in comparison to the results achieved in common solvent-phase reactions, the ones performed in a ball mill are faster and more stereoselective (Table 2.29). Optimal conditions required thiourea catalyst (2.5 mol %) and an excess of nitrostyrene. Similar results were also obtained when the reaction was scaled up and run on a larger scale of 81.0 g.

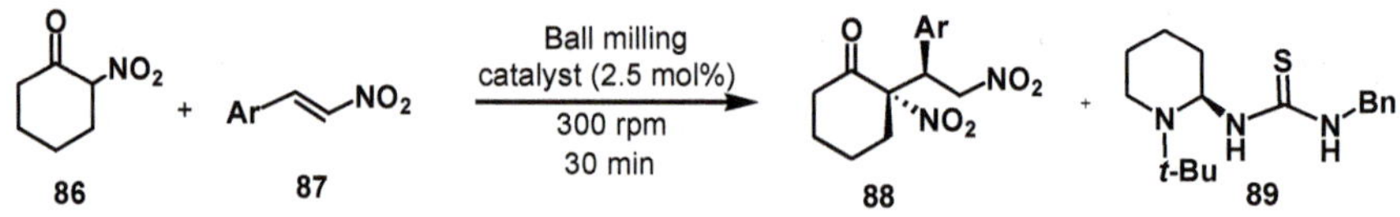

SCHEME 2.33

Asymmetric Michael reaction of α-nitrocyclohexanone with nitrostryrenes.

Table 2.29 Asymmetric Michael Reaction of α-Nitrocyclohexanone With Nitrostryrenes[a]

Entry	Ar	Yield (%)[b]	dr *(anti/syn)*	er (%)
1	2-Furyl	861	94:6	96:4
2	C_6H_5	95	96:4	98:2
3	2-FC_6H_4	91	96:4	97:3
4	4-FC_6H_4	88	94:6	97:3
5	2-ClC_6H_4	91	96:4	99:1
6	4-ClC_6H_4	93	98:2	98:2
7	4-BrC_6H_4	86	96:4	97:3
8	4-$MeOC_6H_4$	80	95:5	97:3
9	4-MeC_6H_4	91	95:5	97:3

dr, *diatereomeric ratio;* er, *enantiomeric ratio.*

[a]*Fritsch Planetary Micro Mill Pulverisette 7, ZrO_2 vessel (12 mL), 4 mm ZrO_2 balls (7.8 g), 300 rpm, 30 min, cyclohexanone:styrene:catalyst 1:1.5:0.025 mol.*

[b]*Isolated products.*

Reproduced from Jörres M, Mersmann S, Raabe G, Bolm C. Organocatalytic solvent-free hydrogen bonding mediated asymmetric Michael additions under ball milling conditions. Green Chem 2013;15:612–6. With permission from the Royal Society of Chemistry.

Hestericovà and Šebesta evaluated a number of chiral thiourea catalysts in two Michael addition reactions under solvent-free ball-milling conditions (Scheme 2.34) [25]. Michael additions of indole and dimethyl malonate to nitrostyrene in the ball mill proceeded with higher enantioselectivities than in solution with several of thiourea catalysts tested (Table 2.30), while the majority of catalysts did not show significant difference in enantiomeric ratios. Lack of solvation was beneficial for enantioselective hydrogen-bonding catalysis in some reactions. In comparison with reactions carried out in solvent, yields obtained in ball mill were generally similar or lower, while, in some cases, enantioselectivities increased, although in the case of catalyst **C2**, much higher enantioselectivity was obtained in solution. Enantioselectivities, yields, and reaction times obtained are inferior to those obtained by Bolm for catalysis of the above described Michael reaction by structurally related chiral thioureas.

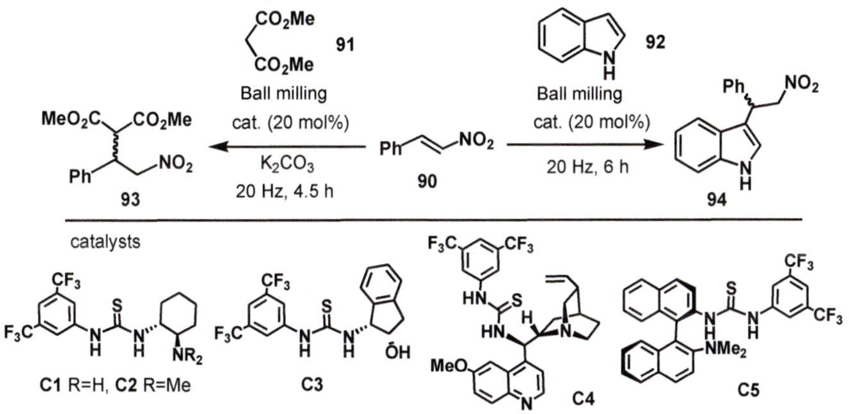

SCHEME 2.34

Asymmetric Michael reaction of nitrostryrene.

Table 2.30 Asymmetric Michael Reaction of Nitrostryrene

	Ball Mill[a]	**Solution**[c]			
Entry	**Cat**	**Yield (%)**[b]	**er/(R/S)**	**Yield (%)**[b]	**er/(R/S)**
Indole					
1	C2	19	11:89	50	60:40
2	C3	79	29:71	56	42:58
3	C4	24	10:90	60	60:40
Dimethyl Malonate					
4	C1	65	40:60	60	50:50
5	C2	78	24:76	90	6:94

Continued

Table 2.30 Asymmetric Michael Reaction of Nitrostryrene—cont'd

	Ball Mill[a]	Solution[c]			
Entry	**Cat**	**Yield (%)[b]**	**er/(R/S)**	**Yield (%)[b]**	**er/(R/S)**
6	C3	85	81:13	44	60:40
7	C5	60	45:55	90	51:49

[a]*Retsch MM400 ball mill, stainless steel vessel (5mL), 5mm ball, 20Hz, 6h, substrate:styrene:catalyst 1:1:0.2mol.*
[b]*Isolated products.*
[c]*CH_2Cl_2, 72h.*
Reproduced from Hestericovà M, Šebesta R. Higher enantioselectivities in thiourea-catalyzed Michael additions under solvent-free conditions. Tetrahedron 2014;70:901–5. Copyright (2014), with permission from Elsevier.

2.6 ALDOL CONDENSATION

Aldol condensation employing vibrating ball mill was firstly reported by Raston and Scott (Scheme 2.35) [26]. However, details are not given to be able to distinguish between results of grinding in a mortar and pestle from ball milling. Veratraldehyde was condensed with 4-phenylcyclohexanone or 1-indanone in the presence of 1 equiv. of NaOH to give condensation products in 77–98% yields. Powdered reagents were ground intermittently, in a mortar and pestle or in a vibrating ball mill for a total of 2 min over a 10-min period.

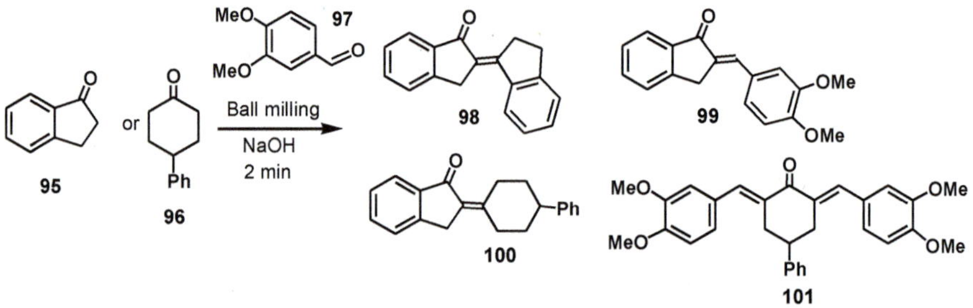

SCHEME 2.35

Aldol reactions of 1-indanone and 4-phenylcyclohexanone.

The solid-state reaction of ninhydrin with dimedone in a ball mill proceeds only as a one-step condensation to give 1:1 product **104** with 100% yield (Scheme 2.36) [27]. In solution, compound was previously obtained in 82% yield as aldol addition product. In solid-state conditions, product **104** was reluctant to eliminate water and form aldol product. Dehydratization was not observed even with gaseous trimethylamine, hydrogen chloride or *p*-toluenesulfonic acid, and this behavior is in contrast to typical condensation reactions of ninhydrin.

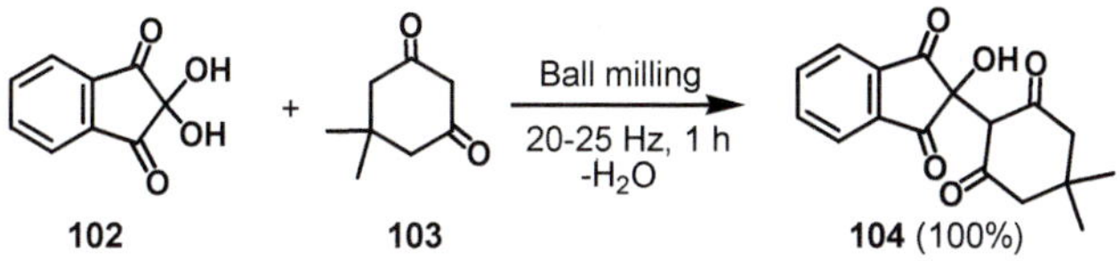

SCHEME 2.36

Solid-state condensations of ninhydrin.

Mitchell and coworkers discovered a novel and simple one-step method for the aluminum-catalyzed self-aldol reaction of valeraldehyde [28]. Ball milling of valeraldehyde **105** with aluminum flakes produced 2-propyl-3-hydroxyl-heptanal **106** (Scheme 2.37). To avoid oxygen or moisture influence on the reaction, stainless steel–milling vial was loaded in a glove box under nitrogen atmosphere. A control experiment without aluminum catalyst for 24 h gave no reaction. This catalytic process was rationalized by reaction mechanism which proposes that the bare aluminum surface produced by milling leads to an acid-catalyzed aldol reaction. The fresh aluminum surface serves to tautomerize the carbonyl to form the enol complex, and activates the carbonyl group of another molecule rendering it electrophilic. Being nucleophilic to the α-carbon, the enol attacks the activated carbonyl compound leading to the formation of the aldol structure.

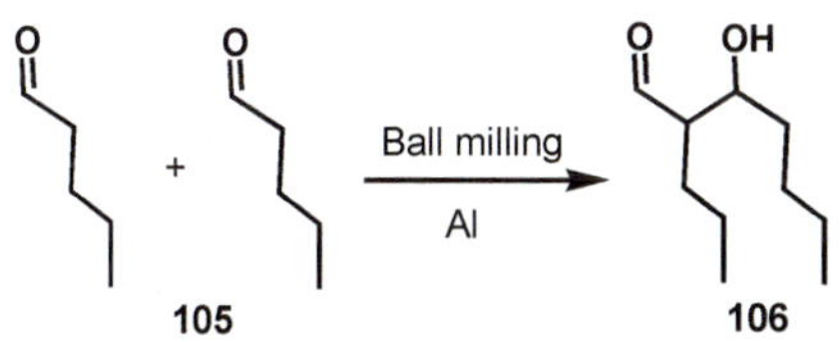

SCHEME 2.37

Self-aldol reaction of valeraldehyde.

2.6.1 ASYMMETRIC ALDOL CONDENSATION

Great deal of effort was given to the development of asymmetric mechanochemical aldol condensations, and several papers were published. For asymmetric aldol reactions under solvent-free conditions in ball mill Tanaka used chiral amine macrocycle-metal(II) complex catalysts [29]. Initial study was carried out on the reaction of cyclohexanone with 4-nitrobenzaldehyde (Table 2.31, entry 1). Optimal reaction conditions were established with (*S,S,S,S,S,S*)-**110** as ligand, and the use of Co ions ($CoBr_2$) resulting in the best enantioselectivity (ee = 93%), *anti–syn* ratio 71:29 and high yield (82%). The solvent-free aldol reaction of 4-nitrobenzaldehyde and cyclohexanone was faster under ball-milling conditions than in a solution or neat liquid. Conventional magnetic stirring in solution using $CuCl_2$ afforded product, but mechanochemical conditions outperformed solution chemistry (*anti–syn* = 70:30) in

Table 2.31 Asymmetric Aldol Reaction of Ketones and Aromatic Aldehydes[a]

Entry	Ketone	Ar	Yield (%)[b]	*anti/syn*	ee (%)
1	Cyclohexanone	4-$NO_2C_6H_4$	82	71:29	93
2	Cyclohexanone	3-$NO_2C_6H_4$	72	69:31	87
3	Cyclopentanone	4-$NO_2C_6H_4$	51	74:26	75
4	Cycloheptanone	4-$NO_2C_6H_4$	42	52:48	26
5	Tetrahydropyranone	4-$NO_2C_6H_4$	55	61:39	69
6	Tetrahydrothiopyranone	4-$NO_2C_6H_4$	54	50:50	60
7	Cyclohexanone	2-ClC_6H_4	14	51:49	22
8	Cyclohexanone	4-ClC_6H_4	42	51:49	75
9	Cyclohexanone	2-BrC_6H_4	12	50:50	88
10	Cyclohexanone	3-BrC_6H_4	20	64:36	85
11	Cyclohexanone	C_6H_5	48	50:50	56
12	Cyclohexanone	4-$OCH_3C_6H_4$	4	54:46	n.d.
13	Acetone	4-$NO_2C_6H_4$	23	–	20

[a]*Fritsch Planetary Micro Mill Pulverisette 7, RT, 24h, 100rpm, ketone:aldehyde:ligand:$CoBr_2$ 2:1:0.2:0.4.*
[b]*Isolated products.*
Reproduced from Tanaka K, Asakura A, Muraoka T, Kalicki P, Urbanczyk-Lipkowska Z. Asymmetric direct aldol reactions catalyzed by chiral amine macrocycle-metal(II) complexes under solvent-free conditions. N J Chem 2013;37:2851–5. With permission from the Royal Society of Chemistry.

13% yield with an enantioselectivity of 79% ee, while neat reaction afforded product (*anti–syn*=60:40) in only 4% yield with an enantioselectivity of 63% ee. Optimal reaction conditions were then applied on the number of aromatic aldehydes and several ketones (Scheme 2.38). Yields obtained in these reactions vary from 14% to 82%, with the best *anti–syn* ratio of products **109** 77:23 and the highest ee 93% (Table 2.32). The ligand (*S,S,S,S,S,S*)-**110** could easily be recovered using column chromatography, and was recycled as a catalyst.

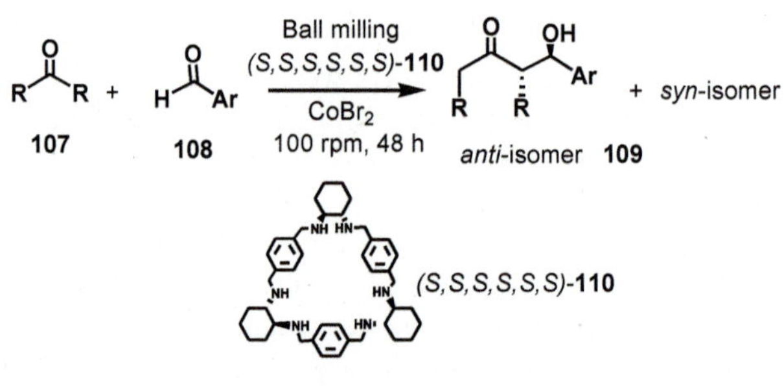

SCHEME 2.38

Asymmetric aldol reaction of ketones and aromatic aldehydes.

In development of solvent-free asymmetric aldol condensations, Nájera carried out reactions using three different methods: conventional magnetic stirring (method A); conventional magnetic stirring with a preformed solution in THF and immediate evaporation of the solvent, which should induce an enhancement of molecule-to-molecule contacts between the reactants, and therefore an acceleration of the reaction rate (method B); and ball-milling conditions (method C) [30]. The optimal conditions for reaction of 4-nitrobenzaldehyde **112** require 2 equiv. of cyclohexanone using 10 mol % of Binam-L-proline-derived catalyst **114** and 20 mol % of benzoic acid (Scheme 2.39). The aldol products **113** were obtained in short time (1–1.5 h) with quantitative conversions, and similar results concerning diastereo- and enantioselectivities using the three methods. In all cases, the *anti*-isomer was obtained with moderate diastereoselectivity (Table 2.32).

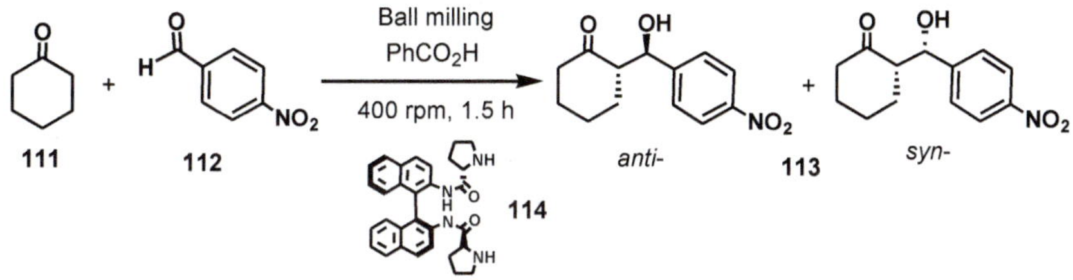

SCHEME 2.39

Asymmetric aldol reaction of 4-nitrobenzaldehyde with cyclohexanone.

Table 2.32 Aldol Reaction of 4-Nitrobenzaldehyde With Cyclohexanone[a]

Entry	Method	*t*/h	Conversion (%)[b]	*anti*/*syn*	ee (%)
1	A	1	99	67:33	88
2	B	1.5	99	72:28	89
3	C	1.5	100	69:31	88

[a]Ketone:aldehyde:ligand:benzoic acid 2:1:0.1:0.2 mol, 25°C. Method A: conventional magnetic stirring. Method B: 4-nitrobenzaldehyde, catalyst, and benzoic acid were dissolved in THF and the solvent was evaporated prior to the addition of cyclohexanone. Method C: ball milling, 300 rpm.
[b]Isolated products.
Reproduced from Guillena G, del Carmen Hita M, Nájera C, Viózquez SF. Solvent-free asymmetric direct aldol reactions organocatalysed by recoverable (S_a)-binam-L-prolinamide. Tetrahedron Asymm 2007;18:2300–4. Copyright (2007), with permission from Elsevier.

Application of the same milling conditions for solid ketones (tetrahydrothiopyranone and *N*-methyl-piperidinone) afforded the corresponding aldol products **115** and **116** in high yields (Table 2.33) [31]. Again, Nájera compared methods B and C and reported that using method C, product ***anti*-115** was achieved with a yield and enantioselectivity higher than that obtained with method B. On the other hand, product ***anti*-116** was obtained with lower yield and enantioselectivity but with slightly better diastereoselectivity than with method B. The use of L-proline (10 mol %) as catalyst under method B conditions gave the corresponding aldol products in much longer reaction time but with similar diastereo- and enantioselectivities for product ***anti*-115**. Under ball-milling conditions using L-proline as catalyst, compound ***anti*-115** was achieved with similar yield and slightly higher de and ee. For *N*-methyl-piperidinone longer reaction time was required, obtaining ***anti*-116** with similar yield, higher de, but lower ee. Finally, acetone was subjected to the reaction with 4-nitrobenzaldehyde, affording the aldol product in only 3 h in 86% yield and 74% ee using method A. Using method B, similar results were obtained but longer reaction time was required. These results were better than those reported by using the ball-milling technique.

Table 2.33 Aldol Reaction of 4-Nitrobenzaldehyde With Ketones[a]

Entry	Method	Major product	*t*/h	Yield (%)[b]	*anti*/*syn*	ee (%)
1	B	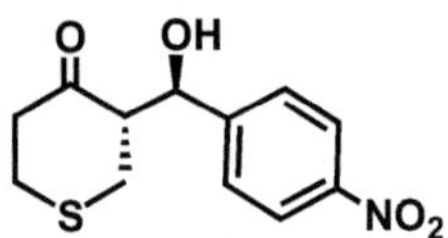 *anti*-**115**	4	64	93:7	80
2	B[c]		48	76	94:6	72
3	C		8	75	88:12	84
4	C[d]		34	79	96:4	90
5	B	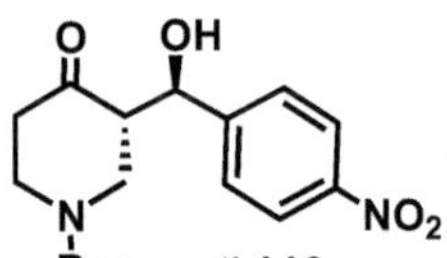 *anti*-**116**	48	69	65:35	76
6	B[c]		192	77	92:8	90
7	C		8	43	73:26	70
8	C[d]		36	50	86:14	55
9	A	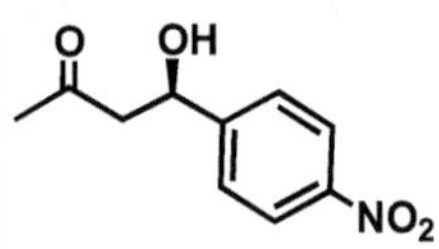	3	86	–	74
10	B		8	88	–	74
11	C		19	73	–	56

[a]*Ketone:aldehyde:ligand:benzoic acid 2:1:0.05:0.1 mol, 25°C. Method A: conventional magnetic stirring. Method B: aldehyde, catalyst, and benzoic acid were dissolved in THF and the solvent was evaporated prior to the addition of ketone. Method C: ball milling, 400 rpm.*
[b]*Isolated products.*
[c]*L-Pro catalyst 10 mol%.*
[d]*Previously published results obtained with L-Pro catalyst.*
Reproduced with permission from Guillena G., del Carmen Hita M., Nájera C., Viózquez S.F. A highly efficient solvent-free asymmetric direct aldol reaction organocatalyzed by recoverable (S)-Binam-L-Prolinamides. ESI-MS evidence of the enamine-iminium formation. J Org Chem 2008;73:5933–43. Copyright (2008), American Chemical Society.

By replacement of Binam-L-proline **114** used in work described above with prolinethioamide catalyst **117**, Nájera obtained better results (Scheme 2.40) [32]. When prolinethioamide catalyst was employed in conjunction with *p*-nitrobenzoic acid as a cocatalyst, reactions were finished in very short time (Table 2.34). The efficacy of catalyst was not improved when the reaction was performed in a ball mill (method C) or under conventional magnetic stirring with a preformed solution in THF and immediate evaporation of the solvent (method B).

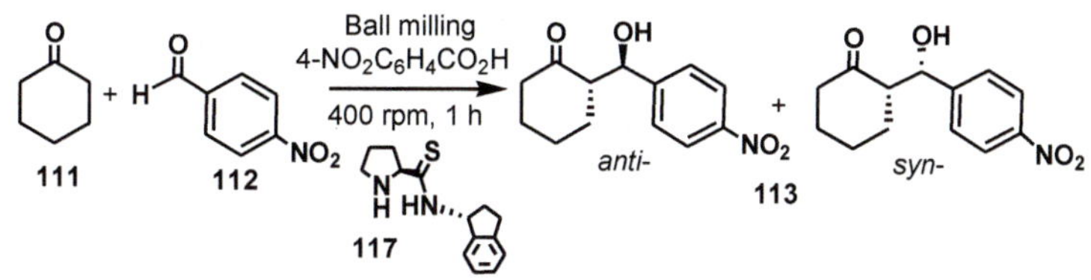

SCHEME 2.40

Aldol reaction of 4-nitrobenzaldehyde with cyclohexanone.

Table 2.34 Aldol Reaction of 4-Nitrobenzaldehyde With Cyclohexanone[a]

Entry	Method	*t*/h	Conversion (%)[b]	*anti*/*syn*	ee (%)
1	A	1	99	89:11	88
2	B	1	98	87:13	74
3	C	1	98	91:9	86

[a]Ketone:aldehyde:ligand:benzoic acid 2:1:0.1:0.2 mol, 25°C. Method A: conventional magnetic stirring. Method B: 4-nitrobenzaldehyde, catalyst, and benzoic acid were dissolved in THF and the solvent was evaporated prior to the addition of cyclohexanone. Method C: ball milling, 400 rpm.
[b]Isolated products.
Reproduced with permission from Almaşi D, Alonso DA, Nájera C. Prolinamides versus prolinethioamides as recyclable catalysts in the enantioselective solvent-free inter- and intramolecular aldol reactions. Adv Synth Catal 2008;350:2467–72. Copyright (2008), Wiley.

Juaristi and coworkers have obtained an improved stereoinduction under solvent-free conditions for aldol reaction of 4-nitrobenzaldehyde with acetone catalyzed by (*S*)-proline-dipeptide **121** (Scheme 2.41) [33]. When aldol reaction was carried out in acetone with 20 mol% of catalyst, the aldol adduct was obtained with only 28% ee. On the other hand, reaction in ball mill in cryogenic conditions at −20°C for 4 h afforded the expected product **120** in 82% yield and with 69% ee, showing superiority of solvent-free process both in terms of required reaction time and enantioselectivity relative to traditional solution conditions (Table 2.36, entries 1 and 2). Ball milling was carried out in milling vial made of Nylamid with one 5 mm stainless steel ball, in apparatus with cooling capability. The optimization of reaction conditions for aldol reaction of 4-nitrobenzaldehyde with cyclohexanone revealed that best results are obtained when using 7 mol% of catalyst at −20°C, after 4 h the *anti*-product being obtained in 95% ee (Scheme 2.41). When the reaction was carried out at higher frequency (3080 rpm) the aldol product was obtained in similar yield but in lower enantiomeric excess (76% vs. 95%), perhaps as a consequence of increased reactor temperature at higher milling speed. Under the optimum reaction conditions (*S*)-proline-(*S*)-phenylalanine methyl ester catalyst **121** was employed in asymmetric aldol reaction between cyclohexanone and cyclopentanone and several benzaldehyde derivatives prepared (Table 2.35). Products were obtained in moderate to high yields and enantioselectivities. Observed results are rationalized by formation of transition state in which a hydrogen bond between amide group of catalyst and the aldehyde controls interaction. In solution, this interaction is weakened by solvation, and therefore enantioinduction is less effective. By contrast, under solvent-free conditions there is no competition between the aldehyde and the solvent for hydrogen bond formation and this interaction is maximized.

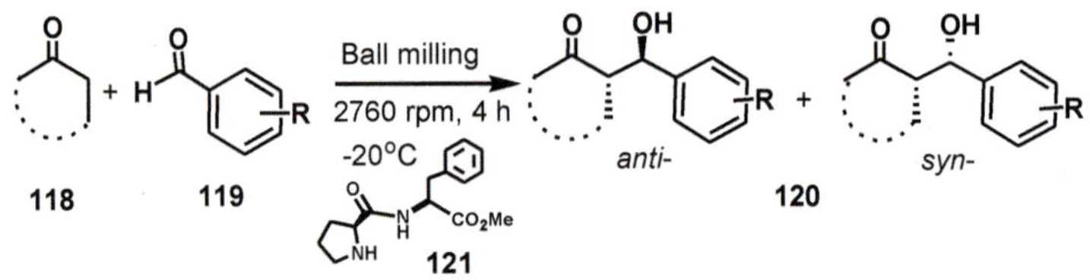

SCHEME 2.41

Aldol reaction of ketones and aromatic aldehydes.

Table 2.35 Aldol Reaction of Ketones and Aromatic Aldehydes[a]

Entry	Ketone	Ar	Yield (%)[b]	*anti/syn*	ee (%)
1	Acetone	4-$NO_2C_6H_4$	82	–	69
2	Acetone	4-$NO_2C_6H_4$	88[c]	–	28
3	Cyclohexanone	4-$NO_2C_6H_4$	92	90:10	95
4	Cyclohexanone	3-$NO_2C_6H_4$	94	88:12	85
5	Cyclohexanone	2-$NO_2C_6H_4$	88	89:11	90
6	Cyclopentanone	4-ClC_6H_4	90	89:11	82
7	Cycloheptanone	3-ClC_6H_4	87	90:10	89
8	Cyclohexanone	2-ClC_6H_4	80	91:9	90
9	Cyclohexanone	4-BrC_6H_4	81	88:12	88
10	Cyclohexanone	3-BrC_6H_4	83	91:9	88
11	Cyclohexanone	C_6H_5	70	88:12	65
12	Cyclohexanone	4-$OCH_3C_6H_4$	62	84:16	57
13	Cyclopentanone	4-$NO_2C_6H_4$	70	31:69	55

[a]*Digital Mixer/Amalgamator, Nylamid cylinder (25 × 10 mm), one 5 mm stainless steel ball, –20°C, 4 h, 2760 rpm, ketone:aldehyde:catalyst 1.1:1:0.07 mol.*
[b]*Isolated products.*
[c]*Acetone solution, 20 mol% catalyst, 24–48 h.*

The detailed evaluation of the reaction parameters, for example, catalyst, amount of catalyst, and equivalents of water and additives, led to optimized reaction conditions, in which catalyst **C1** (Chart 2.1) afforded aldol product in 94% ee, only slightly lower than results in Table 2.35 [34]. An improvement in the stereoselectivity of the product was observed when adding 1.1 equiv. of water to the reaction mixture. These results are consistent with an explanation that water may be involved in interactions in transition state which facilitate the required proximity between the catalyst and aldehyde. Further improvement in catalyst activity was obtained by addition of acidic proton donor additives such as benzoic acid. This property is important for pyrrolidine-based organocatalysts, where the reaction presumably proceeds via an enamine intermediate. Catalysis of aldol reaction between selected ketones and benzaldehydes with dipeptides **C1–C3** under the optimized reaction conditions (3 mol % of catalyst, 1.1 equiv. of water, 5 mol % of benzoic acid) afforded products in high yields and ee's ranging from 55% to 98% (Table 2.36). In these conditions, better enantioselectivities were obtained relative to the reaction catalyzed by 7 mol % of catalyst **C1**.

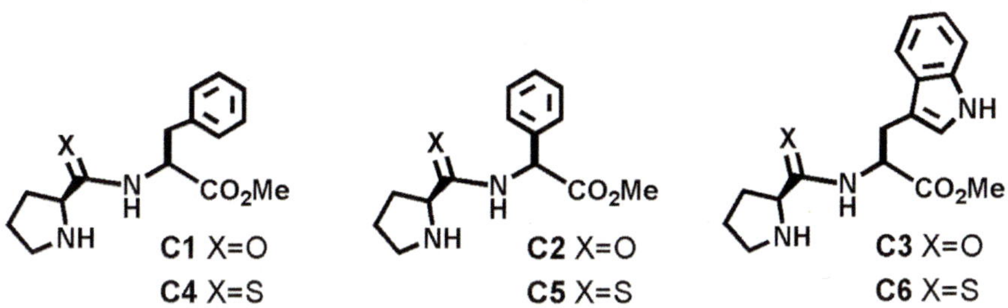

CHART 2.1

Reaction catalysts **C1** to **C6**.

Table 2.36 Aldol Reaction of Ketones and Aromatic Aldehydes[a]

Entry	Ketone	Ar	Yield (%)[b]	*anti/syn*	ee (%)
1	Cyclohexanone	4-$NO_2C_6H_4$	89[c]	93:7	94
2	Cyclohexanone	4-$NO_2C_6H_4$	79[d]	91:9	85
3	Cyclohexanone	4-$NO_2C_6H_4$	88	92:8	>98
4	Cyclohexanone	2-$NO_2C_6H_4$	94	88:12	85
5	Cyclopentanone	2-ClC_6H_4	87	90:10	89
6	Cycloheptanone	4-BrC_6H_4	80	91:9	90
7	Cyclohexanone	3-BrC_6H_4	81	88:12	88
8	Cyclohexanone	4-CNC_6H_4	83	91:9	88
9	Cyclohexanone	4-$CH_3C_6H_4$	70	88:12	65
10	Cyclohexanone	C_6H_5	62	84:16	57
11	Cyclopentanone	4-$NO_2C_6H_4$	70	31:69	55
12	Acetone	4-$NO_2C_6H_4$	65 (80)[c]		58 (69)[c]
13	Acetone	3-$NO_2C_6H_4$	85[c]		28
14	Acetone	2-$NO_2C_6H_4$	68 (75)[c]		69 (67)[c]

[a]*Digital Mixer/Amalgamator, Nylamid cylinder (25 × 10 mm), one 5 mm stainless steel ball, −20°C, 6 h, 2760 rpm, ketone:aldehyde:catalyst C:H_2O:$PhCO_2H$ 1.1:1:0.03:1.1:0.05 mol.*

[b]*Isolated products.*

[c]*Catalyst C1.*

[d]*Catalyst C2.*

Reproduced from Hernández JG, Juaristi E. Efficient ball-mill procedure in the 'green' asymmetric aldol reaction organocatalyzed by (S)-proline-containing dipeptides in the presence of water. Tetrahedron 2011;67:6953–9. Copyright (2011), with permission from Elsevier.

In continuation of synthetic study, Juaristi and coworkers investigated the efficacy of the series of (*S*)-proline–thioamide catalysts **C4–C6** (Chart 2.1) [35]. When aldol reaction of 3-nitrobenzaldehyde with cyclohexanone was carried out, the best results were obtained when using 7 mol % of catalyst **C4**. On the other hand, in ball-milling conditions at −20°C the use of water and acidic additives was beneficial for the increase of diastereoselectivity in favor of the *anti*-diastereomer; however, the enantiomeric excess of the process was not improved. Thus, reactions between cyclic ketones and aromatic were carried out in conditions identical to these in Table 2.35 to afford aldol products in good isolated yields (70–89%) with high diastereo- (92:8 to >98:2 *anti/syn*) and enantioselectivities (82–96% ee) (Table 2.37).

Table 2.37 Aldol Reaction of Ketones and Aromatic Aldehydes[a]

Entry	Ketone	Ar	Yield (%)[b]	*anti/syn*	ee (%)
1	Cyclohexanone	4-$NO_2C_6H_4$	89	98:2	91
2	Cyclohexanone	3-$NO_2C_6H_4$	86	92:8	96
3	Cyclohexanone	2-$NO_2C_6H_4$	80	>98:2	93
4	Cyclohexanone	4-ClC_6H_4	74	>97:3	89
5	Cyclopentanone	3-ClC_6H_4	70	>93:7	87
6	Cycloheptanone	2-ClC_6H_4	78	98:2	90
7	Cyclohexanone	4-BrC_6H_4	82	97:3	88
8	Cyclohexanone	3-BrC_6H_4	77	97:3	82
9	Cyclohexanone	4-CNC_6H_4	80	>97:3	87
10	Cyclohexanone	C_6H_5	63	92:8	86
11	Cyclohexanone	4-$OCH_3C_6H_4$	51[c]	90:10	50
12	Cyclopentanone	4-$NO_2C_6H_4$	76	70:30	66

[a]*Digital Mixer/Amalgamator, 25 × 10 mm Nylamid cylinder, one 5 mm stainless steel ball, −20°C, 6 h, 2760 rpm, ketone:aldehyde:catalyst C4 1.1:1:0.07 mol.*
[b]*Isolated products.*
[c]*8 h.*
Reproduced from Hernández JG, García-López V, Juaristi E. Solvent-free asymmetric aldol reaction organocatalyzed by (S)-proline-containing thiodipeptides under ball-milling conditions. Tetrahedron 2012;68:92–7. Copyright (2012), with permission from Elsevier.

Aldol reaction of isatin with acetone carried out in the same reaction conditions showed that the enantiomeric excesses in aldol product **124** were superior when using ball milling, relative to the reaction under traditional magnetic stirring in solution, regardless of the catalyst employed (Scheme 2.42, Table 2.38). The use of *N*-protected isatins led to an improvement in the enantioselectivity of the aldol product, whereas isatins containing electron-withdrawing groups on the aromatic ring generated the aldol product with higher enantioselectivities (the highest 86% enantiomeric excess).

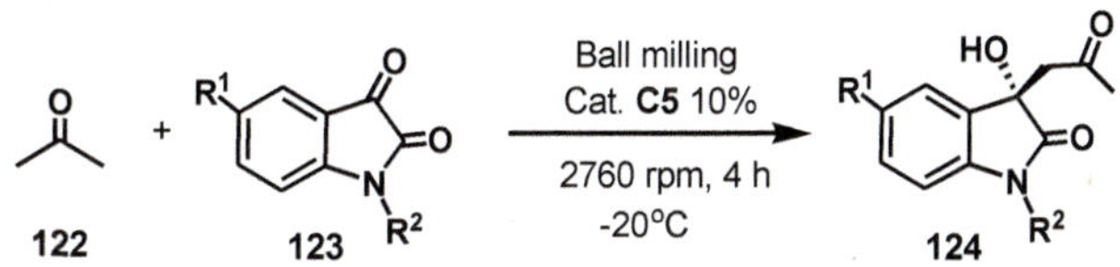

SCHEME 2.42

Asymmetric aldol reaction of isatins and acetone.

Table 2.38 Asymmetric Aldol Reaction of Isatins and Acetone[a]

Entry	R^1	R^2	Yield (%)[b]	ee (%)
1	H	H	33[d] (62)[c]	38 (39)[c]
2	H	H	52[e] (47)[c]	35 (31)[c]
3	H	H	60[f] (72)[c]	50 (24)[c]
4	H	H	30[g]	48
5	H	H	54	56
6	H	CH_3	68	62
7	H	Bn	60	65
8	NO_2	H	65	69
9	Br	H	61	86

[a]*Digital Mixer/Amalgamator, Nylamid cylinder (25×10mm), one 5mm stainless steel ball, −20°C, 4h, 2760rpm, ketone:aldehyde:catalyst C5 1.1:1:0.1mol.*
[b]*Isolated products.*
[c]*In acetone −20°C, 4h.*
[d]*Catalyst C1.*
[e]*Catalyst C2.*
[f]*Catalyst C3.*
[g]*Catalyst C4.*
Reproduced from Hernández JG, García-López V, Juaristi E. Solvent-free asymmetric aldol reaction organocatalyzed by (S)-proline-containing thiodipeptides under ball-milling conditions. Tetrahedron 2012;68:92–7. Copyright (2012), with permission from Elsevier.

Examination of the catalytic activity of the series of catalysts given in Chart 2.2 by Juaristi in cyclohexanone/*p*-nitrobenzaldehyde asymmetric aldol reaction in ball mill revealed that catalysts **C7** and **C9** have similar activities and stereoselecitivities, while the aldol reaction with catalyst **C8** gave low yield, probably due to its limited solubility [36]. By contrast, ester derivative organocatalyst **C7** is an oil that facilitates "wetting" of the reaction mixture and thus makes the aldol reaction more feasible. It was shown that reactions in solution or neat proceeded in good yields, but required 48h, with lower stereoslectivities. The optimal conditions for improved enantioselectivity were obtained with catalyst **C7** (10mol%), employing water and benzoic acid as additives for 30min at room temperature, and these are applied to aldol reaction with series of ketones and aromatic aldehydes (Table 2.39). The best results were observed with the electron-withdrawing nitro derivatives, with high yields in the 87–98% range, *anti/syn* diastereomeric ratios around 90:10, and enantiomeric ratios for the predominant *anti-* isomer up to 93:7. Furthermore, in most cases the observed enantiomeric excess was lower with an increase in reaction time to 2 h, probably as the result of internal heating upon extended reaction times, and it is proven that reversible process does not take place in the reaction conditions.

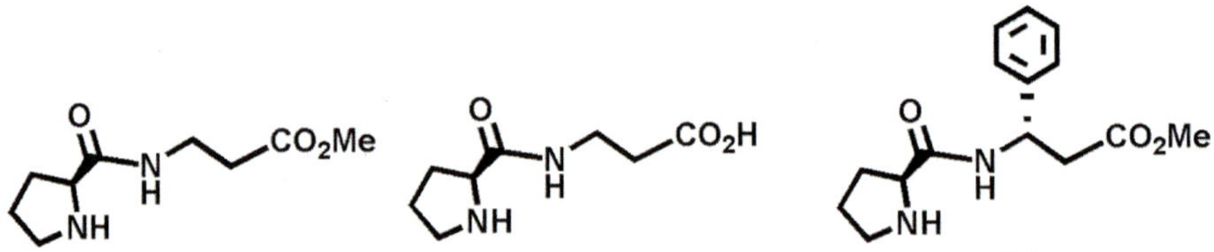

CHART 2.2

Reaction catalysts **C7** to **C9**.

Table 2.39 Aldol Reaction of Ketones and Aromatic Aldehydes[a]

Entry	Ketone	Ar	*t*/h	Yield (%)[b]	dr *anti/syn*	er (%)
1	Cyclohexanone	4-$NO_2C_6H_4$	0.5	94 (99)[c]	90:10 (80:20)[c]	91:9 (84:16)[c]
2		4-$NO_2C_6H_4$	2	93 (95)[d]	89:11(45:55)[d]	89:11 (50:50)[d]
3		3-$NO_2C_6H_4$	0.5	74	89:11	89:11
4		3-$NO_2C_6H_4$	2	98	82:18	n.d.
5		4-BrC_6H_4	0.5	52	91:9	92:8
6		4-BrC_6H_4	2	59	88:12	87:13
7		4-ClC_6H_4	0.5	26	90:10	83:17
8		4-ClC_6H_4	2	51	88:12	98:2
9		2-ClC_6H_4	0.5	37	91:9	95:5
10		2-ClC_6H_4	1	62	89:11	92:8
11		2-ClC_6H_4	2	87	88:12	98:2
12		2-ClC_6H_4	2.5	78	89:11	94:6
13	Cyclopentanone	4-$NO_2C_6H_4$	0.5	97	37:63	76:24
14		4-$NO_2C_6H_4$	2	99	35:65	73:27
15	Acetone	2-$NO_2C_6H_4$	0.5	34	–	76:24
16		2-$NO_2C_6H_4$	2	99	–	77:23

[a]*MM200 ball mill, agate grinding jar (5 mL), 2 × 7 mm agate balls, RT, 15 Hz, ketone:aldehyde:catalyst C7:H_2O:$PhCO_2H$ 1:0.5:0.1:3:0.2 mol.*
[b]*Isolated products.*
[c]*H_2O, 48 h.*
[d]*THF, 10 day.*
Reproduced with permission from Machuca E, Rojas Y, Juaristi E. Synthesis and evaluation of (S)-proline-containing α,β-dipeptides as organocatalysts in solvent-free asymmetric aldol reactions under ball-milling conditions. Asian J Org Chem 2015;4:46–53. Copyright (2015), Wiley.

Bolm et al. reported that (*S*)-proline was a highly efficient catalyst of the aldol reaction between cyclohexanone and 3-nitrobenzaldehyde under solvent-free conditions in ball mill (Scheme 2.43) [37,38]. Complete conversion was achieved after 7 h, and *anti*-aldol product **127** was formed in high yield (94%) and with excellent stereoselectivity (88:12 *anti/syn* ratio; >99 ee). A milling cycle consisting of a 15 min milling period at a rotational speed of 400 rpm, followed by a 5 min pause was used. Higher rotational speeds (>500 rpm) resulted in lower enantiomeric excess values, together with decreased yields mainly due to the formation of byproducts, presumably as a result of the observed temperature increase. Comparative study revealed the beneficial effect of the ball milling with respect to conventional magnetic stirring (Table 2.40). Under solvent-free conditions the aldol products were generally obtained in high yields and with excellent stereoselectivities. By means of the ball mill the aldol reaction proceeded much faster, indicating the advantage of using mechanochemical technique. In the ball-milling process the increased efficiency in mixing, the slightly elevated temperature and the pressure exhibited by the balls upon stirring are beneficial for a more rapid melt formation (phase-change liquefaction to honey-like paste), which corresponds to the observed rate enhancement. Beside liquid ketones, several solid ketones were used (entries 11–14); however, the reactions in the solid–solid system proceeded much slower than in the liquid–solid one, but with comparable stereoselectivities.

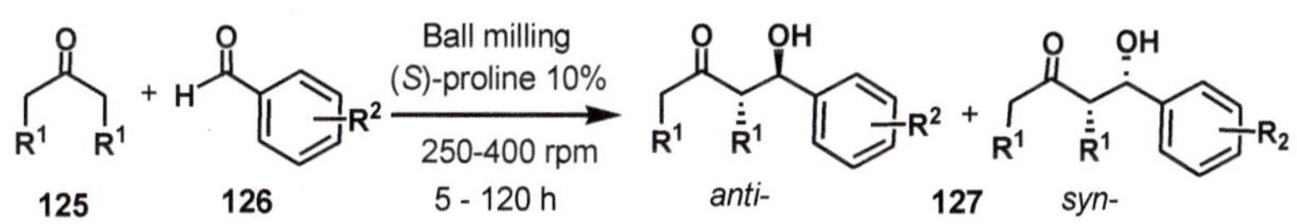

SCHEME 2.43

Asymmetric aldol reaction of ketones and aromatic aldehydes.

Table 2.40 Asymmetric Aldol Reaction of Ketones and Aromatic Aldehydes[a]

Entry	Ketone	Ar	t/h	Yield (%)[b]	*anti*/*syn*	ee (%)
1	Cyclohexanone	4-$NO_2C_6H_4$	5.5	99	89:11	94
2		2-$NO_2C_6H_4$	7	97	93:7	97
3		4-ClC_6H_4	20	87	74:26	75
4		4-ClC_6H_4	72[c]	85	78:22	67
5		2-OCH_3-C_6H_4	36	65	66:34	63
6		C_6H_5	11	53	81:19	45
7	Cyclopentanone	4-$NO_2C_6H_4$	5	53	57:43	95
8		4-$NO_2C_6H_4$	24[c]	90	52:48	87
9		3-$NO_2C_6H_4$	5	93	50:50	90
10	Acetone	4-$NO_2C_6H_4$	19	73	-	56
11	4*R*-*t*Bu-Cyclo-	4-$NO_2C_6H_4$	33	85	91:9	91

Continued

Table 2.40 Asymmetric Aldol Reaction of Ketones and Aromatic Aldehydes[a]—cont'd

Entry	Ketone	Ar	*t*/h	Yield (%)[b]	*anti*/*syn*	*ee* (%)
12	Hexanone	4-$NO_2C_6H_4$	120[c]	58	93:7	89
13	Thiopyranone	4-$NO_2C_6H_4$	36	79	96:4	90
14		3-ClC_6H_4	33	72	95:5	85

[a]*Fritsch Planetary Micro Mill Pulverisette 7, ZrO_2 milling bowl (45 mL), 60×5 mm ZrO_2 balls, 250–400 rpm, sequential intervals of 25 min milling followed by a 5 min pause, ketone:aldehyde:(S)-proline 1.1:1:0.1 mol.*
[b]*Isolated products.*
[c]*Magnetic stirring.*
Reproduced with permission from Rodríguez B, Bruckmann A, Bolm C. A highly efficient asymmetric organocatalytic aldol reaction in a ball mill. Chem Eur J 2007;13:4710–22. Copyright (2007), Wiley.

2.7 MORITA–BAYLIS–HILLMAN REACTION

Significant rate enhancement of the Morita–Baylis–Hillman reaction through solid-state milling has been noted by Mack et al. (Scheme 2.44). [39]. In such conditions, Baylis–Hillman products **130** were obtained in up to >98% yield in as little as 0.5 h by solvent-free reaction of *p*-nitrobenzaldehyde **129** and methyl acrylate **128**. Various bases were tested and 1,4-diazabicyclo[2.2.2]octane (DABCO) showed the best performance (Table 2.41). Other *p*-substituted aromatic aldehydes reacted with methyl acrylate much slower, within 9–45 h and lower yields (28–97%) were obtained. This represents one of the fastest methods of Baylis–Hillman reactions under neat conditions. One of the main drawbacks of this reaction carried out in classical conditions is its slow rate, which has been shown typically to take days to weeks to produce adequate product yields.

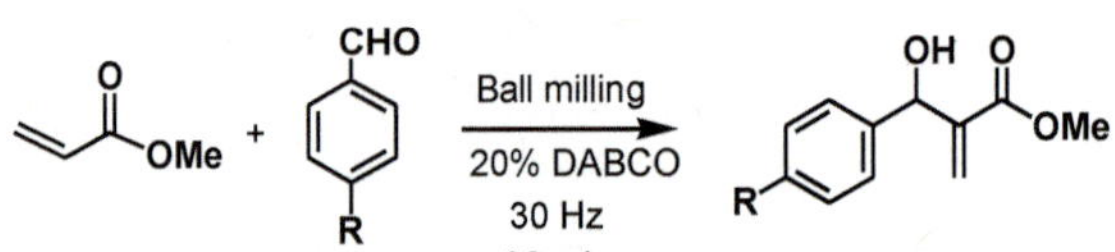

128 129 130

SCHEME 2.44

Morita–Baylis–Hillman reaction.

Table 2.41 Mechanochemical Morita–Baylis–Hillman Reaction of *p*-Substituted Benzaldehydes[a]

Entry	R	Time/h	Yield (%)[b]
1	NO_2	0.5	>98
2	Br	9	97
3	H	39	96
4	Cl	21	54
5	OCH_3	45	28

[a]*SpexCertiprep 8000M mixer mill, 0.5 inch screw-capped stainless steel vial, one 0.125 inch ball, aldehyde: acrylate: DABCO 1:1:0.5 mol.*
[b]*Isolated products.*
Reproduced from Mack J, Shumba M. Rate enhancement of the Morita-Baylis-Hillman reaction through mechanochemistry. Green Chem 2007;9:328–30. with permission from the Royal Society of Chemistry.

2.8 KNOEVENAGEL CONDENSATION

Suzuki and Wada explored the Knoevenagel condensation of malononitrile and methyl cyanoacetate under solvent-free conditions (Scheme 2.45) [40]. Different alkaline earth carbonates and fluospar were used as mild catalysts for the condensation of 2 equiv. of malononitrile and 1 equiv. of aromatic aldehyde to obtain products in high yields, after short milling (30 min, Table 2.42). When methyl cyanoacetate was ball milled with aldehydes, the reaction occurred with the stereoselective formation of (*E*)-cyanocinnamic esters **133**. However, significant oxidative loss of aldehydes was encountered. The addition of a small amount of hydroquinone as antioxidant was found to solve the problem and drastically increase yields.

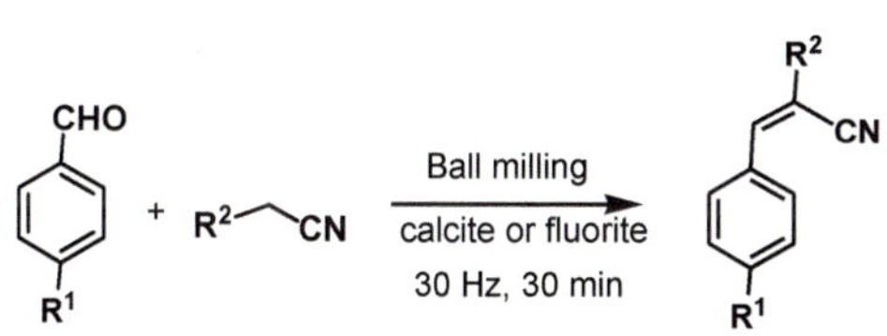

131 132 133

SCHEME 2.45

Knoevenagel condensation of malononitrile and methyl cyanoacetate.

Table 2.42 Knoevenagel Condensation of Malononitrile and Methyl Cyanoacetate[a]

Entry	Mineral Reagent	R¹	R²	Yield (%)[b]
1	Iceland spar	NO_2	CN	96
2	($CaCO_3$)	CN		92
3	from Kansu	Br		81
4		Cl		84
5		Me		97 (66)[c]
6		OH		93
7		H		58 (36)[c]
8		NO_2	CO_2Me	100
9		CN		100
10		CN	CN	87
11	Fluorspar	Cl		99 (30)[c]
12	(CaF_2)	Me		93 (38)[c]
13	from Hunan	H		100 (30)[c]
14		NO_2	CO_2Me	82
15		CN		51[d]
16		Br		82[d]

[a]*Retsch MM200 mixer mill, 30Hz, stainless steel jar (5mL), 2×7mm balls, nitrile:aldehyde 2:1 mol, 30min.*
[b]*Isolated products.*
[c]*Catalytic amount of hydroquinone. Yields obtained in the absence of antioxidant are given in parentheses.*
[d]*NMR yields.*
Reproduced from Wada S, Suzuki H. Calcite and fluorite as catalyst for the Knoevenagel condensation of malononitrile and methyl cyanoacetate under solvent-free conditions. Tetrahedron Lett 2003;44:399–401. Copyright (2003), with permission from Elsevier.

The role of calcite and fluorite used in ball-milling process was proposed. When the minerals are mechanically crushed, the newborn solid surface is activated with the naked ionic species in situ generated. The naked carbonate and fluoride anions act as a strong base capable of deprotonating the active methylene compounds, with the consequent formation of a carbanion stabilized via the coordination with calcium cation, which combines with a carbonyl compound, eventually leading to the Knoevenagel product.

Ondruschka commented later on their observation that the reaction proceeded to high yields when the mineral reagent was mechanically crushed to 1–3 mm pieces prior to use, but no reaction occurred when fine powder was used [43]. It appears more likely that the added minerals did not act as a catalyst but merely served as additional grinding stones during milling of the reactants. When mineral powder was used instead, the reaction mixtures became effectively diluted and the grinding less effective.

Kaupp et al. carried out solvent-free uncatalyzed Knoevenagel condensations and Michael additions of aromatic aldehydes in the solid state and in the melt with quantitative yield [41]. Knoevenagel condensation of equimolar quantities of barbituric acids and aromatic aldehydes was carried out either at room temperature or by heating to 50°C to achieve full conversion within 1 h (Scheme 2.46). The solid powders were dried to give pure products **136** with 100% yield and did not require purification workup. Yields of these condensations carried out in solution were inferior to mechanochemical ones. Condensation of vanillin and barbituric acid (Table 2.43, entry 4) was also carried out in planetary ball mill by Stolle obtaining yield of >97% [42]. Large-scale reactions (200 g) of stoichiometric 1:1 mixtures of *p*-hydroxybenzaldehydes and barbituric acids were carried out in the 2 L steel Simoloyer ball mill (2 kg CR6 balls of 5 mm, 1200 min^{-1} for 1 h, and the products were milled out for 10 min at frequently changing rotor frequencies of 900 and 1200 min^{-1}). For reaction of 1,3-dimethylbarbituric acid the temperature was 18–20°C (with water cooling) and 50°C (without cooling) for barbituric acid. The solids were dried to give pure products with 100% yield and did not require purification workup.

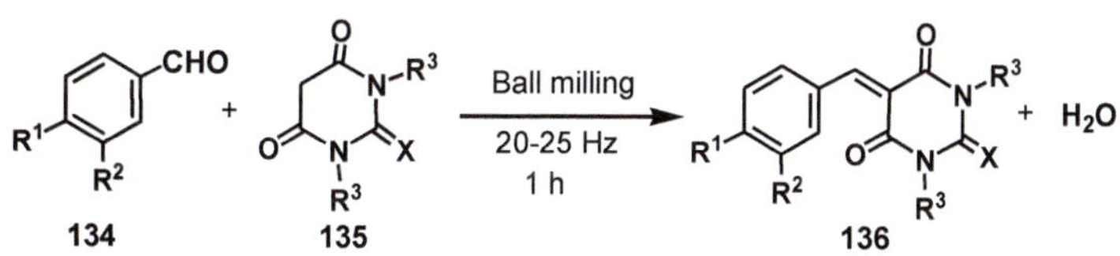

SCHEME 2.46

Uncatalyzed Knoevenagel condensation of barbituric acids and aromatic aldehydes.

Table 2.43 Uncatalyzed Knoevenagel Condensation of Barbituric Acids and Aromatic Aldehydes[a]

Entry	R^1	R^2	R^3	X	Conditions	Yield (%)[b]
1	NMe_2	H	Et	S	1 h, rt	100 (87)[c]
2	OH	H	Me	O	1 h, rt	100 (80–90)[c]
3	OH	H	H	O	1 h, rt to 50°C	100 (95)[c]
4	OH	OMe	H	O	1 h, 50°C	100

[a]*Retsch MM 2000 swing mill, 20–25 Hz, stainless steel double-walled beaker (10 mL) with fittings for circulating coolants, 2 × 12 mm balls, acid:aldehyde 1:1 equiv.*

[b]*Isolated products.*

[c]*In solution, literature values.*

Reproduced from Kaupp G, Naimi-Jamal MR, Schmeyers J. Solvent-free Knoevenagel condensations and Michael additions in the solid state and in the melt with quantitative yield. Tetrahedron 2003;59:3753–60.

The stoichiometric reaction of aromatic aldehydes with Meldrum's acid at 50°C gave an intermediate melt from which products **139** crystallized quantitatively at the reaction temperature (Scheme 2.47). These Knoevenagel condensations were termed "melt reactions with direct crystallization." Similar condensation with dimedone **140** afforded a cascade reaction starting with Knoevenagel condensation, which was followed by elimination and subsequent Michael addition (Scheme 2.48). Unlike the solution synthesis, yields are superior and no wastes was produced.

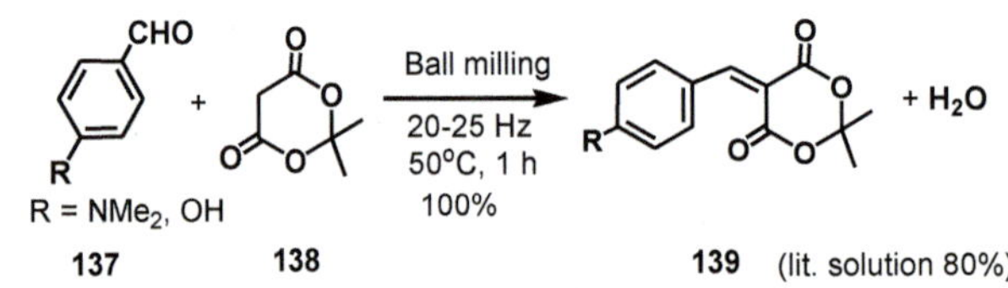

SCHEME 2.47

Knoevenagel condensation of Meldrum's acid and aromatic aldehydes.

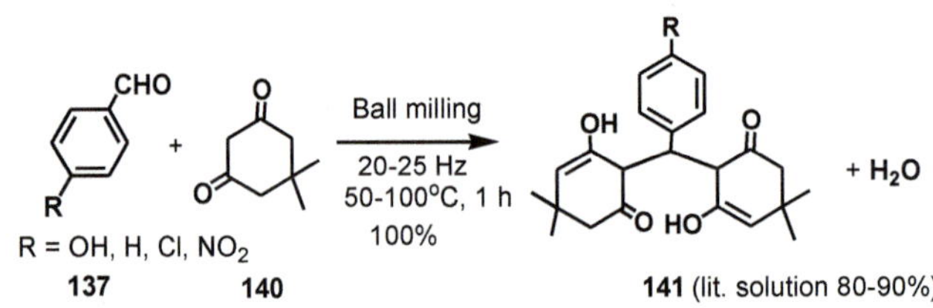

SCHEME 2.48

Cascade Knoevenagel condensation/Michael addition of dimedone and aromatic aldehydes.

Analogous Knoevenagel reaction is an example of use of gaseous reagents in ball mill. Ball milling of aromatic aldehydes with cyanoacetamide were too slow at room temperature and required a basic catalyst such as gaseous trimethylamine to achieve full conversion without the need for purification (Scheme 2.49). In classical melt conditions (150–170°C) quantitative yields were obtained within 1 h.

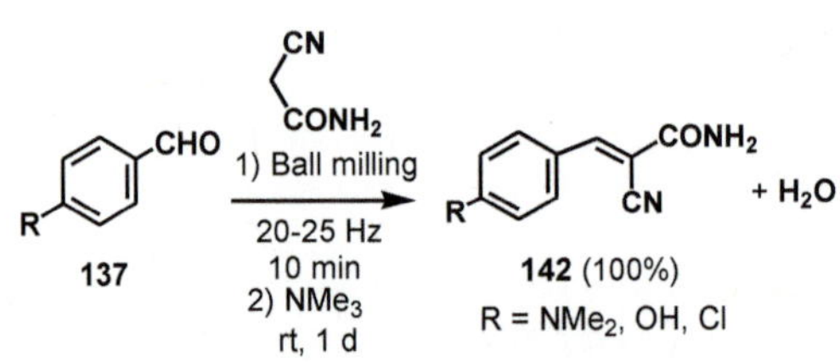

SCHEME 2.49

Knoevenagel condensation of cyanoacetamide with aromatic aldehydes.

Ondruschka et al. carried out solvent-free and waste-free mechanochemically reacted Knoevenagel condensation of aromatic aldehydes with malononitrile (Scheme 2.50) [43]. Obtained vibration mill results were compared with melt microwave conditions and planetary mill (Table 2.44). Knoevenagel reactions by microwave-assisted melts of aldehydes with malononitrile were high yielding (>40%). In comparison, ball-milling condensations showed the influence of reaction time: in short time (10 min), yields were very low, while after 60 min, general improvement of yields was obtained. Furthermore, milling in planetary mill afforded similar yields when reaction was carried out at 400 rpm and the increase of speed to 800 rpm was beneficial to obtain higher yields. The temperature inside the reaction vessel was determined immediately upon completion of the experiment using an IR sensor. The temperatures for the experiments at 400 rpm were always less than 30°C, but were 50–60°C after grinding with 800 rpm. It appears that the overall process of mixing of the reactants and the ensuing reaction for the Knoevenagel reactions is endothermic, absorbing some of the heat that is generated by the grinding. For each experiment in planetary mill, yields were significantly improved when reaction mixtures were allowed standing for 24 h at room temperature. The pre-reacted mixtures were observed to continue to convert in many cases into the desired product after completion of the milling process when left to stand likely due to direct crystallization which can be initiated by the presence of the target product.

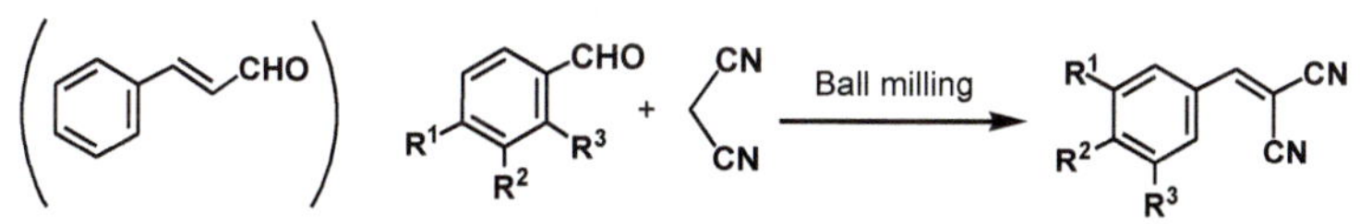

144 143 145 146

SCHEME 2.50

Knoevenagel condensation of malononitrile and aldehydes.

Table 2.44 Knoevenagel Condensation of Malononitrile and Aldehydes

				Yield (%)				
Aldehyde Type	**R¹**	**R²**	**R³**	**BM1[a]**	**BM2[b]**	**PM1[c,e]**	**PM2[d,e]**	**MW[f,g]**
143	H	H	H	5	>99	31 (99)	97 (99)	83
	NO_2	H	H	99		99		95
	$(CH_3)_2N$	H	H	2	71	2 (71)	69 (78)	40
	OH	OCH_3	H	5	2	2 (89)	5 (91)	91
	H	OCH_3	OCH_3	2	98	6 (6)	95 (96)	92
	Cl	H	H	7		10 (76)	>99	96
	H	H	Cl	15		30 (99)	>99	95

Continued

Table 2.44 Knoevenagel Condensation of Malononitrile and Aldehydes—cont,d

				Yield (%)				
Aldehyde Type	**R[1]**	**R[2]**	**R[3]**	**BM1[a]**	**BM2[b]**	**PM1[c,e]**	**PM2[d,e]**	**MW[f,g]**
	OH	H	H	51		77 (95)	96 (99)	
	H	H	OH	23		21 (60)	59 (77)	
144	H	H	H	13	93	9 (21)	95 (95)	84

[a]*BM1 Retsch MM 301 vibration mill, 30 Hz, stainless steel jar (25 mL), 5×10 mm balls, nitrile:aldehyde 1:1 mol, 10 min.*
[b]*BM2 1 h.*
[c]*PM1 Fritsch Pulverisette Planetary Mill, 400 rpm, stainless steel jar (45 mL), 5×15 mm balls, nitrile:aldehyde 1:1 mol, 1 h.*
[d]*PM2 800 rpm, 1 h.*
[e]*Immediately after grinding; in parentheses yields after 24 h.*
[f]*MW 1 h, 150°C, 100 W.*
[g]*GC analysis.*
Reproduced from Trotzki R, Hoffmann MM, Ondruschka B. Studies on the solvent-free and waste-free Knoevenagel condensation. Green Chem 2008;10:767–72. With permission from the Royal Society of Chemistry.

One of the advantages of ball-milling process is that the side products seen in MW reactions were not found during the mechanochemical conversion. Since the mechanochemical conversions were most successful at internally generated temperatures of 50–90°C it appears that the formation of side products requires larger thermal activation above these temperatures. A comparison of total-power consumption using the different methods of conducting Knoevenagel condensation was obtained by measurement with a line-power consumption meter when carrying out the reaction. The power consumption of the MW system running for 1 h at 100 W of MW irradiation is 130 kWh. The power consumption of the milling systems at the same reaction time is only 0.4 and 0.2 kWh for the VM at 30 Hz cycle time and the PM at 800 rpm, respectively. The corresponding power consumption on a molar basis of reactant for 1 h reaction time is 6.5 kW mmol^{-1} for MW system, 0.04 kW mmol^{-1} for the VM and 0.01 kW mmol^{-1} for the PM.

Unusual and novel spiro dihydrofuran structure **148** was obtained in condensation reaction of 5,5-dimethylcyclohexane-1,3-dione (dimedone), aldehyde, dimethylaminopyridine (DMAP), and molecular iodine (Scheme 2.51; Table 2.45) [44]. Wang has found that other bases (DABCO, DBU, Na_2CO_3, Cs_2CO_3, and K_2CO_3) and different promoters (NBS, NCS, CBr_4, Oxone/NaBr, and Oxone/$ZnCl_2$) were less successful.

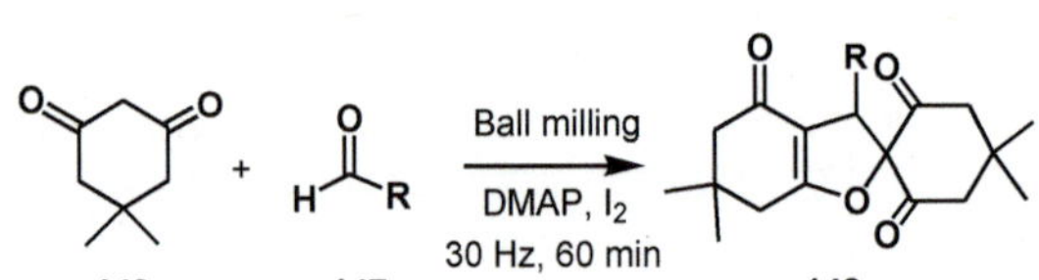

140 147 148

SCHEME 2.51

Reaction of aldehydes with dimedone.

Table 2.45 Reaction of Dimedone With Benzaldehydes[a]

Entry	R	Yield (%)[b]
1	3-$NO_2C_6H_4$	91 (77)[c]
2	4-$NO_2C_6H_4$	92
3	4-CNC_6H_4	88
4	3,4-$Cl_2C_6H_4$	86
5	4-$CHOC_6H_4$	83
6	4-BrC_6H_4	81
7	3,4-$(CH_3)_2C_6H_4$	82
8	3,4-$(OCH_3)_2C_6H_4$	84
9	H	79
10	Et	86
11	n-Pr	85
12	2-Pyridyl	77
13	2-Thienyl	80
14	2-Furyl	85

[a]Retsch MM200 Mixer Mill, 1800 rpm, stainless steel jar (5 mL), 7 mm ball, dimedone:aldehyde: base: I_2 2.4:1:2.5:1.65 mol, 60 min.
[b]Isolated products.
[c]Dioxane, 1 h.
Reproduced with permission from Wang G-W, Gao J. Selective formation of Spiro dihydrofurans and cyclopropanes through unexpected reaction of aldehydes with 1,3-dicarbonyl compounds. Org Lett 2009;11:2385–8. Copyright (2009), American Chemical Society.

When these reaction conditions were employed to oxidative addition of aldehydes with 1,3-indandione **149**, different type of product was obtained, the bispiro-substituted cyclopropanes **150** exclusively and in good yields (Scheme 2.52, Table 2.46). Model reactions carried out in dioxane solvent gave the same products, but after 1 h the yields were lower. The reaction mechanism is thought to start with Knoevenagel condensation, followed by iodination, and intramolecular nucleophilic O-attack with HI elimination to dihydrofurans. When intramolecular nucleophilic C-attack occurs, with subsequent elimination of HI, cyclopropanes were produced.

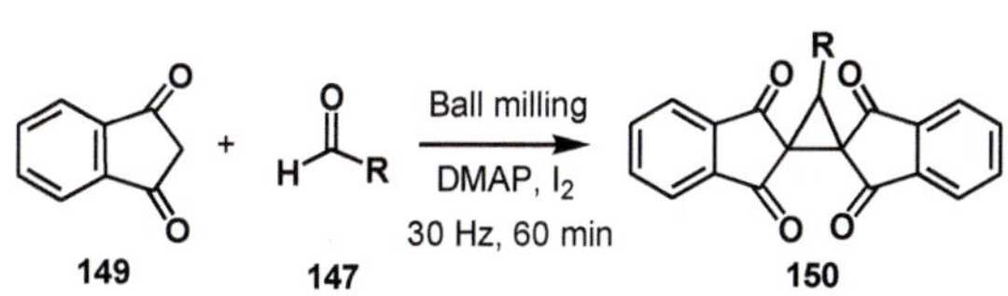

SCHEME 2.52

Reaction of aldehydes with dimedone.

Table 2.46 Reaction of 1,3-Indandione With Benzaldehydes[a]

Entry	R	Yield (%)[b]
1	3-$NO_2C_6H_4$	88 (59)[c]
2	4-$NO_2C_6H_4$	89
3	4-CNC_6H_4	84
4	3,4-$Cl_2C_6H_4$	80
5	4-$CHOC_6H_4$	75
6	4-BrC_6H_4	82
7	4-ClC_6H_4	83
8	C_6H_5	69
9	4-$CH_3C_6H_4$	60

[a]*Retsch MM200 mixer mill, 1800 rpm, stainless steel jar (5 mL), 7 mm ball, dione:aldehyde: base: I_2 2.7:1:2.5:1.65 mol, 60 min.*
[b]*Isolated products.*
[c]*Dioxane, 1 h.*
Reproduced with permission from Wang G-W, Gao J. Selective formation of Spiro dihydrofurans and cyclopropanes through unexpected reaction of aldehydes with 1,3-dicarbonyl compounds. Org Lett 2009;11:2385–8. Copyright (2009), American Chemical Society.

Less common are literature examples in which mechanochemical reaction was carried out at elevated temperature. Naimi-Jamal reported the heating of double-walled ball-mill beaker equipped with fittings for circulating water at 96°C (boiling water as circulant) [45]. One-pot solvent-free synthesis of pyrano[2,3-d]pyrimidine-2,4(1*H*,3*H*)-diones **154** was achieved by simply ball milling a stoichiometric mixture of an aromatic aldehyde, malononitrile, and barbituric acid, without addition of solvent and catalyst (Scheme 2.53). Quantitative yields were obtained (Table 2.47) and products generally did not require purification, the solid products were just dried at 80°C in vacuum and recrystallized, if necessary. Reaction presumably takes place by initial Knoevenagel condensation of aromatic aldehyde with malononitrile to afford the intermediate Michael acceptor, which subsequently reacts with barbituric acid. Tautomerization of Michael adduct is followed by intramolecular cyclocondensation and another tautomerization to afford pyrano[2,3-d]pyrimidine-2,4(1*H*,3*H*)-diones **154**.

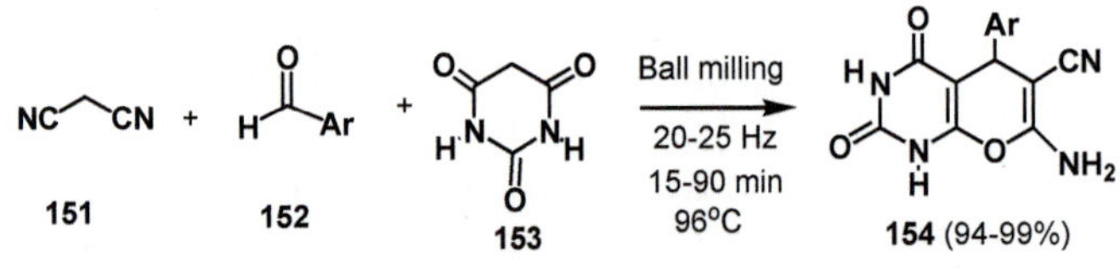

SCHEME 2.53

Synthesis of pyrano[2,3-d]pyrimidine-2,4(1*H*,3*H*)-diones.

Table 2.47 Mechanosynthesis of Pyrano[2,3-d] Pyrimidine-2,4(1*H*,3*H*)-Diones[a]

Entry	Ar	Time (min)	Yield (%)[b]
1	C_6H_5	70	>99
2	2-ClC_6H_4	90	>99
3	4-$ClNC_6H_4$	55	>99
4	2-$NO_2C_6H_4$	60	>99
5	3-$NO_2C_6H_4$	15	>99
6	4-$NO_2C_6H_4$	25	>99
7	4-BrC_6H_4	60	(94)
8	4-$OCH_3C_6H_4$	30	>99

[a]*Retsch MM2000 mixer mill, 20–25 Hz, 96°C, stainless steel beaker (10 mL), 2 × 12 mm balls.*
[b]*Conversions, isolated yield in parenthesis.*
Reproduced from Mashkouri S, Naimi-Jamal MR. Mechanochemical solvent-free and catalyst-free one-pot synthesis of pyrano[2,3-d]pyrimidine-2,4(1H,3H)-diones with quantitative yields. Molecules 2009;14:474–9. With permission from MDPI.

Milling of equimolar amounts of ninhydrin and malononitrile (Retsch MM 2000 mill, stainless steel vial 10 mL, two 12 mm balls) produced after 1 h Knoevenagel product **155** in quantitative yield (Scheme 2.54) [46]. After reaction completed, product did not need any purification. Solution reaction conditions are in variance with solvent-free method which does not require any catalyst. Metwally has shown that heating of ninhydrin and malononitrile in solution (EtOH/AcOH) yielded entirely different product **156**, via initial condensation of two molecules of malononitrile. Manual grinding in mortar afforded after 30 min products in considerably lower yields.

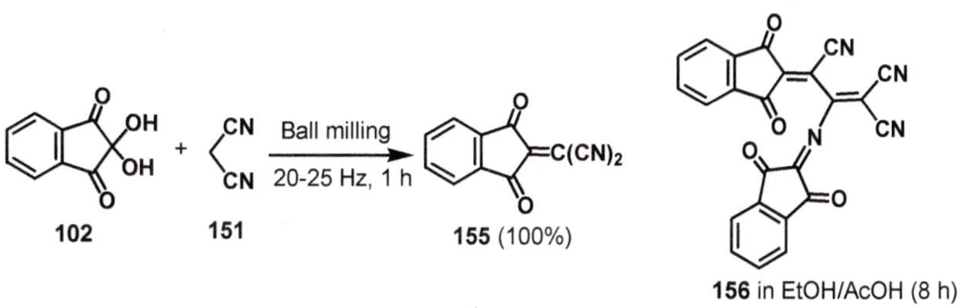

SCHEME 2.54

Knoevenagel reaction of ninhydrin.

Naimi-Jamal et al. reported that anhydrous piperazine is an efficient organocatalyst for the preparation of 2-amino-4*H*-pyrans **158** and tetrahydro-4*H*-chromenes **159** via a three-component condensation under solvent-free ball-milling conditions at ambient temperature [47]. Without the presence of the catalyst, the reaction of *p*-chlorobenzaldehyde, malononitrile, and ethyl acetoacetate did not take place, even by heating at 80°C. However, an excellent yield of products was obtained by ball milling of an equimolar mixture of starting materials at ambient temperature in the presence of 15 mol% piperazine (Table 2.48). After completion of the reaction, the crude solid was stirred with hot water to remove piperazine and filtered. The product was purified by recrystallization, and piperazine was recycled. Replacement of ethyl acetoacetate with 5,5-dimethyl-1,3-cyclohexanedione afforded the tetrahydro-4*H*-chromenes **159** (Scheme 2.55). A broad spectrum of aldehydes was used, and high yields of tetrahydro-4*H*-chromenes were obtained in relatively short reaction times.

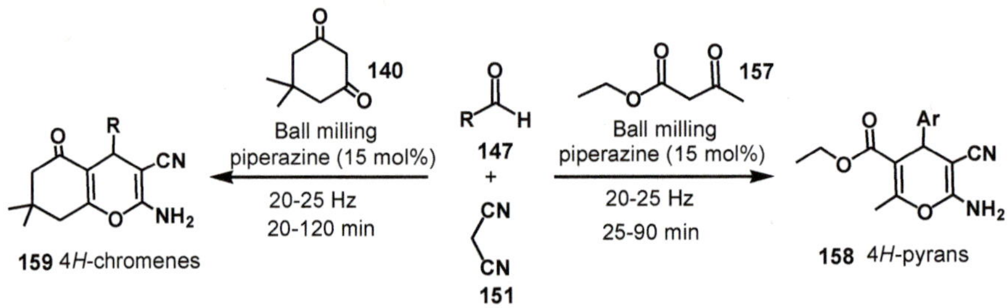

SCHEME 2.55

Multicomponent synthesis of 4*H*-pyrans and 4*H*-chromenes.

Table 2.48 Multicomponent Synthesis of 4*H*-Pyrans and 4*H*-Chromenes[a]

		4H-Pyrans		4H-Chromenes		
Entry	**R**	**Time (min)**	**Yield (%)[b]**	**Entry**	**Time (min)**	**Yield (%)[b]**
1	C_6H_5	25	95	15	20	96
2	4-MeC_6H_4	45	93	16	30	92
3	2-$NO_2C_6H_4$	30	95	17	20	95
4	3-$NO_2C_6H_4$	35	90	18	25	93
5	4-$NO_2C_6H_4$	40	94	19	27	92
6	2-ClC_6H_4	20	97	20	30	94
7	4-ClC_6H_4	50	95	21	25	91
8	4-BrC_6H_4	30	95	22	25	94
9	4-FC_6H_4	40	95	23	30	96
10	4-CNC_6H_4	35	90	24	25	95
11	4-$MeOC_6H_4$	55	91	25	45	90
12	2-furanyl	50	91	26	35	95
13	2-thienyl	90	90	27	45	94
14	4-$(CH_3O_2C)MeOC_6H_4$	50	94			
15	*trans*-$C_6H_5CH{=}CH$			28	100	91
16	$C_6H_5CH_2CH_2$			29	30	93

[a]Retsch MM 2000 horizontal mill, stainless steel vessel (20 mL), 3 × 12 mm balls, 20–25 Hz, aldehyde: malononitrile:ethyl acetoacetate:piperazine 1:1:1:0.15 mol.
[b]Isolated yields.
Reproduced from Amirnejad M, Naimi-Jamal MR, Tourani H, Ghafuri H. A facile solvent-free one-pot three-component method for the synthesis of 2-amino-4H-pyrans and tetrahydro-4H-chromenes at ambient temperature. Monatsh Chem 2013;144:1219–25. With permission of Springer.

A mechanism for the piperazine-catalyzed formation of 4*H*-chromenes is complex cascade of reactions, starting with piperazine acting as a base which activates malononitrile, promoting Knoevenagel condensation, and also formation of an enamine, followed by Michael condensation, proton transfer, intermolecular cyclization via a nucleophilic addition of the enolate oxygen to the nitrile group (hetero-Thorpe–Ziegler), and finally hydrolysis and tautomerization.

2.9 BARBIER ALLYLATION

Suzuki and coworkers reinvestigated the bismuth-mediated allylation of aldehydes with allyl bromide under ball-milling conditions (Scheme 2.56) [48]. The Barbier-type allylation of aromatic aldehydes took place smoothly under this completely dry conditions, giving the expected homoallylic alcohols **162** in moderate to good yield (Table 2.49). For this purpose bismuth shots were used as milling balls. Less-reactive allyl chloride also reacted in ball-milling conditions, while the chloride usually fails to react in solution. Prolonged ball milling led to a gradual decline of product yield, probably due to the decomposition of the initial product to a diene.

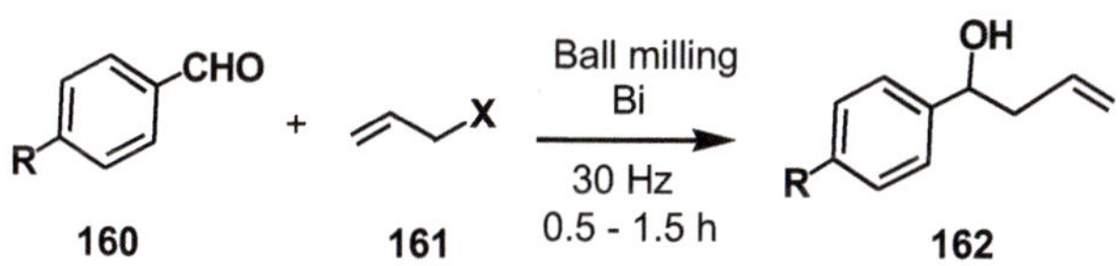

SCHEME 2.56

Synthesis of homoallyl alcohols.

Table 2.49 Synthesis of Homoallyl Alcohols by Barbier-Type Allylation Reaction[a]

Entry	R	X	Aldehyde/Halide/Bi Ratio	Time/h	Yield (%)[b]
1	H	Br	1/2/8	0.5	81
2	H	Br	1/2/8	1.5	88
3	H	Br	1/1.5/8	1.5	94
4	Me	Br	1/1.5/8	0.5	69
5	CN	Br	1/1.5/8	0.5	80
6	CN	Br	1/1.5/8	1.5	91
7	NO_2	Br	1/1.5/8	0.5	84
8	NO_2	Br	1/1.5/8	1.5	95
9	NO_2	Cl	1/1.5/8	1.5	34

[a]*Retsch mixer mill MM200, 30 Hz, small stainless steel balls.*
[b]*Isolated yield.*
Reproduced from Wada S, Hayashi N, Suzuki H. Noticeable facilitation of the bismuth-mediated Barbier-type allylation of aromatic carbonyl compounds under solvent-free conditions. Org Biomol Chem 2003;1:2160–3. With permission from the Royal Society of Chemistry.

Allylation of aromatic aldehydes typically required more time (0.5–1.5 h), while allylation of aromatic ketones (*p*-substituted acetophenones) was complete within 0.5 h (Scheme 2.57, Table 2.50). Aromatic ketones react sluggishly or fail to undergo the Barbier-type allylation in solution; however, they undergo the reaction with moderate ease under the solvent-free conditions. Pinacols that often accompany the Barbier-type reaction of ketones in solution were not detected. Reaction conditions tolerate various functionalities and sensitive functional moieties such as nitro and cyano groups remained intact throughout the reaction. The physical form of the metallic bismuth does not seem to have much influence on the results (shot and powder). In some instances, nonreactive co-milling agents, such as calcite grains, were added to the reaction mixture to solve the problem of formation of very sticky reaction mixture in which the stainless steel balls and bismuth shot clung together to form a doughy lump in the reaction vessel.

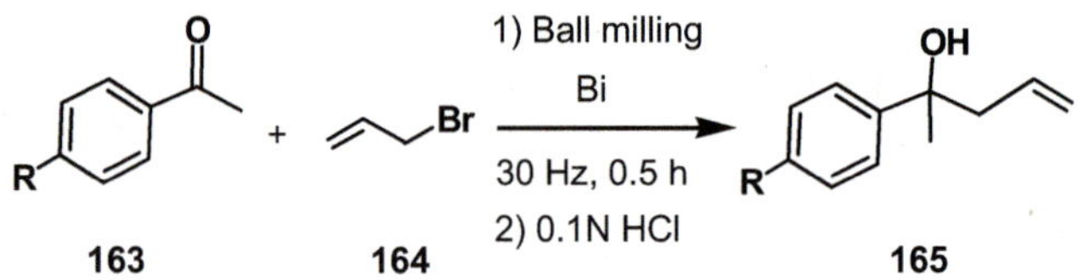

SCHEME 2.57

Barbier-type allylation of aromatic ketones.

Table 2.50 Barbier-Type Allylation of Aromatic Ketones[a]

Entry	R	Bi Form	Ketone/Halide/Bi Ratio	Yield (%)[b]
1	H	Shot	1/1.5/4	44
2	H	Shot	1/1.5/8	46
3	H	Powder	1/1.5/8	56
4	NO_2	Shot	1/1.5/4	36
5	NO_2	Shot	1/1.5/8	70
6	NO_2	Powder	1/1.5/8	51
7	CN	Shot	1/1.5/4	40
8	Br	Powder	1/1.5/8	69
9	Cl	Powder	1/1.5/8	53
10	Me	Powder	1/1.5/8	50
11	OMe	Shot	1/1.5/4	20[c]
12	OMe	Powder	1/1.5/4	31[c]
13	OMe	Powder	1/1.5/8	36[c]

[a]*Retsch mixer mill MM200, 30 Hz, 0.5 h, small stainless steel balls.*
[b]*Isolated yield.*
[c]*NMR yield.*
Reproduced from Komatsu K, Wang G-W, Murata Y, Tanaka T, Fujiwara K. Mechanochemical synthesis and characterization of the fullerene dimer C_{120}. J Org Chem 1998;63:9358–66. With permission from the Royal Society of Chemistry.

The indium- and tin-mediated allylation of aromatic aldehydes works similarly in solvent-free conditions, although the byproducts also form. The advantageous use of bismuth over indium and tin in the chemoselective Barbier-type allylation was manifested in the allylation of 4-nitrobenzaldehyde and 4-nitroacetophenone. Bismuth afforded the expected alcohols in good to moderate yields, whereas indium and tin gave a complex mixtures. Ketones with different steric requirements such as cyclohexanone and benzophenone reacted with moderate ease (cyclohexanone) and sterically more hindered benzophenone with more difficulty (Scheme 2.58). Reactivity of these ketones is markedly different from the solution conditions. Cyclohexanone undergoes reaction under certain conditions, while benzophenone often fails to react, or gives pinacol rearrangement. However, the difference in reactivity is not large in solvent-free conditions, while steric factor seem to be more pronounced in solution reactions.

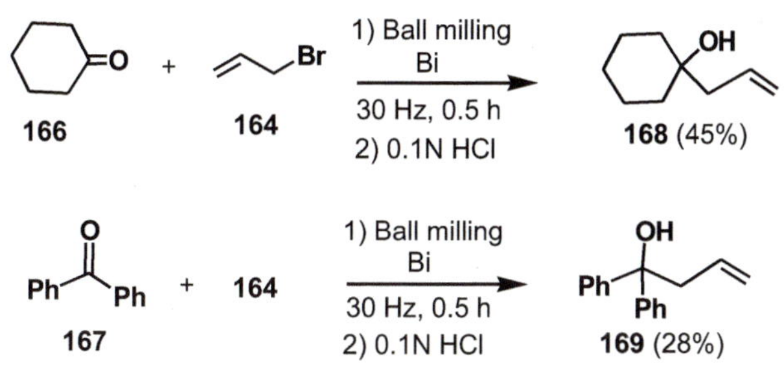

SCHEME 2.58

Barbier-type allylation of ketones.

2.10 GEWALD REACTION

Base-catalyzed multicomponent reaction of the alkyl–aryl ketone, the α-cyanoester, and sulfur carried out in ball mill led to the formation of substituted thiophenes (Scheme 2.59). Using this protocol, Mack et al. developed one-step/one-pot mechanochemical variant of Gewald reaction in which the first step is Knoevenagel condensation between the ketone and the α-cyanoester [49]. By employment of equimolar amounts of reagents and morpholine as base, thiophenes were prepared in low to moderate yields (12–41%, Table 2.51). A comparison of reactivity with solution reaction (heating in oven at 120°C) showed similar results. Heat-assisted ball-milling experiments in which the vial was heated with heat gun accommodated to ball mill showed an increase in rates of reaction, but yields remained about the same. The reaction can be also conducted in a conventional oven under neat conditions using a catalytic amount of base.

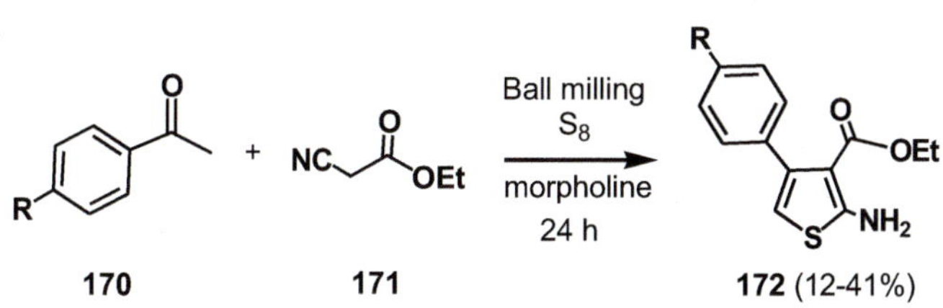

SCHEME 2.59

Synthesis of 2-aminothiophenes.

Table 2.51 Synthesis of 2-Aminothiophenes by Gewald Reaction

		Yield (%)[b]			
Entry	**R**	**Ball Mill[a] 24h**	**Ball Mill[c] 0.5h, Heat Gun**	**Oven 120°C, 1h**	**Oven 120°C, 1h, Cat. Morpholine**
1	H	41	14	29	19
2	NO_2	32	53	43	43
3	Br	30	25	27	19
4	Cl	27	22	26	20
5	CN	27	32	31	36
6	CH_3CONH-	12	15	25	5

[a]*Spex certiprep mixer/mill 8000M; ½″×2″ stainless steel vial, one 1/8″ ball bearing; ketone: ethylcyanoacetate: S: morpholine 1:1:1:1 ratio.*
[b]*Isolated yields.*
[c]*Vial was continuously heated by the heat gun to a temperature of 130°C, at 18Hz (1080rpm) for 30min.*
Reproduced from Shearouse WC, Shumba MZ, Mack J. A solvent-free, one-step, one-pot gewald reaction for alkyl-aryl ketones via mechanochemistry. Appl Sci 2014;4:171–9. With permission from MDPI.

Other amine bases (triethylamine, pyridine, and piperidine) produced yields similar to what is reported for morpholine; however, products were not formed when sodium bicarbonate, potassium carbonate, or sodium hydroxide was used as the base. The final step in the mechanism of the Gewald reaction is thought to be protonation, preceded by the initial Knoevenagel condensation followed by attack of elemental sulfur and cyclization. Adding the reagents at 120°C using 10% morpholine provided product yields similar to other experiments demonstrating the catalytic nature of the reaction. Under solvent-free conditions the proton in the protonation step probably comes from the conjugate acid of morpholine.

The effect of reagent to vial ratio (ie, the ratio of the amount of reagents that fill the volume of the vial) on the rate and yield of mechanochemical reaction was established using equimolar amounts of *p*-nitroacetophenone and reagents at 10%, 25%, 50%, 75%, and 95% of vial capacity. Regardless of the reagent-to-vial ratio each reaction gave similar yields. The size that the reagents fill the milling vessel has been postulated as an important factor to consider. In contrast to reactions in solution, uniformity in ball-milled reactions is not obtained immediately. For the low reagent-to-vial ratio, a longer time is needed to all reactants to react together. When the ratio is too high then there is not enough empty volume in the vial to allow moving the balls everywhere in the reaction and bring the energy needed to trigger the reaction. For multicomponent reactions the reagent-to-vial ratio might be more crucial because these reactions depend on a series of bimolecular reactions before the product is formed. Enlargement of the reagent-to-vial ratio would increase the reaction rate and a higher yield.

2.11 PINACOL REACTION

Pinacol reaction enhanced by mechanical agitation was studied by Wang and coworkers [50]. Mechanochemical reaction of a series aromatic ketones and aldehydes in the presence of $Zn/ZnCl_2$ afforded pinacols **174** and reduced alcohol products **175** (Scheme 2.60). This reaction is one example showing that the higher rotation speed does not necessary produce best results and that too much energy given to the system could be counterproductive. Different activation methods were compared (eg, high speed–vibration mill, mixer mill, and mortar grinder) showing that when reactions were performed in high speed–vibration mill or mixer mill, mixtures of pinacol and reduced alcohol were obtained. Surprisingly, despite mortar grinder works much slower and operates at 1.5 Hz, yields of pinacols were significantly better (73–90%). In these milling conditions reaction was selective, and alcohol side product was obtained in smaller amount. These observations might be rationalized by the reversibility of pinacol products under high-energy milling and their disproportionation to ketone and alcohol, which was shown by ball milling of one of the isolated products for 30 min. Decomposition of the product in an association–dissociation equilibrium was noted in high speed–vibrational milling of C_{60} reactions (chapter Applications of Ball Milling in Nanocarbon Material Synthesis) [51]. Among three grinding techniques, the results from the mechanochemical reactions of aromatic ketones and aldehydes by automated mortar grinder are most close to those by hand grinding [52].

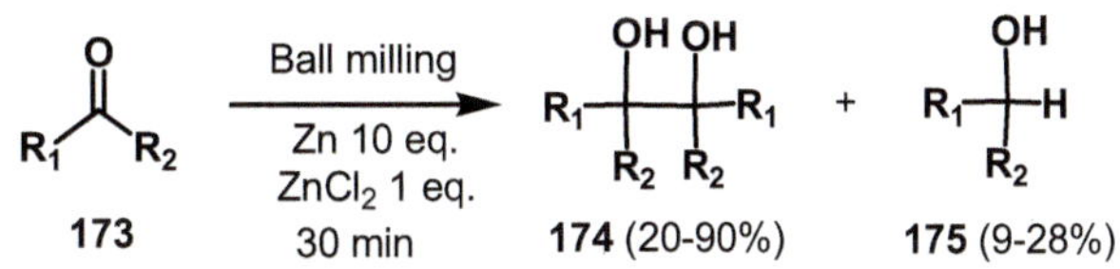

Substrate	Conditions	174	175
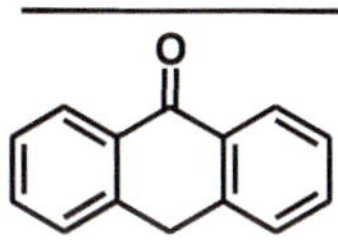	A) High speed vibration mill, 60 Hz	57	28
	B) Retsch MM200 mixer mill, 30 Hz	58	28
	C) Retsch RM100 mortar grinder, 1.5 Hz	79	9
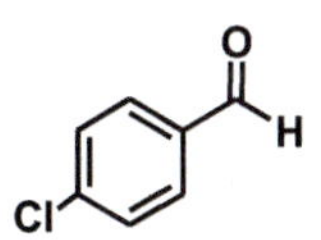	A	20	10
	B	24	13
	C	73	14
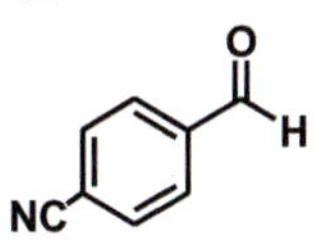	A	50	0
	B	57	0
	C	89	0

SCHEME 2.60

Pinacol reaction of aromatic carbonyl compounds.

2.12 GRIGNARD AND MCMURRY REACTIONS

Harrowfield and coworkers have shown that ball-milling technique can be used to produce Grignard reagents with at least comparable efficiency to conventional solution methods [53]. Reactions of 1-chloro- and 1-bromo-naphthalene with an excess of magnesium (4:1 ratio) in a Spex 8000 vibratory mill (10×12.7 mm stainless steel balls) proceed smoothly (Scheme 2.61).

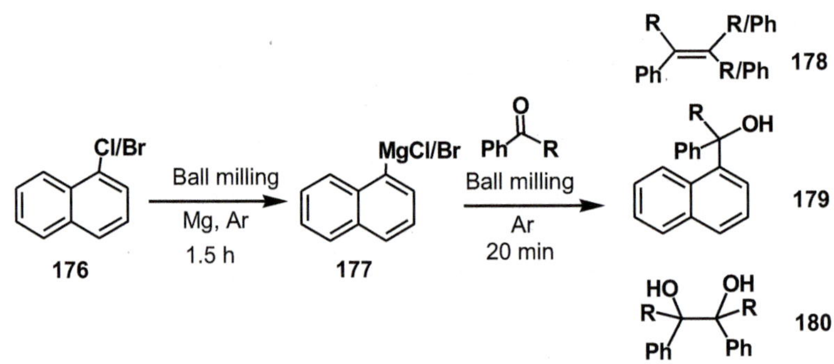

SCHEME 2.61

Grignard and McMurry reactions in a ball mill.

Estimated yields of the Grignard species of c. 95% are similar to those of solvent-based preparations. Alcohols were produced by subsequent mechanochemical quenching of the reaction mixtures with benzophenone or acetophenone in 74–80% yields. In experiments where benzophenone or frozen acetophenone were added to the mill and grinding was continued for another 20 min complex product mixtures were obtained due to the occurrence of McMurry reactions, although the products of such pathways were minor components of the mixtures. The products included those of coupling (to give "dimeric" pinacol, alkene, and alkane species), as well as simple reduced species. Benzophenone provided 1,1,2,2-tetraphenylethanediol, tetraphenylethene, and 1,1,2,2-tetraphenylethane as major products, along with small quantities of diphenylmethanol and diphenylmethane. Acetophenone provided principally *cis*- and *trans*-2,3-diphenylbut-2-ene and *meso*- and *rac*-2,3-diphenylbutane-2,3-diol, as well as some 1-phenylethanol, with the yield of the alkenes increasing significantly with milling time. Isolated yields of the alkenes were as high as 40% from benzophenone and 15% from acetophenone. Substitution of Mg by other metals (Al, Fe, Ti, and Zn) led to much slower reactions. Only coupling (McMurry) products with Ti and Zn were readily detected.

2.13 GRIGNARD–ZEREWITINOFF-LIKE REACTION

Birke et al. showed that small particle–size metals (produced during milling react at room temperature in a single process) rapidly and selectively dechlorinate polychlorinated biphenyls (Scheme 2.62) [54]. Reduction to the parent hydrocarbons takes place in the presence of a hydrogen donors. Formation of a Grignard reagent and its subsequent protonation by hydrogen donors (Zerewitinoff-like reaction) has rarely been performed in one-pot reaction and requires elevated temperatures to obtain high yields. By conducting reaction in a ball mill (by grinding magnesium turnings in the presence of organic halides), a highly selective one-pot reductive dehalogenation of polyhalogenated aromatics takes place. The advantage of the milling procedure is that mechanical impact constantly applied on the metal prevents the poisoning of the reactive surface during the reaction, continuously produces fresh supply of highly accessible metal surfaces, and removes impurities adsorbed on the metal surface. It is proposed that activated metal serves as the reducing agent which delivers electrons for breaking down the carbon–halogen bond in the initial reaction step via electron transfer, producing a radical or a carbanion and inorganic halide. This transfer occurs at the metal surface while the halogenated molecule is adsorbed to the activated surface. In a second step, the hydrogen donor transfers hydrogen atoms or protons, to replace the stripped-off halogen atom.

Dechlorination reactions were carried out in Retsch S100 ball mill, using steel grinding jar (50 mL) and 3×20 mm balls, by deploying external cooling jackets. Sand was used as grinding auxiliary with molar ratio of 1,3-DCB:*n*-butylamine 1:40 dissolved in toluene which was employed as a diluting agent to lower reaction rates. Full disappearance of DCB was observed within 50 min. Some abrasion of steel was noted during milling and experiments with added small amounts of the steel components (iron and chromium) showed that they do not impact the course of the reaction.

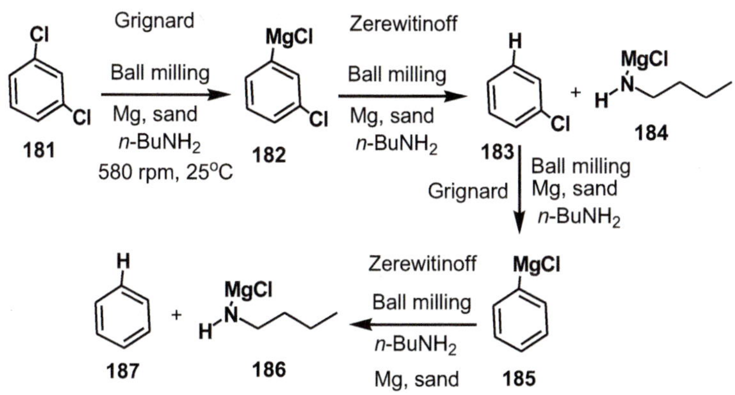

SCHEME 2.62

Reaction mechanism for one-pot reductive MC dechlorination of DCB to benzene.

2.14 WITTIG REACTION

Balema et al. prepared two stable phosphorus ylides **189** in nearly quantitative yields by solvent-free mechanochemical grinding of phosphonium salts **188** and an excess of anhydrous K_2CO_3 (Scheme 2.63) [55]. Prepared phosphorus ylides were subjected, without isolation to subsequent milling with ketones to achieve Wittig products **191** in high yields (Table 2.52). Forced air-cooling of the vial was employed to prevent its heating during ball-milling experiments. The mechanochemical preparation of stabilized phosphorus ylides and the solvent-free Wittig reaction was also successfully carried out as a "one-pot" process starting with triphenylphosphine, an organic halogenide and an organic carbonyl compound in the presence of K_2CO_3 to afford alkenes **194** in high yields (entry 8). There are two major differences between the conventional and the mechanochemically induced Wittig reactions. First, triphenylphosphoranes can be generated mechanochemically using K_2CO_3, while their preparation in a solution requires much stronger bases. Second, while Wittig reaction between triphenylphosphoranes and aromatic carbonyl compounds in a solution produces preferably the *Z*-stilbenes or mixtures with nearly equal content of the *Z*- and *E*-isomers, the mechanochemical technique favors more thermodynamically stable *E*-stilbenes (Scheme 2.64).

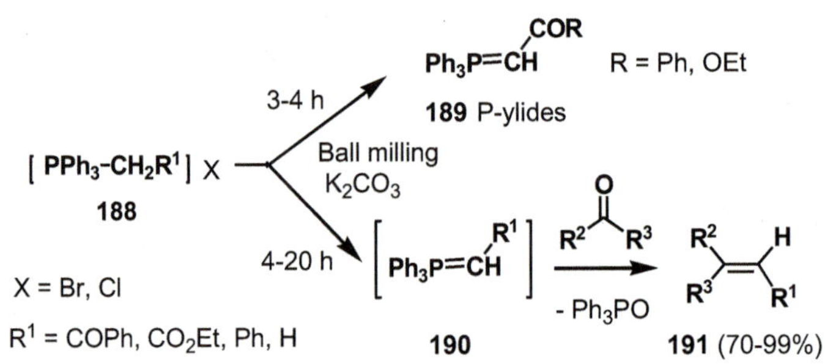

SCHEME 2.63

Solvent-free Wittig reaction.

Table 2.52 Mechanochemical Wittig Reaction[a]

Entry	R	R^1	R^2	R^3	*E:Z* Ratio	Milling Time/h	Yield[b] (%)
1	Ph				–	3	99
2	OEt				–	4	96
3		Ph	H	2-Naphthyl	1.6:1	7	85
4		Ph	H	4-BrC_6H_4	2:1	8	92
5		Ph	Me	4-BrC_6H_4	3.4:1	14	70
6		H	H	2-Naphthyl	–	20	73

Table 2.52 Mechanochemical Wittig Reaction[a]—cont'd

Entry	R	R^1	R^2	R^3	*E:Z* Ratio	Milling Time/h	Yield[b] (%)
7	Ph/Ph				–	4	99
8[c]		4-BrC_6H_4	H	2-Naphthyl	3.5:1	8	93

[a]*Spex 8000 mill, stainless steel vial, ball (21 g), He atmosphere, forced air cooling.*
[b]*Isolated.*
[c]*One-pot procedure.*
Reproduced with permission from Balema VP, Wiench JW, Pruski M, Pecharsky VK. Mechanically induced solid-state generation of phosphorus ylides and the solvent-free wittig reaction. J Am Chem Soc 2002;124:6244–5. Copyright (2002), American Chemical Society.

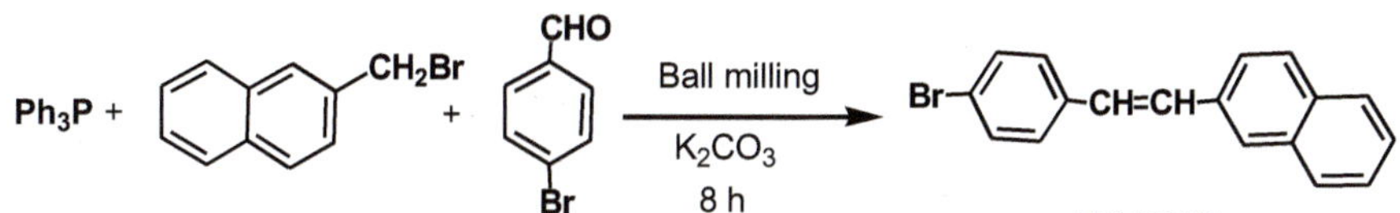

192 193 194 (93%)

SCHEME 2.64

One-pot solvent-free Wittig reaction.

Mechanochemical synthesis of stilbenes **198** by Wittig reaction employing triphenylphosphine-functionalized polystyrene was carried out by Shearouse and Mack (Scheme 2.65) [56]. The Wittig reaction is troubled by the difficulties in separation of the product from triphenylphosphine oxide by-product. Performing the Wittig reaction under ball-milling conditions (Spex 8000 M mixer, stainless steel vessel (3 mL), 5 mm ball) using functional resin–enabled separation of triphenylphosphine oxide resin by filtration, allowing the olefin product to be isolated without the need for column chromatography. Traditionally, nonpolar, organic solvents such as benzene or dichloromethane were used to swell functional resins, allowing reagents to diffuse through the cross-linked, functional portions of the resin. In contrast, swelling was found to be less important when using ball-milled functional resin particles (FRPs). When triphenylphosphine-functionalized polystyrene was used in the solvent-free, ball-milled synthesis of stilbene, only 17% of the aldehyde was converted to olefin. Ethyl acetate and ethanol significantly enhanced the reaction over traditionally used nonpolar "swelling" solvents, suggesting that under LAG conditions, the added "solvent" was not swelling the FRPs, as nonswelling "solvents" outperformed traditional swelling "solvents," instead of swelling the FRPs, they undergo conformational changes. Optimized reaction conditions in ball mill were applicable to a wide variety of substrates (aldehyde:bromide ratio 1:1.4, Scheme 2.65). Stabilized ylides provided *E*-diastereoselectivity, while semi-stabilized ylides gave a 50:50 mixture of *E*- and *Z*-diastereomers. LAG conditions showed no electronic effect on diastereoselectivity as *E*/*Z* ratios do not follow the ones observed in solution chemistry, suggesting that diastereoselectivity here is dependent on the solvent environment, as well as on the interactions of the reagents with the polystyrene backbone of the resin.

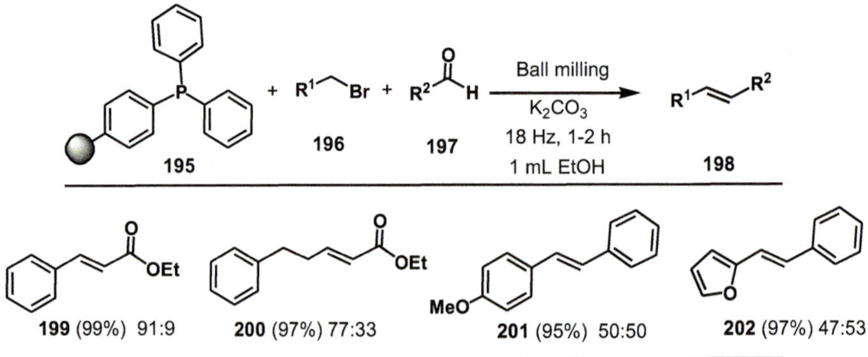

SCHEME 2.65

Synthesis of stilbenes by Wittig reaction.

2.15 HORNER–WADSWORTH–EMMONS REACTION

The Horner–Wadsworth–Emmons (HWE) version of the Wittig reaction focuses on the use of more stabilized phosphonate ylides **203**. Solvent-free method for making olefinic amino esters **205** from an aldehyde compound and a phosphonate derivative of glycine was established by Lamaty (Scheme 2.66) [57]. For successful reaction in planetary ball mill, an excess of glycine phosphonate **203** was required, in the presence of an excess of inorganic carbonate base (Cs_2CO_3 or K_2CO_3). Optimized reaction conditions could be applied to variety of aldehydes, both aromatic and aliphatic to obtain products in high yield after 7 h of milling (Table 2.53). Most of the reactions exhibited very high *Z/E* selectivity. In contrast to aldehydes, ketones (benzophenone and 2-propanone) were unreactive. Ball-milling HWE reaction is advantageous over reported solvent-free reactions in which very often an excess of liquid reactant serves as the solvent, while solvent synthesis usually requires the use of an excess of organic base and a low temperature for the deprotonation step.

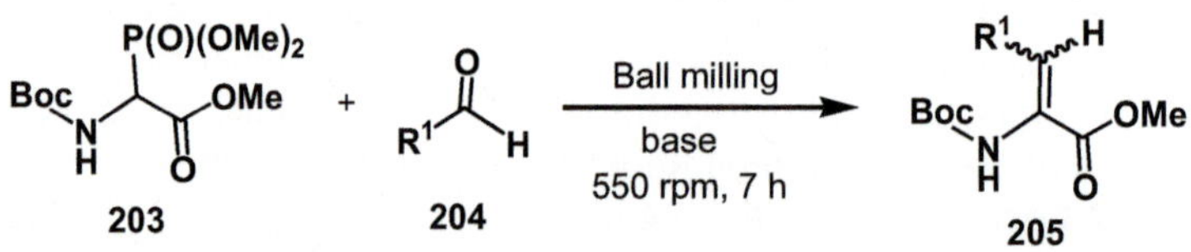

SCHEME 2.66

Horner–Wadsworth–Emmons reaction of glycine phosphonate.

Table 2.53 Horner–Wadsworth–Emmons Reaction of Glycine Phosphonate[a]

Entry	R^1	Base	Conversion (%)	Yield (%)[b]	*Z/E* Ratio
1	2,4-di(MeO)C_6H_3	Cs_2CO_3	100	95	100/0
2	2-BrC_6H_4	K_2CO_3	96	85	100/0
3	2-Pyridyl	K_2CO_3	100	88	100/0
4	2-Furyl	Cs_2CO_3	100	89	100/0
5	2-Naphthyl	K_2CO_3	100	76	82/18
6	*n*-Propyl	K_2CO_3	96	80	100/0
7	Benzyl	K_2CO_3	96	61	66/44
8	Neopentyl	K_2CO_3	96	91	61/39
9	*Tert*-butyl	K_2CO_3	35	n.d.	100/0

[a]*Retsch Planetary Mill PM200, stainless steel jar (12mL), 5×7mm balls, 550rpm, substrate:aldehyde:base 1:0.9:1.8.*
[b]*Isolated yields.*
Reproduced from Baron A, Martinez J, Lamaty F. Solvent-free synthesis of unsaturated amino esters in a ball-mill. Tetrahedron Lett 2010;51:6246–9. Copyright (2010), with permission from Elsevier.

2.16 ASYMMETRIC ALKYLATION

Asymmetric carbon–carbon bond formation reaction under solvent-free conditions was carried out by Bolm et al. in ball mill [58]. Here, nickel(II) complex **206** was used as a chiral auxiliary in alkylation with various bromides (Scheme 2.67). Optimized reaction conditions were set to increase the stereoselectivity; however, the desired monoalkylated product **207** was often accompanied by small amount of doubly alkylated side product **208**. Two different bases were used (NaOMe/$MgSO_4$ or Cs_2CO_3) and grinding of nickel complex **206** with bromides for 30–75 min afforded alkylation products in moderate to high yields and with complete stereoselectivity (selected examples, Table 2.54).

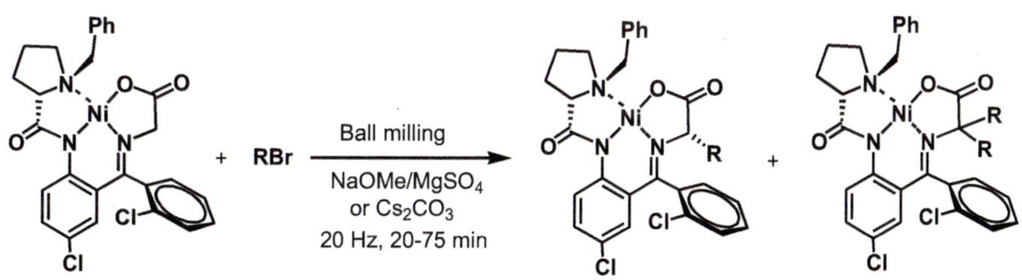

206 207 208

SCHEME 2.67

Asymmetric alkylation of a nickel complex.

Table 2.54 Asymmetric Alkylation of a Nickel Complex[a]

Entry	R	Time (min)	Method	207 Yield (%)[b]	dr of 207	208 Yield (%)[b]
1	$C_6H_5CH_2$	45	A	76	>98:2	10
2	$4\text{-}BrC_6H_4CH_2$	45	A	89	>98:2	2
3	$3\text{-}FC_6H_4CH_2$	45	A	86	>98:2	1
4	$3\text{-}NO_2C_6H_4CH_2$	60	A	81	>98:2	6
5	$3\text{-}MeOC_6H_4CH_2$	30	A	92	>98:2	0
6	$C_6H_5CH_2$	45	B	92	>98:2	1
7	$4\text{-}CF_3C_6H_4CH_2$	75	B	91	>98:2	1
8	$PhCH{=}CHCH_2$	30	B	54	98:2	0

[a]*Retsch MM400 mixer mill, stainless steel vessel and balls, 20 Hz, substrate:RBr 1:1.5 equiv. Method A: NaOMe:MgSO$_4$ 20:1 equiv. Method B: Cs$_2$CO$_3$ 5 equiv.*
[b]*Isolated.*
Reproduced with permission from Jörres M, Aceña JL, Soloshonok VA, Bolm C. Asymmetric carbon-carbon bond formation under solventless conditions in ball mills. ChemCatChem 2015;7:1265–9. Copyright (2015), Wiley.

Synthetic utility of solvent-free approach was demonstrated by development of aldol addition reaction (Scheme 2.68). Grinding of nickel complex **206** with *n*-butyraldehyde using NaOMe as the base afforded aldol product **209** in 76% yield (single stereoisomer). Further example was given by one-pot, two-step asymmetric alkylation/1,3-dipolar cycloaddition click reaction sequence. Nickel complex reacted in ball mill with propargyl bromide in the presence of NaOMe/$MgSO_4$ to provide alkylation product, which was in the second step treated with benzyl azide, copper acetate (as catalyst), sodium ascorbate (as additive) and silica gel (as milling auxiliary). Such procedure yielded 1,2,3-triazole product **210** in 88% yield as a single diastereomer (Scheme 2.68).

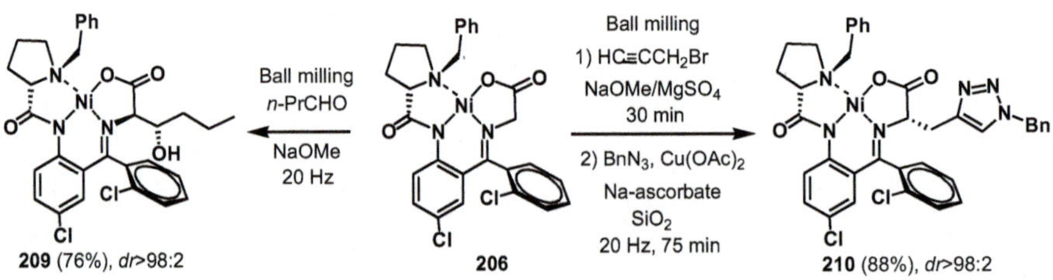

SCHEME 2.68

Asymmetric aldol and alkylation/click reactions of a nickel complex.

Another synthetic approach was used by Lamaty et al. in asymmetric alkylation of Schiff bases in ball mill (Scheme 2.69) [59]. In optimized reaction conditions, grinding of Schiff base of glycine ester **213** with organic halides in the presence of base (KOH) and chiral phase–transfer catalyst **215** (10 mol%) gave within 1–2 h a complete conversion and excellent yields, with enantiomeric excesses varying from 36% to 75% (selected results, Table 2.55). These enantiomeric excesses are lower than those obtained in solution. Full conversion to products and use of equimolar amounts of reagents simplify the isolation procedure to extraction and washing with water and catalyst could be recovered by extraction of the aqueous layer. Authors had an interesting observation that in contrast to classical phase-transfer catalysis, the catalyst was not necessary for the reaction and good results are obtained in similar reaction time. Absence of PTC should not favor the formation of one enantiomer; nevertheless, the enantiomeric excesses that were observed are high and might be explained by the existence of different phases, which determine the step sequence in which the reaction takes place.

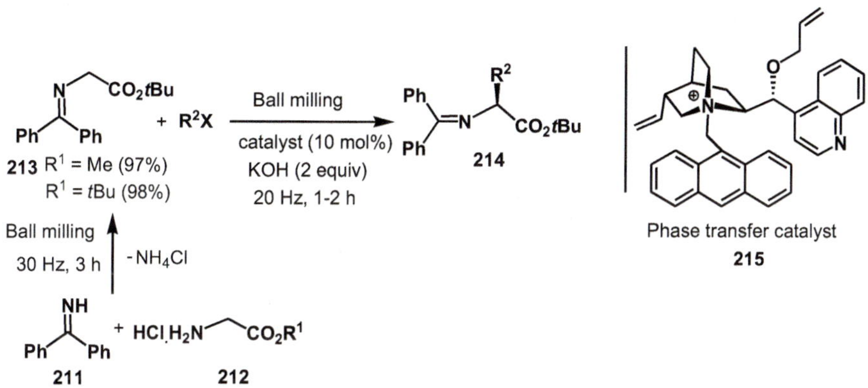

SCHEME 2.69

Asymmetric alkylation of Schiff bases.

Required Schiff base substrates were also prepared in solvent-free ball-mill conditions by mechanochemical transimination reaction. Grinding of an equimolar amount of the methyl ester of glycine hydrochloride and benzophenone imine for 3 h exclusively produced Schiff base (Scheme 2.60). Aqueous washing eliminated ammonium chloride, and the pure product was obtained without recrystallization in nearly quantitative yield. The reaction time was reduced to 3 h compared with one night for the synthesis in solution.

Table 2.55 Asymmetric Alkylation of Schiff Bases[a]

Entry	Electrophile	R^2	Time/h	ee (%)	Yield (%)[b]
1	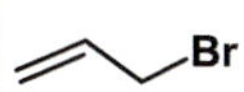	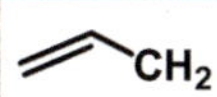	1	75	97
2	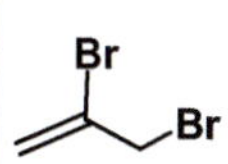	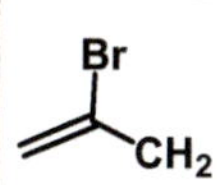	1	61	93

Continued

Table 2.55 Asymmetric Alkylation of Schiff Bases[a]—cont'd

Entry	Electrophile	R^2	Time/h	ee (%)	Yield (%)[b]
3	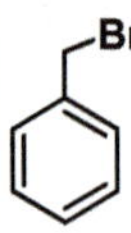	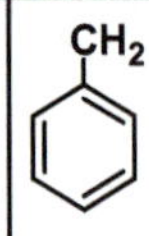	1	59	96
4		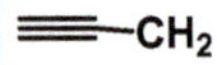	1	92	70
5	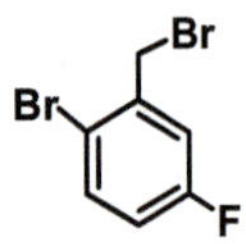	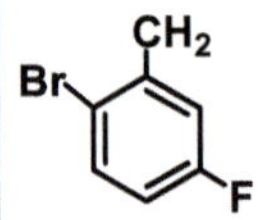	1	96	53
6	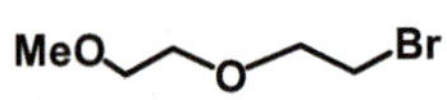	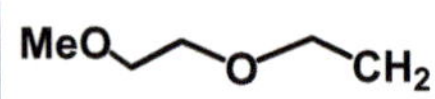	2	93	36

[a]Retsch MM400 mixer mill, stainless steel vessel (5 mL), 2×5 mm balls, 20 Hz substrate:RBr:KOH:catalyst 1:1:2:0.1 equiv.
[b]Isolated.
Reproduced with permission from Nun P, Pérez V, Calmès M, Martinez J, Lamaty F. Preparation of chiral amino esters by asymmetric phase-transfer catalyzed alkylations of schiff bases in a ball mill. Chem Eur J 2012;18:3773–9. Copyright (2012), Wiley.

2.17 OLEFIN CROSS-METATHESIS

Friščić et al. have shown that olefin cross-metathesis employing ruthenium catalysts could be transferred to ball-mill conditions (Scheme 2.70) [60]. The second-generation Hoveyda–Grubbs catalyst **219** (0.5–2 mol%) was the most efficient of the used catalysts, and Teflon-milling jars had to be used since the stainless steel jars gave irreproducible results associated with catalyst activity reduction by impact of steel. The escape of ethylene side product was enabled by non-hermetically sealing of the milling jars. LAG with THF improved yields of less-reactive substrates possessing COOR substituents, and the yields were further enhanced by addition of NaCl as grinding auxiliary (Table 2.56). Metathesis products **217** were isolated by catalyst removal by milling with aqueous L-Cys or disodium EDTA, followed by ethyl acetate or acetone wash and filtration. Reactions conducted with a solid auxiliary were first washed with water to remove the salt additive.

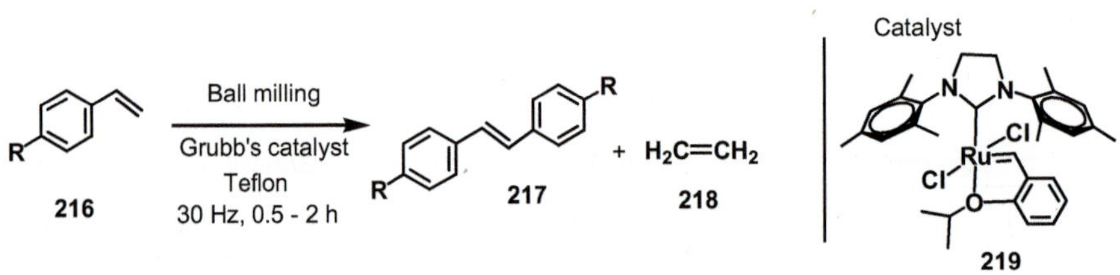

SCHEME 2.70

Olefin cross-metathesis.

Table 2.56 Olefin Cross-Metathesis[a]

Entry	R^1	Catalyst/ mol%	Liquid/μL	Grinding Auxiliary	Time/h	Yield (%)[b]
1	H	0.5	–	–	0.5	90
2	$COOCH_3$	1	THF (50)	–	1.5	40
3	COOH	2	THF (50)	–	5	49
4	$COOCH_3$	1	–	NaCl	1.5	90
5	COOH	2	EtOAc (50)	NaCl	5	73

[a]Retsch MM400 Teflon vial (10 mL), 10 mm stainless steel ball, 30 Hz.
[b]Isolated.
Reproduced with permission from Do J-L, Mottillo C, Tan D, Štrukil V, Friščić T. Mechanochemical ruthenium-catalyzed olefin metathesis. J Am Chem Soc 2015;137:2476–9. Copyright (2015), American Chemical Society.

Ring-closing metathesis (RCM) using ruthenium catalyst **219** also afforded cyclopentene and dihydropyrroles **221** in high yields (Scheme 2.71, Table 2.57). Synthesis of dinitro dihydropyrrole product is particularly notable, as an earlier synthesis gave only a 3.2% yield after 1 day at 45°C. Further studies showed that RCM of an ionic reactant BF_3 was not achieved by neat-milling or by LAG conditions, except when highly polar propylene carbonate was added. Authors demonstrated that mechanochemical CM and RCM could be scaled up 10-fold (3 g) scale. The optimized reactions were readily adapted to this scale by increasing the catalyst loading by only 50%.

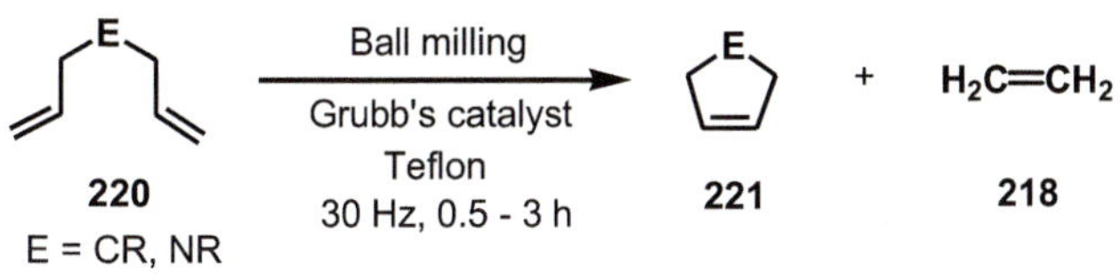

SCHEME 2.71

Ring olefin cross-metathesis.

Table 2.57 Ring Olefin Cross-Metathesis[a]

Entry	E	Catalyst/ mol%	Liquid/μL	Grinding Auxiliary	Time/h	Yield (%)[b]
1	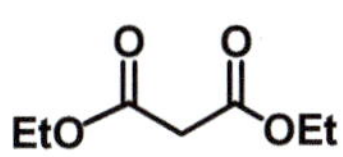	0.25	EtOAc (50)	NaCl	0.5	94
2	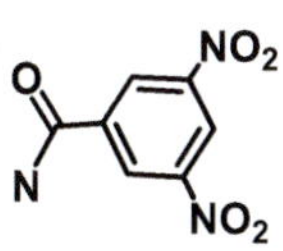	0.5	EtOAc (50)	NaCl	3	92

Continued

Table 2.57 Ring Olefin Cross-Metathesis[a]—cont'd

Entry	E	Catalyst/ mol%	Liquid/μL	Grinding Auxiliary	Time/h	Yield (%)[b]
3	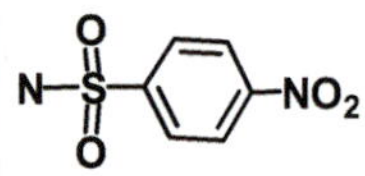	0.5	EtOAc (50)	NaCl	3	94
4	$BF_4^- + H_2N$	2	Propylene carbonate	NaCl	1.5	96

[a]Retsch MM400 Teflon vial (10 mL), 10 mm stainless steel ball, 30 Hz.
[b]Isolated.
Reproduced with permission from Do J-L, Mottillo C, Tan D, Štrukil V, Friščić T. Mechanochemical ruthenium-catalyzed olefin metathesis. J Am Chem Soc 2015;137:2476–9. Copyright (2015), American Chemical Society.

2.18 CROSS-DEHYDROGENATIVE COUPLING

A high-yielding synthesis of racemic 2,2′-dihydroxy-1,1′ binaphthyl **223** was achieved by solid-phase oxidative coupling of 2-naphthol in ball mill catalyzed by $FeCl_3 \cdot 6H_2O$ in the presence of sodium chloride as milling auxiliary (Scheme 2.72) [61]. Fritsch Planetary Mill Pulverisette 7 was applied for milling. New synthetic protocol is an alternative and practical method to existing procedures. Conventional synthesis in solution has disadvantages such as the use of large amounts of CH_2Cl_2 in high dilute conditions, or lower yield in THF, while for the solid-state dimerization it is difficult to control the heating of powders to 50°C.

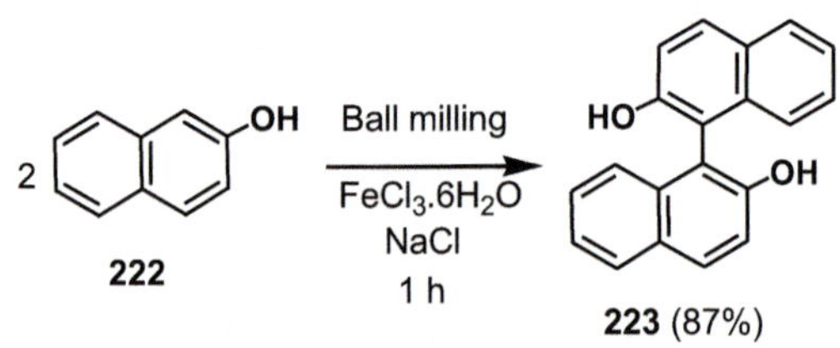

SCHEME 2.72

Dimerization of 2-naphthol.

By building their own vibration ball mill, Shayesteh and coworkers obtained 100% conversion in just 12 min at 30 Hz for cross-dehydrogenative coupling (CDC) of 2-naphthol [62]. In comparison, the Toda's mortar grinding and above described method by Rasmussen et al. needed 2 and 1 h, respectively. On operation condition of 30 Hz vibration ball mill gives better performance in comparison to other commonly used methods such as ball mill and mortar and pestle, due to higher mechanical energy given to reactants by vibration mill. Oxidative coupling reactions of anthrone and 2,3-dihydroxynaphthalene also afforded products **225** and **227** in short time and in high yields (Scheme 2.73).

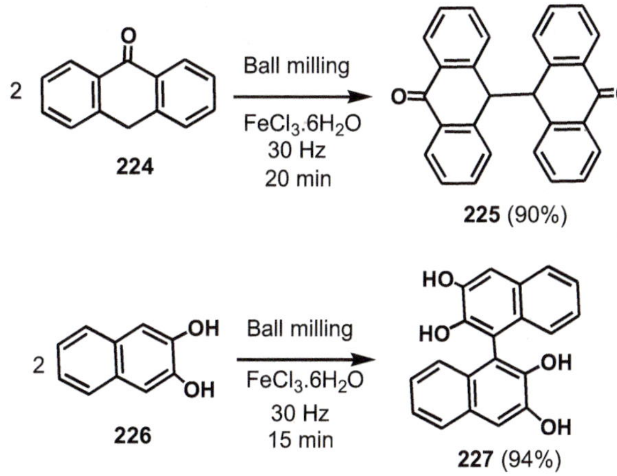

SCHEME 2.73

Dimerizations of anthrone and 2,3-dihydroxynaphthalene.

An efficient DDQ-mediated solvent and metal-free CDC reaction of tetrahydroisoquinoline **228** was developed by Su et al. (Scheme 2.74). [63]. Three types of reactions based on functionalization of sp^3 C—H bonds adjacent to nitrogen atom of tetrahydroisoquinoline proceeded rapidly under ball-milling condition: sp^3 C—H/sp^3 C—H, sp^3 C—H/sp C—H and sp^3 C—H/aryl-sp^2 C—H bond CDC. Nitroalkanes reacted rapidly in ball mill giving products **233** in high yield (Table 2.58, entries 1–4) when reactions were carried out in the presence of silica gel as grinding auxiliary. Comparative reactions carried out in neat solution of nitroalkanes required significantly longer time to achieve similar yields. Malononitrile was also viable participant which reacted with 2-phenyltetrahydroisoquinoline to give comparable yield. Coupling product **234** was afforded in 55% yield along with 10% of cyanated product **235**, formed by oxidative degradation of malononitrile with the cleavage of the C—CN bond by DDQ. Coupling reaction of tetrahydroisoquinolines with alkynes via ball milling are catalyzed by trace amount of copper, which was conveniently provided from copper-milling balls. In short time, in the presence of silica gel as grinding auxiliary, coupling products **230** were obtained in high yields (entries 5–8). Employing similar reaction conditions, indoles **231** were smoothly coupled to tetrahydroisoquinolines at position 3 (entries 9–12), and it was established that even free indole (NH) is suitable for reaction. These mechanochemical conditions are advantageous over traditional solvent-free reactions under heating conditions. Mechanochemical CDC method is also amenable to larger-scale preparation, which was demonstrated to give appreciable yields for nitromethane/indole coupling.

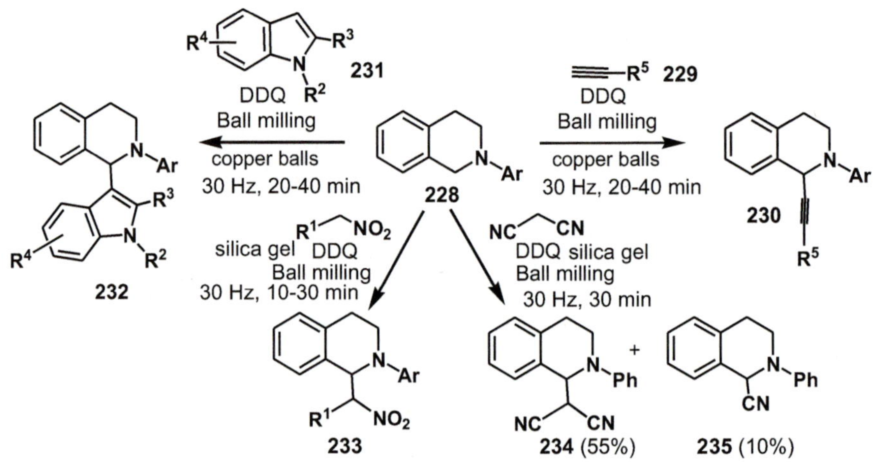

SCHEME 2.74

Cross-dehydrogenative coupling of tetrahydroisoquinolines.

Table 2.58 Cross-Dehydrogenative Coupling of Tetrahydroisoquinolines

Entry	Ar	R[1]	R[2]	R[3]	R[4]	R[5]	Time/min	Yield (%)[b]
Nitroalkanes[a]								
1	C_6H_5	H					10 (120)[c]	85 (89)[c]
2	C_6H_5	CH_3					15 (180)[c]	75 (78)[c]
3	4-$CH_3OC_6H_4$	H					20 (120)[c]	80 (75)[c]
4	4-$CH_3OC_6H_4$	C_2H_5					30 (210)[c]	70 (60)[c]
Alkynes[d]								
5	C_6H_5					C_6H_5	20	78
6	C_6H_5					Pr	30	67
7	C_6H_5					2-Py	40	69
8	4-$CH_3OC_6H_4$					C_6H_5	20	84
Indoles[e]								
9	C_6H_5		H	H	H		30	77 (79)[f]
10	C_6H_5		CH_3	H	H		40	80 (44)[f]
11	C_6H_5		H	H	OCH_3		20	70 (57)[f]
12	4-$CH_3OC_6H_4$		H	H	H		30	67

[a]*High-energy vibrational micro mill, stainless steel vial (10mL), 2×8mm balls, 30Hz, amine:nitroalkane:DDQ 1:2:1, silica gel.*
[b]*Isolated yields.*
[c]*Isolated yields under stirring condition: amine:DDQ 1:1, nitroalkanes (10mL).*
[d]*Amine:alkynes:DDQ 1:1.1:1, silica gel, 2×8mm copper balls.*
[e]*Amine:indoles:DDQ 1:0.8:0.8, silica gel, 2×8mm copper balls.*
[f]*Solvent-free heating synthesis, overnight.*
Reproduced with permission from Su W, Yu J, Li Z, Jiang Z. Solvent-free cross-dehydrogenative coupling reactions under high speed ball-milling conditions applied to the synthesis of functionalized tetrahydroisoquinolines. J Org Chem 2011;76:9144–50. Copyright (2015), American Chemical Society.

Asymmetric variant of CDC coupling reaction of tetrahydroisoquinolines **228** with alkynes by ball milling was carried out in subsequent paper (Scheme 2.75) [64]. In this protocol, copper balls were used in conjunction with chiral PyBox-1 ligand **238** (10 mol %) to afford various α-alkynylamine derivatives **237** in good yields and enantiomeric excesses up to 79% (Table 2.59). Interestingly, copper ball has been proved to be an efficient metal catalyst to coordinate with chiral ligands. After the reaction was completed, the catalyst was recovered simply by removing the copper ball from the reaction media. The ball bearings were used for more than 20 times without significant change in mass. This protocol was also applied to 6,7-dimethoxy-2-aryl- tetrahydroisoquinolines obtaining moderate yields (56–72%) and fair enantiomeric excess (20%).

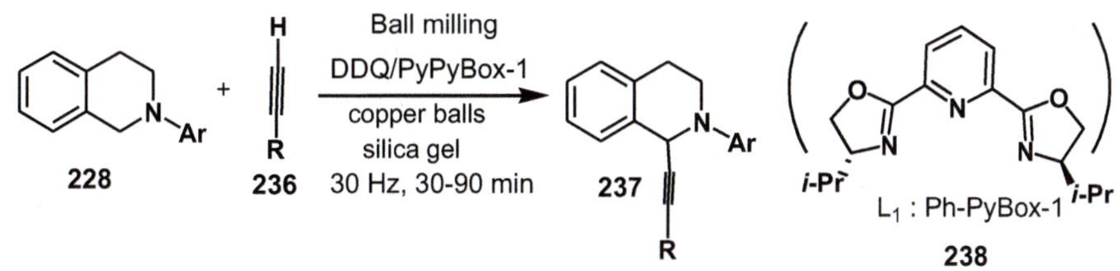

SCHEME 2.75

Asymmetric coupling reaction of tetrahydroisoquinolines with alkynes.

Table 2.59 Asymmetric Coupling Reaction of Tetrahydroisoquinolines With Alkynes[a]

Entry	R¹	R²	Time/ min	Yield (%)[b]	ee (%)
1	4-$CH_3OC_6H_5$	C_6H_5	40	69	74
2	4-$CH_3OC_6H_5$	4-FC_6H_5	30	75	55
3	4-$CH_3OC_6H_5$	CO_2CH_3	40	60	55
4	C_6H_5	C_6H_5	50	74	43
5	C_6H_5	$CO_2CH_2CH_3$	50	67	56
6	C_6H_5	2-Py	90	60	44
7	2-$CH_3OC_6H_5$	4-FC_6H_5	40	65	34
8	2-$CH_3OC_6H_5$	CO_2CH_3	30	75	79
9	4-$CH_3C_6H_5$	C_6H_5	30	75	38
10	4-$CH_3C_6H_5$	CO_2CH_3	40	72	62

[a]*High-energy vibrational micro mill QM-3A 1200 rpm, stainless steel jar (50 mL), 2×12 mm copper balls, substrate:alkyne:DDQ: PyBox-1 1:1.1:1:0.1, silica gel.*

[b]*Isolated yields.*

Reproduced from Yu J, Li Z, Jia K, Jiang Z, Liu M, Su W. Fast, solvent-free asymmetric alkynylation of prochiral sp³ C—H bonds in a ball mill for the preparation of optically active tetrahydro isoquinoline derivatives. Tetrahedron Lett 2013;54:2006–9. Copyright (2013), with permission from Elsevier.

2.19 REFORMATSKY REACTION

Some of substrates required for study of Achmatowicz oxidative rearrangement (chapter Oxidations and Reductions) were prepared in solid-state conditions. For instance, König prepared ethyl 3-(furan-2-yl)-3-hydroxypropanoate **241** by Reformatsky reaction from furfural ethyl-4-bromoethanoate employing Zn/NH_4Cl as reducing agent (Scheme 2.76). Planetary ball mill (agate beaker and balls) was employed, with Zn:NH_4Cl:furfural ratio of 10:4:1 [65].

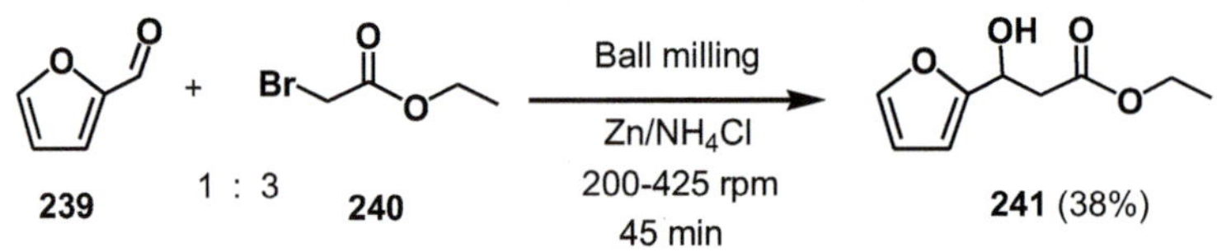

SCHEME 2.76

Reformatsky reaction of furfural.

2.20 CASCADE REACTIONS

2.20.1 PORPHYRIN SYNTHESIS

Simple two-step mechanochemical synthesis of porphyrins was developed by Hamilton and coworkers [66]. It involves the milling of an aromatic aldehyde and pyrrole in the presence of an acid catalyst, followed by the second milling step with an oxidizing agent to produce porphyrins (Scheme 2.77). After condensation milling step, oxidizing agent (1.5 equiv.) was added to the grinding jar and mixture further ball milled. A variety of oxidants and grinding agents were tried to optimize reaction conditions; however, the yields in oxidation step were unsatisfactory, not exceeding 10% (Table 2.60). Interestingly, milling with some oxidizing agents produces the porphyrin in a shorter time than simple exposure to air. Addition of grinding auxiliaries has a different effect for two oxidizing agents tested: in the case of oxone, sodium chloride works much better than magnesium sulfate, while for sodium ethoxide magnesium sulfate is beneficial, on the other hand sodium chloride retards the reaction.

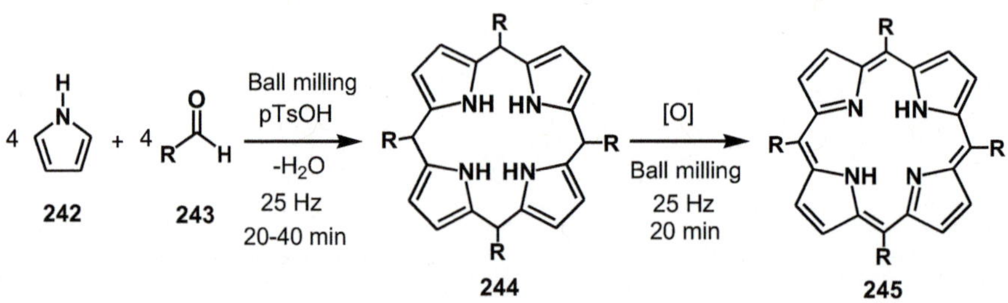

SCHEME 2.77

Two-step porphyrin synthesis.

Table 2.60 Two-Step Mechanochemical Synthesis of Porphyrins[a]

Oxidizing Agent	Grinding Agent	Milling Time (min) Condensation	Milling Time (min) Oxidation	Yield[b] (%)
None	None	20	20	0
DDQ	None	20	20	0
I_2	None	20	20	1.6
HNO_3	None	20	20	11.5
Sodium perborate	None	20	20	8.3
NaOEt	None	20	20	8.2
NaOEt	None	20	480	6.7
NaOEt	Silica	20	20	5.2
NaOEt	NaCl	20	20	0
NaOEt	$MgSO_4$	20	20	9.3
Oxone	None	20	20	0
Oxone	None	20	480	8.1
Oxone	Silica	20	20	5.3
Oxone	NaCl	20	20	4.7
Oxone	$MgSO_4$	20	20	0

[a]*Retsch MM200 mill, 25 Hz, stainless steel grinding jar (10 mL), 2×5 mm balls.*
[b]*Determined by UV–vis spectroscopy.*
Reproduced from Shy T, Mackin P, Orvieto AS, Gharbharan D, Peterson GR, Bampos N, et al. The two-step mechanochemical synthesis of porphyrins. Faraday Discuss 2014;170:59–69. With permission from the Royal Society of Chemistry.

Better results for oxidation step were obtained when oxidation was carried out in solution (Table 2.61). In this procedure, cyclocondensation of pyrrole and aromatic aldehydes in the presence of acid catalyst (pTsOH) were carried mechanochemically, followed by DDQ oxidation in solution (chloroform, 2 h).

Table 2.61 Mechanochemical Synthesis of Porphyrins, With Oxidation in Solution Using DDQ[a]

RCHO; R=	Milling Time (min)	Yield[b] (%)
Ph	20	27.9
2-ClC_6H_4	20	32.7
3-ClC_6H_4	20	21.9
4-ClC_6H_4	30	20.0
2-FC_6H_4	30	28.1

Continued

Table 2.61 Mechanochemical Synthesis of Porphyrins, With Oxidation in Solution Using DDQ[a]—cont'd

RCHO; R=	Milling Time (min)	Yield[b] (%)
3-FC_6H_4	20	25.0
4-FC_6H_4	30	9.7
α-naphthyl	30	2.7
β-naphthyl	40	31.3

[a]Retsch MM200 mill, 25 Hz, stainless steel grinding jar (10 mL), 2×5 mm balls.
[b]Determined by UV–vis spectroscopy.
Reproduced from Shy T, Mackin P, Orvieto AS, Gharbharan D, Peterson GR, Bampos N, et al. The two-step mechanochemical synthesis of porphyrins. Faraday Discuss 2014;170:59–69. With permission from the Royal Society of Chemistry.

2.20.2 NINHYDRIN CONDENSATIONS

Another example of cascade reaction carried out in ball mill is given by Kaupp [27]. The solid enaminoester (3-methylcrotonate) reacts quantitatively with ninhydrin in stoichiometric ball-mill runs to give the heterocyclic building block **248** with α-hydroxyketone and *N/O*-semiaminal functionality (Scheme 2.78). The cascade process consists of vinylogous substitution, cyclization, and 1,3-hydrogen shift. For comparison, solution reaction gives the product in 82% yield.

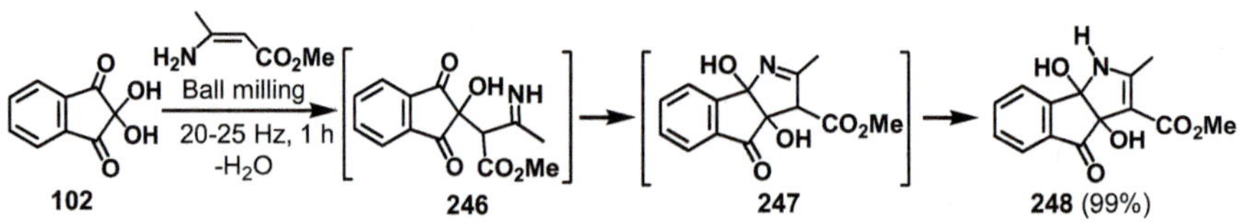

SCHEME 2.78

Solid-state condensations of ninhydrin.

2.21 ARYLAMINOMETHYLATION

Highly reactive methylene iminium chlorides **250** were used by Kaupp for carbon–carbon formation in arylaminomethylations of 2-naphthol in a ball mill (Scheme 2.79) [67]. Due to susceptibility of iminium chlorides to moisture, reagents were handled under an argon atmosphere. After milling, electrophilic aromatic substitution products **252** were dissolved in dihloromethane and recrystallized. The yields were 45–60% due to intermediate melting to viscous oils that could not yet be avoided.

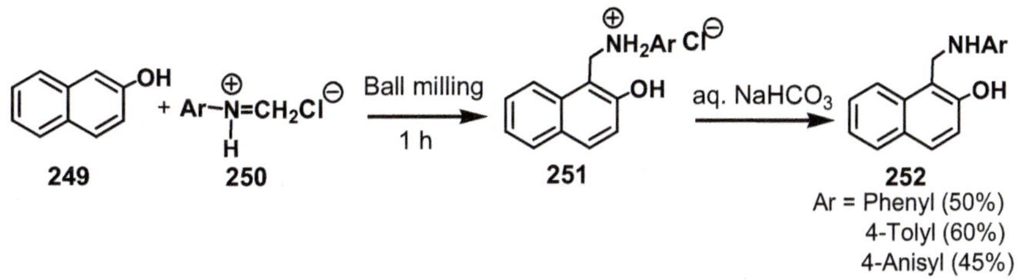

SCHEME 2.79

Arylaminomethylation of 2-naphthol.

2.22 NAPHTHOPYRAN SYNTHESIS

Two-step mechanochemical synthesis of naphthopyrans was effected by Wang et al. (Scheme 2.80). [68]. Under solvent-free ball-milling conditions, the nucleophilic addition of terminal alkynes to carbonyl compounds takes place with the aid of potassium hydroxide base to afford propargylic alcohols **255** in moderate to high yields (Table 2.62). One or more equivalents of KOH were required, and addition of 18-crown-6 significantly improved reaction yields. Mechanochemical procedure is advantageous in terms of short reaction time in comparison with reaction of methyl ketone in organic solvent where small amount of by-product was usually formed due to the competitive aldol condensation, in which methyl ketone must be added slowly to prevent the side reaction. However, under solvent-free conditions, good yields could be readily achieved even when the reactants were added in one pot. Other bases were tested in conjunction with 18-crown-6. When K_2CO_3 along with 18-crown-6 or CH_3ONa, DBU, or DABCO was used, no target product was formed. When a catalytic amount of KOH/18-crown-6 was used, only moderate yield of the product could be obtained.

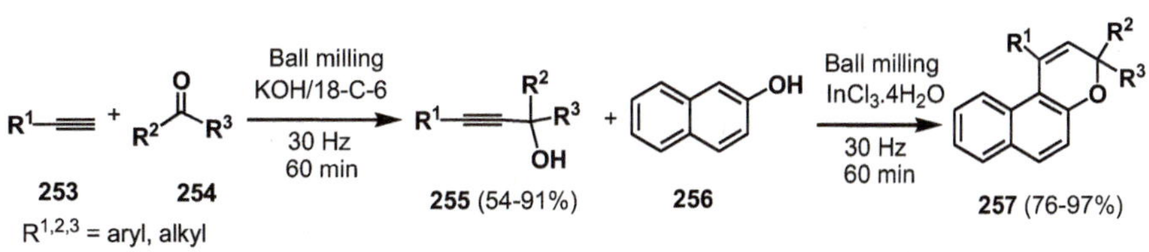

SCHEME 2.80

Naphthopyran synthesis.

Table 2.62 Nucleophilic Additions of Terminal Alkynes With Carbonyl Compounds[a]

Entry	R^1	R^2	R^3	Yield (%)[b]
1	Ph	Ph	Ph	97
2	Ph	4-$OCH_3C_6H_4$	4-$NO_2C_6H_4$	93
3	Ph	Ph	4-$OCH_3C_6H_4$	81
4	Ph	Fluorenone		92
5	Ph	2-F-fluorenone		90
6	Ph	Cyclohexanone		89
7	Ph	Ph	CH_3	83
8	Ph	2-pyrazinyl	CH_3	81
9	Ph	Isatin		91
10	2-Thienyl	Ph	Ph	96
11	2-Thienyl	Fluorenone		94
12	2-Thienyl	Cyclohexanone		91
13	2-Pyridyl	Ph	Ph	91
14	*n*-hexyl	Ph	Ph	54

[a]*Retsch MM200 mixer mill, 1800 rpm (30 Hz), 1 h, stainless steel jar (10 mL), one 7 mm milling ball, alkyne:ketone:KOH:18-crown-6 1.1:1:1:1.1 mol.*
[b]*Isolated yield.*
Reproduced from Dong Y-W, Wang G-W, Wang L. Solvent-free synthesis of naphthopyrans under ball-milling conditions. Tetrahedron 2008;64:10148–54. Copyright (2008), with permission from Elsevier.

Propargylic alcohols **255** were then subjected to indium trichloride catalyzed cyclization with 2-naphthol to obtain naphthopyrans **256** under the same ball-milling conditions (Scheme 2.80, Table 2.63). Other Lewis acids ($ZnCl_2$ and $SnCl_4$) afforded lower yields. The main advantages of mechanochemical procedure in comparison with the literature methods are milder reaction conditions, higher yields, and shorter reaction time without side reactions. The reason for the efficiency of the solvent-free procedure is postulated by an enhanced reaction rate resulted from ultimately high concentrations of reactants without using solvent. Reaction consists of formation of propargyl-naphthol ether followed by Claisen rearrangement and subsequent enolization.

Table 2.63 Cyclizations of Propargylic Alcohols With 2-Naphthol[a]

Entry	R^1	R^2	R^3	Yield (%)[b]
1	Ph	Ph	Ph	97
2	Ph	4-$OCH_3C_6H_4$	4-$NO_2C_6H_4$	90
3	Ph	Ph	4-$OCH_3C_6H_4$	97
4	Ph	Fluorenone		93
5	Ph	2-F-fluorenone		97
6	Ph	cyclohexanone		76

Table 2.63 Cyclizations of Propargylic Alcohols With 2-Naphthol[a]—cont'd

Entry	R^1	R^2	R^3	Yield (%)[b]
7	Ph	Ph	CH_3	92
8	Ph	2-pyrazinyl	CH_3	0
9	Ph	Isatin		23 (82)[c]
10	2-Thienyl	Ph	Ph	93
11	2-Thienyl	Fluorenone		93
12	2-Thienyl	cyclohexanone		82
13	2-Pyridyl	Ph	Ph	0
14	*n*-hexyl	Ph	Ph	95

[a]*Retsch MM200 mixer mill, 1800 rpm (30 Hz), 1 h, stainless steel jar (10 mL), one 7 mm ball, alcohol:2-naphthol:$InCl_3.4H_2O$: 1:1.1:0.2 mol.*

[b]*Isolated yield; 1.5 equiv. of catalyst.*

Reproduced from Dong Y-W, Wang G-W, Wang L. Solvent-free synthesis of naphthopyrans under ball-milling conditions. Tetrahedron 2008;64:10148–54. Copyright (2008), with permission from Elsevier.

REFERENCES

[1] Tullberg E, Peters D, Frejd T. The Heck reaction under ball-milling conditions. J Organomet Chem 2004;689:3778–81.

[2] Tullberg E, Schacher F, Peters D, Frejd T. Solvent-free Heck-Jeffery reactions under ball-milling conditions applied to the synthesis of unnatural amino acids precursors and indoles. Synthesis 2006:1183–9.

[3] Zhu X, Liu J, Chen T, Su W. Mechanically activated synthesis of (*E*)-stilbene derivatives by high-speed ball milling. Appl Organomet Chem 2012;26:145–7.

[4] Declerck V, Colacino E, Bantreil X, Martinez J, Lamaty F. Poly(ethylene glycol) as reaction medium for mild Mizoroki–Heck reaction in a ball-mill. Chem Commun 2012;48:11778–80.

[5] Hermann GN, Becker P, Bolm C. Mechanochemical Rhodium(III)-catalyzed C-H bond functionalization of acetanilides under solventless conditions in a ball mill. Angew Chem Int Ed 2015;54:7414–7.

[6] Nielsen SF, Peters D, Axelsson O. The Suzuki reaction under solvent-free conditions. Synth Commun 2000;30:3501–9.

[7] Klingensmith LM, Leadbeater NE. Ligand-free palladium catalysis of aryl coupling reactions facilitated by grinding. Tetrahedron Lett 2003;44:765–8.

[8] Schneider F, Ondruschka B. Mechanochemical solid-state suzuki reactions using an in situ generated base. ChemSusChem 2008;1:622–5.

[9] Bernhardt F, Trotzki R, Szuppa T, Stolle A, Ondruschka B. Solvent-free and time-efficient Suzuki-Miyaura reaction in a ball mill: the solid reagent system KF-Al_2O_3 under inspection. Beilstein J Org Chem 2010;6. http://dx.doi.org/10.3762/bjoc.66.7.

[10] Schneider F, Stolle A, Ondruschka B, Hopf H. The Suzuki-Miyaura reaction under mechanochemical conditions. Org Proc Res Dev 2009;13:44–8.

[11] Schneider F, Szuppa T, Stolle A, Ondruschka B, Hopf H. Energetic assessment of the Suzuki-Miyaura reaction: a curtate life cycle assessment as an easily understandable and applicable tool for reaction optimization. Green Chem 2009;11:1894–9.

[12] Cravotto G, Garella D, Tagliapietra S, Stolle A, Schüßler S, Leonhardt SES, Ondruschka B. Suzuki cross-couplings of (hetero)aryl chlorides in the solid-state. N J Chem 2012; 36:1304–7.
[13] Fulmer DA, Shearouse WC, Medonza ST, Mack J. Solvent-free Sonogashira coupling reaction *via* high speed ball milling. Green Chem 2009;11:1821–5.
[14] Thorwirth R, Stolle A, Ondruschka B. Fast copper-, ligand- and solvent-free Sonogashira coupling in a ball mill. Green Chem 2010;12:985–91.
[15] Stolle A, Ondruschka B. Solvent-free reactions of alkynes in ball mills: It is definitely more than mixing. Pure Appl Chem 2011;83:1343–9.
[16] Schmidt R, Thorwirth R, Szuppa T, Stolle A, Ondruschka B, Hopf H. Fast, ligand- and solvent-free synthesis of 1,4-substituted Buta-1,3-diynes by Cu-catalyzed homocoupling of terminal alkynes in a ball mill. Chem Eur J 2011;17:8129–38.
[17] Waddell DC, Mack J. An environmentally benign solvent-free Tishchenko reaction. Green Chem 2009;11:79–82.
[18] Chen L, Betsegaw E, Lemma BE, Rich JS, Mack J. Freedom: a copper-free, oxidant-free and solventfree palladium catalysed homocoupling reaction. Green Chem 2014; 16:1101–3.
[19] Zhang Z, Dong Y-W, Wang G-W, Komatsu K. Mechanochemical Michael reactions of chalcones and azachalcones with ethyl acetoacetate catalyzed by K_2CO_3 under solvent-free conditions. Chem Lett 2004;33:168–9.
[20] Zhang Z, Dong Y-W, Wang G-W, Komatsu K. Highly efficient mechanochemical reactions of 1,3-Dicarbonyl compounds with chalcones and Azachalcones catalyzed by potassium carbonate. Synlett 2004:61–4.
[21] Jia C, Chen D, Zhang C, Zhang Q, Cao B, Zhao Z. Mechanosynthesis of γ-nitro dicarbonyl compounds via $CaCl_2$-catalyzed Michael addition. Tetrahedron 2013;69:7320–4.
[22] Veverková E, Poláčková V, Liptaková L, Kázmerová E, Mečiarová M, Toma Š, Šebesta R. Organocatalyst efficiency in the michael additions of aldehydes to nitroalkenes in water and in a ball-mill. ChemCatChem 2012;4:1013–8.
[23] Wang Y-F, Chen R-X, Wang K, Zhang B-B, Li Z-B, Xu D-Q. Fast, solvent-free and hydrogen-bonding-mediated asymmetric Michael addition in a ball mill. Green Chem 2012;14:893–5.
[24] Jörres M, Mersmann S, Raabe G, Bolm C. Organocatalytic solvent-free hydrogen bonding mediated asymmetric Michael additions under ball milling conditions. Green Chem 2013;15:612–6.
[25] Hestericovà M, Šebesta R. Higher enantioselectivities in thiourea-catalyzed Michael additions under solvent-free conditions. Tetrahedron 2014;70:901–5.
[26] Raston CL, Scott JL. Chemoselective, solvent-free aldol condensation reaction. Green Chem 2000;2:49–51.
[27] Kaupp G, Naimi-Jamal MR, Schmeyers J. Quantitative reaction cascades of ninhydrin in the solid state. Chem Eur J 2002;8:594–600.
[28] Heintz AS, Gonzales JE, Fink MJ, Mitchell BS. Catalyzed self-aldol reaction of valeraldehyde via a mechanochemical method. J Mol Catal A Chem 2009;304:117–20.
[29] Tanaka K, Asakura A, Muraoka T, Kalicki P, Urbanczyk-Lipkowska Z. Asymmetric direct aldol reactions catalyzed by chiral amine macrocycle-metal(II) complexes under solvent-free conditions. N J Chem 2013;37:2851–5.
[30] Guillena G, del Carmen Hita M, Nájera C, Viózquez SF. Solvent-free asymmetric direct aldol reactions organocatalysed by recoverable (S_a)-binam-L-prolinamide. Tetrahedron Asymm 2007;18:2300–4.

[31] Guillena G, del Carmen Hita M, Nájera C, Viózquez SF. A highly efficient solvent-free asymmetric direct aldol reaction organocatalyzed by recoverable (*S*)-Binam-L-Prolinamides. ESI-MS evidence of the enamine-iminium formation. J Org Chem 2008; 73:5933–43.
[32] Almaşi D, Alonso DA, Nájera C. Prolinamides *versus* prolinethioamides as recyclable catalysts in the enantioselective solvent-free inter- and intramolecular aldol reactions. Adv Synth Catal 2008;350:2467–72.
[33] Hernández J, Juaristi E. Asymmetric aldol reaction organocatalyzed by (*S*)-Proline-Containing Dipeptides: Improved stereoinduction under solvent-free conditions. J Org Chem 2011;76:1464–7.
[34] Hernández JG, Juaristi E. Efficient ball-mill procedure in the 'green' asymmetric aldol reaction organocatalyzed by (*S*)-proline-containing dipeptides in the presence of water. Tetrahedron 2011;67:6953–9.
[35] Hernández JG, García-López V, Juaristi E. Solvent-free asymmetric aldol reaction organocatalyzed by (*S*)-proline-containing thiodipeptides under ball-milling conditions. Tetrahedron 2012;68:92–7.
[36] Machuca E, Rojas Y, Juaristi E. Synthesis and evaluation of (*S*)-proline-containing α,β-dipeptides as organocatalysts in solvent-free asymmetric aldol reactions under ball-milling conditions. Asian J Org Chem 2015;4:46–53.
[37] Rodríguez B, Bruckmann A, Bolm C. A highly efficient asymmetric organocatalytic aldol reaction in a ball mill. Chem Eur J 2007;13:4710–22.
[38] Rodríguez B, Rantanen T, Bolm C. Solvent-free asymmetric organocatalysis in a ball mill. Angew Chem Int Ed 2006;45:6924–6.
[39] Mack J, Shumba M. Rate enhancement of the Morita-Baylis-Hillman reaction through mechanochemistry. Green Chem 2007;9:328–30.
[40] Wada S, Suzuki H. Calcite and fluorite as catalyst for the Knoevenagel condensation of malononitrile and methyl cyanoacetate under solvent-free conditions. Tetrahedron Lett 2003;44:399–401.
[41] Kaupp G, Naimi-Jamal MR, Schmeyers J. Solvent-free Knoevenagel condensations and Michael additions in the solid state and in the melt with quantitative yield. Tetrahedron 2003;59:3753–60.
[42] Burmeister CF, Stolle A, Schmidt R, Jacob K, Breitung-Faes S, Kwade A. Experimental and computational investigation of Knoevenagel condensation in planetary ball mills. Chem Eng Technol 2014;37:857–64.
[43] Trotzki R, Hoffmann MM, Ondruschka B. Studies on the solvent-free and waste-free Knoevenagel condensation. Green Chem 2008;10:767–72.
[44] Wang G-W, Gao J. Selective formation of Spiro dihydrofurans and cyclopropanes through unexpected reaction of aldehydes with 1,3-dicarbonyl compounds. Org Lett 2009;11:2385–8.
[45] Mashkouri S, Naimi-Jamal MR. Mechanochemical solvent-free and catalyst-free one-pot synthesis of pyrano[2,3-d]pyrimidine-2,4(1H,3H)-diones with quantitative yields. Molecules 2009;14:474–9.
[46] Etman HA, Metwally HM, Elkasaby MM, Khalil AM, Metwally MA. Green, two components highly efficient reaction of ninhydrin with aromatic amines, and malononitrile using ball-milling technique. Am J Org Chem 2011;1:10–3.
[47] Amirnejad M, Naimi-Jamal MR, Tourani H, Ghafuri H. A facile solvent-free one-pot three-component method for the synthesis of 2-amino-4*H*-pyrans and tetrahydro-4*H*-chromenes at ambient temperature. Monatsh Chem 2013;144:1219–25.

[48] Wada S, Hayashi N, Suzuki H. Noticeable facilitation of the bismuth-mediated Barbier-type allylation of aromatic carbonyl compounds under solvent-free conditions. Org Biomol Chem 2003;1:2160–3.
[49] Shearouse WC, Shumba MZ, Mack J. A solvent-free, one-step, one-pot gewald reaction for alkyl-aryl ketones via mechanochemistry. Appl Sci 2014;4:171–9.
[50] Wang G-W, Liu L. Mechanochemical reaction of aromatic ketones and aldehydes in the presence of Zn-$ZnCl_2$. Chin Chem Lett 2004;15:587–90.
[51] Komatsu K, Wang G-W, Murata Y, Tanaka T, Fujiwara K. Mechanochemical synthesis and characterization of the fullerene dimer C_{120}. J Org Chem 1998;63:9358–66.
[52] Tanaka K, Kishigimi S, Toda F. A new method for coupling aromatic aldehydes and ketones to produce.alpha.-glycols using zinc-zinc dichloride in aqueous solution and in the solid state. J Org Chem 1990;55:2981–3.
[53] Harrowfield JM, Hart RJ, Whitaker CR. Magnesium and aromatics: mechanically-induced grignard and McMurry reactions. Aust J Chem 2001;54:423–5.
[54] Birke V, Schütt C, Ruck WKL. Small particle size magnesium in one-pot grignard-zerewitinoff-like reactions under mechanochemical conditions: on the kinetics of reductive dechlorination of persistent organic pollutants (POPs). In: Geiger CL, Carvalho-Knighton K, editors. Environmental applications of Nanoscale and Microscale reactive metal particles. ACS Symposium series, Washington DC, vol. 1027. 2009. p. 39–54. [Chapter 3].
[55] Balema VP, Wiench JW, Pruski M, Pecharsky VK. Mechanically induced solid-state generation of phosphorus ylides and the solvent-free wittig reaction. J Am Chem Soc 2002;124:6244–5.
[56] Shearouse WC, Mack J. Diastereoselective liquid assisted grinding: 'cracking' functional resins to advance chromatography-free synthesis. Green Chem 2012;14:2771–5.
[57] Baron A, Martinez J, Lamaty F. Solvent-free synthesis of unsaturated amino esters in a ball-mill. Tetrahedron Lett 2010;51:6246–9.
[58] Jörres M, Aceña JL, Soloshonok VA, Bolm C. Asymmetric carbon-carbon bond formation under solventless conditions in ball mills. ChemCatChem 2015;7:1265–9.
[59] Nun P, Pérez V, Calmès M, Martinez J, Lamaty F. Preparation of chiral amino esters by asymmetric phase-transfer catalyzed alkylations of schiff bases in a Ball mill. Chem Eur J 2012;18:3773–9.
[60] Do J-L, Mottillo C, Tan D, Štrukil V, Friščić T. Mechanochemical ruthenium-catalyzed olefin metathesis. J Am Chem Soc 2015;137:2476–9.
[61] Rasmussen MO, Axelsson O, Tanner D. A practical procedure for the solid-phase synthesis of racemic 2,2′-Dihydroxy-1,1′-binaphthyl. Synth Commun 1997;27:4027–30.
[62] Shayesteh K, Moghaddas J, Haghighi M, Gholamhasan Imanzadeh G. A mechanochemical reaction for highly efficient preparation of 1,1′-bi-2-naphthol from 2-naphthol. Sci Res Essay 2009;4:1201–4.
[63] Su W, Yu J, Li Z, Jiang Z. Solvent-free cross-dehydrogenative coupling reactions under high speed ball-milling conditions applied to the synthesis of functionalized tetrahydroisoquinolines. J Org Chem 2011;76:9144–50.
[64] Yu J, Li Z, Jia K, Jiang Z, Liu M, Su W. Fast, solvent-free asymmetric alkynylation of prochiral sp^3 C-H bonds in a ball mill for the preparation of optically active tetrahydro isoquinoline derivatives. Tetrahedron Lett 2013;54:2006–9.

[65] Falenczyk C, Pölloth B, Hilgers P, König B. Mechanochemically initiated Achmatowicz rearrangement. Synth Commun 2014;45:348–54.
[66] Shy T, Mackin P, Orvieto AS, Gharbharan D, Peterson GR, Bampos N, Hamilton TD. The two-step mechanochemical synthesis of porphyrins. Faraday Discuss 2014;170:59–69.
[67] Kaupp G, Schmeyers J, Boy J. Iminium salts in solid-state syntheses giving 100% yield. J Prakt Chem 2000;342:269–80.
[68] Dong Y-W, Wang G-W, Wang L. Solvent-free synthesis of naphthopyrans under ball-milling conditions. Tetrahedron 2008;64:10148–54.

CHAPTER

3 Carbon–Nitrogen Bond-Formation Reactions

CHAPTER OUTLINE

Mechanochemical Organic Synthesis. http://dx.doi.org/10.1016/B978-0-12-802184-2.00003-0

3.1 1,2-DIONE/DIAMINE CONDENSATION

The first account on 1,2-dione/diamine condensation carried out mechanochemically was given by Zefirov, who investigated solid-phase synthesis of benzo[*b*]phenazine **3** by reaction of phenylenediamine **1** with 2,3-naphthoquinone **2** (Scheme 3.1) [1]. Synthesis was carried out in an SVM-0.4 vibrational ball mill (120 mL chamber volume). Cascade reaction mechanism consists of the four reactions: substitution, elimination, cyclization, and elimination.

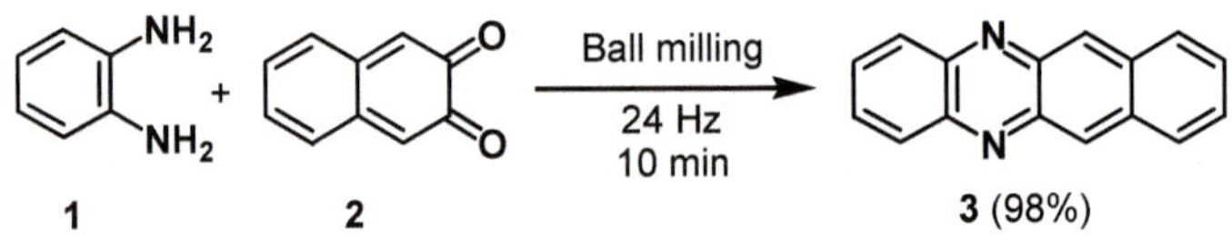

SCHEME 3.1

1,2-Dione/diamine condensation.

Kaupp et al. recognized the utility of 1,2-dione/diamine condensation for preparation of various nitrogen-heterocyclic products [2]. Various *o*-phenylenediamines (all solids) **4a–c** reacted quantitatively by milling with ninhydrin to give the indenoquinoxaline ketones **6a–c** (Scheme 3.2). Retsch MM2000 swing mill with a 10 mL stainless steel double-walled beaker with fittings for circulating coolants in conjunction with two 12 mm stainless steel balls was employed. Ball milling was performed usually at room temperature (without circulating liquid the temperature did not rise above 30°C). Water or methanol of the appropriate temperature was circulated for heating or cooling. Quantitative yields required no purification. Similar reactions proceed in solution with yields from 61% to almost quantitative. Identical results were obtained by Metwally for *o*-phenylenediamine condensation in the same ball-milling apparatus (for 1 h), while solution reaction required reflux for longer time, with lower yield (5 h, 60% yield) [3].

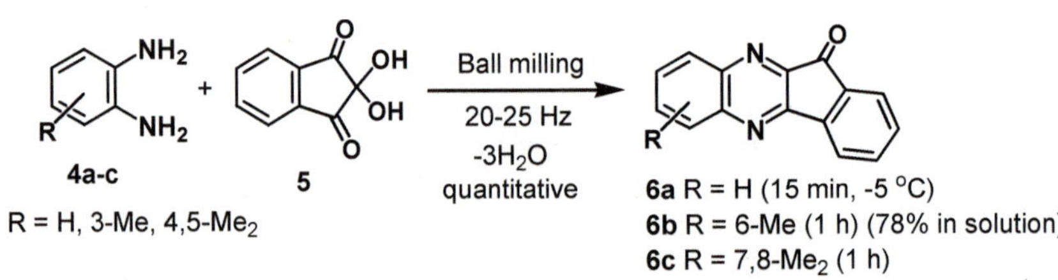

SCHEME 3.2

1,2-Dione/diamine condensation.

Further mechanochemical 1,2-dione/diamine condensation reactions were described by Kaupp and Naimi-Jamal in their subsequent article [4]. Substituted quinoxalines **9** were obtained by ball milling of equimolar amounts of *o*-phenylenediamines **7** and benzyls **8** (Scheme 3.3, Table 3.1). Milling beaker was heated at different temperatures through double walls with fittings for circulating coolants. Quantitative conversion did not require any isolation, and products were simply dried. Solution reactions gave much lower yields and additional waste was produced by the necessary workup in solvents.

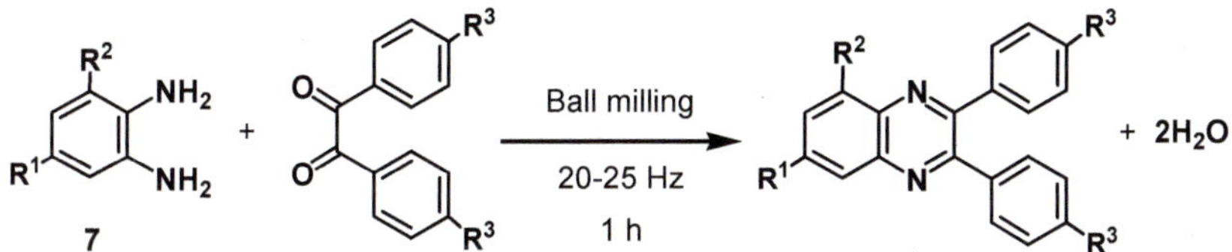

7 8 9

SCHEME 3.3

1,2-Dione/diamine condensation.

Table 3.1 Mechanochemical Condensation of *o*-Phenylenediamines and Benzyls[a]

Entry	R^1	R^2	R^3	T (°C) Ball Mill	T (°C) Melt	*Time* (min) Ball Mill	*Time* (min) Melt	Yield (%)[b] Ball Mill	Yield (%)[b] Melt	Yield (%)[b] In Solution
1	H	H	H	25	100	30	20	100	100	62
2	H	H	OCH_3	75	130	60	30	100	100	85
3	CH_3	H	OCH_3	80	130	60	30	100	100	
4	H	CH_3	OCH_3	75		30		100		

[a]*Retsch MM2000 swing mill, 20–25 Hz stainless steel jar (10 mL) and 2 × 12 mm milling balls, double-walled beaker with fittings for circulating coolants; diamine:dione 1:1 mol.*
[b]*NMR yields.*

Mechanochemical condensation between *o*-phenylenediamine **1** and 2-hydroxy-1,4-naphthoquinone **10** produced gave benzo[*a*]phenazin-5-ol **12** in 100% yield in solid-state 1:1 runs in 15 min at 70°C (Scheme 3.4). In melt conditions (30 min at 120°C), quantitative yield was also obtained, while solution chemistry gave lower yield (77%) after a complicated workup.

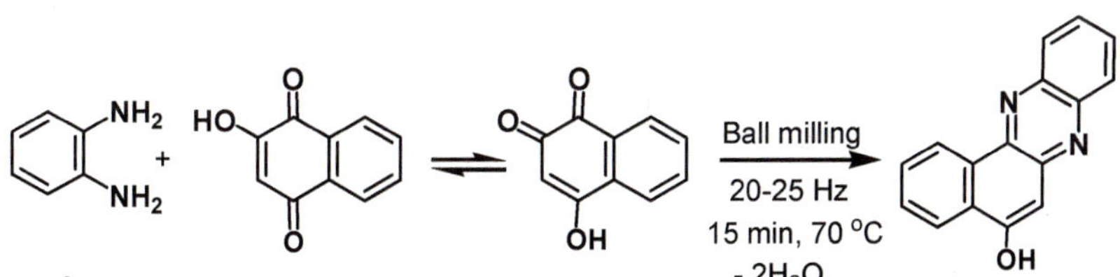

1 10 11 12

SCHEME 3.4

1,2-Dione/diamine condensation.

Condensation between *o*-phenylenediamine and 2-oxoglutaric acid **12** by ball milling for 10 min at room temperature gave a 90% yield of 3-(3-oxo-3,4-dihydroquinoxalin-2-yl)propionic acid **18**, which could not be increased by further milling for 1 h (Scheme 3.5). However, when the mixture obtained by milling was heated to 120–125°C for 30 min, a quantitative yield of products was obtained. Mechanosynthesis compared favorably with the lower yields obtained in solution conditions. Reactions of *o*-phenylenediamines **13** and alloxane hydrate **14** at room temperature quantitatively produced 3-oxo-3,4-dihydroquinoxalin-2-yl carbonyl ureas **15** in four cascade process consisting of substitution, elimination, cyclization, and ring opening. On the other hand, the milling of the stoichiometric mixture of *o*-phenylenediamine hydrochloride and alloxane hydrate afforded a mixture of 3-oxo-3,4-dihydroquinoxalin-2-yl carbonyl urea **16** and alloxazine **17**, while reaction in acidic solution was almost quantitative.

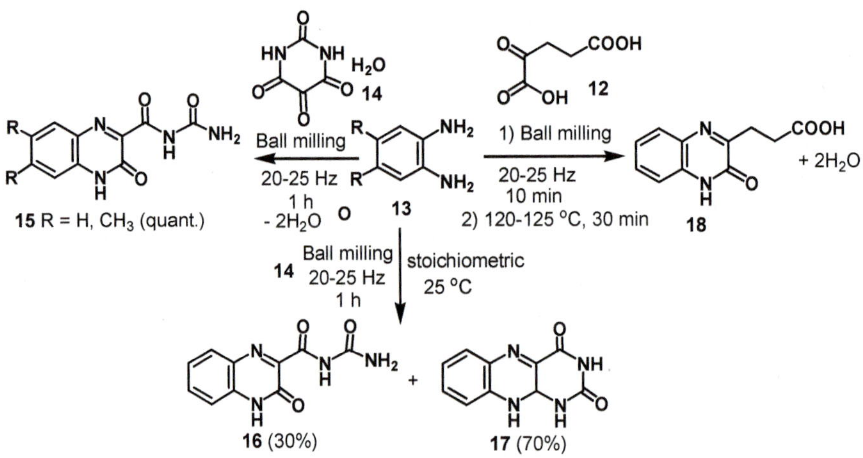

SCHEME 3.5

1,2-Dione/diamine condensation.

Carlier et al. have investigated similar mechanochemical condensation of equimolar amounts of phenanthrene-9,10-dione **20** with aromatic 1,2-diamines **19a–c** which furnished dibenzo[*a,c*]phenazines **21–22** in moderate to high yield (Scheme 3.6) [5]. For this transformation, to obtain reasonable yields, long reaction time was needed (4 h). In the reaction of 9,10-phenanthrene-9,10-dione with 2,3-diaminopyridine, corresponding dibenzo[*f,h*]pyrido[3,2-*b*]quinoxaline **23** was produced in high yield. These molecules are classically synthesized by a solvent route in refluxing ethanol or acetic acid for several hours. Fritsch Pulverisette 0 vibrating ball mill was used with a single agate milling ball (180 g). Mechanistic study of 9,10-phenanthrenequinone/phenylene diamine condensation under milling conditions supports mechanism which involves an initial formation of quinone imine intermediate. Generated water molecule forms hydrogen bonds as an intramolecular bridge and acts as a catalyst for the completion of the reaction by elimination of the second molecule of water, favored by the stabilization in the product by aromatization [6].

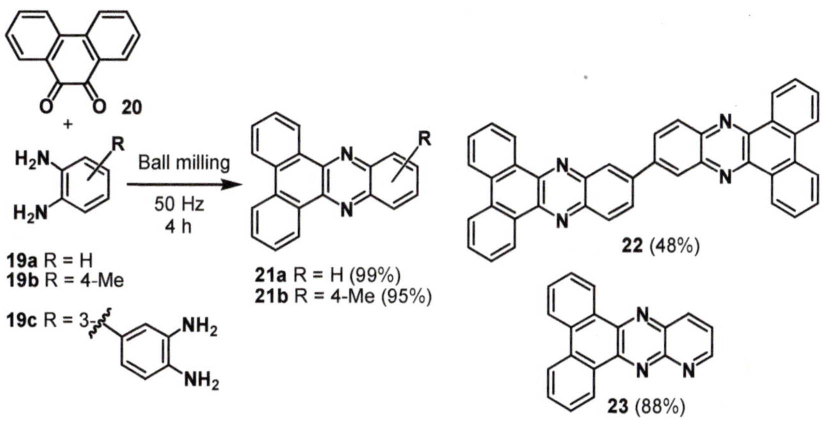

SCHEME 3.6

1,2-Dione/diamine condensation of phenanthrene-9,10-dione.

Synthesis of dibenzo[*a,c*]phenazine illustrated in Scheme 3.6 is in good accordance with our own results on nonoptimized 1,2-diamine/dione condensations (Table 3.2) [7,8]. It suggests that optimized reaction conditions (prolonged reaction time) gave product in higher yield (entry 1). Similarly, condensations of other aromatic diones 1,10-phenanthroline-5,6-dione **26** and porphyrin-2,3-dione **28** were carried out in mechanochemical conditions. Furthermore, instead of aromatic 1,2-diones, ninhydrin has been successfully employed in condensations with aromatic diamines (entries 5–7). Reactions could be also carried out in solution, however in many cases in low yields and contaminated with complex mixture of side products. These results show the advantageous use of ball milling in reaction where double condensation product **33** was obtained, which was not possible to prepare by classical solution reaction of ninhydrin and phenylenetetramine hydrochloride **32**.

3.2 SYNTHESIS OF IMINES

Kaupp and Toda performed solid–solid reactions by grinding together equimolar amounts of anilines and aldehydes in a mortar and keeping the mixture at room temperature for several hours or days. For instance, for quantitative conversion chloro imine required 120h, while methoxy/nitro imine needed 24h at 50°C (Scheme 3.7) [9]. The rate of reaction considerably increased when these reactions are performed in ball mills. Thus, chloro and methoxy/nitro imines **36a–c** can be quantitatively obtained in less than 30min at room temperature when stoichiometric mixtures of the starting materials are ball milled. Larger runs could be performed in ball mills under heat control.

Table 3.2 1,2-Dione/Diamine Condensations[a]

Entry	1,2-Dione	1,2-Diamine	Product	Yield (%)[b]
1	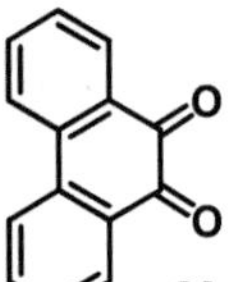 **20**	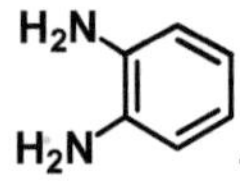 **1**	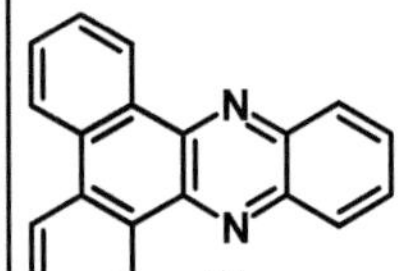 **21a**	16
2	20	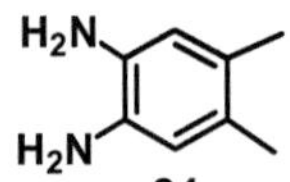 **24**	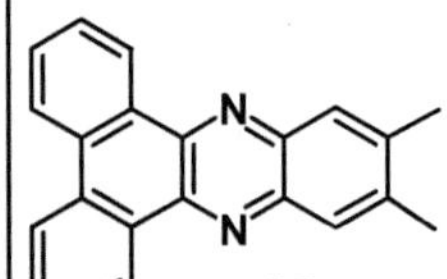 **25**	14
3	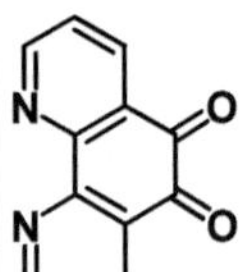 **26**	1	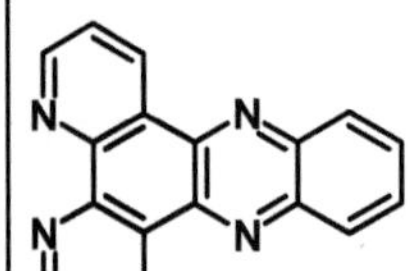 **27**	25
4	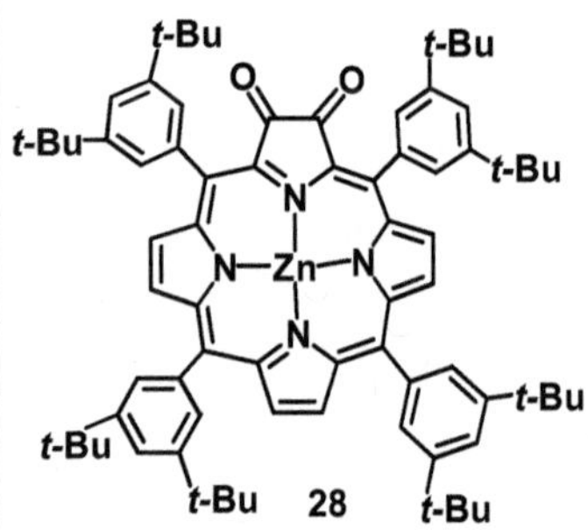 **28**	1	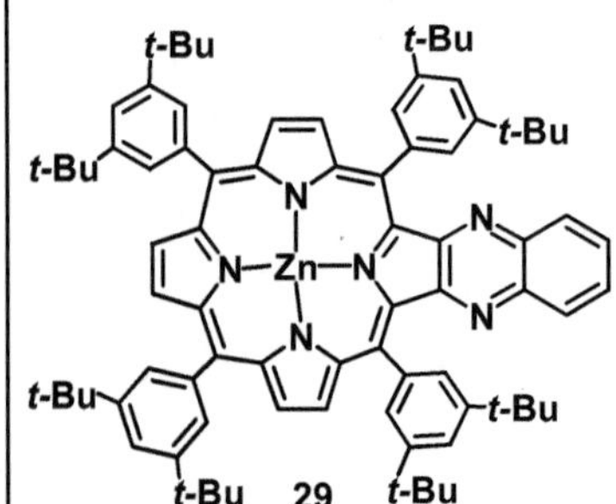 **29**	11
5	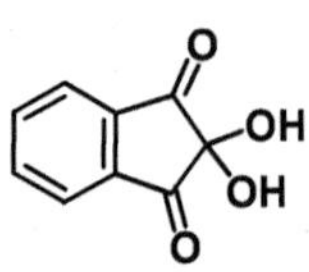 **5**	1	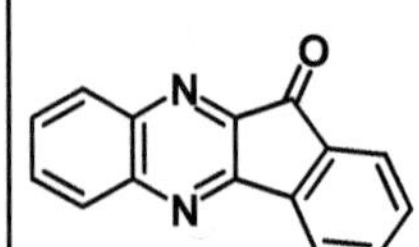 **30**	17 [8]
6	5	24	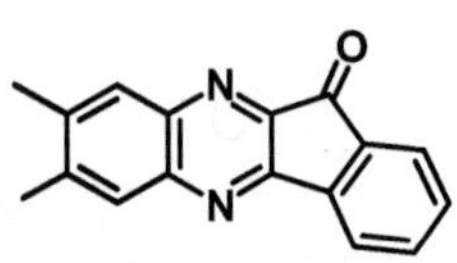 **31**	22 [8]
7	5	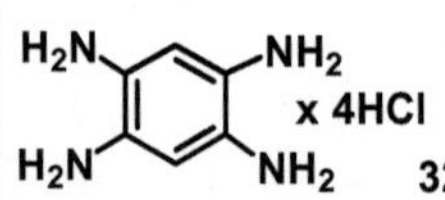 **32**	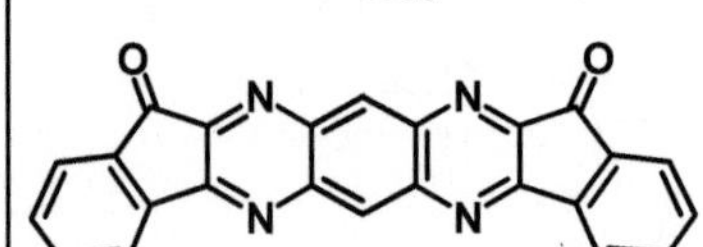 **33**	13 [8]

[a]*Home-made vibration ball mill, 30 min, 60 Hz, stainless steel capsule, one 5 mm stainless steel ball, equimolar amounts of reagents.*
[b]*Estimated by NMR.*

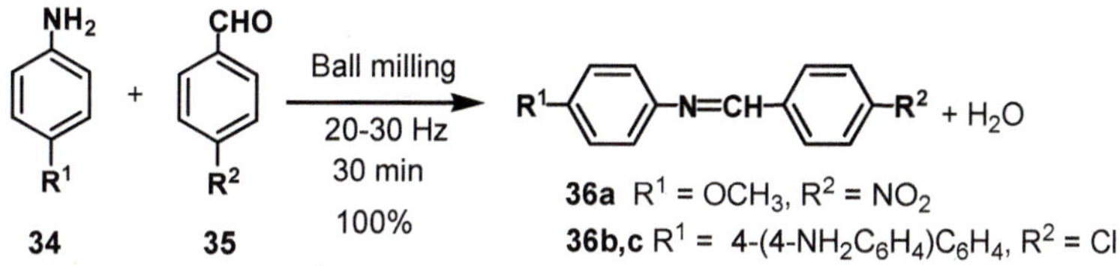

SCHEME 3.7

Mechanosynthesis of imines.

Imine-based, salen-type ligands **39** were synthesized by ball milling of equimolar amounts of diamines and aldehydes (Scheme 3.8). James reported that in optimized reaction conditions ligands were produced in quantitative yields within 1 h of milling and without the need for purification (Table 3.3). Drying in oven was the sole product manipulation [10]. For the synthesis of salenH_2, the starting materials (ethylene diamine, and salicylaldehyde) are both liquids. Such a reaction was also performed without using solvent by simply mixing these two liquids with conventional stirring. It led to partial reaction, while mechanochemical mixing affords full conversion, which was also superior to solution synthesis (alcohol, reflux 1–2 h). Additional advantage is that mechanochemical method offers a full stoichiometric control of reaction. Monoamine required for preparation of unsymmetrical ligands could be prepared quantitatively after 60 min of milling of equimolar amounts of aldehyde and 1,2-diaminobenzene. Preparation of the bulky *tert*-butyl chiral ligand ("Jacobsen" ligand) required liquid-assisted grinding LAG using 30 μL of methanol (1.75 M equiv.), since partial reaction took place after 30 min of neat milling at 25 Hz.

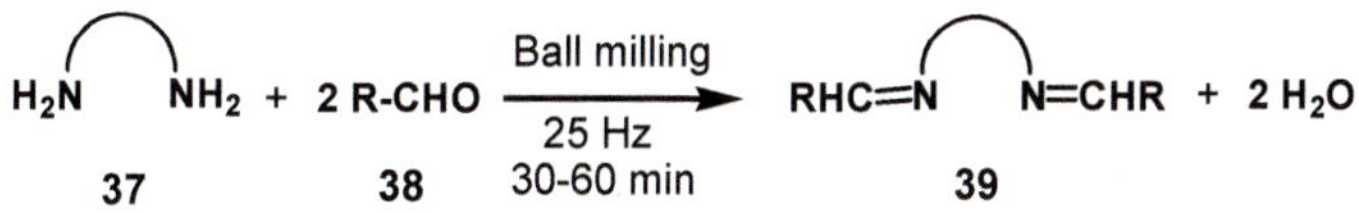

SCHEME 3.8

Mechanosynthesis of salen-type ligands.

Table 3.3 Mechanosynthesis of Salen-type Ligands[a]

Entry	RCHO	Diamine	Conditions
1	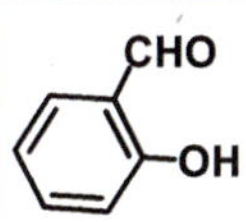	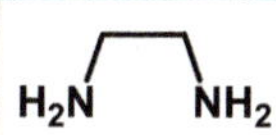	30 min
2		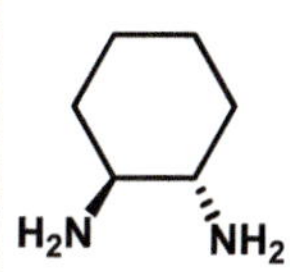	30 min

Continued

Table 3.3 Mechanosynthesis of Salen-type Ligands[a]—cont'd

Entry	RCHO	Diamine	Conditions
3		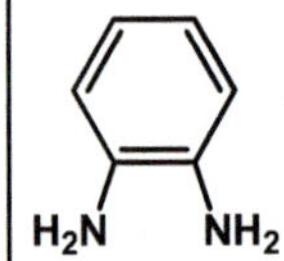	60 min
4	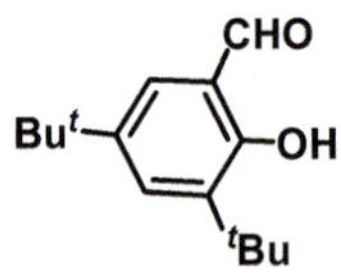	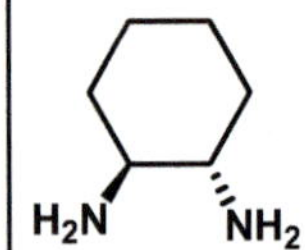	30 min LAG MeOH

[a]Retsch MM400 vibrational ball mill, stainless steel capsule (25 mL), one ball (15 mm, 13.6 g), 25 Hz, aldehyde:amine 2:1.

[b]Estimated by NMR.

Reproduced from Ferguson M, Giri N, Huang X, Apperley D, James SL. One-pot two-step mechanochemical synthesis: ligand and complex preparation without isolating intermediates. Green Chem 2014;16:1374–82, with permission from the Royal Society of Chemistry.

In continuation, one-pot two-step sequential synthesis of salen tetradentate complexes **43** was developed, where metal salts (ZnO, $Ni(OAc)_2{\cdot}4H_2O$ and $Cu(OAc)_2{\cdot}H_2O$) were added directly to the reaction vessel after preparation of salenH_2 **42** as described above, without isolating or drying of the ligand. In each case, milling was performed for a further 30 min at 25 Hz for complexation and LAG conditions were applied using methanol (1.1 M equiv.) for quantitative complexation (Scheme 3.9). Finally, both steps of the reaction were performed by simply milling together ethylenediamine, salicylaldehyde, and ZnO directly for 60 min, without the addition of solvent, to afford Zn(salen). This enhancement of reaction may be associated by the fact that since the jar did not need to be opened for addition of reagents, the water by-product was not able to evaporate and thus stayed in the vessel accelerating the reaction.

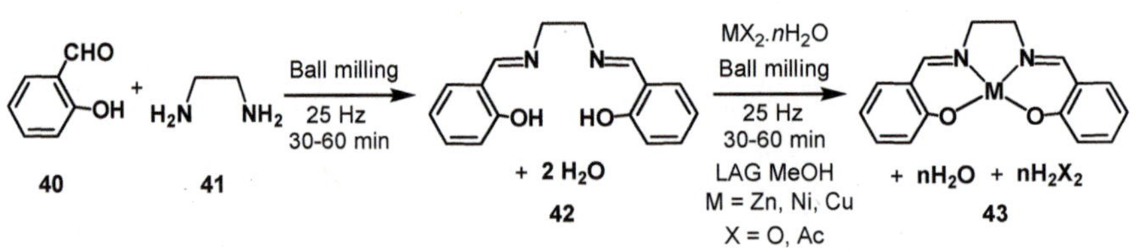

SCHEME 3.9

One-pot two-step synthesis of salen complexes.

Metwally carried out several substitutions of hydroxyl group of ninhydrin in ball-milling apparatus to afford imine products in quantitative yields after 1 h milling (Retsch MM2000, stainless steel vial (10 mL), two 12 mm balls)

(Scheme 3.10) [3]. Equimolar amounts of either benzidine or *p*-toluidine were used (in synthesis of imine **46** and *bis*-imine **47**, respectively) and after reaction completed, products did not need any purification. On the other hand, solution reaction required reflux for longer time, with lower yield in the case of benzidine (5 h, 60% yield). Manual grinding in mortar afforded after 30 min products in considerably lower yields.

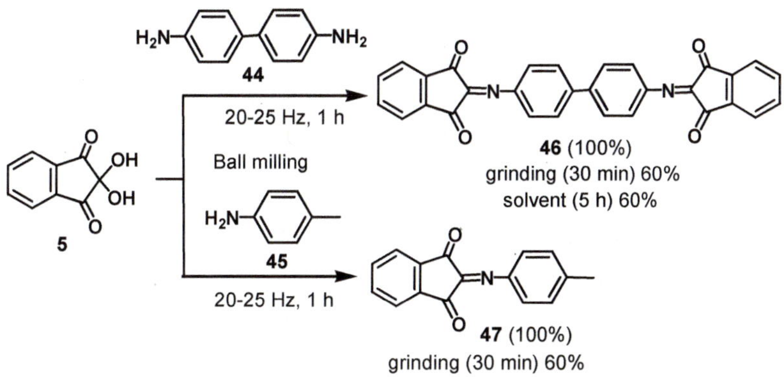

SCHEME 3.10

Ninhydrin substitutions.

LAG of an equimolar mixture of 2-hydroxy-1-naphthaldehyde and 5-aminoisophthalic acid in Retsch MM200 grinder mill afforded imine product, 5-[(2-oxo-2H-naphthalen-1-ylidenemethyl)-amino]-bezene-1,3-dicarboxylic acid (Scheme 3.11). Kaitner et al. noted the dependence of the solvent used for LAG on the reaction time needed for full conversion: milling with methanol, pyridine, and DMF took 50, 40, and 20 min, respectively [11]. Solution-phase synthesis in methanol at room temperature was complete in 30 min. PXRD and X-ray diffraction analysis revealed that variation of solvents used for LAG led to imine product (methanol), or its solvates (pyridine and DMF).

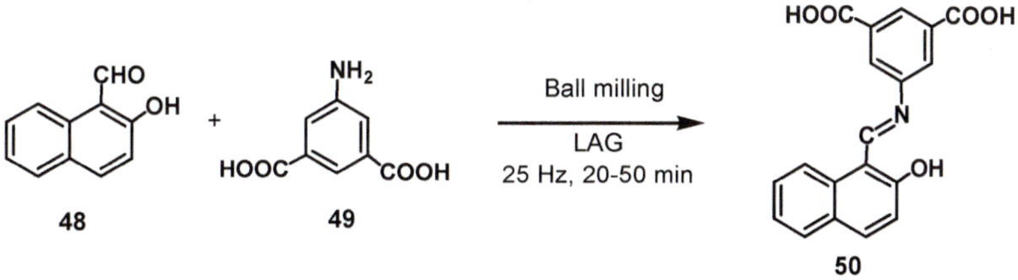

SCHEME 3.11

Imine synthesis by liquid-assisted grinding.

3.3 SYNTHESIS OF (THIO)SEMICARBAZONES

Imino bond was also constructed when thiosemicarbazone was used as an amine component. For instance, 2,5-dimethoxybenzaldehyde thiosemicarbazone **53** was prepared in high yield by ball milling of 2,5-dimethoxybenzaldehyde **51** and thiosemicarbazide **52** in a QM-3B shaking mill (Scheme 3.12). Yuan et al. used planetary and shaking ball mills for mechanosynthesis of thiosemicarbazone [12]. Among the two types of ball mills, the shaking type provides higher mechanical energy for driving a complete reaction and pure product was obtained. This method did not require any solvents or salts. In planetary ball mill, the starting materials remained unreacted even after 1 h of milling. The use of liquid- (or ion)-assisted grinding in the presence of EtOH, MeOH, DMF, or anhydrous Na_2SO_4, led to incomplete reactions.

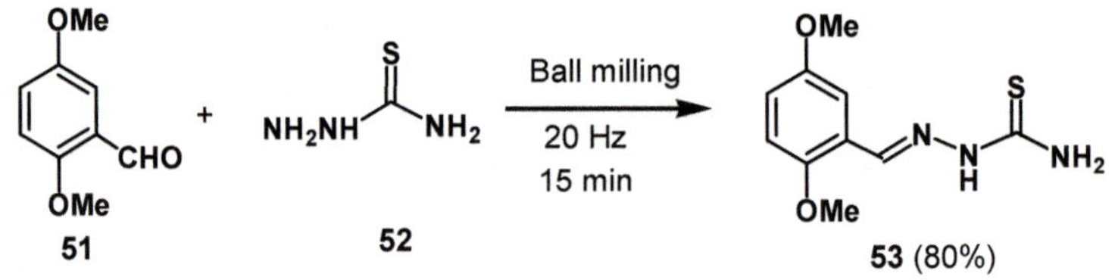

SCHEME 3.12

Mechanosynthesis of thiosemicarbazone.

Analogous synthesis of semicarbazones under solvent-free conditions was carried out in ball mill by Naimi-Jamal et al. They reported an efficient synthesis by milling of stoichiometric amounts of an aldehyde (or ketone, including cyclohexanone and (+)-carvone) and semicarbazide hydrochloride **55** in Retsch MM200 mill (10 mL stainless steel vessel, 2×12 mm balls) (Scheme 3.13) [13]. Grinding was followed by treatment with aqueous $NaHCO_3$ to afford essentially pure products **56** in quantitative yield (Table 3.4). The method is superior to the other methods for synthesis: solution, microwave, and mortar and pestle grinding. For instance, benzaldehyde reaction in solution (methanol and NaOAc) provided the corresponding semicarbazone in 78% yield.

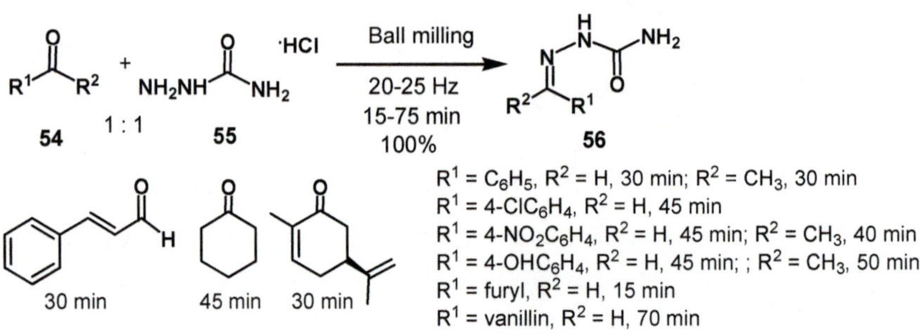

SCHEME 3.13

Synthesis of semicarbazones.

Difference in reactivity of aldehydes and ketones was used for selective protection of aldehydes in the presence of a ketone functional group. Chemoselectivity is demonstrated by two competitive reactions. When a 1:1 mixture of *p*-hydroxy- or *p*-nitrobenzaldehyde and *p*-hydroxy- or *p*-nitroacetophenone was allowed to react with 1.0 equiv. of semicarbazide hydrochloride for 45 min, the aldehydes were transformed quantitatively to their corresponding semicarbazones **59**, whereas the ketones remain unreacted (Scheme 3.14).

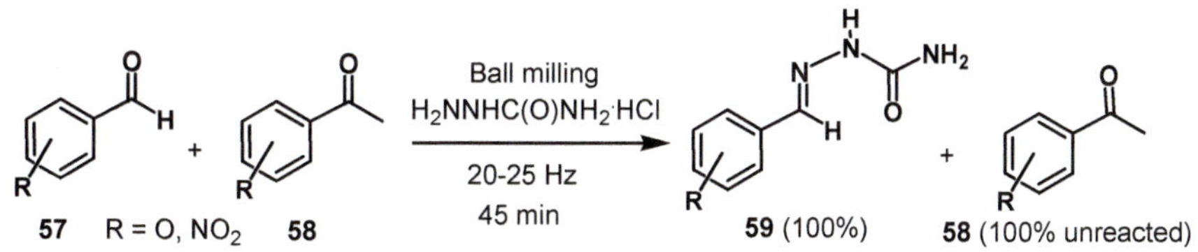

SCHEME 3.14

Chemoselective synthesis of semicarbazones.

3.4 SYNTHESIS OF OXIMES

Naimi-Jamal, Kaupp et al. stoichiometrically ball milled aldehydes and ketones with hydroxylamine hydrochloride to obtain the hydrated oxime salts **61**, from which the free oximes **62** were obtained in quantitative yield after treatment with base (Scheme 3.15) [14]. The only waste produced in quantitative reactions was stoichiometric NaCl in water. Milling times were 5–10 min by heating at 45–50°C, and 30 min at 60–65°C, whereas benzophenone required milling for 90 min at 140°C for complete reaction (selected results, Table 3.4). Without heating of the mill, the milling times ranged from 1 to 2 h. This result underlines the superiority of the heating/cooling device of the mill. Oxime hydrochlorides were simply washed with a solution of $NaHCO_3$ and water to obtain the free oximes, and this technique gave superior results to the previous literature examples.

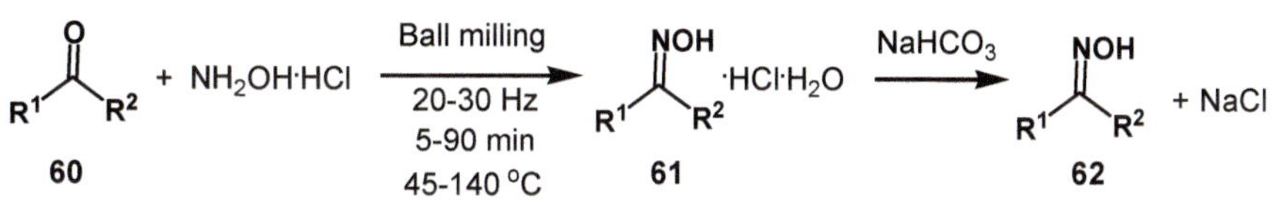

SCHEME 3.15

Synthesis of oximes.

3.5 SYNTHESIS OF HYDRAZONES

Using the same milling equipment, mechanochemical reactions of aldehydes with 2,4-dinitrophenylhydrazine **64** at 65–70°C afforded 2,4-dinitrophenylhydrazones **65** in quantitative yields within 10–20 min (Scheme 3.16) [14]. The less-reactive ketones required milling at higher temperature and milling times up to 1 h for quantitative

Table 3.4 Synthesis of Oximes Under Ball-Milling Conditions[a]

Entry	R^1	R^2	T (°C)	Time (min)	Yield (%)[b]
1	H	C_6H_5	45–50	10	100
2	H	4-HO-C_6H_4	25–30	120	95
			45–50	10	100
3	H	4-Cl-C_6H_4	25–30	60	100
			45–50	10	100
4	H	2-Cl-C_6H_4	45–50	10	100
5	H	4-Br-C_6H_4	45–50	10	100
6	H	4-CN-C_6H_4	45–50	10	100
7	H	4-NO_2-C_6H_4	45–50	5	100
8	H	4-CH_3-C_6H_4	45–50	10	100
9	H	4-N$(CH_2)_2$-C_6H_4	45–50	10	100
10	H	2-OCH_3-C_6H_4	45–50	10	100
11	H	*i*Pr	45–50	10	100
12	H	$CH_3CH{=}CH$	45–50	10	100
13	H	2-Furyl	45–50	10	100
14	H	$PhCH{=}CH_2$	45–50	10	100
15	H	$CH_3(CH_2)_6$	45–50	10	100
16	CH_3	C_6H_5	60–65	30	100
17	CH_3	2-OCH_3-C_6H_4	65–70	30	75
18	CH_3	$CH_3(CH_2)_4$	65–70	30	30
19	C_6H_5	C_6H_5	140	90	100

[a]*Retsch MM200 swing mill, 20–25Hz, stainless steel jar (10mL) with fittings for circulating liquids, 2×12mm milling balls, aldehyde:hydroxylamine 1:1 mol.*
[b]*Isolated yield.*

reactions (selected results, Table 3.5). To minimize the potential explosiveness of dry 2,4-dinitrophenylhydrazine, desensitized wetted crystals were used. After milling, all the 2,4-dinitrophenylhydrazones were in the pure form, while only two of the products were recrystallized from ethanol. When these reactions are performed in solution, strong acid catalysis is required but they occur rapidly in ball mill without the catalyst.

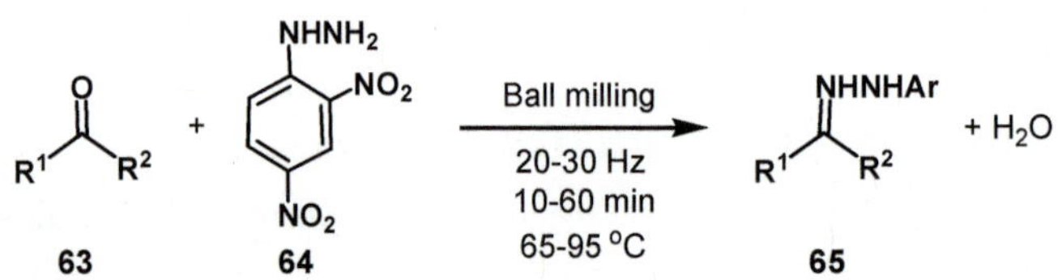

SCHEME 3.16

Synthesis of 2,4-dinitrophenylhydrazones.

Table 3.5 Synthesis of 2,4-Dinitrophenylhydrazones Under Ball-Milling Conditions[a]

Entry	R^1	R^2	T (°C)	Time (min)	Yield (%)[b]
1	H	C_6H_5	65–70	10	100
2	H	4-HO-C_6H_4	65–70	10	100
3	H	4-Cl-C_6H_4	65–70	10	100
4	H	4-NO_2-C_6H_4	65–70	20	100
5	H	4-CH_3-C_6H_4	65–70	10	100
6	H	4-$N(CH_2)_2$-C_6H_4	65–70	10	100
7	H	2-OCH_3-C_6H_4	65–70	10	100
8	H	*i*Pr	65–70	10	100
9	H	$CH_3CH{=}CH$	25–30	15	100
10	H	2-Thiophenyl	65–70	10	100
11	H	$PhCH{=}CH_2$	65–70	10	100
12	H	$CH_3(CH_2)_6$	65–70	20	100
13	CH_3	C_6H_5	80–85	40	100
14	CH_3	4-OCH_3-C_6H_4	80–85	60	58
15	CH_3	$CH_3(CH_2)_4$	75–80	15	100
16	C_6H_5	C_6H_5	90–95	60	0

[a]*Retsch MM200 swing mill, 20–25 Hz, stainless steel jar (10 mL) with fittings for circulating liquids, 2 × 12 mm balls, aldehyde: 2,4-dinitrophenyhydrazine 1:1 mol.*
[b]*Isolated yield.*
Copyright (2009), Wiley. Used with permission from Mokhtari et al. ChemSusChem 2009;2:248–54.

The different reactivities of aldehydes and ketones with hydroxylamine hydrochloride were used as a method for protecting aldehydes by oximation in the presence of a ketone functional group. The competitive reactions with 1:1 mixtures of *o*-nitro- or *p*-nitrobenzaldehyde and *o*-nitro- or *p*-nitroacetophenone when milled with 1 equiv. of $NH_2OH{\cdot}HCl$ for 60 min at 25°C afforded only the transformation of aldehydes to their corresponding oximes, whereas the ketones did not react. In a similar manner the difference in temperatures needed for the conversion of ketones into the 2,4-dinitrophenylhydrazones was used for chemoselective reaction of aldehydes. As an example, milling of 1:1 mixture of *p*-nitrobenzaldehyde and *p*-nitroacetophenone with 2,4-dinitrophenylhydrazine and water at 70°C for 50 min produced quantitative transformation of aldehyde to its 2,4-dinitrophenylhydrazone, whereas the ketone remained unaffected.

Mechanosynthesis was also used by Lamaty et al. to prepare a set of Boc-, Bz-, Fmoc-, and tosyl-hydrazones **68** from aldehydes by employment of Retsch MM200 mill (5 mL stainless steel jar, 2×5 mm balls). Aromatic, heteroaromatic, and alkyl aldehydes readily reacted under solvent-free conditions with stoichiometric amounts of hydrazines obtaining products quantitatively (selected results are given in Scheme 3.17) [15]. When the reaction was complete, the product was recovered as a solid directly in the jar and dried. The mechanochemical activation was shown to be essential to obtain a quantitative conversion: with only a magnetic stirring in round-bottom flask, the highest conversion in hydrazone after 24h was only 41%. This mechanochemical method is applicable to larger scale which was demonstrated with the preparation of 1.7g of hydrazone in 120min and using the same grinding jars. Some ketones could be also transformed to hydrazones by applying this procedure. Usually condensations on aldehydes are quite easy, but it is not always the case with ketones, as heating the reaction mixture is often required. In contrast, mechanosynthesis with ketones was carried out at room temperature with full conversion.

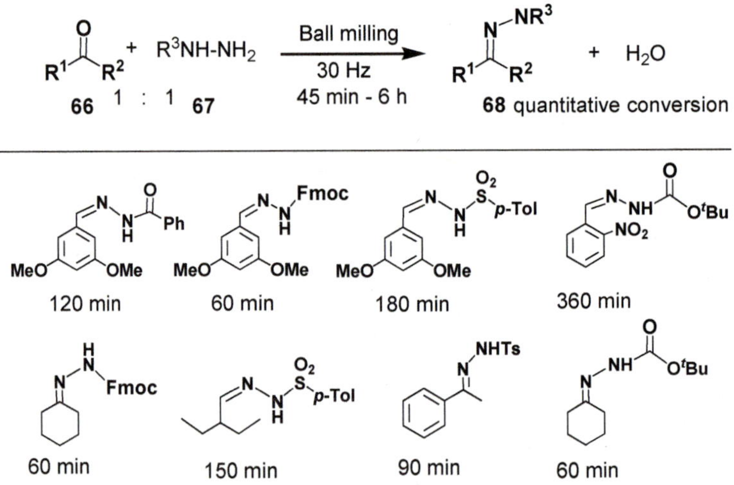

SCHEME 3.17

Preparation of hydrazones from hydrazines and aldehydes.

Phenolic hydrazone derivatives **71** were effectively prepared in analogous manner by grinding phenolic aldehydes and amines, without any reagent or additive by Baltas et al. (Scheme 3.18) [16]. Ball milling of stoichiometric amounts of reagents for 4h afforded products **71** in high conversions; many of them were quantitative (Table 3.6). Further improvement in conversion was obtained in some reactions by longer milling time (8h). After the grinding time the powder was recovered to be analyzed without any purification and no side product was detected. It is notable that unusually large single stainless steel ball was used, weighing 500g. The mechanochemical route was found to be more efficient when compared the results with the classical ones under reflux and heat.

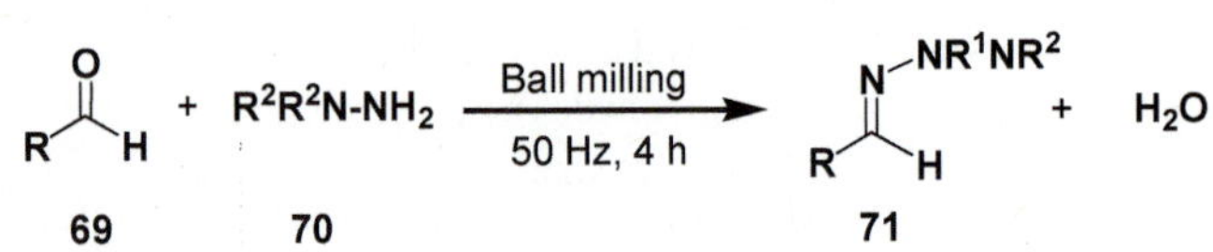

SCHEME 3.18

Preparation of hydrazones from hydrazines and aldehydes.

Table 3.6 Preparation of Hydrazones From Hydrazines and Aldehydes[a]

	Hydrazines				Conversion	(%)[b]
Aldehydes	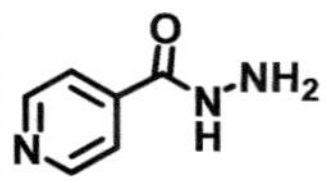	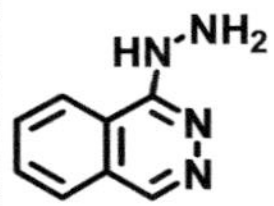	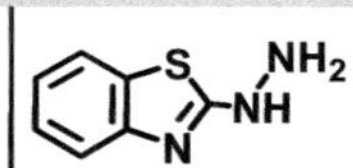	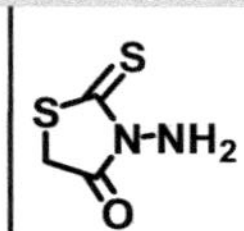	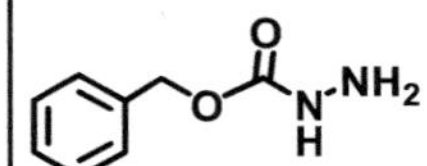	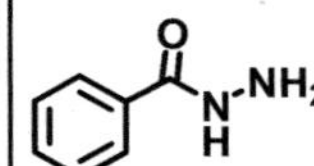
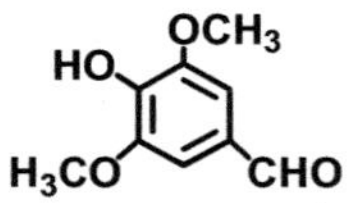	90 (97)[c]	99 (75)	92 (97)	88 (70) ≥99[d]	≥99 (90)	≥99
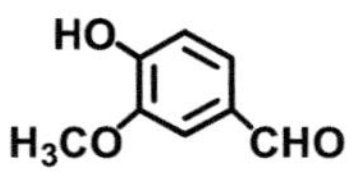	90 (90)	99 (90)	94 (99)	84 (37) ≥99[d]	≥99 (99)	≥99
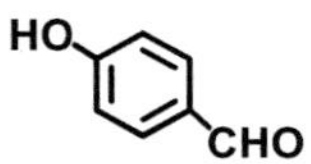	95	99	95	93 ≥99[d]	≥99	≥99
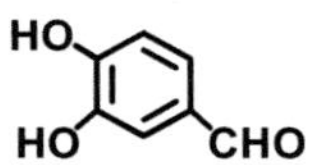	88 (81)	99 (81)	98 (74)	86≥99[d]	≥99 (99)	≥99

[a]*Fritsch vibratory ball mill Pulverisette 0, milling vessel (9.5cm diameter), 50mm stainless steel ball (500g), 50Hz, 4h, hydrazine:aldehyde 1:1mol.*
[b]*By NMR.*
[c]*Yields (%) from classical methods, reflux 4–8h.*
[d]*Conversions, grinding 8h.*
Reproduced from Oliveira PFM, Baron M, Chamayou A, André-Barrés C, Guidetti B, Baltas M. Solvent-free mechanochemical route for green synthesis of pharmaceutically attractive phenolhydrazones. RSC Adv 2014;4:56736–42, with permission from the Royal Society of Chemistry.

Liquid–solid and solid–solid condensations of aromatic aldehydes with anthranilhydrazide **72** carried out by Fülöp et al. in a mixing ball mill produced hydrazones **74** (Scheme 3.19) [17]. The use of equimolar amounts of reactants yielded hydrazones quantitatively after 2h of milling (Table 3.7). It was noted that during the milling time the temperature of the reaction vessel reached 60–70°C. Problems with the presence of iron particles abraded from the stainless steel balls during the long milling process were circumvented by use of ZrO_2-milling balls. Under these conditions, various hydrazones could be synthesized with high conversion (up to 99%) and high isolated yields (Table 3.7). Separation of the material was achieved by suspension of the reaction mixtures in *n*-hexane, and collection of the solid products by filtration. Similarly to the analogous reactions performed in solution, the solvent-less mechanochemical condensations of **72** with 1 equiv. of aromatic aldehydes led to the corresponding hydrazones. The formation of the 1:2 products of quinazolinone-type **73** was below the limit of detection (<1%).

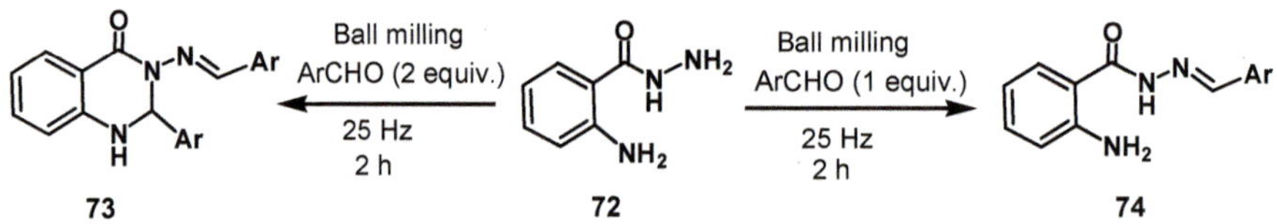

SCHEME 3.19

Synthesis of hydrazones and quinazolinones.

Table 3.7 Mechanochemical Synthesis of Hydrazones and Quinazolinones[a]

	Hydrazones 74		Quinazolinones 73		Ph-quinazolinones 76	
Ar	Entry[c]	Yield (%)[b]	Entry[d]	Yield (%)[b]	Entry[c]	Yield (%)[b]
4-$NO_2C_6H_4$	1-1	93	1-2	–	1-3	90 (96)[e]
4-ClC_6H_4	2-1	93	2-2	90	2-3	86 (94)
4-MeC_6H_4	3-1	91	3-2	92	3-3	86 (93)
4-$MeOC_6H_4$	4-1	90	4-2	94	4-3	90 (95)
3,4-$(MeO)_2C_6H_3$	5-1	92	5-2	92	5-3	90 (94)
4-$Me_2NC_6H_4$	6-1	95	6-2	-	6-3	90 (97)
2-naphthyl	7-1	93	7-2	91	7-3	42 (50)
4-PhC_6H_4	8-1	92	8-2	90	8-3	30 (50)
2-thienyl	9-1	90	9-2	92	9-3	94 (97)

[a]*Retsch MM400 mixer mill, 25Hz, 25°C, two ZrO_2 balls 15mm in diameter, stainless steel jar.*
[b]*Isolated yields.*
[c]*Equimolar amounts of reactants.*
[d]*Excess of aldehyde.*
[e]*Conversion, estimated from ^1H NMR.*
Reproduced from Chem Heterocycl Comp 2015;50:1463–9, with permission of Springer.

However, the uncatalyzed solid-state milling reactions of anthranilhydrazide with a twofold excess of aldehydes in a vibrational mill under the above conditions furnished the double-condensed quinazolinone products **73** in high yields (>90%). In the cases of some aldehydes (Ar=4-$NO_2C_6H_4$, or 4-$Me_2NC_6H_4$), only hydrazones were formed instead of quinazolinones. Unsymmetrical phenyl quinazolinone products **76** were obtained starting from phenyl anthranilhydrazide (2-amino-*N′*-phenylbenzohydrazide) **75**. Almost quantitative conversions and yields were obtained, except in the cases of naphthyl and biphenyl aldehydes (Scheme 3.20, Table 3.7 entries 1-3 to 9-3). The moderate conversions in these cases could be explained by the steric effects of the bulky groups.

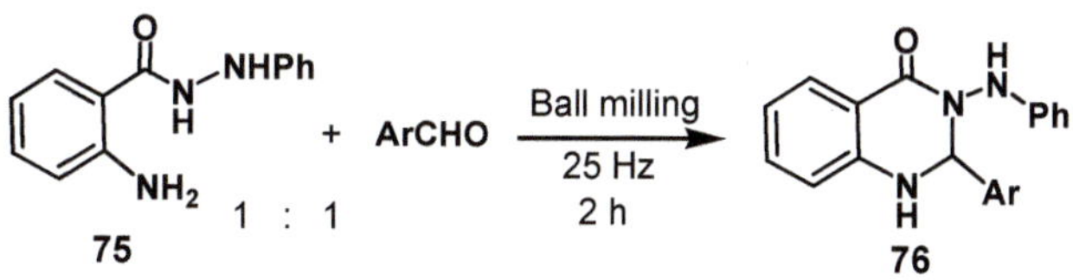

SCHEME 3.20

Synthesis of *N′*-phenyl-quinazolinones.

Another example of mechanochemical preparation of hydrazones is given by Kaupp et al. [18]. They found that the use of common acid catalysis may be avoided in the quantitative synthesis of hydrazones, and if the reaction partners are used in a 1:1 ratio, no waste is formed in the 100% yield reactions. The formation of iminium salts is avoided in this approach. Ball milling of benzhydrazide **77** and the solid aldehydes for 1 h gave pure hydrazones **81** (Scheme 3.21). Milling of benzhydrazide with isatin requires 3 h for a complete reaction to give the corresponding hydrazone **79** as a single regioisomer (in solution yield is 75%).

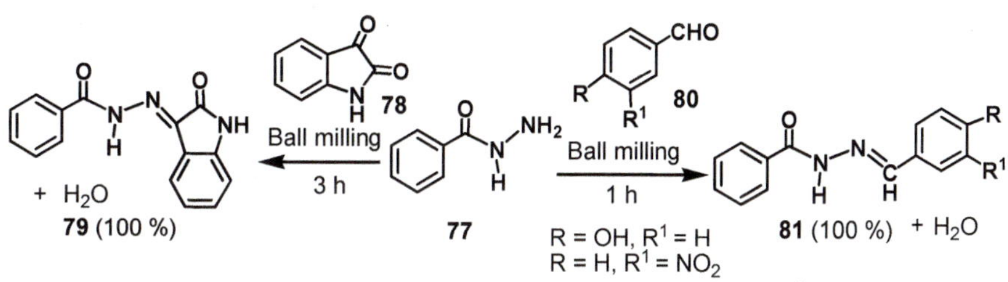

SCHEME 3.21

Synthesis of benzoylhydrazones.

3.6 SYNTHESIS OF AZINES

Kaupp and Schmeyers subjected to ball milling the solid aldehydes and ketones with the hydrazine–hydroquinone complex **83** in a 2:1 ratio to obtain azines **84** at room temperature (Scheme 3.22) [19]. Experiments were carried out in Retsch MM200 with 10 mL stainless steel beakers and two balls (6.5 g). Solid-state reactions of aldehydes with hydrazine monohydrochloride remain largely incomplete; while all of the aldehydes underwent complete reaction with hydrazine–hydroquinone complex in 1 h, the less-reactive 4-nitroacetophenone required a longer milling time (3 h). These solid-state double condensations are more effective than the reported solution reactions, and hydroquinone was easily extracted from the reaction mixture with water and recycled.

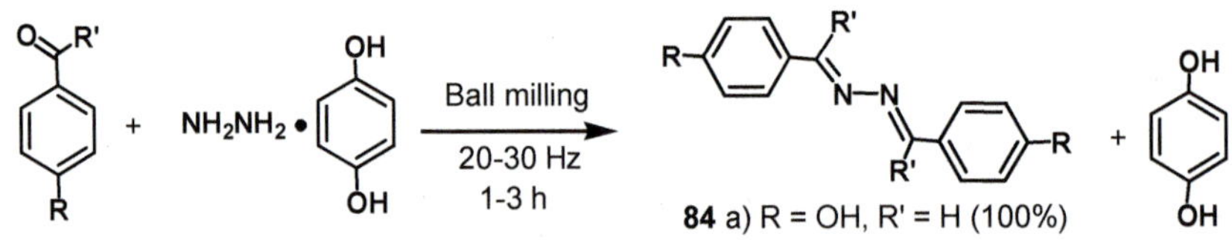

82 **83** **84** a) R = OH, R' = H (100%) b) R = Cl, R' = H (94%) c) R = $(NCH_3)_2$, R' = H (92%) d) R = NO_2, R' = CH_3 (91%) **85**

SCHEME 3.22

Solid-state synthesis of azines.

Solid reactants were also used for preparation of methylene azines in ball mill. The methylene azine **88** was quantitatively obtained in mechanochemical methylenation reaction from the solid hydrazone **86** using solid paraformaldehyde as a source of formaldehyde (Scheme 3.23) [18].

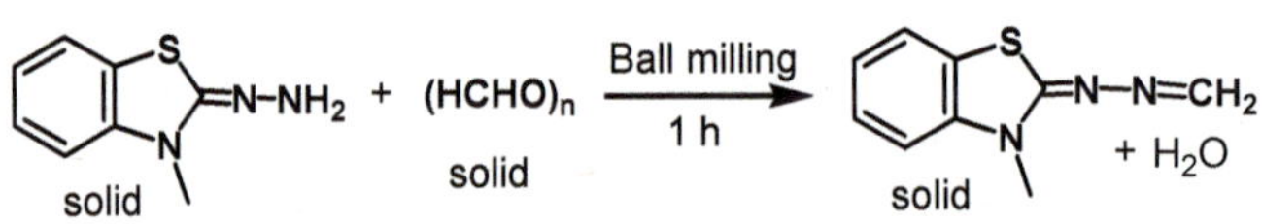

86 **87** **88** (100%)

SCHEME 3.23

Synthesis of methylene azine.

3.7 AMIDE AND THIOAMIDE BOND FORMATION

Different synthetic methods which were used for amide and thioamide bond formation in solution synthesis were transferred in solid-state ball-milling conditions. These include peptide-coupling reagents, addition of cumulenes, oxidative amination, and acylation.

3.7.1 COUPLING OF ACTIVATED ACIDS AND AMINES

Solvent-free mechanochemical procedure for the synthesis of di- and tripeptides starting from Boc-protected α-amino acid *N*-carboxyanhydrides (Boc-AA-NCA), α-amino acid alkyl ester hydrochlorides (HCl·H-AA-OR) and $NaHCO_3$ was developed by Lamaty et al. (Scheme 3.24) [20]. Here, urethane-protected α-amino acid

N-carboxyanhydrides (UNCAs) **89** are used as the activated forms of amino acids for peptide synthesis, which renders the use of classical peptide-coupling reagents. Ball milling of equimolar amounts of UNCA and amino esters **90** provided di- and tripeptides **91** with high conversion and yields (Table 3.8). Employed ball-milling conditions were found to be non-racemizing since no epimerization during the course of milling was detected. This straightforward, high-yielding method has been used in synthesis of protected aspartame by ball milling of Boc-Asp(OtBu)-NCA and HCl·H-Phe-OMe (entry 13). Fully solvent-free method does not use organic solvents, neither in the coupling reaction, nor for protecting group removal. During the course of reaction volatile side product (CO_2) and water-soluble inorganic salts were formed, which are removed by simple precipitation of the final product from water.

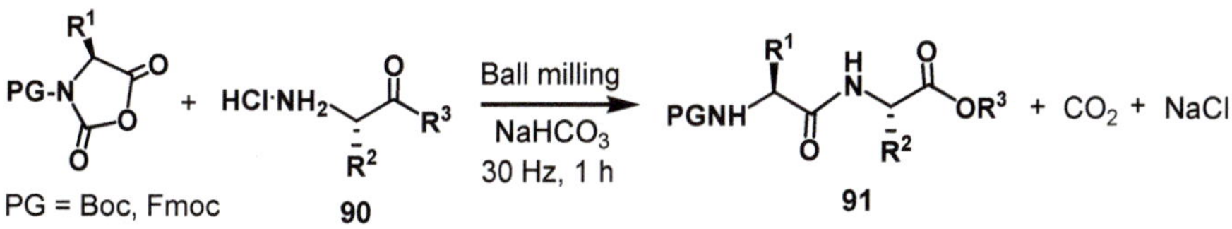

SCHEME 3.24

Synthesis of dipeptides and tripeptides.

Table 3.8 Synthesis of Dipeptides and Tripeptides (Selected Examples)[a]

Entry	UNCA	Amino Ester	Dipeptide/ Tripeptide	Conversion (%)[c]	Yield (%)[b]
1	Boc-Val-NCA	HCl·H-Leu-OMe	Boc-Val-Leu-OMe	100	87
2		HCl·H-Leu-O*t*Bu	Boc-Val-Leu-O*t*Bu	97	85
4		HCl·H-Ala-O*t*Bu	Boc-Val-Ala-O*t*Bu	100	100
5		HCl·H-Phe-OMe	Boc-Val-Phe-OMe	100	88
6	Fmoc-Val-NCA	HCl·H-Leu-OMe	Fmoc-Val-Leu-OMe	90	–
7		HCl·H-Ala-OMe	Fmoc-Val-Ala-OMe	100	76
8	Boc-Phe-NCA	HCl·H-Leu-OMe	Boc-Phe-Leu-OMe	85	–
9		HCl·H-Ala-OMe	Boc-Phe-Ala-OMe	99	79
10		HCl·H-Ala-O*t*Bu	Boc-Phe-Ala-O*t*Bu	100	73
11		HCl·H-Phe-OMe	Boc-Phe-Phe-OMe	58	55
12	Boc-Val-NCA	HCl·H-Ala-OMe	Boc-Val-Ala-Gly-OMe	100	89
13	Boc-Asp (O*t*Bu)-NCA	HCl·H-Phe-OMe	Boc-Asp(O*t*Bu)-Phe-OMe	100	97

[a]*Retsch 200 mixer mill, 30 Hz, stainless steel jar (10 mL) 1 × 10 mm ball, UNCA:amino ester:NaHCO$_3$ 1:1:1.5 mol, 1 h.*
[b]*Isolated yield.*
[c]*Estimated by NMR.*

Lamaty et al. used liquid-assisted ball milling (ie, LAG) to overcome problems associated with scale up of peptide synthesis using Boc-Phe-NCA and HCl·H-Leu-OMe reagents in conjunction with $NaHCO_3$ [21]. Increase of milling load used in their previous article (see above) led to decreased yield, and formation of Boc-Phe-OH **95** resulting from Boc-Phe-NCA **92** hydrolysis (Scheme 3.25). This result was explained by the formation of highly viscous and sticky reaction media that leads to mass transfer limitations when excessive material was in the reactor, thereby obstructing the global speed of dipeptide formation. This hypothetical mass-transfer limitation may also account for the formation of **95** resulting from hydrolysis of **92**. Low-toxicity ethyl acetate was used for LAG ($\eta = 1.4\,\mu L\,mg^{-1}$), which fully suppressed hydrolysis reaction and afforded product in 95% yield. After reaction completion, mixture was diluted with EtOAc and washed with sodium carbonate and diluted acid.

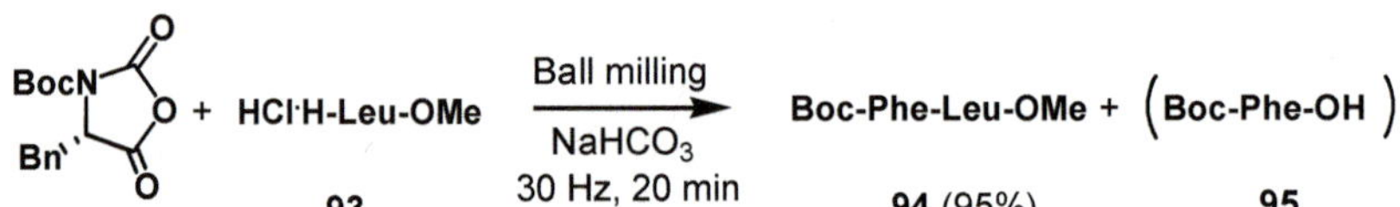

92 Boc-Phe-NCA 93 94 (95%) 95

SCHEME 3.25

Synthesis of Boc-Phe-Leu-OMe.

Optimal reaction conditions were used for ball milling of stoichiometric amounts of variety of Boc-protected α-amino acid *N*-carboxyanhydrides (Boc-AA-NCA) **92** or Boc-protected α-amino acid *N*-hydroxysuccinimide esters (Boc-AA-OSu) **95** with α-amino acid alkyl ester salts (Schemes 3.26 and 3.27, Tables 3.9 and 3.10). In this way, di- to pentapeptides **97** and **100** were produced in high yield and environmentally benign manner. For some of the reactions, *t*BuOAc was used as a grinding auxiliary. Furthermore, it was established that no racemization could be observed as the diastereomeric excess of some products was checked and was found to be superior to 98%. After completion of Boc-AA-OSu reactions, mixture was milled for 5 min at 30 Hz with aqueous NaOH solution, diluted with EtOAc, and washed with sodium carbonate and diluted acid. Pentapeptide Boc-Tyr(Bn)-Gly-Gly-Phe-Leu-OBn product obtained mechanochemically was successfully converted to Leu-enkephalin by classical methods.

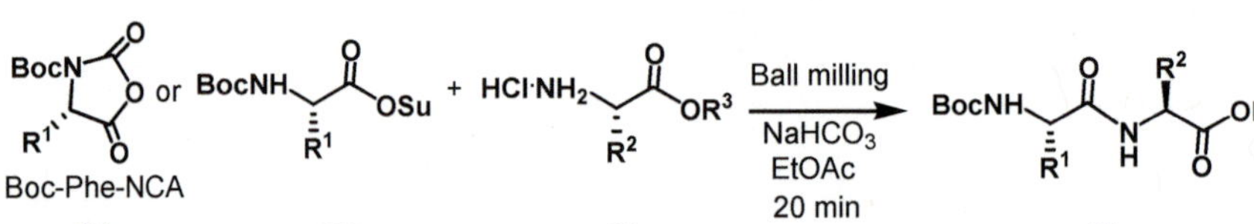

92 95 96 97

SCHEME 3.26

Synthesis of dipeptides.

Table 3.9 Synthesis of Dipeptides[a]

Entry	Boc-AA-NCA or Boc-AA-OSu	AH·H-AA-OR 95	Dipeptides 97	Yield (%)[b]
1	Boc-Leu-NCA	*p*-TsOH·H-Leu-OBn	Boc-Leu-Leu-OBn	93
2	Boc-Val-NCA	HCl·H-Phe-OMe	Boc-Val-Phe-OMe	84
3	Boc-Leu-NCA	HCl·H-Pro-OBn	Boc-Leu-Pro-OBn	90
4	Boc-Tyr(Bn)-OSu	HCl·H-Leu-OMe	Boc-Tyr(Bn)-Leu-OMe	98[c]
5	Boc-Phe-OSu	HCl·H-Phe-OMe	Boc-Phe-Phe-OMe	92
6	Boc-Phe-OSu	HCl·H-Leu-OMe	Boc-Phe-Leu-OMe	83
7	Boc-Phe-OSu	*p*-TsOH·H-Leu-OBn	Boc-Phe-Leu-OBn	90[d] 80[e]

[a]*Retsch 200 mixer mill, 30 Hz, stainless steel jar (10 mL), 1 × 10 mm ball, Boc-AA-NCA:amino acid.HCl: $NaHCO_3$ 1:1:1 mol, EtOAc ($1.1 < \eta < 1.4\ \mu L\,mg^{-1}$), 20 min, $22.0 < ML < 28.4\ mg\,mL^{-1}$.*
[b]*Isolated yield.*
[c]*1 h.*
[d]*$ML = 28.4\ mg\,mL^{-1}$, $\eta = 1.1\ \mu L\,mg^{-1}$.*
[e]*$ML = 188.8\ mg\,mL^{-1}$, $\eta = 1.1\ \mu L\,mg^{-1}$, 2 h.*
Reproduced from Bonnamour J, Métro T-X, Martinez J, Lamaty F. Environmentally benign peptide synthesis using liquid-assisted ball-milling: application to the synthesis of Leu-enkephalin. Green Chem 2013;15:1116–20, with permission from the Royal Society of Chemistry.

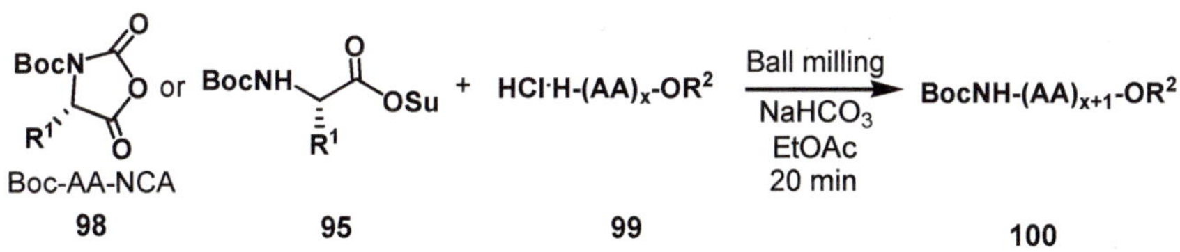

SCHEME 3.27

Synthesis of tri-, tetra-, and pentapeptides.

Table 3.10 Synthesis of Tri-, Tetra-, and Pentapeptides[a]

Entry	Boc-AA-NCA or Boc-AA-OSu	HCl·H-(AA)$_x$-OR	Product 100	Yield (%)[b]
1	Boc-Leu-NCA	HCl·H-Leu-Leu-OBn	Boc-Leu-Leu-Leu-OBn	79
2	Boc-Phe-OSu	HCl·H-Phe-Phe-OMe	Boc-Phe-Phe-Phe-OMe	86
3	Boc-Lys(Boc)-OSu	HCl·H-Gly-Pro-OBn	Boc-Lys(Boc)-Gly-Pro-OBn	79
4	Boc-Trp-NCA	HCl·H-Val-Phe-OMe	Boc-Trp-Val-Phe-OMe	86
5	Boc-Gly-OSu	HCl·H-Gly-Phe-OMe	Boc-Gly-Gly-Phe-OMe	74
6	Boc-Gly-OSu	HCl·H-Phe-Leu-OBn	Boc-Gly-Phe-Leu-OBn	93[c] 94[d]

Continued

Table 3.10 Synthesis of Tri-, Tetra-, and Pentapeptides[a]—cont'd

Entry	Boc-AA-NCA or Boc-AA-OSu	HCl·H-(AA)$_x$-OR	Product 100	Yield (%)[b]
7	Boc-Gly-OSu	HCl·H-Gly-Phe-Leu-OBn	Boc-Gly-Gly-Phe-Leu-OBn	96[e] 90[f]
8	Boc-Tyr(Bn)-OSu	HCl·H-Gly-Gly-Phe-Leu-OBn	Boc-Tyr(Bn)-Gly-Gly-Phe-Leu-OBn	88[g]

[a]*Retsch 200 mixer mill, 30 Hz, stainless steel jar (10mL), 1×10mm ball, Boc-AA-NCA:amino acid.HCl: $NaHCO_3$ 1:1:1 mol, EtOAc ($1.3<\eta<1.6\,\mu L\,mg^{-1}$), 20min, $6.6<ML<8.4\,mg\,mL^{-1}$.*
[b]*Isolated yield.*
[c]*$ML=22.9\,mg\,mL^{-1}$.*
[d]*$ML=152.7\,mg\,mL^{-1}$, 40min, tBuOAc.*
[e]*$ML=22.2\,mg\,mL^{-1}$, $\eta=1.1\,\mu L\,mg^{-1}$.*
[f]*$ML=60.6\,mg\,mL^{-1}$, $\eta=1.1\,\mu L\,mg^{-1}$, 40min, tBuOAc.*
[g]*1 h.*
Reproduced from Bonnamour J, Métro T-X, Martinez J, Lamaty F. Environmentally benign peptide synthesis using liquid-assisted ball-milling: application to the synthesis of Leu-enkephalin. Green Chem 2013;15:1116–20, with permission from the Royal Society of Chemistry.

Lamaty's strategy to mechanochemical green synthesis of α,β- and β,β-dipeptides was applied by Juaristi (Scheme 3.28) [22]. The coupling of urethane-protected β-amino acid *N*-carboxyanhydride (UNCA) derivatives **101** with hydrochloride salts **102/103** derived from α- and β-amino esters was carried out in the presence of $NaHCO_3$ (1.5 equiv.) in high-energy ball mill. The change in mass of the reaction capsule was monitored during the milling until the difference in mass was constant, indicating that the reaction had ended (no more CO_2 was being liberated). In general, the required reaction time was 2h. Then the mixture was dissolved in EtOAc to yield the dipeptide **104/105** of sufficient purity in high yields (Table 3.11), without racemization or epimerization.

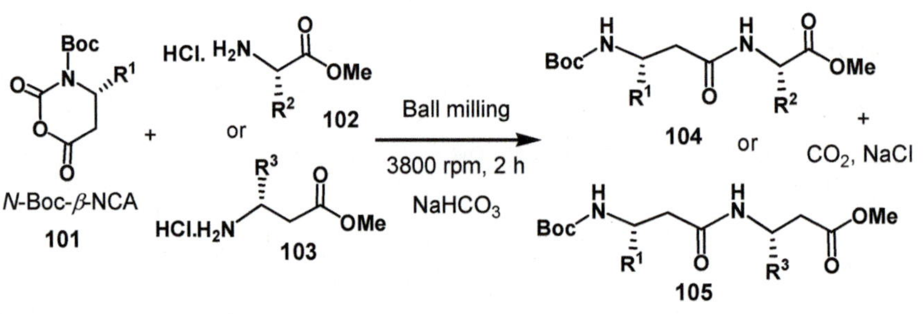

SCHEME 3.28

Synthesis of dipeptides.

Table 3.11 Synthesis of Dipeptides[a]

Entry	*N*-Boc-β-NCA	Base Amino Methyl Ester Hydrochloride	Dipeptide *N*-Boc-aa-aa-OMe	Yield (%)[b]
1	β-ala	Ala	β-ala-Ala	88
2	β-ala	Val	β-ala-Val	82
3	β-ala	Leu	β-ala-Leu	87
4	β-ala	(*S*)-β^3-hPhg	β-ala-(*S*)-β^3-hPhg	91
5	β-ala	L-His-OMe	β-ala-L-His-OMe	91[c]
6	(*S*)-β^3-hPhg	β-ala	(*S*)-β^3-hPhg-β-ala	93
7	(*S*)-β^3-hPhg	(*S*)-β^3-hPhg	(*S*)-β^3-hPhg-(*S*)-β^3-hPhg	94
8	(S)-β^3-chg-(CO_2CH_3)	Ala	(S)-β^3-chg-(CO_2CH_3)-Ala	79
9	(S)-β^3-chg-(CO_2CH_3)	β-ala	(S)-β^3-chg-(CO_2CH_3)-β-ala	91

[a]*Digital mixer/amalgamator, 25×10mm Nylamid cylinder, one 5mm stainless steel ball, 3800rpm, 2h, NCA:amino acid HCl:$NaHCO_3$ 1:1:1.5.*
[b]*Isolated products.*
[c]*$NaHCO_3$ 2 equiv.*
Reproduced from Hernández J, Juaristi E. Green synthesis of α,β- and β,β-dipeptides under solvent-free conditions. J Org Chem 2010;75:7107–11, copyright (2010), with permission from American Chemical Society.

Due to their relative insensitivity toward hydrolysis, *N*-hydroxysuccinimidyl (NHS) esters were also employed as activated esters for acylation of amines in ball mill by Vyle et al. (Scheme 3.29) [23]. Selected amines were transformed to amides employing variety of phenylazobenzoic acid-NHS esters **107** in the presence of DMAP, which provided a significantly faster reaction than DBU and a slight rate enhancement compared with 1,4-diazabicyclo[2.2.2]octane (DABCO). In mere 10min of milling products **108** were obtained in high yields (Table 3.12). Fmoc-protected amines were also obtained under ball-milling conditions using Fmoc-OSu/DMAP, with addition of EtOAc (LAG). Sodium carbonate buffers are commonly employed in the acylation of amines by NHS esters requiring a mixed aqueous organic solvent system with typically long reaction times, and the isolations are often associated with NHS-ester hydrolysis. Chemoselective acylation of these polyfunctional substrates in ball mill was rapid, completely avoided the use of usual solvents and buffers, required minimal workup, and thereby avoided undesirable *E/Z* photoisomerization often associated with azobenzene derivatives. Remarkable increase in reactivity of ball milling in comparison to conventional conditions as illustrated by the time for the reaction to complete between *p*-nitrobenzoyl-NHS and aniline at room temperature was reported to be 60days. Using ball milling, reaction times of 20 or 120min were required to effect the complete reaction to the corresponding anisidide or naphthylamide, respectively.

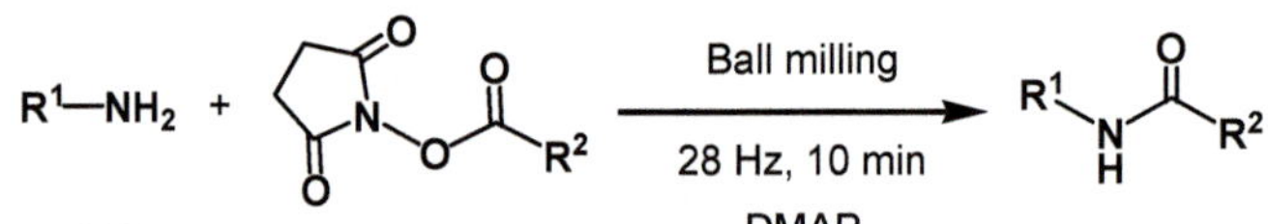

106 107 108

SCHEME 3.29

Acylation of amines.

Table 3.12 Acylation of Amines[a]

Entry	R^1	R^2	Yield (%)[b]
1	Propargyl	p-(Ph—N=N)C_6H_4	94
2		m-(Ph—N=N)C_6H_4	92
3		o-(Ph—N=N)C_6H_4	90
4	D-threoninol	p-(Ph—N=N)C_6H_4	79[c]
5		m-(Ph—N=N)C_6H_4	77[c]
6		o-(Ph—N=N)C_6H_4	83[c]
7	5′-O-(4,4′-dimethoxytrityl)- 2′-Amino-2′-deoxyuridine	p-(Ph—N=N)C_6H_4	90[c]
8	D-threoninol	Fmoc	79
9	p-MeOC_6H_4	p-(Ph—N=N)C_6H_4	78
10	1-Naphthyl	p-(Ph—N=N)C_6H_4	65

[a]*Retsch mixer mill MM400, stainless steel jar (25 mL), 15 mm ball, 28 Hz, 10 min, amine:NHS:DMAP 2:1:1.6 eq.*
[b]*Isolated yields.*
[c]*LAG, EtOAc.*
Reproduced from Ravalico F, James SL, Vyle JS. Synthesis of nucleoside analogues in a ball mill: fast, chemoselective and high yielding acylation without undesirable solvents. Green Chem 2011;13:1778–83, with permission from the Royal Society of Chemistry.

Štrukil et al. employed ball milling for the amide bond formation by one-pot, one-step reaction of aromatic carboxylic acids **109** and aromatic amines **110** in the presence of water-soluble carbodiimide reagent (EDC hydrochloride) [24]. Milling was carried out in very short reaction times to afford *N*-substituted amide derivatives **111** in quantitative yields when anhydrous nitromethane was added to assist grinding (LAG) (Scheme 3.30, Table 3.13). Full conversion allowed simple product isolation by aqueous workup (stirring of reaction mixture with water and filtration). Other standard coupling reagents used in peptide synthesis (DCC, DIC, and CDI) were less efficient. Utilization of EDC reagent was well suited to avoid any use of organic solvent for reaction and purification, since urea side product formed in the course of reaction could be simply washed off with water. Structural characterization of products by PXRD and solid-state NMR completed totally solvent-free organic synthesis.

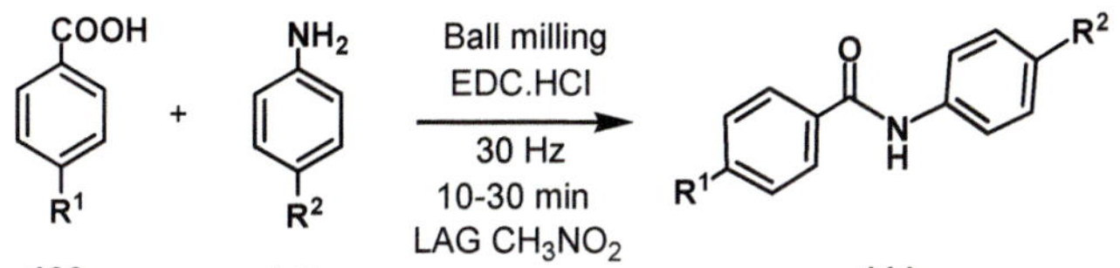

109 110 111

SCHEME 3.30

Carbodiimide-mediated synthesis of peptide bond.

Table 3.13 Carbodiimide Coupling of Aromatic Acids and Amines[a]

Entry	R_1	R_2	Time (min)	Yield (%)[b]
1	Ph	OCH_3	10	95
2		CH_3	10	97
4		F	10	95
4	1-Naphthyl	OCH_3	20	88
5		H	20	85
6		F	20	80/85[c]
7		Cl	20	80/79[c]
8	2-Naphthyl	CH_3	10	96
9		Cl	10	95

[a]*Retsch MM400 mixer mill, 30Hz, stainless steel jar (10mL), one 12mm ball, LAG: nitromethane ($\eta = 0.25\,\mu L\,mg^{-1}$), acid:amine:EDC.HCl 1:1:1 mol.*
[b]*Isolated yield aqueous workup.*
[c]*Isolated yield, column chromatography.*
Reproduced from Štrukil V, Bartolec B, Portada T, Đilović I, Halasz I, Margetić D. One-pot mechanosynthesis of aromatic amides and dipeptides from carboxylic acids and amines. Chem Commun 2012;48:12100–2, with permission from the Royal Society of Chemistry.

Reaction was also applied on coupling of aliphatic carboxylic acids and amines (Table 3.14), and the chiral bis-amides were likewise obtained from diamines without the loss of chirality, suggesting that mechanochemical conditions are applicable for the synthesis of chiral amides. In addition, several dipeptides **114** were prepared (Scheme 3.31). Racemization was not noticed which is the usual drawback of conventional solution-based procedures (Table 3.15). In dipeptide synthesis 2 equiv. of DMAP were used as the base to deprotonate the unprotected amino acid, and as the activator of EDC·HCl and sodium chloride were added as the grinding auxiliary.

Table 3.14 Carbodiimide Coupling of Acids and Amines[a]

Entry	Acid	Amine	Conditions	Time (min)	Yield (%)[b]
1	Succinic	Aniline[d]	NG	10	77
2	Fumaric	*p*-anisidine[d]	LAG	30	79
3	Benzoic	Morpholine	LAG	10	93[c]

Continued

Table 3.14 Carbodiimide Coupling of Acids and Amines[a]—cont'd

Entry	Acid	Amine	Conditions	Time (min)	Yield (%)[b]
4	Benzoic	Thiomorpholine	LAG	10	92[c]
5	Benzoic[d]	(1*R*,2*R*)-cyclohexyldiamine	LAG	30	98
6	Benzoic[d]	(1*R*,2*R*)-1,2-diphenylethylenediamine	LAG	30	88
7	Boc-Ala-OH	*p*-anisidine	LAG	30	87
8	Boc-Ala-OH	4-Chloroaniline	LAG	30	88

[a]*Retsch MM400 mixer mill, 30Hz, stainless steel jar (10mL), one 12mm ball, LAG nitromethane (η=0.25 μL mg^{-1}), acid:amine:EDC.HCl 1:1:1 mol.*
[b]*Isolated yield.*
[c]*after column chromatography.*
[d]*2 equiv.*
Reproduced from Reproduced from Štrukil V, Bartolec B, Portada T, Đilović I, Halasz I, Margetić D. One-pot mechanosynthesis of aromatic amides and dipeptides from carboxylic acids and amines. Chem Commun 2012;48:12100–2, with permission from the Royal Society of Chemistry.

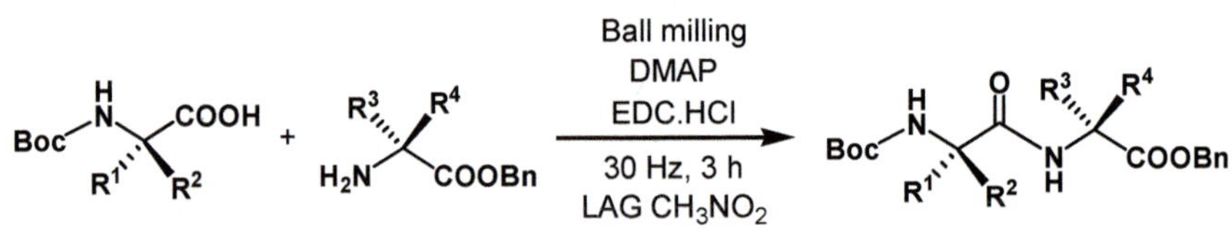

112 **113** **114**

SCHEME 3.31

Mechanosynthesis of dipeptides.

Table 3.15 Carbodiimide Synthesis of Dipeptides[a]

Entry	Acid	Amine	R^1	R^2	R^3	R^4	Yield (%)[b]
1	Boc-Gly-OH	Gly-OBn	H	H	H	H	70
2	Boc-L-Ala-OH	L-Ala-OBn	CH_3	H	H	CH_3	80
3	Boc-L-Ala-OH	Gly-OBn	CH_3	H	H	H	78
4	Boc-D-Ala-OH	Gly-OBn	H	CH_3	H	H	79
5	Boc-D-Ala-OH	L-Ala-OBn	H	CH_3	H	CH_3	81

[a]*Retsch MM400 mixer mill, 30Hz, stainless steel jar (10mL), one 12mm milling ball, LAG nitromethane (η=0.25 μL mg^{-1}), acid:amine:EDC.HCl:DMAP 1:1:1:2mol, 180min, NaCl grinding auxiliary 20 equiv.*
[b]*Isolated yield.*
Reproduced from Reproduced from Štrukil V, Bartolec B, Portada T, Đilović I, Halasz I, Margetić D. One-pot mechanosynthesis of aromatic amides and dipeptides from carboxylic acids and amines. Chem Commun 2012;48:12100–2, with permission from the Royal Society of Chemistry.

Lamaty et al. used another common peptide-coupling reagent *N,N'*-carbonyldiimidazole (CDI) in the mechanosynthesis of amides. They succeeded in the development of complete solvent-free process for amide bond formation

(Scheme 3.32) [25]. In this one-pot, two-step process, amides are prepared by initial milling of carboxylic acids **115** with CDI for 5 min, followed by the addition of the corresponding amine and milling for 10 min. After the completion of the reaction, water is added and mixture milled for further 5 min to remove by-products (imidazole, acid, and amine). Collection of product by filtration completes the fully solvent-free synthesis. By such procedure, a variety of amides **118** was obtained in high yield (Table 3.16). This methodology was not limited to the acylation of primary amine hydrochlorides; *O*-benzyl hydroxylamine, phenylhydrazine, and dibenzylamine hydrochlorides were also acylated. Furthermore, no racemization occurred in the reaction of enantiopure amines or carboxylic acids showing high enantioselectivity.

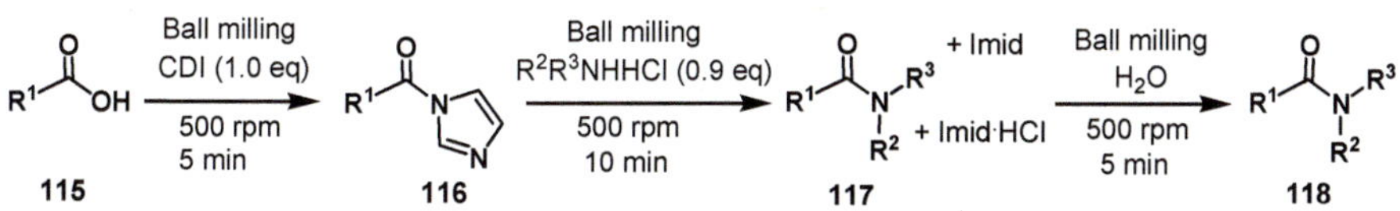

SCHEME 3.32

CDI-mediated synthesis of amides.

Table 3.16 CDI-Mediated Synthesis of Amides (Selected Examples)[a]

Entry	R^1	R^2	R^3	Yield (%)[b]
1	$PhCH_2CH_2$	Bn	H	94
2	*n*-pentyl	Bn	H	99
3	Ad	Bn	H	77
4	*o*-$BrC_6H_4CH_2$	Bn	H	69
5	*o*-CH_3-*m*-NO_2-$C_6H_3CH_2$	Bn	H	91
6	Ph	$PhCH_2CH_2$	H	96
7	Ph	2,4-$(OCH_3)_2$-$C_6H_3CH_2$	H	84
8	Ph	$PhCH_2O$	H	64
9	Ph	PhNH	H	83
10	$PhCH_2CH_2$	Bn	Bn	44[c]
11	Ph	(*S*)-*N*-1-Phenylethyl	H	68 (99 ee)
12	Ph	(*S*)-Phe-Ala-O^{t}Bu	H	78[d] (98 ee)
13	(*S*)-Boc-Phe-Ala	(*S*)-*N*-1-Phenylethyl	H	49[d] (>95 de)

[a]*Retsch PM100 planetary mill, 500 rpm, stainless steel jar (12 mL), 50 × 5 mm milling balls, acid:CDI:amine 1:1:0.9 mol, 5 + 10 + 5 min.*
[b]*Isolated yield.*
[c]*6 h.*
[d]*30 min.*
Reproduced from Métro T-X, Bonnamour J, Reidon T, Sarpoulet J, Martinez J, Lamaty F. Mechanosynthesis of amides in the total absence of organic solvent from reaction to product recovery. Chem Commun 2012;48:11781–3, with permission from the Royal Society of Chemistry.

Anilines were also effectively amidated by this procedure. Since aniline-derived acyl-imidazole intermediates are slow reacting considering addition, the removal of the solvent has an accelerating effect (Scheme 3.33, Table 3.17). The difference is most pronounced in the case of a less-reactive substrate (entry 4).

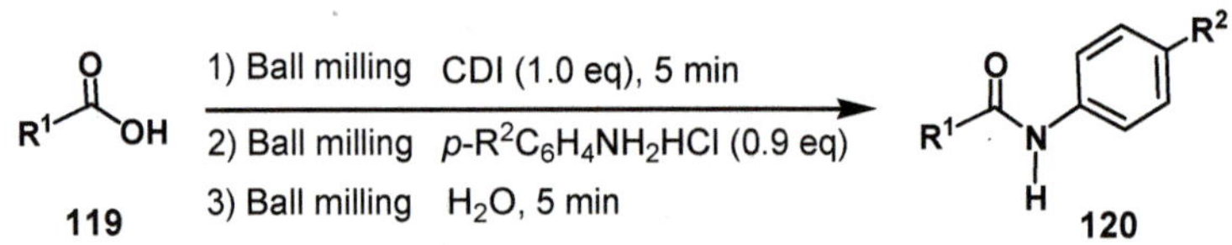

SCHEME 3.33

CDI-mediated synthesis of aniline amides.

Table 3.17 CDI-Mediated Synthesis of Aniline Amides[a]

Entry	R^1	R^2	Ball Milling	Yield (%)[b]	Solvent	Conversion (%)
1	Ph(CH$_3$)CH	C_6H_5	10 min	91	3 h[c]	86
2	Ph	C_6H_5	10 min	93	1 h[d]	>90
3	Ph	4-CF$_3$-C$_6$H$_4$	30 min	82	29 h[d]	>90
4	Ph	4-CN-C$_6$H$_4$	3.5 h	78	>143 h[d]	>90

[a]*Retsch PM100 planetary mill, 500 rpm, stainless steel jar (12 mL), 50×5 mm milling balls, acid:CDI:amine 1:1:0.9 mol.*
[b]*Isolated yield.*
[c]*Boiling EtOAc [0.39M].*
[d]*50°C, NMP [0.41M].*
Reproduced from Métro T-X, Bonnamour J, Reidon T, Sarpoulet J, Martinez J, Lamaty F. Mechanosynthesis of amides in the total absence of organic solvent from reaction to product recovery. Chem Commun 2012;48:11781–3, with permission from the Royal Society of Chemistry.

The synthesis of *N*-benzylbenzamide was scaled up to a 4.5 g (using 250 mL stainless steel jar containing 50 balls with 10 mm diameter). Initial reaction of benzoic acid with CDI finished in 5 min, and addition of benzylamine hydrochloride finished in 10 min to afford product in 96% yield. The CDI coupling was also applied to solvent-free preparation of Teriflunomide, an active pharmaceutical ingredient. The amide bond was synthesized by mechanochemical treatment of 5-methyl-4-isoxazolocarboxylic acid **121** with CDI for 20 min, followed by reaction with 4-(trifluoromethyl)aniline hydrochloride for 5 h to obtain Leflunomide **122** (Scheme 3.34). For such a long milling, the jar within the planetary mill was rotated at 500 rpm, with 1 min break every 10 min, with inversion of the rotation direction after each break. Reaction mixture was then milled for 5 min with water, and synthesis finalized by isoxazole ring opening by classical stirring with hydrochloric acid to afford Teriflunomide **123** in 81% yield.

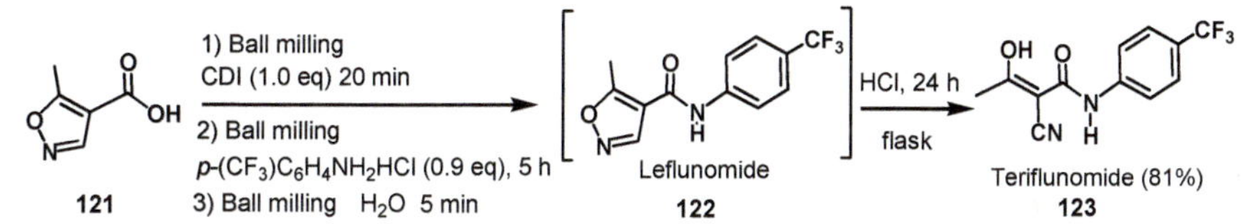

SCHEME 3.34

CDI-mediated synthesis of Teriflunomide.

The comparison of environmental impact of this synthetic approach with other reagents and conditions for the synthesis of benzylamide showed that mechanochemical CDI-mediated procedure with aqueous workup (taking into account yield, cost, safety, technical setup, temperature and reaction time, workup, and purification) has the highest Eco-scale score [26]. Furthermore, studies of nature of the milling material on reaction efficiency and metal contamination from wear of the balls and container revealed that the lowest contamination was obtained when using a planetary ball mill with the jar and balls made of agate. Contaminants could be effectively eliminated by solubilization of the product in ethyl acetate followed by filtration either through cotton or through a PTFE filter.

N,N′-carbonyldiimidazole (CDI) was also the reagent of choice by Kartha et al. for carboxylic acid activation in amide bond mechanosynthesis in glycopeptides and ureas [27]. Ball-milling method involves a pre-activation step with CDI (3 equiv.) followed by an amidation step, and both steps are performed in the same pot (Scheme 3.35). First, planetary ball milling with CDI resulted in the transformation of 1,12-dodecanedioic acid **124** to activated dicarbodiimidazolyl intermediate **125**. After introduction of amine, milling was continued to obtain two different products: glycosylated *bis*-amide **127** and urea **128**. The *p*-aminophenyl glycoside **126** required for the reaction was also prepared in a planetary ball mill from bromide by reaction with *p*-nitrophenolate generated from *p*-nitrophenol and K_2CO_3 (in 90% yield), followed by catalytic hydrogenation (Scheme 3.36). Products were isolated by simple aqueous workup and crystallization.

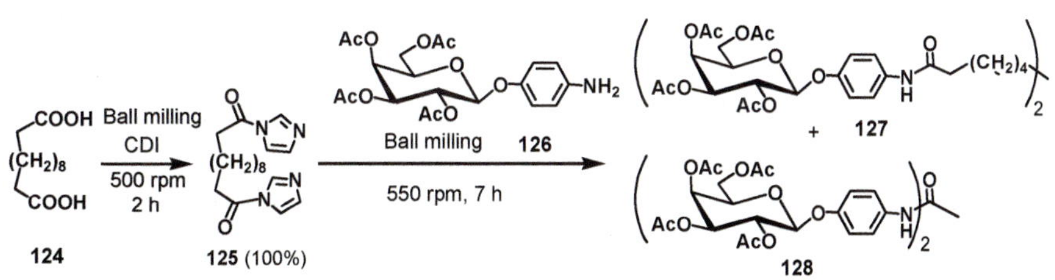

SCHEME 3.35

CDI-mediated synthesis of galactose cross-linked diamide **127**.

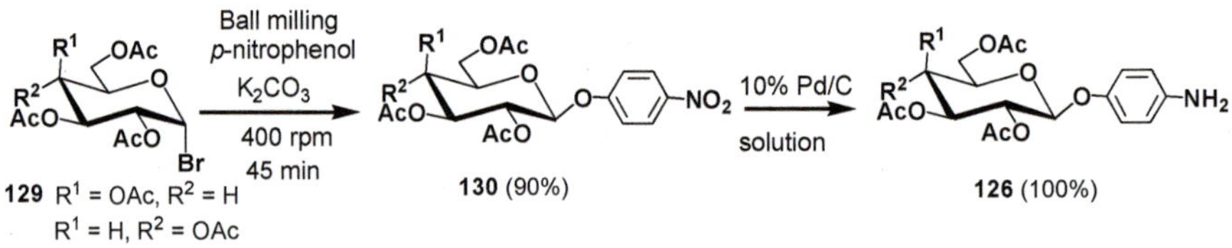

SCHEME 3.36

Synthesis of *p*-aminophenyl glycoside **126**.

Mechanochemical preparation of galactose-based cross-linked diamides required the diacetone galactose–derived amine **133**, which was obtained from the corresponding tosylate **131** (Scheme 3.37). In the first reaction step, nucleophilic substitution carried out by a mechanochemical method (Retsch MM400 mixer mill), afforded 6-azido-6-deoxygalactose derivative **132** as the sole product. It was isolated by simple aqueous workup in 95% yield and reduced in the following step to amine **133** by classical hydrogenation. In the final reaction step, amine was converted to mixture of diamide **134** and urea **135** by ball milling with acylated imidazole **125**.

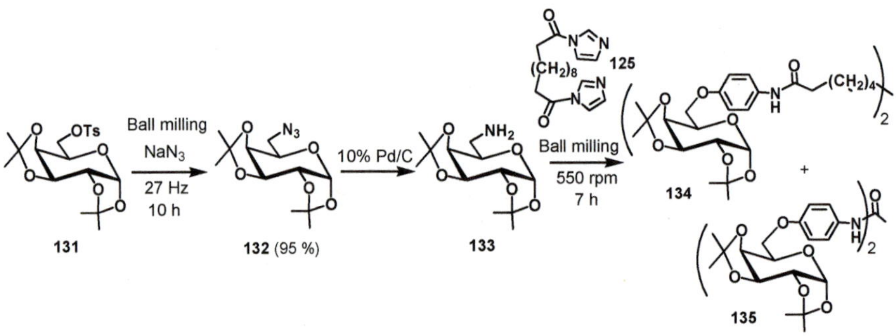

SCHEME 3.37

Synthesis of diacetone galactose azide.

For the synthesis of glycopeptides from Boc/Fmoc-protected amino acids, the initial carboxylic acid activation step was performed by milling at lower frequency of 450 rpm to avoid deprotection (Scheme 3.38). Ball milling of amine **137** with amino-protected lysine **136** formed the corresponding amide **138** and urea **139** in good overall yield.

The exclusive formation of 1,3-disubstituted urea **137** was achieved by grinding amine with CDI under solvent-free conditions. When 4-*O*-glucosyl aniline **140** was ball milled with CDI (3 equiv.), the respective 1,3-disubstituted urea **139** was obtained in 94% yield after a simple aqueous workup (Scheme 3.38). Under these environmentally friendly conditions, other sugar-linked anilines also reacted efficiently within 3–5 h to yield the respective ureas in high yields (80–90%).

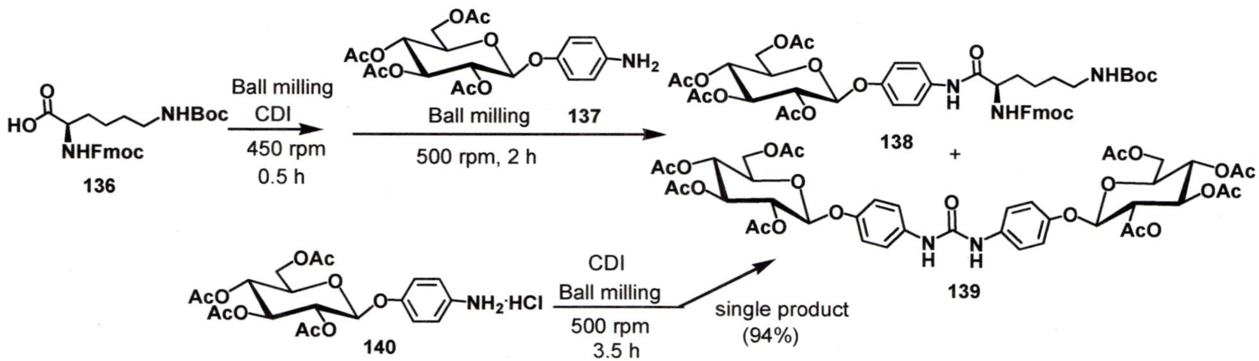

SCHEME 3.38

Synthesis of amide and urea glycopeptides.

Acyl azide was also used as activated carboxylic acid reagent for amide bond construction. In situ Raman spectroscopy was employed in the study of the reaction between 4-nitrobenzoyl azide **141** and 1,4-diaminobenzene **142** yielding *N*-(4-aminophenyl)-4-nitrobenzamide **143** (Scheme 3.39). Halasz et al. have found that in mechanochemical conditions reaction mechanism takes place via nucleophilic substitution on the carbonyl group of an acyl azide, while the other possible mechanism (Curtius rearrangement yielding an isocyanate intermediate which reacts with the amine to give the urea derivative) was not detected [28]. Possibility of the Curtius rearrangement of this substrate was verified by heating the solution of acyl azide where isocyanate was formed; however, the mild external stimuli applied during milling at room temperature was insufficient to trigger this reaction.

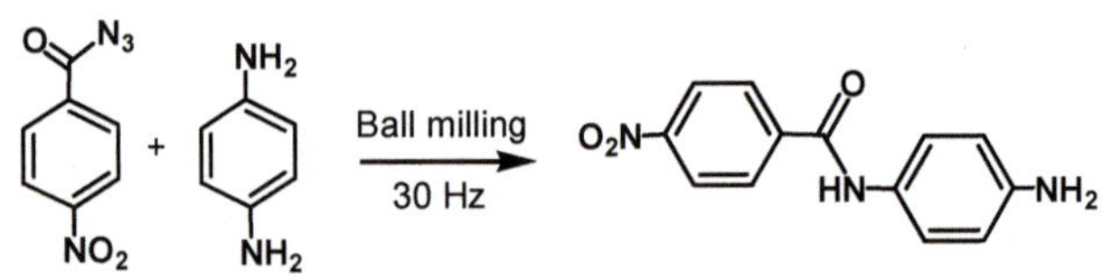

141 **142** **143**

SCHEME 3.39

Nucleophilic substitution on the carbonyl group of an acyl azide.

The effect of various additives differing in proticity, polarity, and acidity/basicity properties showed that neat grinding resulted in slow and incomplete formation of product, while LAG was more efficient (Table 3.18). Except for chloroform which had little effect, liquid additives dramatically increased the reaction rate, and DMF being the most effective. Product formation was faster when the liquid additive was more basic. Addition of basic salts such as sodium acetate in the reaction mixture along with ethanol (ILAG) showed an acceleration relative to pure ethanol, while slightly acidic ammonium nitrate did not show any change. Solubilities of reactants and products could not be correlated with observed reaction acceleration. Finally, strong base DABCO accelerated reaction the most. These findings indicate that the nucleophilic substitution reaction on a carbonyl group is catalyzed by base and proton abstraction is an important mechanistic step. A comparison of mechanochemical conditions with solution chemistry (in ethanol) showed much faster reaction with higher conversion.

Table 3.18 Reaction Times Using Various Additives[a]

Additive	Reaction Time[b] (min)	Additive	Reaction Time[b] (min)
LAG (liquid)		**ILAG (ethanol, salt)**	
DMF	4	DABCO	6.5
Ethanol	8 (2 h, 59%)[c]	NaOAc	7.2

Table 3.18 Reaction Times Using Various Additives[a]—cont'd

Additive	Reaction Time[b] (min)	Additive	Reaction Time[b] (min)
Acetone	10	NH_4NO_3	8
Acetonitrile	16.5		
Chloroform	30[d]		
Neat	60[d]		

[a]*Retsch MM301 ball mill, 30 Hz, poly(methylmetha)acrylate (PMMA) milling vessels (14 mL) with two 7 mm diameter stainless steel balls (ball weight 1.3 g), azide:amine 1:1 mol; LAG (η = 0.18 µL mg^{-1}).*
[b]*Reaction completion was determined by in situ Raman spectroscopy.*
[c]*Ethanol, 2 h, RT.*
[d]*Incomplete reaction.*
Reproduced from Tireli M, Juribašić M, Kulcsá M, Cindro N, Gracin D, Biliškov N, et al. Mechanochemical reactions studied by in situ Raman spectroscopy: base catalysis in liquid-assisted grinding. Chem Commun 2015;51:8058–61, with permission from the Royal Society of Chemistry.

3.7.2 ADDITION OF CUMULENES TO AMINES

Swinburne and Steed reported the mechanochemical synthesis of urea **146** which was achieved by grinding isocyanate **145** and amine **144** in a ball mill (Retsch MM200 with 25 mL vessel) in high yield [29]. No purification of the powder was performed, and obtained purity is similar to solution-state synthesis after purification (Scheme 3.40). This mechanochemical reaction is significantly faster than the typical solution-based procedure and demonstrates a high yielding, facile route to urea derivatives.

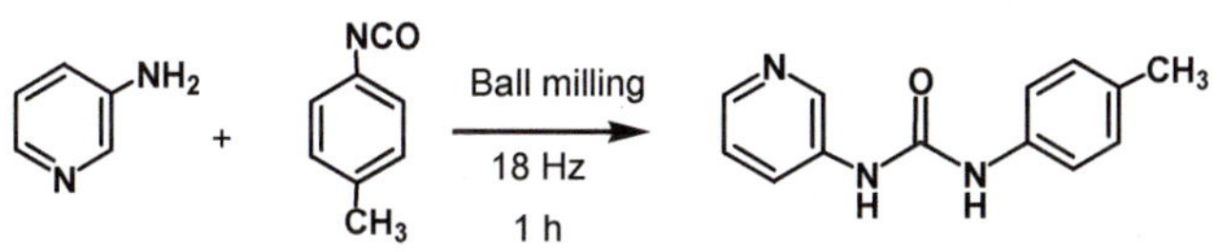

SCHEME 3.40

Synthesis of urea from amines and isocyanate.

Pyridyl ureas obtained by this protocol were used in the next synthetic step, formation of pyridinium salts, a tripodal anion receptors **149** by *N*-alkylation (Scheme 3.41). In all cases the solid reactants consisting of a pyridine derivative **148** and bromomethylated triethylbenzene derivative **147** were milled for 1 h. LAG technique was used in all cases (acetonitrile), except for compound R=H (Br^-) where a liquid was used as a reagent. The solid-state synthesis appears to be significantly cleaner than the solution-state method with significantly less impurity. Addition of methanol, filtration, and precipitation of product with NH_4PF_6 were employed for purification. In most cases, the mechanochemical technique is able to compete with the solution-based yield with gains in simplicity, time, and removal of one solvent–based step. Using the same general procedure for 1,10-bis(3,5-bis(bromomethyl)-2,4,6-triethylbenzyl)-4,4′-bipyridine-1,1′diium bromide **150** and 1-pyridin-3-yl-3-*p*-tolyl-urea **151**, tetrapodal anion receptor **152** was prepared in 27% yield (Scheme 3.42).

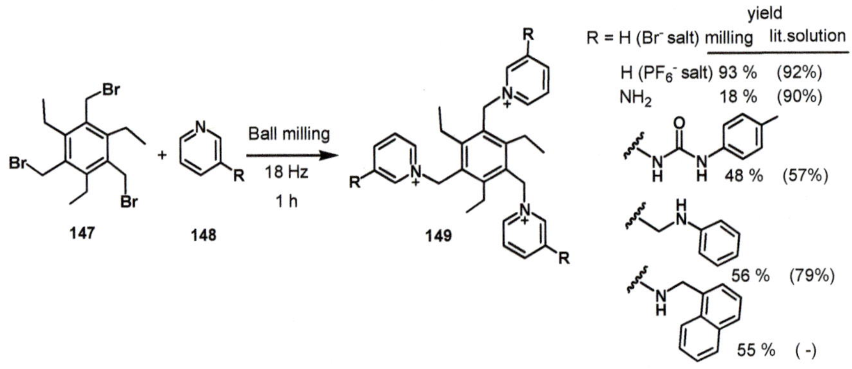

SCHEME 3.41

Synthesis of tripodal anion receptors.

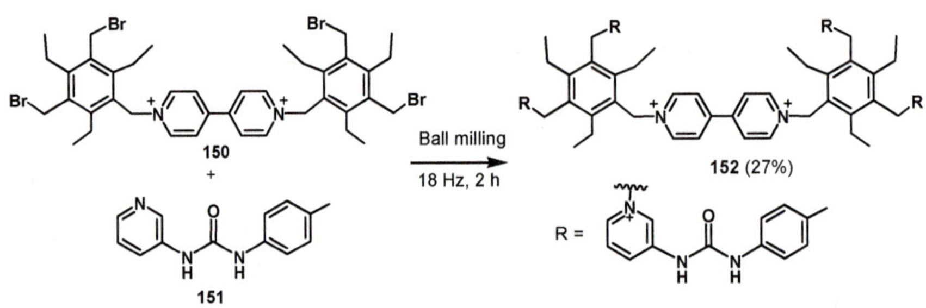

SCHEME 3.42

Synthesis of tetrapodal anion receptor.

Employment of potassium cyanate for mechanochemical synthesis of ureas was described by Colacino et al. [30]. Reaction carried out in the planetary mill with KNCO/H_2O, without addition of any supplementary base afforded pure phenylalanyl urea **154** in high yield (97%) without need of purification (Scheme 3.43). From an environmental point of view, regarding atom and solvent economy, this procedure for the preparation of urea derivative was very advantageous. In contrast, an excess of potassium cyanate and heating were necessary to afford phenylalanylurea in solution in 78% yield.

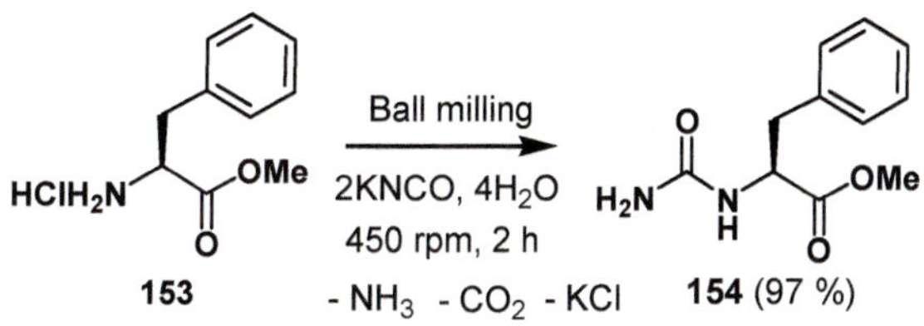

SCHEME 3.43

Synthesis of phenylalanylurea.

Procedure was employed in one-pot, two-step mechanochemical preparation of a library of substituted hydantoins **159**, as well as six-membered heterocycles **160**, from variety of α-amino methyl esters **155**, quaternary amino methyl esters or β-amino methyl esters **156** (Scheme 3.44). These substrates were easily converted into their urea derivatives **157–158** in the first step within 2h of milling, while the conditions for the second step had to be adjusted according to the starting amino ester, particularly concerning the nature of the base used for cyclization (Table 3.19). Products were isolated by grinding with ammonium chloride and ethyl acetate at 450rpm for 10min to eliminate inorganic salts and extracted with ethyl acetate, without the need for purification. No by-product was formed, the only waste being inorganic salts (KCl or CsCl), water, gases (CO_2 and NH_3), and MeOH (easily eliminated by evaporation). Six-membered heterocycles, dihydrouracil, and dihydrothymine were also synthesized starting from the corresponding β-amino ester precursors (entries 9 and 10) in moderate yields.

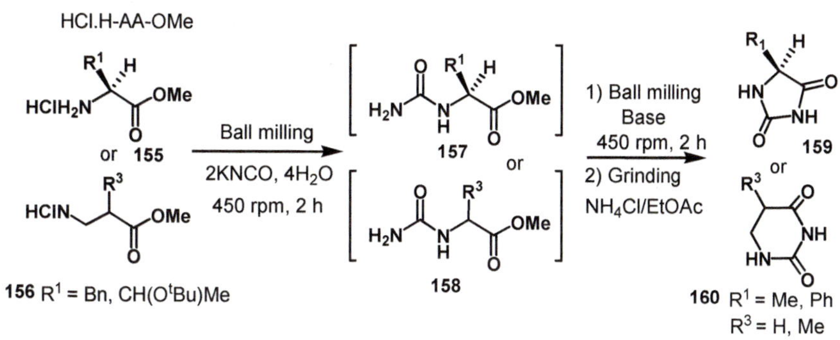

SCHEME 3.44

Synthesis of hydantoins and six-membered heterocycles.

This method was successfully applied to the sterically hindered diphenylglycine methyl ester hydrochloride (entry 8) to prepare the antiepileptic drug phenytoin **22**. In this synthesis, trimethylsilyl isocyanate (TMS-NCO), a more reactive alternative to potassium cyanate was used in the first step. The decomposition of TMS-NCO

Table 3.19 Synthesis of Hydantoins and Six-Membered Heterocycles[a]

Entry	Substrate HCl·H-AA-OMe	Step 1 Time (h)	Step 1 Yield (%)[b]	Step 2 Base Equivalent	Step 2 Time (h)	Step 2 Yield (%)[b]
Hydantoins 159						
1	L-HCl·H-Phe-OMe	1.5	97	K_2CO_3/1	2	80
2	D-HCl·H-Phe-OMe	2		K_2CO_3/Et_3N 1/1	2	70
3	L-HCl·H-Tyr(O^tBu)-OMe	2	100	Cs_2CO_3/3	2	53
4	L-HCl·H-Met-OMe	2		Cs_2CO_3/3	2	91
5	L-HCl·H-Ala-OMe	2	79	K_2CO_3/1	2	53
6	D-HCl·H-Leu-OMe	2	93	K_2CO_3/1	2	56
7	L-HCl·H-Ser(O^{t}Bu)-OMe	2		K_2CO_3/1	2	100
8	L-HCl·H-Gly(Ph)$_2$-OMe	8[c]		Cs_2CO_3/3	2	84
Six-membered Heterocycles 160						
9	HCl·H-β-Ala-OMe	2		Cs_2CO_3/3	2	47
10	HCl·H-3-Aib-OMe	2		Cs_2CO_3/3	2	57

[a]*Retsch PM100 planetary ball mill, stainless steel jar (12 mL), 25 × 5 mm balls, 450 rpm, 2 h, amino ester.HCl:KNCO:H_2O 1:3:4 equiv.*
[b]*Isolated yields.*
[c]*Trimethylsilyl isocyanate added in portions at start, after 3 and 6 h.*
Reproduced from Konnert L, Reneaud B, de Figueiredo RM, Campagne J-M, Lamaty F, Martinez J, et al. Mechanochemical preparation of hydantoins from amino esters: application to the synthesis of the antiepileptic drug phenytoin. J Org Chem 2014;79:10132–42, copyright (2014), with permission from American Chemical Society.

in the presence of water accounts for the in situ generation of hexamethyldisilazane (HMDS) base in the presence of HCl, leading also to the formation of the amino nucleophile. Water was essential to the reaction allowing the generation of the base as well as, in a second step, the hydrolysis of the *N*-protected trimethylsilyl ureido derivative to afford product in 84% yield. Phenytoin is usually synthesized from the condensation of benzil and urea according to the Biltz reaction, and when this reaction was carried out in ball mill (30 Hz, 2 h), phenytoin was obtained in low 10% yield. The major products of the reaction were the 3α,6α-diphenylglycoluril and the 1,3-bis(2-oxo-1,2-diphenylethylidene)urea coming from the double condensation of benzil with urea, showing that the mechanochemical activation switched the selectivity toward the formation of glycoluril and double condensation product as the main products, while pinacol rearrangement leading to phenytoin is the favorite pathway in solution.

Thiourea derivatives **164** were prepared by Kaupp et al. by mechanochemical reactions in which stoichiometric quantities of solid *o*-phenylene-di-isothiocyanate **161** reacts with the solid aromatic amines **162** in a ball-mill at room temperature (Retsch MM2000 mill with a 10 mL stainless steel beaker and balls) [31]. The products, 1-(aminoaryl-thiocarbonyl)-benzimidazolidine-2-thiones **163** were presumably formed via thiourea intermediate and intramolecular cyclization in 100% yield (Scheme 3.45). After completion of milling, no workup or isolation was required.

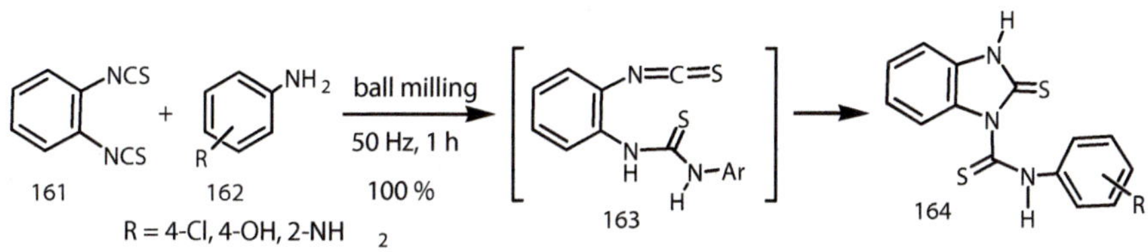

SCHEME 3.45

Synthesis of benzimidazolidine-2-thiones.

Facile mechanochemical synthesis of thioureas was developed by Štrukil et al. (Scheme 3.46) [32]. Formation of new thioamide C—N bond was achieved by reaction of aromatic isothiocyanates **166** with aromatic amines (Table 3.20). Addition of small amount of methanol (LAG) was beneficial and the high yields were obtained regardless on the reactant physical state. The whole synthesis was carried out without the use of bulk organic solvents, that is solvent-free mechanosynthesis was quantitative, which voids the use of solvent for purification. Mechanochemical reaction

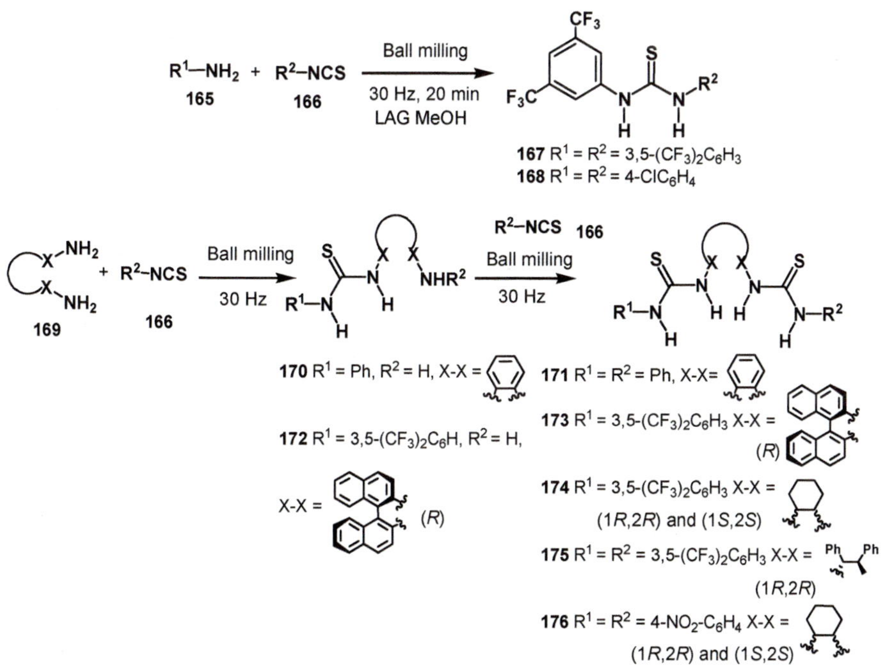

SCHEME 3.46

Synthesis of thioureas from amines and isothiocyanates.

conditions can be readily optimized to form either the 1:1 or the 1:2 thiourea product, demonstrating that reaction stoichiometry can be more efficiently controlled under these conditions than in solution-based syntheses. In solution an equimolar amounts of reagents form complex mixture of mono- and bis-thioureas (entries 11 and 14). The reaction procedure can be readily scaled up to at least the 1-g scale, although the reaction time needed to be extended to achieve the same conversion. It was also shown that the enantiomerically pure reagents could be transformed into chiral *bis*-thioureas without the loss of chirality (entries 4–10 and 14). In the case of sterically more hindered *ortho*-phenylenediamine, significantly longer time was required for completion of *bis*-thiourea reaction (entry 12).

Table 3.20 Thiourea Synthesis From Amines and Isothiocyanates[a]

Entry	Product	Reactant State	Method[b]	Time (min)	Yield (%)[h]
1	**167**	Liquid/liquid	LAG[c]	30	>99 (84)[i]
2	**167**	Liquid/liquid	LAG (1 g scale)[c]	180	98
3	**168**	Liquid/solid	LAG[d]	20	>99
4	**(1*R*,2*R*)-174**	Liquid/solid	Neat/LAG[e]	20	>99 (94)[i]
5	**(1*R*,2*R*)-174**	Liquid/solid	Neat (1 g scale)[e]	180	98
6	**(1*R*,2*S*)-174**	Liquid/solid	Neat/LAG[e]	20	>99
7	**(1*R*,2*R*)-175**	Liquid/solid	Neat/LAG[e]	20	>99 (81)[i]
8	**(1*R*,2*R*)-176**	Solid/solid	Neat/LAG[e]	30	>99 (89)[i]
9	**(1*S*,2*S*)-176**	Solid/solid	Neat/LAG[e]	30	>99
10	**(*R*)-173**	Liquid/solid	LAG[f]	90	>99 (85)[i]
11	**170**	Liquid/solid	Neat[g]	20	99 (81, **170**:**171** 29:1)[j]
12	**171**	Liquid/solid	LAG[e]	180	>99
13	**171**	Liquid/solid	LAG (1 g scale)[e]	540	99 (**170**:**171** 7:1)[j]
14	**(*R*)-172**	Solid/solid	Neat[c]	30	37+**(*R*)-173** (21) (72+**(*R*)-173** (16))[j]

[a]*Retsch MM400 mixer mill, 30 Hz, stainless steel jar (10 mL), one 12 mm milling ball.*
[b]*LAG MeOH (η=0.25 µL mg^{-1}).*
[c]*Amine:isothiocyanate 1:1 mol.*
[d]*Amine:isothiocyanate 1:1.05 mol.*
[e]*Amine:isothiocyanate 1:2 mol.*
[f]*Amine:isothiocyanate 1:2.1 mol.*
[g]*Amine:isothiocyanate 1.1:1 mol.*
[h]*NMR/HPLC yield.*
[i]*Literature solution synthesis.*
[j]*In dichloromethane, 24 h, RT.*

In the follow-up article, Eckert-Maksić and Friščić applied the amine/iso(thio)cyanate reaction to the synthesis of *ortho*-mono(thio)ureas **180** (Scheme 3.47) [33]. Excellent control of the stoichiometric composition of the product in mechanochemical synthesis was achieved. Thiourea products were obtained selectively and in almost quantitative yield when reagents were used in equimolar amounts and small amount of solvent was added to assist grinding (ie, LAG) (Table 3.21, entries 1–4). Quantitative formation of single product allowed simple collection by scraping off the walls of the grinding jar. The synthesis of symmetrical bis(thio)ureas by the amine-iso(thio)cyanate coupling in the 1:2 M ratio required longer reaction time (entries 7–10). Finally, the mixed, unsymmetrical bis-(thio)ureas were prepared by milling of equimolar amounts of monothioureas **178** with the corresponding iso(thio)cyanates **179** (entries 10–13). Good stoichiometric control allowed the preparation of mixed thiourea/urea products in quantitative yield (entries 15–18). Comparison of mechanochemical results with classical solution conditions showed significant improvement of yields.

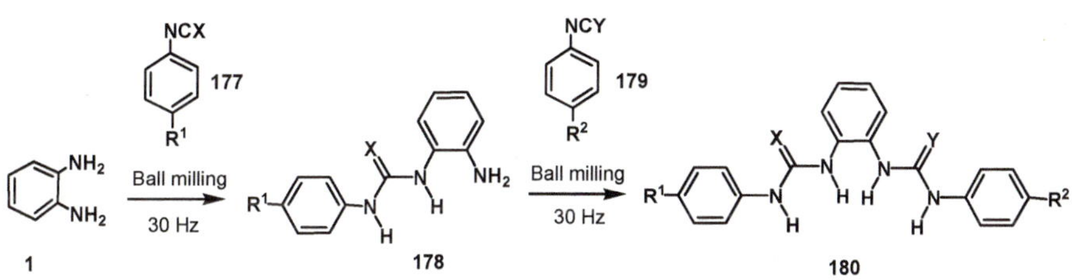

SCHEME 3.47

Mechanochemical synthesis of *ortho*-(thio)ureas.

Table 3.21 Synthesis of *ortho*-(Thio)ureas[a]

Entry	R^1	R^2	X,Y	Conditions	Time (min)	Yield (%)[b]
Mono-*ortho*-(Thio)ureas						
1	OCH_3	–	S, –	LAG	30	≥98
2	H	–	S, –	Neat	30	≥95
3	Cl	–	S, –	LAG	30	≥99
4	NO_2	–	S, –	LAG	30	≥98 (78)[c]
5	H	–	O, -	Neat	30	78[d] (99)[e]
Bis-*ortho*-(Thio)ureas						
6	OCH_3	OCH_3	S,S	LAG	540	≥95
7	H	H	S,S	LAG	180	≥99
8	Cl	Cl	S,S	LAG	180	≥99
9	NO_2	NO_2	S,S	LAG	180	≥99 (54)[c]
10	OCH_3	Cl	S,S	LAG	180	≥99

Continued

Table 3.21 Synthesis of *ortho*-(Thio)ureas[a]—cont'd

Entry	R^1	R^2	X,Y	Conditions	Time (min)	Yield (%)[b]
11	OCH_3	NO_2	S,S	LAG	180	≥99
12	NO_2	Cl	S,S	LAG	180	≥97
13	H	NO_2	S,S	LAG	30/180	68/≥ 99
14	H	H	O,O	Neat	30	≥99 (96)[c]
15	OCH_3	H	S,O	Neat	180	≥99
16	H	H	S,O	Neat	180	≥99
17	Cl	H	S,O	Neat	180	≥99
18	NO_2	H	S,O	Neat	180	≥99

[a]*Retsch MM400 mixer mill, 30Hz, stainless steel jar (10mL), one 12mm milling ball, LAG methanol (η=0.25 μL mg^{-1}), amine:isothiocyanate 1:1 mol.*
[b]*Isolated yield.*
[c]*Literature solution synthesis.*
[d]*NMR yield.*
[e]*Solution synthesis.*
Reproduced from Štrukil V, Margetić M, Igrc MD, Eckert-Maksić M, Friščić F. Desymmetrisation of aromatic diamines and synthesis of non-symmetrical thiourea derivatives by click-mechanochemistry. Chem Commun 2012;48:9705–7, with permission from the Royal Society of Chemistry.

The selective desymmetrization of the sterically less hindered *para*-phenylenediamine (***p*-pda**) **181** was more difficult to achieve, presumably due to increased steric hindrance (Scheme 3.48). Milling of a neat equimolar mixture of *p*-methoxyphenyl isothiocyanate **182** and **181** resulted in only 65% conversion to thiourea **183**, along with 17% of *bis*-thiourea **184** (entry 1). Optimized reaction conditions required LAG and sodium chloride as grinding auxiliary (Table 3.22). Smaller size of balls and the use of Teflon as material for grinding jars and balls also increased the yield. These conditions were not successful for the *p*-nitrophenyl derivative, yielding a mixture of the mono- and bis-thiourea, which might be explained by higher reactivity of *p*-nitrophenyl isothiocyanate. Milling of mono-thiourea (R^1=OCH_3) and *p*-chloro or *p*-nitrophenyl isothiocyanate quantitatively yielded the corresponding nonsymmetrical products **184** (R^1=OCH_3, R^2=Cl) and (R^1=OCH_3, R^2=NO_2), respectively. Environmentally benign isolation procedure for mono-urea is just scraping off the product from the walls of the grinding jar and for *bis*-thioureas is simple washing of products with water and filtration.

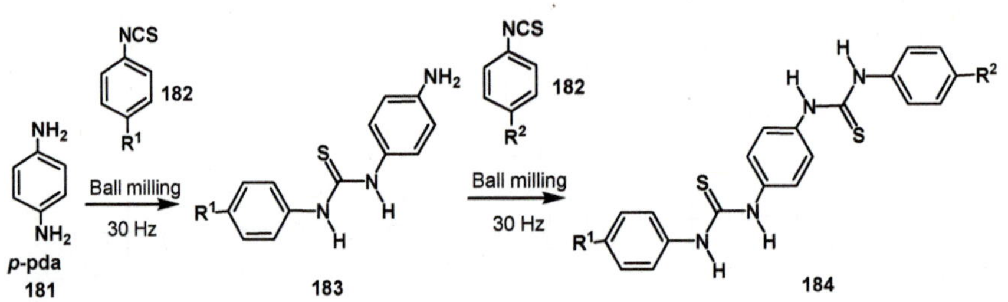

SCHEME 3.48

Mechanochemical synthesis of *para*-thioureas and bisthioureas.

Table 3.22 Mono-Thiourea Synthesis From *p*-Phenylenediamine and *p*-(OMe)Isothiocyanate[a]

Entry	Grinding Material[b]	Conditions[c]	Yield (%)[d]
1	Steel	Neat	65
2	Steel	LAG—methanol	70
3	Steel	LAG—water	36
4	Steel	LAG—DMF	70
5	Steel	LAG—cyclohexane	74
6	Steel	LAG—EtOAc	88 (78)[g]
7	Steel	LAG—EtOAc, 8mm ball	91
8	Teflon	Neat	67
9	Teflon	LAG—EtOAc	93
10	Teflon	LAG—EtOAc,[e] m(NaCl) = 10 × m(**181**)	95
11	Teflon	LAG—EtOAc,[e] m(NaCl) = 20 × m(**181**)	≥96
12	Teflon	LAG—EtOAc,[f] m(NaCl) = 20 × m(**181**)	≥97

[a]*Retsch MM400 mixer mill, 30Hz, 30min.*
[b]*Stainless steel jar (10mL), one 12mm milling ball or Teflon jar (10mL), one 10mm Teflon grinding ball.*
[c]*LAG (η = 0.25 µL mg^{-1}), amine:isothiocyanate 1:1 mol.*
[d]*NMR yield.*
[e]*Diluted with NaCl.*
[f]*Diluted with NaCl (η = 0.8 µL mg^{-1}).*
[g]*In ethyl acetate, 24h.*
Reproduced from Štrukil V, Margetić M, Igrc MD, Eckert-Maksić M, Friščić F. Desymmetrisation of aromatic diamines and synthesis of non-symmetrical thiourea derivatives by click-mechanochemistry. Chem Commun 2012;48:9705–7, with permission from the Royal Society of Chemistry.

Effective control of reaction outcome of the ball-milling reaction by stoichiometry of added reagents was also achieved by Zhang et al. [34]. Synthesis of thioureas was carried out using aromatic amines, carbon disulfide, and potassium hydroxide in process where isothiocyanate reagent **187** was firstly generated (Scheme 3.49). By this protocol isothiocyanate **187** and thiourea **188** were obtained in high yields (Table 3.23). In these conditions, isothiocyanates were formed via transient dithiocarbamates **186**, and further react with an amine giving thioureas. When an excess of CS_2 was applied, within 40min isothiocyanate was prepared exclusively. As the ratio of CS_2 to amine changes to equimolar, single thiourea product is obtained (entries 1–3). Employment of these observations led to development of simple mechanochemical synthesis of thioureas using large excess of CS_2 (entries 7–9). Symmetrical thioureas were prepared in high yield by one-step milling process using equimolar amounts of amine and CS_2 (entries 4–6). Reaction mixtures were isolated by simple treatment with aqueous HCl and filtration to obtain products in moderate to high yields, resembling the influence of electronic nature of substituents on reactivity. Finally, two-step ball-milling process was used for synthesis of unsymmetrical thioureas.

In the first step isothiocyanates were prepared by the procedure described above, and then without isolation 1 equiv. of second amine was added and milling continued. Upon completion, treatment of reaction mixture with aqueous hydrochloric acid and filtration afforded unsymmetrical products in high yields (entries 10–13).

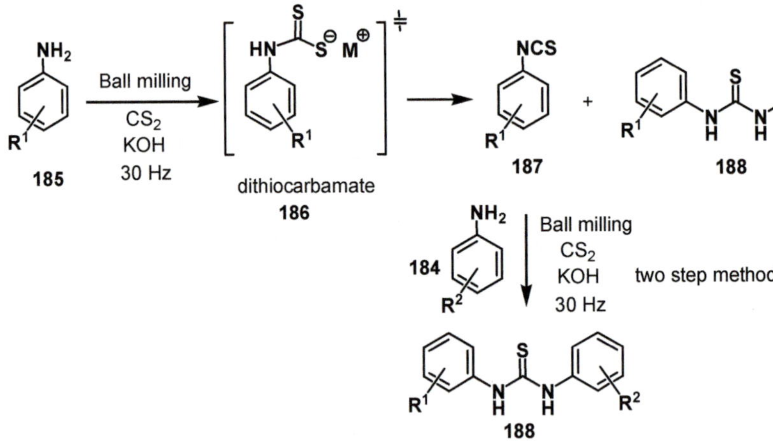

SCHEME 3.49

Preparation of isothiocyanates and thioureas.

Table 3.23 Preparation of Isothiocyanates and Thioureas[a]

Entry	R^1	R^2	CS_2 Equivalent	Time	Yield (%)[b]
Isothiocyanates and Thioureas					
1	4-CH_3	4-CH_3	5	40 min	**187** 98%[c]
2	4-CH_3	4-CH_3	3	40 min	**187** 58% + **188** 27%[c]
3	4-CH_3	4-CH_3	1	40 min	**187** trace + **188** 93%[c]
One-Step Method[d] Symmetrical Thioureas					
4	4-CH_3	4-CH_3	1[e]	40 min	**188** 93%
5	4-OCH_3	4-OCH_3	1[e]	40 min	**188** 95%
6	4-NO_2	4-NO_2	1[e]	90 min	**188** 43%
Isothiocyanates					
7	2-OCH_3	2-OCH_3	5	40 min	**187** 91%
8	2-Cl	2-Cl	5	70 min	**187** 71%
9	4-NO_2	4-NO_2	5	90 min	**187** 61%
Two-Step Method[d] Unsymmetrical Thioureas					
10	4-CH_3	2-CH_3	5	40 + 35 min	**188** 89%

Table 3.23 Preparation of Isothiocyanates and Thioureas[a]—cont'd

Entry	R^1	R^2	CS_2 Equivalent	Time	Yield (%)[b]
11	4-OCH_3	4-Cl	5	40+60 min	**188** 90%
12	4-OCH_3	4-Br	5	40+60 min	**188** 94%
13	3-CH_3	4-OCH_3	5	45+60 min	**188** 91%

[a]*Retsch MM400 stainless steel vial (25 mL), 7 mm ball, 30 Hz, amine1:KOH 1:1 equiv.*
[b]*Isolated.*
[c]*HPLC analysis.*
[d]*Step 1 amine 1:KOH 1:1 equiv. Step 2 amine 2 1 equiv.*
[e]*Amine 1:KOH 1:1.2 equiv.*
Reproduced from Zhang Z, Wu H-H, Tan Y-J. A simple and straightforward synthesis of phenyl isothiocyanates, symmetrical and unsymmetrical thioureas under ball milling. RSC Adv 2013;3:16940–4, with permission from the Royal Society of Chemistry.

High potential of mechanochemistry to study reactive intermediates which are not isolable in solution reactions was explored by Štrukil and Friščić. Advantageous use of milling in preparation and study of reactive intermediates was demonstrated by thiourea synthesis from anilines and bis(benzotriazolyl)methanethione **189** (Scheme 3.50) [35]. In solution, this reaction proceeds via reactive aryl *N*-thiocarbamoylbenzotriazole intermediate **189**, which is not isolable, but quickly fragments into isothiocyanate **193** and benzotriazole **194**. In contrast, by solid-state milling (LAG with MeCN), *N*-thiocarbamoylbenzotriazoles were isolated as stable solids in high yields. Remarkably, using *p*-phenylene diamine, a stable *bis-N*-thiocarbamoylbenzotriazole was prepared **190**. The formation of these intermediates was monitored

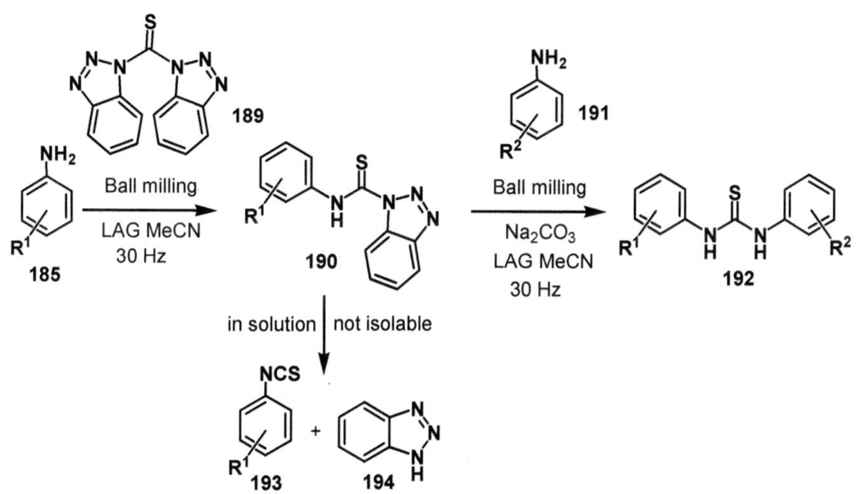

SCHEME 3.50

Synthesis of *N*-thiocarbamoylbenzotriazoles and thioureas.

in situ by Raman spectroscopy, and their fragmentation demonstrated in base-catalyzed milling reactions. The optimized milling conditions were applied to synthesis of variety of symmetrical and unsymmetrical thioureas **192** (selected examples, Table 3.24). Synthesis of unsymmterical thiourea products was achieved by two-step, one-pot procedure: the first step was ball milling of anilines with bis(benzotriazolyl) methanethione **189** (LAG MeCN) for 10 min. Upon completion, MeCN was allowed to evaporate, then K_2CO_3 and new amount of MeCN (LAG) added and milling continued to afford thioureas **192** in quantitative yields.

Table 3.24 Synthesis of *N*-Thiocarbamoylbenzotriazoles and Thioureas[a]

Entry	R¹	R²	Yield (%)[b]
***N*-Thiocarbamoylbenzotriazoles[f]**			
1	4-Br	–	99
2	4-Cl	–	98
3	2-NH_2	–	98
4	4-OCH_3	–	99
Thioureas[g]			
5	4-Br	4-Br	97[c]
6	4-CH_3	4-CH_3	98
7	4-OCH_3	4-Cl	97[c,d]
8	4-Br	2-NH_2	98[d,e]

[a]*Retsch MM400 Teflon vial (10 mL), 10 mm stainless steel ball, 30 Hz, 10 min, LAG with CH_3CN (η = 0.25 mL mg^{-1}).*
[b]*Isolated.*
[c]*45 min LAG.*
[d]*Two-step synthesis.*
[e]*30 min LAG.*
[f]*Amine 1:bis(benzotriazolyl) methanethione 1:1 equiv.*
[g]*Amine 1:amine 2:K_2CO_3 1:1:1.*

Amino-substituted bis-*N*-thiocarbamoylbenzotriazole **196** was readily obtained by milling of 2 equiv. of bis(benzotriazolyl)methanethione **189** with 1 equiv. of *para*-phenylenediamine. On the other hand, analogous synthesis starting from *ortho*-phenylenediamine and 1 equiv. of **189** yielded bisthione **198** as a result of an intramolecular cyclization (Scheme 3.51).

Similarly to *N*-thiocarbamoylbenzotriazoles, their *bis*-variants represent masked aryl isothiocyanates useful in thiourea mechanosynthesis. They allow the synthesis of bisthioureas: *bis*-*N*-thiocarbamoylbenzotriazole **199** with *p*-anisidine and Na_2CO_3 gave bisthiourea **203** quantitatively after 3 h (Scheme 3.52). Alternatively, the same

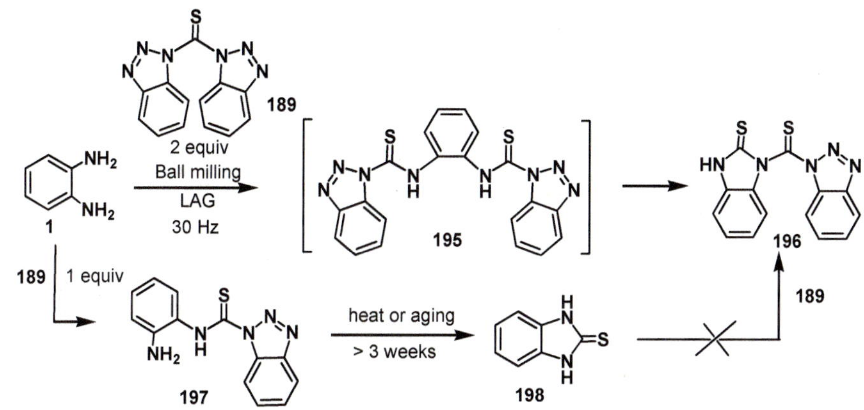

SCHEME 3.51

Mechanochemical reaction of *ortho*-phenylenediamine.

product was obtained in 98% yield by milling of 2 equiv. of *N*-thiocarbamoylbenzotriazole **200** and *para*-phenylenediamine with Na_2CO_3 for 45 min.

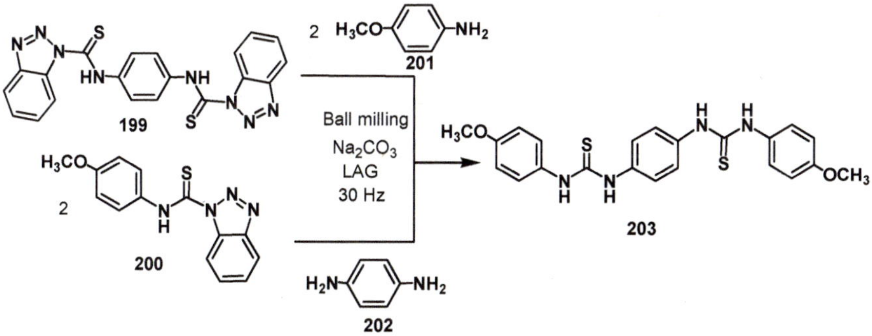

SCHEME 3.52

N-thiocarbamoylbenzotriazoles as masked isothiocyanates.

Mechanosynthesis of sulfonyl(thio)ureas starting with sulfonamide was developed by Friščić et al. (Scheme 3.53) [36]. Two synthetic approaches were developed: stoichiometric base-assisted and copper-catalyzed coupling of sulfonamides and iso(thio)cyanates. For stoichiometric base-assisted coupling, two-step mechanosynthesis of sulfonyl(thio)ureas involved in the first-step deprotonation of sulfonamide **204** with K_2CO_3, followed in the second step by milling with 1 equiv. of iso(thio) cyanate. Full conversion of substrates was achieved, and pure products **205–208** were obtained by simple aqueous wash, thus achieving truly solvent-free organic synthesis.

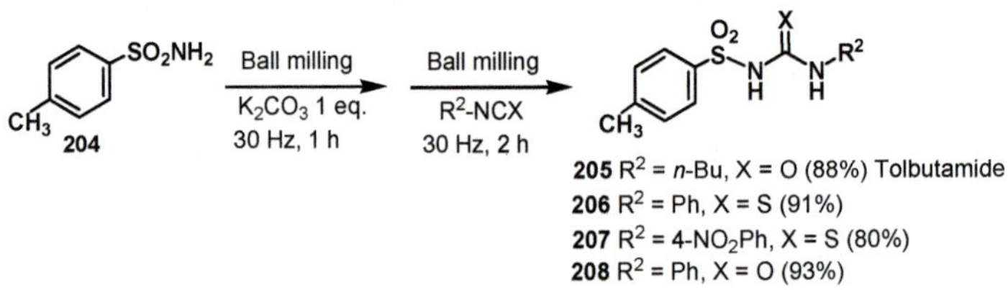

SCHEME 3.53

Synthesis of sulfonyl(thio)ureas.

Direct copper-catalyzed coupling of sulfonamides and isocyanates was achieved with the highest yield when CuCl (5 mol%) catalyst was used in LAG (nitromethane) milling (Scheme 3.54). Within 2 h of grinding, sulfonylureas **211** were produced in high yield, and workup included ball milling for 10 min with water and Na_2H_2EDTA to remove catalyst (Table 3.25).

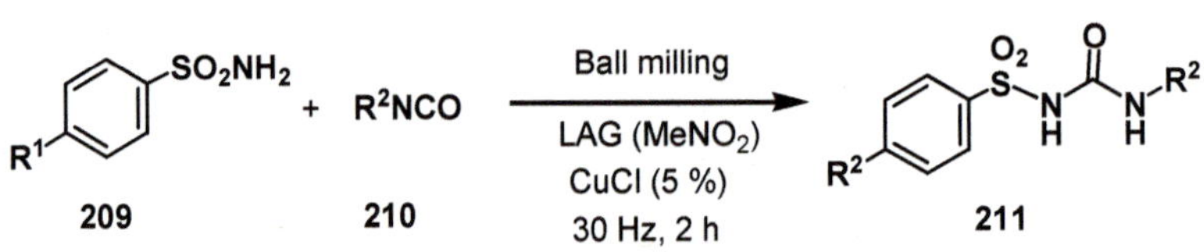

SCHEME 3.54

Synthesis of sulfonylureas.

Table 3.25 Synthesis of Sulfonylureas[a]

Entry	R^1	R^2	Yield (%)[b]
1	Me	*n*-Bu	90
2	Me	Cy	88
3	Me	*n*-Pr	86
4	Cl	*n*-Pr	92
5	Cl	*n*-Bu	92
6	Cl	Cy	91

[a]Retsch MM400 mixer mill, stainless steel vessel (10 mL), 10 mm ball, 30 Hz, 2 h.
[b]Isolated.
Reproduced from Tan D, Štrukil V, Mottillo C, Friščić T. Mechanosynthesis of pharmaceutically relevant sulfonyl-(thio)ureas. Chem Commun 2014;50:5248–50, with permission from the Royal Society of Chemistry.

Mechanosynthetic protocol optimized for sulfonylureas was also applied to preparation of antidiabetic drug glibenclamide **216** (Scheme 3.55). Synthesis was carried out in two reaction steps: first, coupling of benzoic acid derivative **212** and *p*-(2-aminoethyl) benzenesulfonamide **213** afforded sulfonamide precursor **214**, which was subjected to copper-catalyzed coupling with cyclohexylisocyanate. For this step, excess of isocyanate was required, due to formation of the dicyclohexylurea side product.

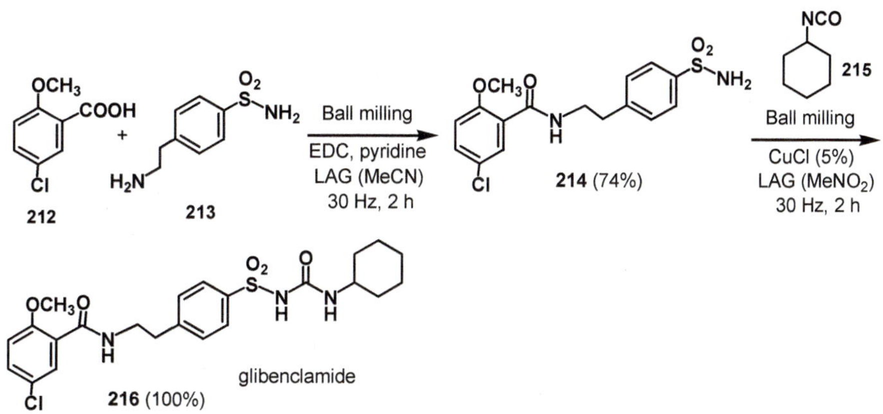

SCHEME 3.55

Synthesis of glibenclamide.

3.7.3 OXIDATIVE AMIDATION

Efficient amide bond formation by oxidative amidation was reported by Wang and Gao [37]. Reactions of aromatic aldehydes with aromatic amines in the presence of oxone and anhydrous magnesium sulfate gave clearly the corresponding amides **219** (Scheme 3.56). Various oxidants and additives were used to optimize the reaction conditions (oxone, iodine, $K_2S_2O_8$). As the imines are presumably formed in the initial step of reaction, the presence of magnesium sulfate as dehydrating agent was beneficial and oxone showed the best oxidation activity. Anilines substituted with either electron-donating or electron-withdrawing groups exhibited similar activities (Table 3.26). When aldehydes bearing electron-withdrawing groups were used, the corresponding amides were obtained in moderate to good yields. The steric hindrance of the substituted groups on the phenyl rings of aldehydes was deleterious to the reaction, leading to decreased yields. Compared with the traditional solution reaction (entries 3 and 4), this novel solvent-free and metal catalyst–free procedure makes the synthesis of amides more efficient, low cost, and ecofriendly.

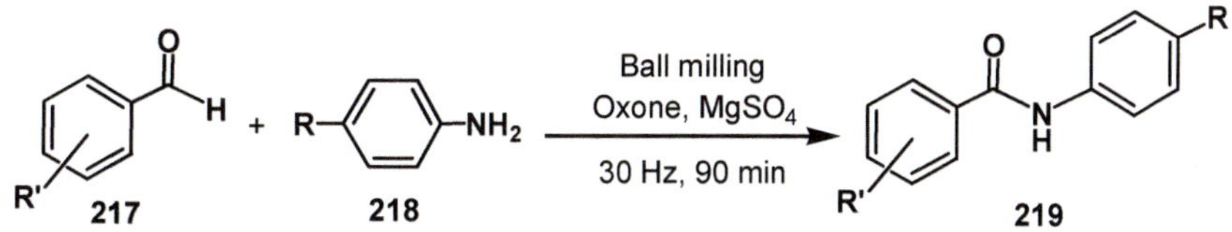

SCHEME 3.56

Oxidative amidation of aromatic aldehydes with amines.

Table 3.26 Oxidative Amidation of Aldehydes With Anilines Under Ball-Milling conditions[a]

Entry	R	R′	Yield (%)[b]
1	3-$NO_2C_6H_4$	4-$CH_3C_6H_4$	75
2	4-$NO_2C_6H_4$	4-$CH_3C_6H_4$	78
3	4-CNC_6H_4	4-$CH_3C_6H_4$	65
4	2-$NO_2C_6H_4$	4-$CH_3C_6H_4$	38
5	3-$NO_2C_6H_4$	4-BrC_6H_4	71
6	4-$NO_2C_6H_4$	4-BrC_6H_4	66
7	4-CNC_6H_4	4-BrC_6H_4	60
8	3,4-$C_{l2}C_6H_3$	4-BrC_6H_4	65
9	2-$NO_2C_6H_4$	4-BrC_6H_4	45
10	3-$NO_2C_6H_4$	4-ClC_6H_4	62
11	2-thienyl	4-$CH_3C_6H_4$	42

[a]*Retsch MM200 mixer mill, 1800 rpm (30 Hz), 1.5 h, stainless steel jar (5 mL), one 7 mm ball, aldehyde:aniline:$MgSO_4$:Oxone 1:1:1:2 mol.*

[b]*Isolated yield.*

Reproduced from Giao J, Wang G-W. Direct oxidative amidation of aldehydes with anilines under mechanical milling conditions. J Org Chem 2008;73:2955–8, copyright (2008), with permission from American Chemical Society.

3.7.4 NUCLEOPHILIC ADDITION OF CARBOXYLATE

Employment of gaseous reagent in ball-mill synthesis was reported by Pinhas et al. In contrast to literature conditions, mechanochemical conversion of aziridines **219** to oxazolidinones by reaction with CO_2 was carried out without the need for catalysts, at room temperature and at low pressure (Scheme 3.57) [38]. Substrate and an excess of dry ice were simply closed in reaction vessel and milled for 17 h. Quantitative yield and high regioselectivity were obtained by this simple procedure with variously substituted aziridines (Table 3.27). For comparison, uncatalyzed reaction of aziridine (R^1 = Bn, R^2 = CH_3) carried out in solvent gave significantly lower yield (THF, 12 h at 4 atm, 40% of regiosimeric products **221** and **222**). Multistep reaction mechanism starts with nucleophilic carboxylate (CO_2) attack on nitrogen, forming amide bond in **220**, followed by the second nucleophilic attack on aziridine ring with subsequent ring opening and reclosure.

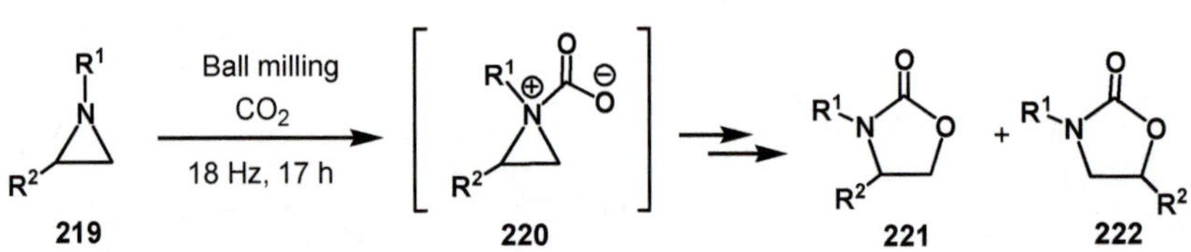

SCHEME 3.57

Conversion of aziridines to oxazolidinones.

Table 3.27 Conversion of Aziridines to Oxazolidinones[a]

Substrate	Product	Yield (%)[b]	Substrate	Product	Yield (%)[b]
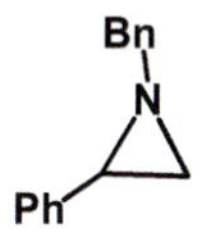	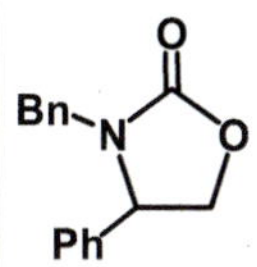	100	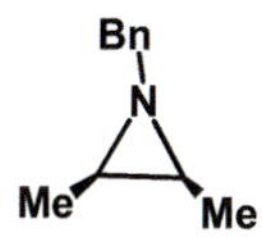	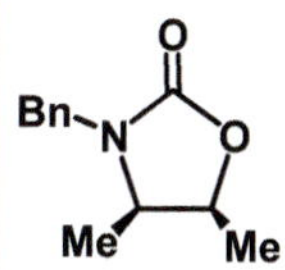	100
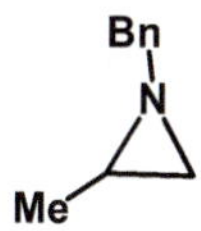	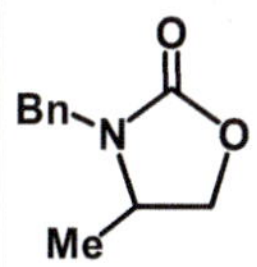	100[c]	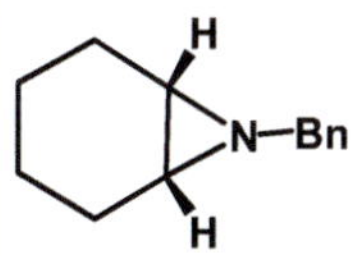	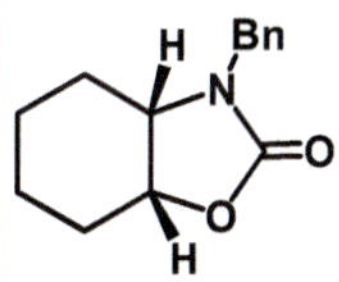	100
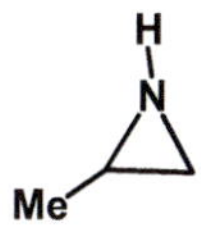	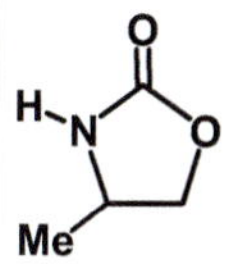	100	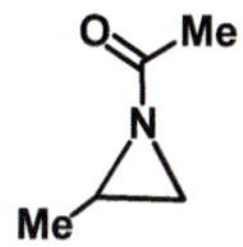	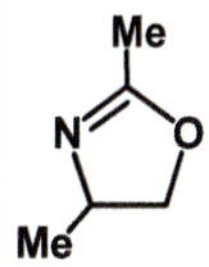	100

[a]*Spex 8000M mixer, stainless steel vessel (3.5 mL), 3.2 mm ball, 18 Hz, 17 h.*
[b]*Isolated yields.*
[c]*A:B regioisomers 93:7.*
Reproduced from Phung C, Ulrich RM, Ibrahim M, Tighe NTG, Lieberman DL, Pinhas AR. The solvent-free and catalyst-free conversion of an aziridine to an oxazolidinone using only carbon dioxide. Green Chem 2011;13:3224–9, with permission from the Royal Society of Chemistry.

3.7.5 CARBAMATE NITROGEN PROTECTION

Sikchi and Hultin reported a simple mechanochemical method for nitrogen protection of nucleoside derivatives (Scheme 3.58) [39]. For this purpose, simple Pyrex glass apparatus with Pyrex glass balls was used. A loose fit of glass apparatus was used, which prevents pressure buildup as gaseous CO_2 released during the reactions.

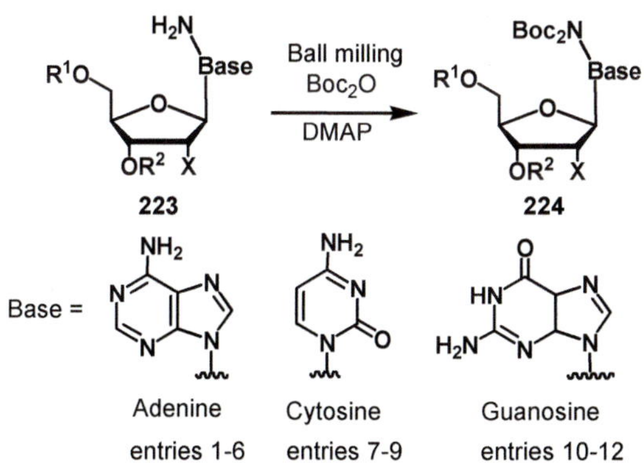

SCHEME 3.58

N-protection of *O*-protected nucleoside derivatives.

This simple mill provided a relatively gentle grinding action, which was nevertheless quite adequate for these reactions. Ball milling of *O*-protected nucleosides **223** with an excess of di-*tert*-butyl dicarbonate (Boc_2O) and 4-*N*,*N*-(dimethylamino)pyridine (DMAP) resulted in *N*-protection in 90–99% yield (Table 3.28). Acetyl esters, silyl ethers, and acetonide derivatives were unaffected by these conditions. The reactions were complete within 1–7h, affording a single product **224** in very high yield except in the case of the guanosine derivatives, where *N*-protection was accompanied with *O*-protection.

Table 3.28 *N*-Protection of *O*-Protected Nucleoside Derivatives Under Ball-Milling Conditions[a]

Entry	R[1]	R[2]	X	Conditions	Time (h)	Yield (%)[b]
1	Ac	Ac	OAc	A	6	90
2	TBDMS	TBDMS	OTBDMS	B	2	99
3	Ac	$C(Me)_2$	$OC(Me)_2$	B	4	96
4	TBDMS	$C(Me)_2$	$OC(Me)_2$	B	2	99
5	Ac	Ac	H	B	1	99
6	TBDMS	TBDMS	H	B	1	99
7	TBDMS	TBDMS	OTBDMS	C	1	99
8	TBDMS	TBDMS	H	C	1	99
9	Ac	Ac	H	C	2	50
10	Ac	Ac	OAc	D	7	25[c]
11	TBDMS	TBDMS	OTBDMS	D	6	40[c]
12	TBDMS	TBDMS	H	E	6	70[c]

[a]*Pyrex glass apparatus, 9mm Pyrex balls. Method A: DMAP:Boc_2O, 1.2:4 mol. Method B: DMAP:Boc_2O, 0.3:3 mol. Method C: DMAP:Boc_2O, 0.8:4 mol. Method D: DMAP:Boc_2O, 1:5mol. Method E: DMAP:Boc_2O, 2.4:6.*
[b]*Isolated yield.*
[c]*O-protection also occurs.*
Reproduced from Sikchi SA, Hultin PG. Solventless protocol for efficient bis-n-Boc Protection of adenosine, cytidine, and guanosine derivatives. J Org Chem 2006;71:5888–91, copyright (2006), with permission from American Chemical Society.

Further *N*-protections were carried out on nucleosides with transiently protected hydroxyl groups **226**. Transient protection of the sugar hydroxyls using the trimethylsilyl (TMS) group was achieved mechanochemically by addition of 3–6 equiv. of HMDS, DMAP, and 3 mol percent of TMSOTf to promote silylation (Scheme 3.59, Table 3.29). DMAP itself was ineffective to achieve solvent-free silylation. In the second reaction step, Boc_2O was added directly to the reaction mixture, and ball milling continued to afford Boc-protected nucleoside amine functionality **227**. Finally, *O*-silyl groups were removed by standard solution procedure yielding the bis-*N*-Boc adenosine and cytidine products **228** in good yields (Table 3.29).

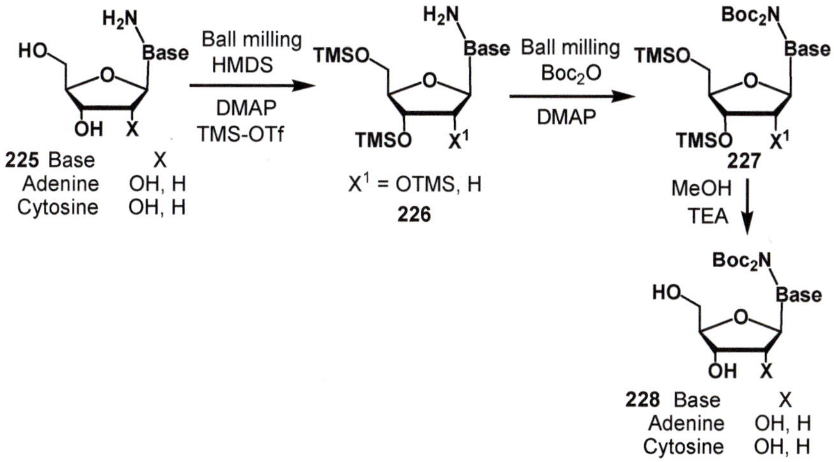

SCHEME 3.59

Transient silylation and *N*-protection of nucleoside derivatives.

Table 3.29 Transient Silylation and *N*-Protection of Nucleoside Derivatives[a]

			Conditions				
Entry	**Base**	**X**	**Silylation[c]**	**Time (h)**	**Boc rxn[d]**	**Time (h)**	**Yield (%)[b]**
1	Adenine	OH	6:0.18:1.2	3	5:0.5	6–7	60
2	Adenine	H	4:0.12:0.4	3	5:1	6–7	72
3	Cytosine	OH	4:0.12:0.4	2	4:0	4	65
4	Cytosine	H	3:0.06:0.3	1	4:0.4	4	50

[a]*Pyrex glass apparatus, 9mm Pyrex balls.*
[b]*Isolated yield.*
[c]*Molar ratio HMDS:TMSOTf:DMAP.*
[d]*Molar ratio Boc_2O:DMAP.*
Reproduced from Sikchi SA, Hultin PG. Solventless protocol for efficient bis-n-Boc Protection of adenosine, cytidine, and guanosine derivatives. J Org Chem 2006;71:5888–91, copyright (2006), with permission from American Chemical Society.

Selective nitrogen protection of α-amino acids was achieved in one-pot/two-step procedure in a ball mill (Scheme 3.60). Colacino et al. developed low-cost ecofriendly method which allows easy and simple preparation of *N*-protected amino acids in short reaction time at room temperature [40]. Different carbamate-protecting groups were utilized: 9-fluorenylmethyloxycarbonyl (Fmoc), *tert*-butyloxycarbonyl (Boc), and benzyloxycarbonyl (Z). The first step is the transient α-carboxylic acid protection as the internal salt **230**, which was achieved by grinding in the presence of an inorganic base (K_2CO_3) and NaCl as grinding auxiliary. Then, in the second step, protection reagent (1 equiv., Boc_2O, benzyloxysuccinimide (ZOSu), or 9-fluorenylmethyloxysuccinimide (FmocOSu)) is added to reaction vessel and milling continued. The final product was recovered by acidic workup, while the by-products were removed in the aqueous phase (inorganic salts or residual α-amino acid, hydroxysuccinimide) and by simple evaporation (*t*-butanol or CO_2), without further product purification. The methodology was applicable to various α-amino acids (selection of data, Table 3.30). The highest yield were obtained for Boc protection, while Z- and Fmoc-protection protocols afforded lower, but still respectable yields (>75%). Efforts to increase yields for Fmoc protection using various mill apparatuses and milling parameters did not succeed.

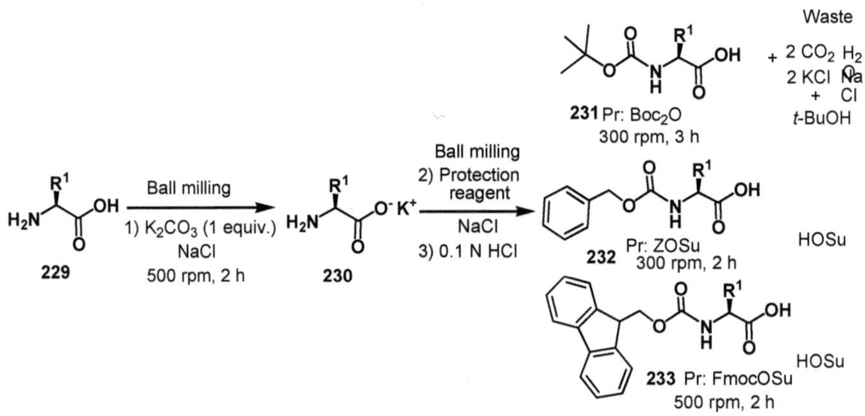

SCHEME 3.60

N-protection of amino acids.

Table 3.30 *N*-Protection of Amino Acids[a]

		Boc[c]	Z[d]	Fmoc[e]
Entry	**Amino acid**	**Yield (%)[b]**	**Yield (%)[b]**	**Yield (%)[b]**
1	H-Phe-OH	100	75	87[f]
2	H-Tyr(Me)-OH	95	78	95
3	H-Ser(Bzl)-OH	96	88	76
4	H-Thr(Bzl)-OH	85	87	
5	H-Glu(Bzl)-OH	68	95	
6	H-His(Bzl)-OH	72	97	
7	H-Trp(For)-OH	95	73	

[a]*Retsch planetary mill PM100, stainless steel jar (12mL), 24×5mm balls, 300rpm, 2h, substrate:K_2CO_3 1:1, NaCl.*
[b]*Isolated yields.*
[c]*Step 2: Boc_2O 1 equiv., 300rpm, 3h.*
[d]*Step 2: ZOSu 1 equiv., 300rpm, 2h.*
[e]*Step 2: FmocOSu 1 equiv., 500rpm, 2h.*
[f]*650rpm.*
Reproduced from Konnert L, Gauliard A, Lamaty F, Martinez J, Colacino E. Solventless synthesis of n-protected amino acids in a Ball mill. ACS Sustain Chem Eng 2013;1:1186–91, copyright (2013), with permission from American Chemical Society.

This synthetic methodology was extended to selective *N*-protection of α- and β-amino esters **234** and **235** and procedure simplified by change of reagents (Scheme 3.61) [41]. Mechanochemical introduction of carbamate-based protecting groups (Fmoc, Z, and Boc) was carried out by Colacino et al. with 1 equiv. of commonly used protection reagents (Fmoc-Cl, Z-Cl, and Boc_2O) in the presence of inorganic base ($NaHCO_3$). Except for β-amino esters yields were similar as for the usual synthetic methods in solution (Table 3.31). Final products were recovered by a simple workup, based on a precipitation/filtration without need of chromatographic purification for most of cases, contrary to synthesis in solution. Great improvement of yield (three times more than in solution) was obtained with Boc protection of the hindered substrate H-Aib-OMe suggesting that mechanochemical activation could be particularly suitable with poorly reactive (hindered) substrates. This solvent-free protocol is environmentally friendly, using stoichiometric amounts of reagents, while the only waste was easily removable water, sodium chloride, CO_2, and *t*-butyl alcohol.

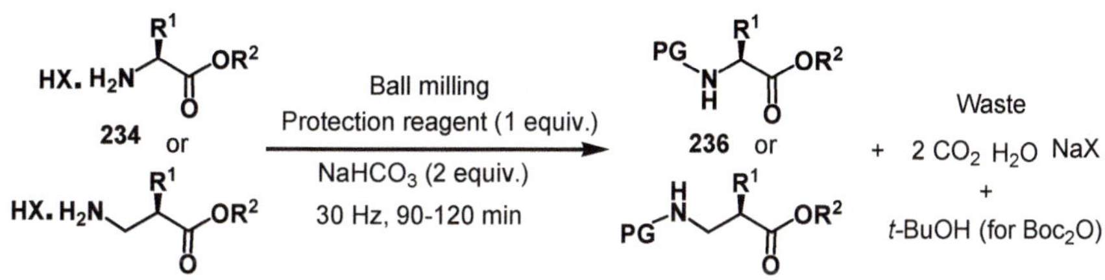

SCHEME 3.61

N-protection of α- and β-amino esters.

Table 3.31 *N*-Protection of α- and β-Amino Esters[a]

Entry	Substrate	Reagent	Time (min)	Yield (%)[b]
1	H-Gly-OMe	Fmoc-Cl	90	96
2	H-Phe-OMe		90	95
3	H-Leu-O^{t}Bu		90	87
4	H-Gly-OMe	Z-Cl	90	80
5	H-Gly-O^{t}Bu		120	96
6	H-Leu-O^{t}Bu		90	67[c]
7	H-Phe-OMe	Boc_2O	90	91
8	H-Pro-OMe		90	81
9	H-Aib-OMe		120	68

[a]*Retsch vibratory mill MM200, stainless steel jar (10 mL), 2×5 mm balls, 30 Hz, substrate:reagent:$NaHCO_3$ 1:1:2.*
[b]*Isolated yields.*
[c]*One 5 mm ball.*
Reproduced from Konnert L, Lamaty F, Martinez J, Colacino E. Solventless mechanosynthesis of n-protected amino esters. J Org Chem 2014;79:4008–17, copyright (2014), with permission from American Chemical Society.

3.7.6 SYNTHESIS OF IMIDES

One of the pioneering organic mechanochemical reaction was selective solid-state reaction of phthalic anhydride **238** and aromatic amine **239** reported by Boldyrev et al. (Scheme 3.62) [42]. When stoichiometric milling of phthalic anhydride and sulfathiazole **239** was carried out, the antibacterial sulfonamide phthalazole (phthalamic acid) **240** was obtained exclusively. The C—N bond formation stops at the 1:1 stage, and bis-amide or imide side products, which usually accompany reaction in the solution or in the melt were not found. Addition of dehydrating agents ($CaCl_2$, $MgSO_4$, and Na_2SO_4) did not facilitate the formation of an imide. Reaction was carried out in a thermostated planetary mill centrifuge at room temperature. It was found that the reaction rate is significantly increased with decreasing diameter of the milling balls. However, increasing the mass of the grinding media to 200 g essentially eliminates the difference of reaction rates of their diameter. It was explained that during the activation of reactants by smaller diameter balls occur greater interactions at larger mechanical contact areas of reacting particles per unit time. Assumption was confirmed by the fact that the rate of the reaction is proportional to the number of grinding balls. Benzoic acid also reacts with sulfathiazole in identical mechanochemical conditions.

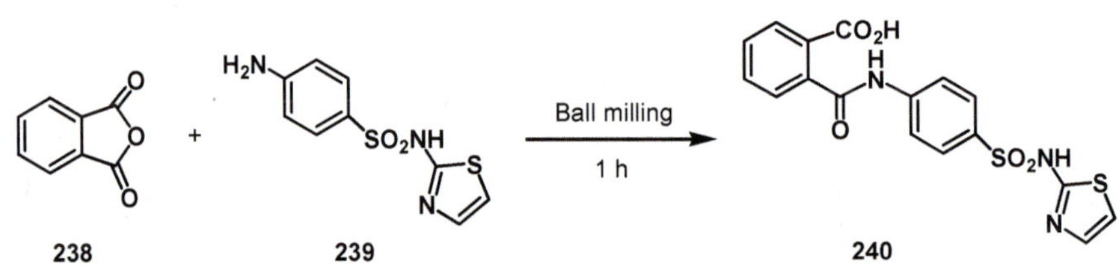

SCHEME 3.62

Synthesis of phthalamic acid.

When phthalic anhydride and 4-toluidine were subjected to the solid–solid reaction by ball milling in a 1:1 ratio, a quantitative yield of the imide **242** was obtained (Scheme 3.63). Kaupp et al. have found that mechanochemical solid–solid reaction of anhydride and amine differs from the solid–gas reaction of tetrachlorophthalic anhydride and gaseous ethylamine which gives different type of product, the amidic carboxylic salt [31].

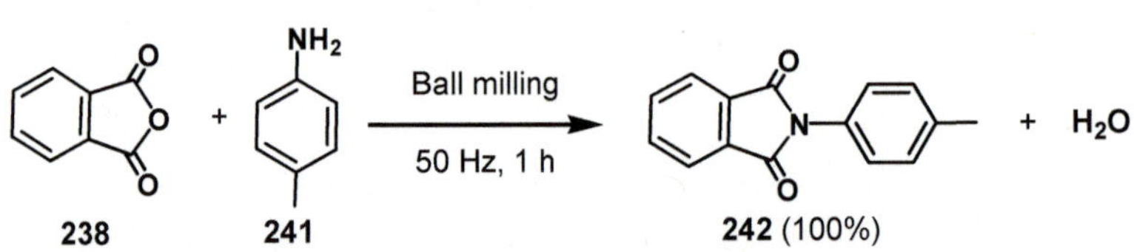

SCHEME 3.63

Synthesis of phthalimide.

3.8 SYNTHESIS OF NITRONES

Lamaty and collaborators have carried out solvent-free synthesis of nitrones in a ball mill [43]. Equimolar amounts of various aldehydes and *N*-substituted-hydroxylamines were used in conjunction with sodium bicarbonate (Scheme 3.64). For stoichiometric reaction between benzaldehyde and *N*-methylhydroxylamine hydrochloride, small amount of dichloromethane was used. This reaction proceeded as a one-step condensation to give the corresponding nitrones in nearly quantitative yields, without the need to exclude air and moisture, in a shorter reaction time than by classical methods, but with comparable yields, where excess of reagents under inert atmosphere are reacted at room temperature or under heating for 1 or 2 days (selected results, Table 3.32). Furthermore, the *E,Z*-isomeric ratio of the products obtained in ball-milling conditions were identical to those already reported in literature: all the nitrones were in the more stable *Z*-form, with the exception of *N*-(naphth-2-ylmethylidene) methylamine *N*-oxide, known to be in the *E*-form. Microwave-assisted reactions gave lower yields than ball mill.

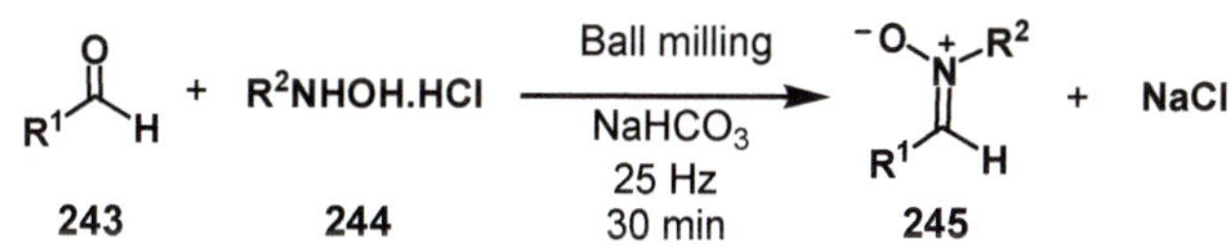

SCHEME 3.64

Synthesis of nitrones.

Table 3.32 Synthesis of Nitrones[a]

Entry	R^1	R^2	Time (h)	Yield (%)[b] BM	Yield (%)[b] MW[c]
1	C_6H_5	CH_3	1	82	72
2	C_6H_5	Bn	1	88	80
3	C_6H_5	$C(CH_3)_3$	2	100	74
4	4-CN-C_6H_4	CH_3	0.5	93	64
5	4-NO_2-C_6H_4	CH_3	0.5	94	
6	4-I-C_6H_4	CH_3	1	100	
7	3-I-C_6H_4	CH_3	1	94	
8	2-I-C_6H_4	CH_3	2	100	
9	4-CH_3-C_6H_4	Bn	2	88	65
10	4-OCH_3-C_6H_4	CH_3	2	91	
11	4-Pyridyl	$C(CH_3)_3$	2	98	
12	2-Furyl	Bn	2	94	
13	2-Thiophenyl	CH_3	2	92	

Continued

Table 3.32 Synthesis of Nitrones[a]—cont'd

Entry	R^1	R^2	Time (h)	Yield (%)[b] BM	Yield (%)[b] MW[c]
14	2-Benzofuranyl	CH_3	0.5	100	
15	2-Naphthyl	Bn	2	100	
16	3-Quinolyl	CH_3	2	95	
17	*n*-Hexyl	CH_3	1	90	

[a]*Retsch MM200 mixer mill, 30 Hz, stainless steel vial (10 mL), 2×7 mm balls, aldehyde: N-substituted hydroxylaminehydrochloride:NaHCO$_3$ 1:1:1 mol.*
[b]*Isolated yields.*
[c]*Microwave: 400 W, 120°C, 1 min.*
Reproduced from Colacino et al. Tetrahedron 2008;64:5569–76, copyright (2008), with permission from Elsevier.

Scale up of reaction in the case of stoichiometric reaction between 4-cyanophenyl aldehyde (5.0 mmol) and *N*-methylhydroxylamine hydrochloride gave limited success. Milling for 30 min showed the formation of nitrone in 88% yield. However, the same experiment on 10 mmol scale showed the lower substrate conversion (68%), which might be a result of an inefficient mixing, showing that the 10 mL jar was not adapted for such a scale.

The effect of three milling operating conditions on the variation of external temperature of the vessel measured during the milling experiment for a period of 30 min at 30 Hz showed that the starting temperature of the sample is about 30°C and rises to 44°C (experiment with empty vessel) or 40°C (experiments on 0.5 and 5 mmol scale) at the end of the test.

3.9 SYNTHESIS OF ENAMINES AND ENAMINE KETONES

Addition of amines to acetylene carboxylates **246** in ball mill was studied by Stolle and Thorwirth (Scheme 3.65) [44]. Employing solvent-free protocol in planetary ball mill in conjunction with quartz sand as inert grinding auxiliary, a number of enamines **247** were prepared in high yield (Table 3.33). No further catalysts or additives was required to accomplish reaction within 5 min. Obtained *E/Z* ratios vary with amine applied: in the case of dicarboxylates, mainly the *E*-isomers have been identified. In contrast, reactions with propiolates and anilines yielded favorably the *Z*-isomer, whereas secondary alkyl amines furnished the *E*-isomer preferably. Simple stirring of DMAD and aniline in flask afforded only 80% conversion and chemoselectivity was lower (*E/Z* ratio of 87:13) in comparison to ball milling. A common method for synthesis is the reaction in solvent, for which was found to have a strong influence on the *E/Z* ratio of the product, while this influence is not present in solvent-free milling.

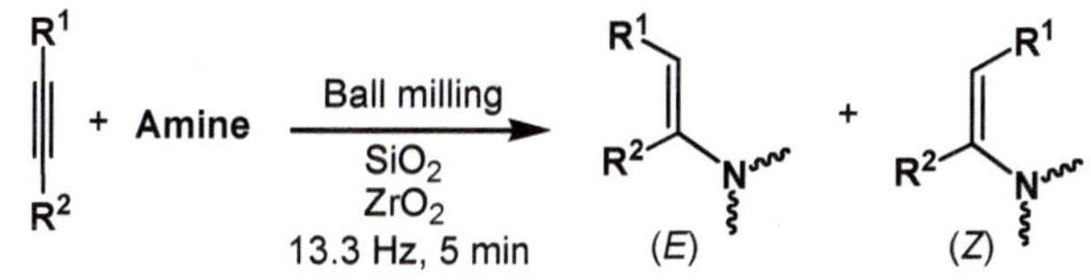

246 **247**

SCHEME 3.65

Addition of amines to acetylenes.

Table 3.33 Addition of Amines to Acetylenes[a]

Entry	R^1	R^1	Amine	Yield (%)[b]	Ratio *E/Z*
1	CO_2Me	CO_2Me	Aniline	91 (80)[c]	97:3 (87:13)[c]
2			*p*-toluidine	87	98:2
3			*o*-toludine	88	96:4
4			*m*-anisidine	89	97:3
5			Piperidine	82	>99:1 < 1
6			Morpholine	84	>99:1 < 1
7			Diisoproply-amine	94	>99:1 < 1
8	CO_2Et	CO_2Et	Aniline	92	97:3
9			*p*-toluidine	93	97:3
10	CO_2Me	H	Aniline	89	7:93
11			*p*-toluidine	85	8:92
12			Piperidine	81	>99:1 < 1
13	CO_2Et	H	*p*-toluidine	92	7:93
14			Piperidine	87	7:93

[a]*Fritsch planetary ball mill Pulverisette 7, ZrO_2-milling beaker (45 mL), 6 × 15 mm ZrO_2 balls, 13.3 Hz, 5 min, substrate:amine 1:1 mol, SiO_2.*
[b]*Isolated products.*
[c]*Stirring in flask.*
Reproduced from Thorwirth et al. Synlett 2011;15:2200–2, with permission from Thieme Publishers.

Kaupp et al. described that the 1,3-dicarbonyl compounds such as 1,3-cyclohexanedione, dimedone, and dehydracetic acid reacted in the solid state with aniline derivatives to obtain enamine ketones **250** and **252** in quantitative yields (Scheme 3.66) [18]. Advantages over the solvent synthesis were identified: it is unnecessary to use acid catalysis, solvents, or liquid phases. A further benefit of avoiding iminium salt intermediates in such synthesis is the lack of workup requirements.

Related mechanochemical protocol was published by Li and collaborators who synthesized a number of β-enamino ketones (esters) **255** in ball mill using acyclic 1,3-diones for substrates without catalyst (Scheme 3.67) [45]. Initial study of reaction

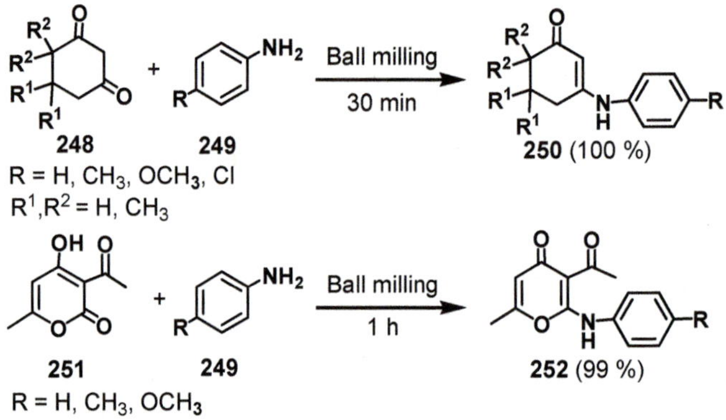

SCHEME 3.66

Synthesis of enamine ketones.

of acetylacetone and ethanolamine identified the optimal reaction conditions (Retsch MM301 ball mill, stainless steel container 9×26 mm, 30 Hz), which were subsequently employed in preparation of series of β-enamino ketones. High yields of products were obtained within 2 h of milling of equimolar amounts of reagents. In general, reactions with acetylacetone were faster and many amines required 30 min to complete, while synthesis of enamine esters with ethyl acetoacetate needed grinding for 1–2 h.

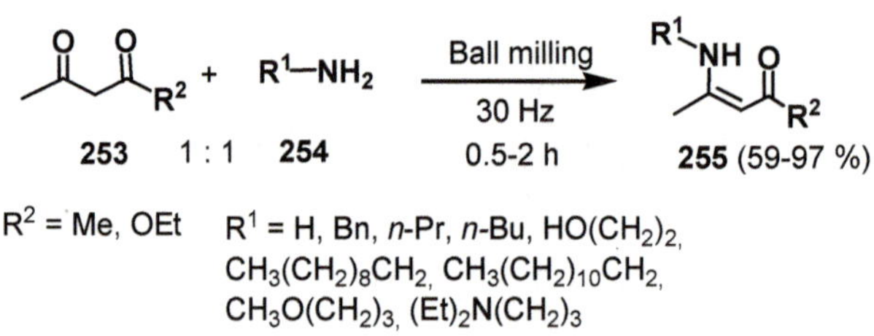

SCHEME 3.67

Synthesis of enamine ketones (esters).

3.10 ALKYLATION OF NITROGEN

Mack et al. carried out *N*-alkylation reaction in ball mill (8000M SpexCertiprep Mixer Mill, stainless steel) [46]. Reaction of 4-bromobenzyl bromide **257** with urea and sodium hydroxide yielding di-(4-bromobenzyl) urea **258** proved that chemical reactions under high-speed ball-milling conditions could be manipulated in order to increase the nucleophilicity of an otherwise weakly nucleophilic substance (Scheme 3.68). Here the hydroxide deprotonates the urea, increasing the nucleophilicity of the nitrogen. Without NaOH, there was a negligible conversion to a dialkyl urea product.

Hydrazones prepared mechanochemically by Lamaty (Section 3.5) were further *N*-alkylated in ball mill [15]. Optimal reaction conditions were grinding of mixture

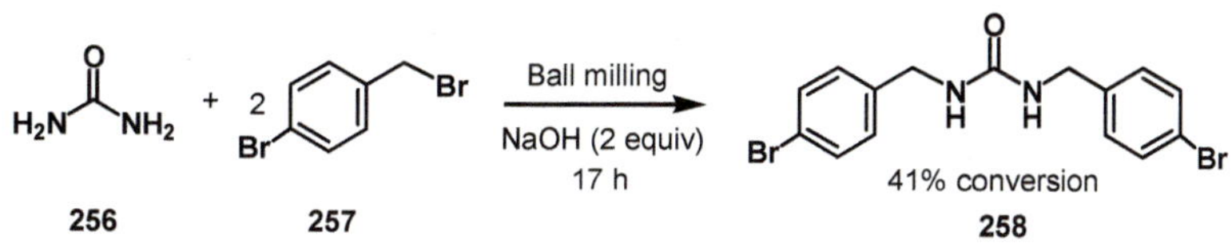

SCHEME 3.68

N-alkylation of urea.

of hydrazone **259**, allyl or phenyl bromide, and Cs_2CO_3 as base in 1:3:3 M ratio during 1 h (in Retsch MM200 mill, 5 mL stainless steel jar, 2×5 mm balls). Simple washing with water afforded the pure products **261** in high yield (Scheme 3.69). For this reaction, different substituents at hydrazones are tolerated.

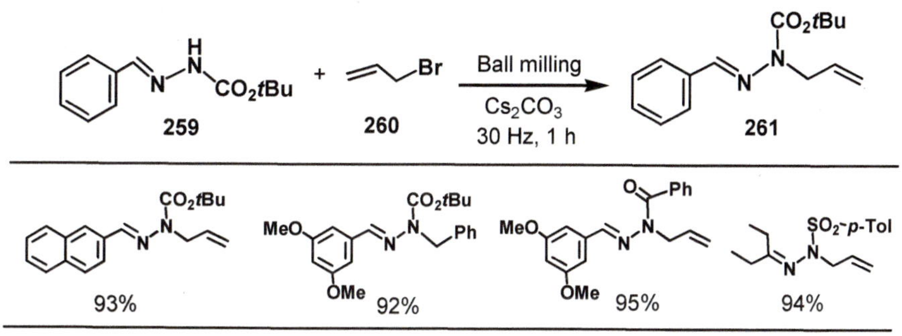

SCHEME 3.69

N-alkylation of hydrazones.

Another mechanochemical *N*-alkylation was effected by Kaupp et al., providing a quaternary iminium salts [18]. *N*-alkylated quaternary iminium salts **263** and **264** were quantitatively prepared from solid imines with solid alkylating agents such as trimethyloxonium or triphenylcarbenium tetrafluoroborate by ball milling within 1 h (Scheme 3.70). The gaseous dimethyl ether that was formed in the synthesis escaped through a Teflon gasket that was set at 50 N m torque. Similarly, ball milling of the 1:1 mixture of benzimidazole and tritylchloride give the salt **267** quantitatively.

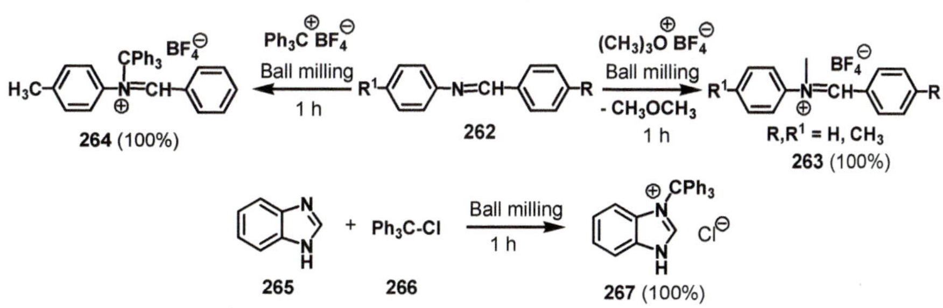

SCHEME 3.70

Synthesis of *N*-alkylated quaternary iminium salts.

3.11 SYNTHESIS OF BIGUANIDES AND GUANIDINES

A series of biguanides was prepared employing nonconventional and environmentally friendly methods by Margetić and Maksić [47]. The C=N bond formation was achieved by addition of tetramethylguanidine **268** to carbodiimides **269**, without the use of catalysts (Scheme 3.71). A comparison of various methods, including mechanochemistry shows that microwave reactions are the best in respect to reaction time and conversion into products (Table 3.34). Since reaction is facilitated by heating, mechanosynthesis of biguanides **270** afforded much lower yields than thermal reactions. The heat generated by ball milling was not adequate to promote reaction. The optimization of ball-milling conditions for the synthesis of biguanide showed in one case better conversion in the planetary mill, compared to mixer ball mill (entry 1). In addition, the influence of size and number of milling balls used in planetary mill was established: smaller size balls and smaller number of balls decreased the conversion (yield): 40×3 mm (26%); 30×3 mm (18%); 40×2 mm (26%); 30×2 mm (18%).

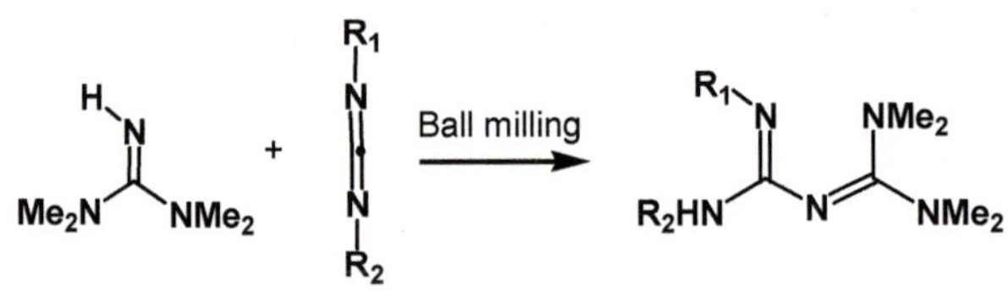

268 269 270

SCHEME 3.71

Mechanochemical synthesis of biguanides.

Table 3.34 Ball-Milling Synthesis of Biguanides[a]

			Yields (%)[b]				
Entry	**R_1**	**R_2**	**BM-1**	**BM-2**	**MW**	**Solution**	**Ultrasound**
1	*i*-Pr	*i*-Pr	15	40	>99 (56)	(94)	53
2	Chex	Chex		5	95 (63)	(62)	43
3	Et	$Me_2N(CH_2)_3$	Traces		>98 (89)	90 (77)	(44)
4	$Me_2N(CH_2)_3$	$Me_2N(CH_2)_3$	<5		>98 (43)	98 (34)	50 (33)
5	$MeO(CH_2)_3$	$MeO(CH_2)_3$	44		>95 (80)	96 (54)	68 (18)
6	$Me_2N(CH_2)_3$	Ph	95 (82)		98 (95)	99 (93)	91 (74)
7	$MeO(CH_2)_3$	*p*-OMePh	94 (86)		99 (92)	99 (93)	86 (75)

[a]*BM-1: Retsch MM400 mixer mill, 30 Hz, 2 h stainless steel jar (10 mL) and one 12 mm ball, NaCl grinding auxiliary; BM-2: Retsch PM200 Planetary Mill, 500 rpm, 2 h, stainless steel jar (10 mL) and 50×3 mm milling balls, NaCl grinding auxiliary; MW: microwave, neat 90–120°C, 5–90 min; solution: THF 90°C, 1–2 h; ultrasound: neat, 50°C, 1 h; guanidine:carbodiimide 1:1.3 mol.*
[b]*NMR yield, isolated yields in parentheses.*
Reproduced from Glasovac et al. Synlett 2013;24:2540–4, with permission from Thieme Publishers.

A comparison of mechanosynthesis with other synthetic methods showed the superiority of microwave-assisted method and solution conditions over mechanochemical and ultrasound reactions for the less-reactive carbodiimides. More-reactive carbodiimides provide full conversion under all conditions, while best yields were obtained in MW and solution reactions (entries 6 and 7).

Analogous mean of formation of the guanidine C═N bond was achieved mechanochemically by Friščić et al. by coupling of arylsulfonamides **271** and carbodiimides **272** (Scheme 3.72) [48]. Guanidines **273** were prepared in high yields when alkyl carbodiimides were applied (Table 3.35). However, TMS and *p*-tolyl carbodiimides did not afford product, and *p*-tolyl carbodiimide reacted only when CuCl was added as catalyst in combination with nitromethane for LAG (entry 6). Environmentally friendly organic synthesis was achieved with solvent-free workup: reaction mixture was milled with and Na_2H_2EDTA for short time (10 min, 25 Hz) and the product simply collected by filtration.

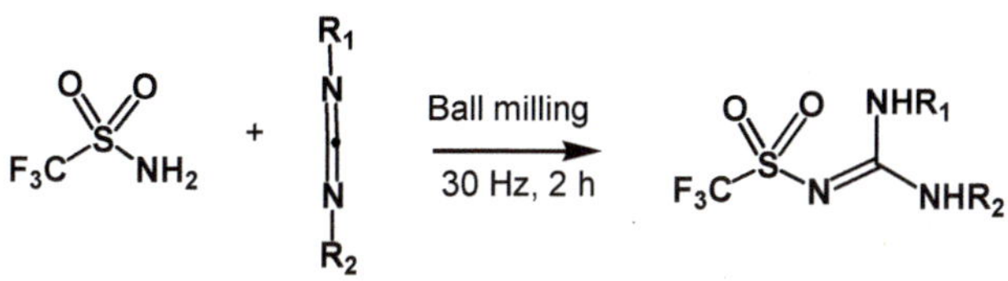

271 272 273

SCHEME 3.72

Mechanochemical synthesis of guanidines.

Table 3.35 Mechanochemical Synthesis of Guanidines From Trifluoromethylsulfonamide[a]

Entry	R_1	R_2	Yields (%)[b]
1	Chex	Chex	98
2	*i*-Pr	*i*-Pr	96
3	*t*-Bu	*t*-Bu	92
4	*t*-Bu	Et	99
5	TMS	TMS	–
6	*p*-Tol	*p*-Tol	- (95)[c]

[a]Retsch MM400 mixer mill, 30 Hz, 2 h stainless steel jar (10 mL) and one 7 or 10 mm milling ball; amide:carbodiimide 1:1 mol.
[b]Isolated yields.
[c]LAG nitromethane (η = 0.25 µL mg^{-1}), 10 mol% CuCl catalyst.
Copyright (2014), Wiley. Used with permission from Tan et al. Angew Chem Int Ed 2014;53:9321–4.

While uncatalyzed reaction of sulfonamides and carbodiimides is limited to aliphatic carbodiimides, the CuCl-catalyzed coupling is more general reaction (Scheme 3.73). CuCl-catalyzed coupling of arylsulfonamides **274** with carbodiimides **275** readily proceeded with DCC, DIC, and DPTC obtaining the corresponding guanidines **276** (Table 3.36), but not with sterically hindered carbodiimides

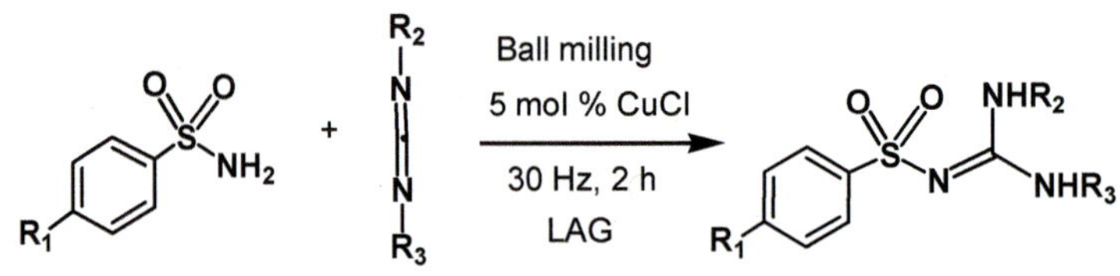

SCHEME 3.73

Copper-catalyzed synthesis of guanidines.

Table 3.36 Copper-Catalyzed Synthesis of Guanidines From Arylsulfonamides[a]

Entry	R_1	R_2	R_3	Yields (%)[b]
1	*p*-Tol	Chex	Chex	98 (0)[g]
2	*p*-Tol	*i*-Pr	*i*-Pr	99
3	*p*-Tol	*p*-Tol	*p*-Tol	98
4	*p*-Cl-C_6H_4	Chex	Chex	91 (96)[h]
5	*p*-Cl-C_6H_4	*i*-Pr	*i*-Pr	95 (8)[g] (87)[h]
6	*p*-Cl-C_6H_4	*p*-Tol	*p*-Tol	88
7	*o*-Cl-C_6H_4	Chex	Chex	94[c] (0)[g]
8	*o*-Cl-C_6H_4	*i*-Pr	*i*-Pr	97
9	*o*-Cl-C_6H_4	*p*-Tol	*p*-Tol	99
10	2-Naphthyl	Chex	Chex	92[f]
11	2-Naphthyl	*i*-Pr	*i*-Pr	99[c,d]
12	*p*-OMe-C_6H_4	Chex	Chex	87[c]
13	*p*-OMe-C_6H_4	*i*-Pr	*i*-Pr	89 (5) [g]
14	*p*-OMe-C_6H_4	*p*-Tol	*p*-Tol	98
15	*p*-NO_2-C_6H_4	Chex	Chex	20[e]
16	*p*-NO_2-C_6H_4	*i*-Pr	*i*-Pr	59[e]

[a]*Retsch MM400 mixer mill, 30 Hz, 2 h stainless steel jar (10 mL) and one 7 or 10 mm milling ball; amide:carbodiimide: CuCl 1:1:0.05 mol, LAG nitromethane (η = 0.25 μL mg^{-1}).*
[b]*Isolated yields.*
[c]*10% CuCl.*
[d]*1.1 equiv. carbodiimide.*
[e]*20% CuCl.*
[f]*10 mol% CuCl, acetone (η = μL mg^{-1}), 4 h.*
[g]*10 mol% CuCl, acetone, reflux 18 h.*
[h]*Multi-gram scale.*

involving *tert*-butyl and trimethylsilyl groups. For the less-reactive *p*-nitrophenylsulfonamide, 20% of CuCl catalyst was needed. Naphthylsulfonamide substrate was also unreactive, and the reaction optimization has indicated that acetone is the most efficient solvent for LAG and longer milling time and catalyst loading of 10% gives the highest conversion (entry 10). In comparison with mechanochemical conditions, synthesis of some of *N*-sulfonylguanidines in acetone afforded very poor yields after 18 h reflux. The applicability of reaction for preparation of large amounts was demonstrated by scale-up of some reactions to gram quantities.

3.12 ADDITIONS TO DOUBLE AND TRIPLE BONDS

Carbon–carbon and carbon–nitrogen double bonds were utilized as an entry point to *N*-functionalization by formation of new carbon–nitrogen bond in mechanosynthetic reactions such as aza-Michael reaction and aminohalogenation of olefins.

3.12.1 AZA-MICHAEL REACTION

The investigation of catalyst-free mechanochemical aza-Michael addition of amines with chalcones by Li et al. led to the formulation of an efficient synthetic protocol (Scheme 3.74) [49]. By applying optimal reaction conditions, a number of aliphatic and aromatic amines **278** in equimolar ratio were reacted with chalcones **277** to obtain azo-Michael products **279** in 44–99% yields (Table 3.37). Compared with conventional methods, the main advantages of the mechanochemical procedure are milder conditions, higher yield, shorter reaction time, and occurrence of no side reactions.

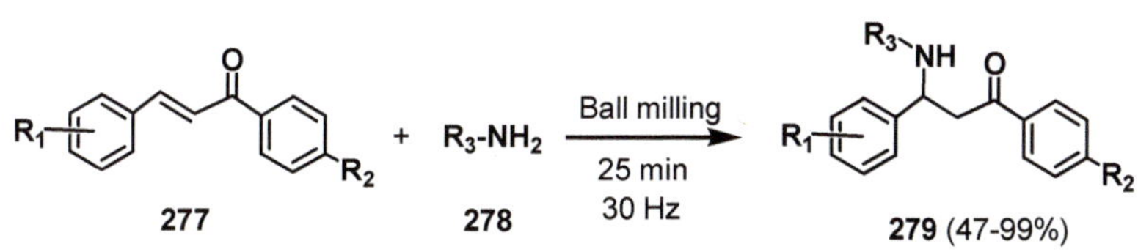

SCHEME 3.74

Aza-Michael addition of amines to chalcones.

Table 3.37 Aza-Michael Addition of Amines to Chalcones[a]

Entry	R_1	R_2	R_3	Yield (%)[b]
1	H	H	Bn	98
2	*m*-Br	H	Bn	99
3	*m*-NO_2	H	Bn	98
4	*p*-CF_3	H	Bn	99
5	*p*-Br	OCH_3	Bn	90
6	*p*-CF_3	OCH_3	Bn	82

Continued

Table 3.37 Aza-Michael Addition of Amines to Chalcones[a]—cont'd

Entry	R_1	R_2	R_3	Yield (%)[b]
7	p-OCH_3	OCH_3	Bn	88
8	p-NO_2	OCH_3	Bn	90
9	H	OCH_3	Bn	94
10	H	H	Bn	83
11	m-NO_2	H	p-CH_3-C_6H_4	59
12	H	H	p-CH_3-C_6H_4	52
13	H	H	p-OCH_3-C_6H_4	47
14	p-CF_3	H	m-Cl-C_6H_4	91
15	m-Br	H	Piperidyl	91
16	H	OCH_3	Piperidyl	89
17	m-NO_2	H	Piperidyl	90
18	p-CF_3	OCH_3	Piperidyl	90

[a]*Retsch MM301 mixer mill, stainless steel capsule, one ball, 30Hz, 25min, chalcone:amine 1:1mol.*
[b]*NMR yields.*
Reproduced from Sci China Chem 2012;55:1252–6, with permission of Springer.

Several highly substituted pyrroles **284** were produced by Kaupp et al. by mechanochemical one-pot reaction of the enamine ketones **282** with *trans*-1,2-dibenzoylethene **280** (Scheme 3.75) [50]. Depending on the reactivity of substrate, different milling temperatures were applied. In addition, ball milling of enamine cyclohexenone **281** under same reaction conditions produced tetrahydroindolone **283** in quantitative yield. Quantitative yields of all products were obtained by heating of reaction mixture after milling for removal of water. In solution, these reactions afforded moderate yields at much higher temperatures (Table 3.38).

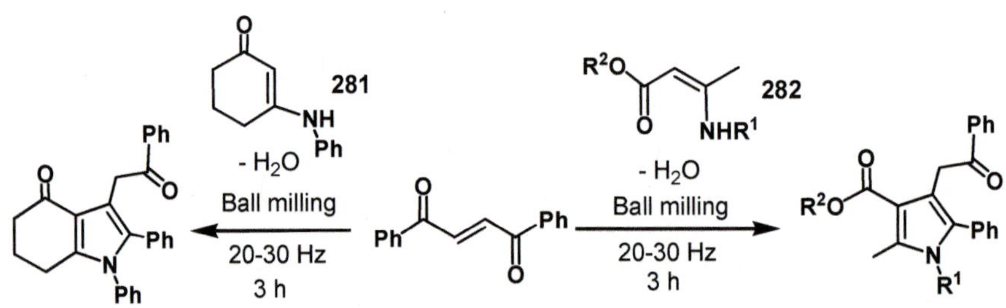

SCHEME 3.75

Mechanochemical synthesis of pyrroles.

Reaction mechanism is postulated to have four sequential steps starting with aza-Michael addition (including hydrogen transfer) followed by imine/enamine rearrangement. The generated amino group adds to the more favorably situated carbonyl

Table 3.38 Mechanochemical Synthesis of Pyrroles[a]

Entry	Pyrrole	R^1	R^2	Time	T (°C)	Yield (%)[b] Milling	Yield (%)[b] Solution[c]
1	**282**	H	CH_3	3	25	100	68
2		CH_3	CH_3	3	25	100	81
3		CH_3	C_2H_5	3	−20	100	78
4		CH_2Ph	C_2H_5	3	0	100	55
5	**283**			3	25	100	46[d]

[a]*Heatable/coolable Retsch MM2000 mixer mill, 20–30 Hz, stainless steel jar (10 mL), two balls (6.5 g), enamine:dibenzoylethene 1:1.*
[b]*Isolated yields.*
[c]*Methanol, reflux 3 h.*
[d]*Xylene, reflux 1.5 h.*

function leading to the formation of a five-membered ring, from which the pyrrole is obtained by elimination of water.

Another example of mechanochemical aza-Michael addition to C=C bond with successive cyclization to nitrogen heterocycle is given by Su et al. They developed an efficient and comparatively green strategy for the synthesis of 1,3,5-triaryl-2-pyrazolines **287** by ball milling (Scheme 3.76) [51]. Cyclization of chalcone and phenylhydrazine in the presence of $NaHSO_4{\cdot}H_2O$ was found to proceed most efficiently with chalcone:phenylhydrazine:$NaHSO_4{\cdot}H_2O$ ratio of 1:2:0.2 in the presence of silica gel as grinding auxiliary. To prevent "overheating" of the reaction mixture, a milling was carried out for 5 min followed by a 5 min pause. Excess of phenylhydrazine was successfully recovered by chromatography and reused without noticeable effect on the yield. The generality of the mechanochemical procedure was demonstrated by employment of the optimal reaction conditions to structurally diverse chalcones **285** and phenylhydrazines **286** (Table 3.39). All reactions were high yielding, in which substrates possessing electron-donating groups are more reactive. The reuse of the phenyl hydrazine resulted in 93%, 92%, 90%, and 87% yield over four recycles (entry 1), showing recyclability of hydrazine excess. Finally, the employment of thiosemicarbazide and benzylideneacetone afforded 1,3,5-trisubstituted-2-pyrazolines possessing one nonaromatic substituent (entries 16 and 18).

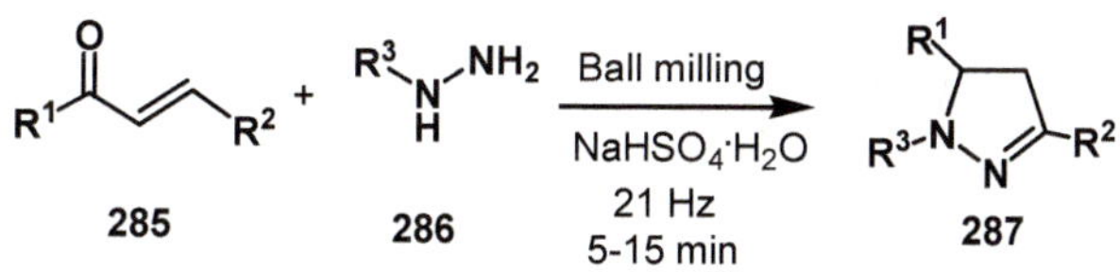

SCHEME 3.76

Synthesis of 1,3,5-triaryl-2-pyrazolines.

Table 3.39 Mechanochemical Synthesis of 1,3,5-Triaryl-2-Pyrazolines[a]

Entry	R^1	R^3	R^3	Time (min)	Yield (%)[b]
1	C_6H_5	C_6H_5	C_6H_5	5	95[c]
2	C_6H_5	C_6H_5	4-ClC_6H_5	5	91
3	4-ClC_6H_5	C_6H_5	C_6H_5	5	92
4	4-ClC_6H_5	C_6H_5	4-ClC_6H_5	5	90
5	C_6H_5	4-ClC_6H_5	C_6H_5	5	90
6	C_6H_5	4-ClC_6H_5	4-ClC_6H_5	15	89
7	4-ClC_6H_5	4-ClC_6H_5	C_6H_5	5	92
8	4-ClC_6H_5	4-ClC_6H_5	4-ClC_6H_5	5	91
9	4-$CH_3OC_6H_5$	4-ClC_6H_5	C_6H_5	5	90
10	C_6H_5	4-$CH_3OC_6H_5$	C_6H_5	5	92
11	4-$CH_3OC_6H_5$	4-$CH_3OC_6H_5$	C_6H_5	5	93
12	4-$O_2NC_6H_5$	C_6H_5	C_6H_5	15	85
13	C_6H_5	4-$O_2NC_6H_5$	C_6H_5	15	82
14	C_6H_5	C_6H_5	$NH_2C{=}S$	15	80
15	C_6H_5	2-furyl	C_6H_5	15	76
16	C_6H_5	CH_3	C_6H_5	15	62

[a]*Planetary centrifugal mill AGO-2 (acceleration: 60g), stainless steel vial 35mL, 5mm balls (75g), substrate:hydrazine:$NaHSO_4$ 10:2:0.2 mol, silica gel grinding auxiliary 10g.*
[b]*Isolated yields.*
[c]*Recycled Ph hydrazine 93%, second recycle 92%, third recycle, fourth recycle 87%.*
Reproduced from Zhu X, Li Z, Jin L, Xu L, Wu Q, Su W. Mechanically activated synthesis of 1,3,5-triaryl-2-pyrazolines by high speed ball milling. Green Chem 2009;11:163–5, with permission from the Royal Society of Chemistry.

An identical synthetic strategy was employed by Zhang et al. for the efficient synthesis of 3,5-diphenyl-1*H*-pyrazoles under ball-milling conditions (Scheme 3.77) [52]. Two-step, one-pot procedure starts with the grinding of chalkone and slight excess of

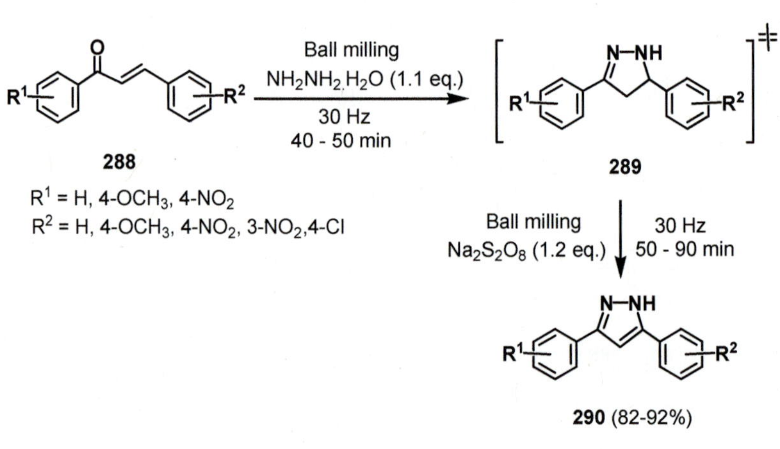

SCHEME 3.77

Synthesis of 3,5-diphenyl-1*H*-pyrazoles.

hydrazine hydrate in Retsch MM400 mill (25 mL stainless steel vial, 7 mm ball). Upon completion of aza-Michael reaction and subsequent cyclization (after 40–50 min), without the isolation of pyrazoline **289**, sodium persulfate oxidant (in slight excess) was introduced in milling jar and second milling carried out. 1*H*-pyrazoles **290** were formed after 50–90 min in yields higher than 80% and nitro group–containing substrates were less reactive and required the longest milling times. Environmentally friendly process employing persulfate is completed by isolation of products by washing with ethanol/water. This mechanochemical solvent-free procedure is superior to solution synthesis in which mixture of pyrazoline and 1*H*-pyrazole was formed.

Application of solid-state chemistry for quantitative multistep cascades in a ball mill is also demonstrated by reaction of enamine ketone **291** with 1,2-dibenzoylethene **292** (Scheme 3.78). Pyrrole derivative **293** was obtained by Kaupp et al. in quantitative yield through four reaction steps (vinylogous Michael addition, imine/enamine rearrangement, cyclization, and elimination), without the use of acid catalysts [18].

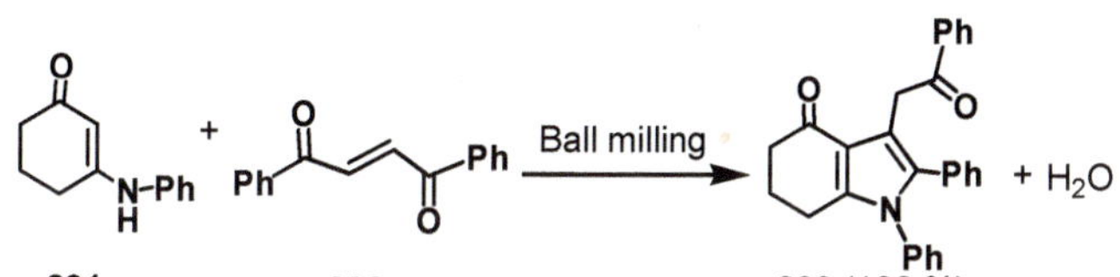

291 **292** **293** (100 %)

SCHEME 3.78

Cascade reaction of enamine ketone.

3.12.2 AMINOHALOGENATION OF OLEFINS

3.12.2.1 Aminochlorination of Olefins

Mechanochemical aminochlorination of electron-deficient olefins with chloramine-T offers an access to highly functionalized systems (Scheme 3.79) [53]. Variety of additives was tested in the reaction of 1,3-diphenylpropenone to find out that reactions promoted by $PhI(OAc)_2$ (50 mol%) are the most efficient to produce vicinal chloroamino compounds **295**. Hydrates of metal chlorides (Cu, Fe, Co, Ni, Cr, Ce, Sn, and In) were less active and CuI nonreactive. Optimized reaction conditions were then employed for various electron-deficient olefins **294** (Table 3.40). Comparison with reactions carried out in organic solvents showed that the solvent-free mechanochemical reactions were the most efficient in terms of product yield and reaction time (entry 1).

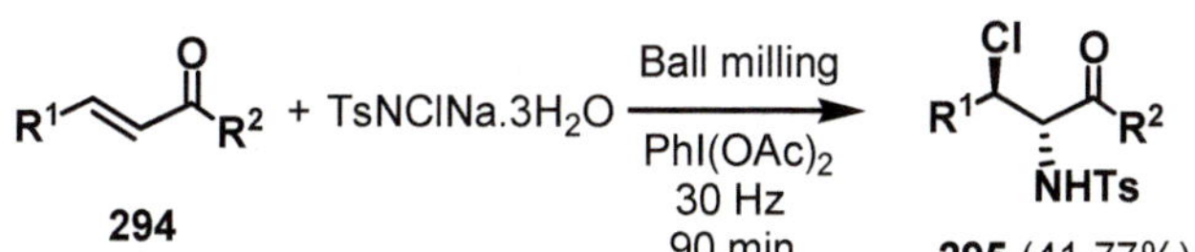

294 **295** (41-77%)

R^1,R^2 = aliphatic, aromatic

dr (*anti/syn* 75/25-99/1)

SCHEME 3.79

Aminochlorination of electron-deficient olefins.

Table 3.40 Aminochlorination of Electron-Deficient Olefins Under Ball-Milling Conditions[a]

Entry	R^1	R^2	Yield (%)[b]	dr (*anti/syn*)
1	Ph	Ph	77	91/9
2	4-$MeOC_6H_4$	Ph	59	75/25
3	4-ClC_6H_4	Ph	75	>99/1
4	2-ClC_6H_4	Ph	63	>99/1
5	3,4-$Cl_2C_6H_3$	Ph	73	>99/1
6	4-$NO_2C_6H_4$	Ph	63[c]	>99/1
7	Ph	4-$MeOC_6H_4$	72	01/9
8	4-ClC_6H_4	4-$MeOC_6H_4$	78	90/10
9	Ph	4-ClC_6H_4	70	>99/1
10	4-ClC_6H_4	4-ClC_6H_4	75	>99/1
11	Ph	Me	41	96/4
12	Et	Ph	0	–
13	*i*-Bu	Me	0	–
14	Ph	OMe	65	>99/1
15	Ph	NEt_2	52	>99/1

[a]*Retsch MM200 mixer mill, 1800 rpm (30 Hz), 1.5 h, stainless steel jar (5 mL), one 7 mm ball, olefin:chloramine-T: $PhI(OAc)_2$ 1:2:0.5 mol.*
[b]*Isolated yield.*
[c]*3 h, chloramine-T 4 mol.*
Copyright (2007), Wiley. Used with permission from Wang et al. Adv Synth Catal 2007;349:1977–82.

3.12.2.2 Aminobromination of Olefins

When chloramine-T used in previous aminochlorination reaction was replaced with *p*-tosylsulfonamide/NBS reagent, corresponding aminobromination product **297** was obtained from 1,3-diphenylpropenone, but also accompanied with dibrominated product **298** (Scheme 3.80, entry 1, Table 3.41) [54]. Different Lewis acids and change of their molarity afforded various reaction outcomes. The highest chemoselectivity toward aminobromination product was obtained by increasing $PhI(OAc)_2$ loading to 75 mol% (entries 7 and 8), while dibromination reaction was not visible. On the other hand, the use of CuI, $InCl_3{\cdot}4H_2O$, or $FeCl_3{\cdot}6H_2O$ shifted the chemoselectivity toward dibromo product (entries 2 and 4).

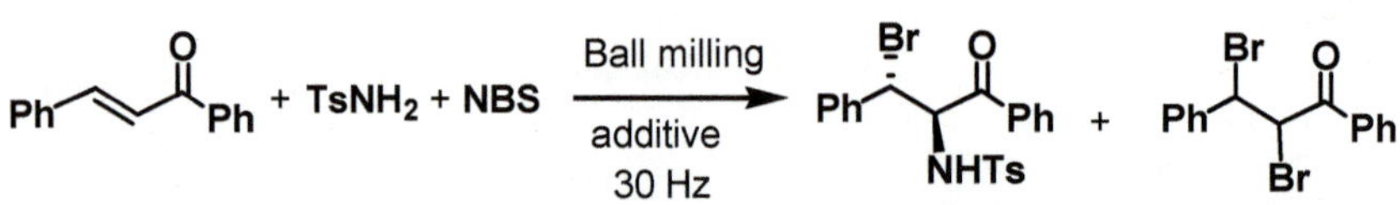

296 297 298

SCHEME 3.80

Aminobromination of 1,3-diphenylpropenone.

Table 3.41 Aminobromination of 1,3-Diphenylpropenone ($R^1 = R^2 = Ph$)[a]

Entry	Additive	Additive Loading (mol%)	Yield (%)[b] 297	Yield (%)[b] 298	dr (*anti/syn*)
1	$PhI(OAc)_2$	25	63	25	91/9
2	CuI	25	0	57	–
3	$MnSO_4{\cdot}H_2O$	25	7	20	92/8
4	$FeCl_3{\cdot}6H_2O$	25	0	38	–
5	$InCl_3{\cdot}4H_2O$	25	0	62	–
6	$PhI(OAc)_2$	50	75	10	92/8
7	$PhI(OAc)_2$	75	83	Trace	92/8
8	$PhI(OAc)_2$	75	69[c]	Trace	92/8
9	$PhI(OAc)_2$	75	67[d]	11	92/8

[a]*Retsch MM200 mixer mill, 1800 rpm (30 Hz), 1.5 h, stainless steel jar (5 mL), one 7 mm ball, 1,3-diphenylpropenone:*$TsNH_2$*:NBS 1:2:1.5 mol.*
[b]*Isolated yield.*
[c]*NBS 1 equiv.*
[d]$TsNH_2$ *1 equiv.*
Reproduced from Wu X-L, Xia J-J, Wang G-W. Aminobromination of olefins with TsNH2 and NBS as the nitrogen and bromine sources mediated by hypervalent iodine in a ball mill. Org Biomol Chem 2008;6:548–53, with permission from the Royal Society of Chemistry.

Application of optimized conditions to various electron-deficient olefins afforded aminobromination products **300** in 48–83% yields and with high diastereomeric ratios (Scheme 3.81, selected results Table 3.42).

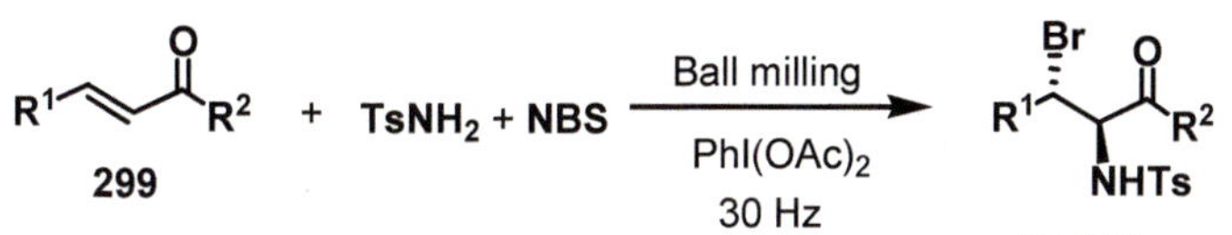

299

R^1, R^2 = aliphatic, aromatic

(±) 300

SCHEME 3.81

Aminobromination of electron-deficient olefins.

Table 3.42 Aminobromination of Electron-Deficient Olefins Under Ball-Milling Conditions[a]

Entry	R^1	R^2	Time (h)	Yield (%)[b]	dr (*anti/syn*)
1	Ph	Ph	1.5	83	92/8
2	$4\text{-}CH_3C_6H_4$	Ph	1.5	61	92/8
3	$2\text{-}ClC_6H_4$	Ph	1.5	79	94/6
4	$3,4\text{-}Cl_2C_6H_3$	Ph	1.5	76[c]	93/7
5	$4\text{-}NO_2C_6H_4$	Ph	3	73[c]	95/5
6	$4\text{-}ClC_6H_4$	$4\text{-}ClC_6H_4$	1.5	71[c]	94/6

Continued

Table 3.42 Aminobromination of Electron-Deficient Olefins Under Ball-Milling Conditions[a]—cont'd

Entry	R^1	R^2	Time (h)	Yield (%)[b]	dr (*anti*/*syn*)
7	Ph	4-$CH_3OC_6H_4$	1.5	83	92/8
8	Ph	CH_3	1.5	70	92/8
9	4-ClC_6H_4	OEt	1.5	57	89/11
10	Ph	$N(C_2H_5)_2$	1.5	69	>99/1

[a]*Retsch MM200 mixer mill, 1800 rpm (30 Hz), stainless steel jar (5 mL), one 7 mm milling ball, alkene:$TsNH_2$:NBS: $PhI(OAc)_2$ 1:2:1.5:0.75 mol.*
[b]*Isolated yield.*
[c]*$TsNH_2$ 4 equiv., $PhI(OAc)_2$ 1 equiv.*
Reproduced from Wu X-L, Xia J-J, Wang G-W. Aminobromination of olefins with TsNH2 and NBS as the nitrogen and bromine sources mediated by hypervalent iodine in a ball mill. Org Biomol Chem 2008;6:548–53, with permission from the Royal Society of Chemistry.

Interestingly, when olefin **301** with a strong electron-donating group substituted on the phenyl ring of the double bond was employed in the reaction, a reversed regioselectivity was observed and product **302** obtained (Scheme 3.82).

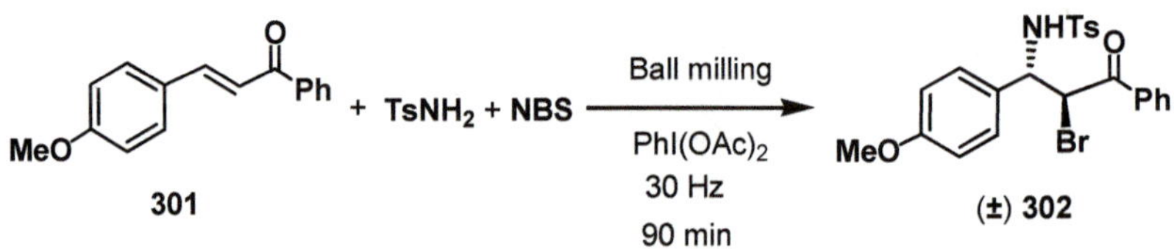

SCHEME 3.82

Aminobromination of olefin **301**.

Simple olefins such as styrene were also reactive in mechanochemical conditions and two regioisomeric aminobrominated products were obtained (Scheme 3.83).

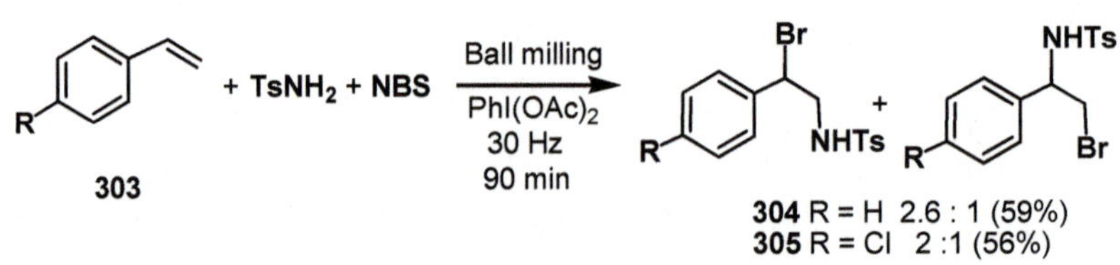

SCHEME 3.83

Aminobromination of styrenes.

Mechanochemical aminobromination of 1,3-diphenylpropenone with other sulfonamides showed that the electron-withdrawing substituents retard the reaction (Scheme 3.84, Table 3.43).

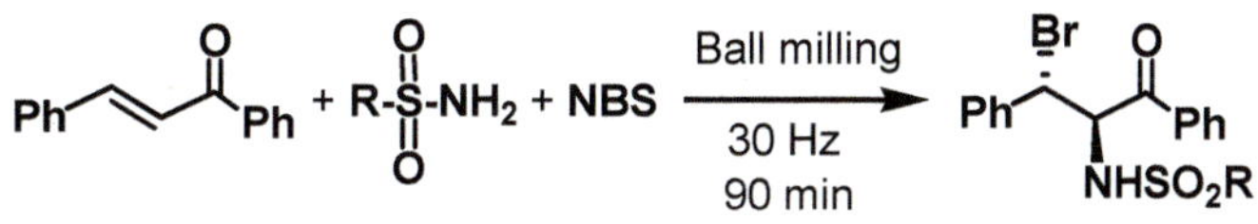

306 307 (±) 308

SCHEME 3.84

Aminobromination of 1,3-diphenylpropenone.

Table 3.43 Aminobromination of 1,3-Diphenylpropenone With sulfonamides[a]

Entry	R	Yield (%)[b]	dr (*anti/syn*)
1	4-$CH_3C_6H_4$	83	92/8
2	Ph	85	92/8
3	4-$NO_2C_6H_4$	37	89/11
4	CH_3	69	89/11

[a]*Retsch MM200 mixer mill, 1800 rpm (30 Hz), stainless steel jar (5 mL), one 7 mm ball, alkene:sulfonamide:NBS: $PhI(OAc)_2$ 1:2:1.5:0.75 mol.*
[b]*Isolated yield.*
Reproduced from Wu X-L, Xia J-J, Wang G-W. Aminobromination of olefins with TsNH2 and NBS as the nitrogen and bromine sources mediated by hypervalent iodine in a ball mill. Org Biomol Chem 2008;6:548–53, with permission from the Royal Society of Chemistry.

3.12.3 INTRAMOLECULAR HYDROAMINATION

N-addition to alkyne triple bond was also utilized in the synthesis of nitrogen-containing heterocycles. Lewis-catalyzed cyclization (intramolecular hydroamination) of 2-(phenylethynyl)aniline **309** yielding 2-phenyl-1*H*-indole in ball mill was studied by Stolle et al. (Scheme 3.85) [55]. Amidst catalysts tested, zinc bromide was found to be the most satisfactory with good selectivity for indole product, and stoichiometric amounts are required for high yields. Sodium chloride was used as a grinding auxiliary, because it is not participating in the reaction, it has a high stiffness and a high melting point, hence eutectic melts will not occur while the reaction is running. Hardened steel vessel and balls produced the highest yields and selectivity in comparison with copper and ZrO_2-milling materials. Two ways to carry out the cyclization were compared: the fast way (45 min, 800 rpm) results in a slightly lower selectivity and lower yield (Table 3.44), while the slow way (8 h, 200 rpm) afforded a higher yield; however, the milling time had to be significantly increased. Most synthetic procedures for intramolecular hydroamination in solution utilize toluene at reflux, whereas ball milling proceeds in the solid state and the bulk temperatures did not raise above 80°C. A drawback of the present method is the need to employ stoichiometric amounts of $ZnBr_2$.

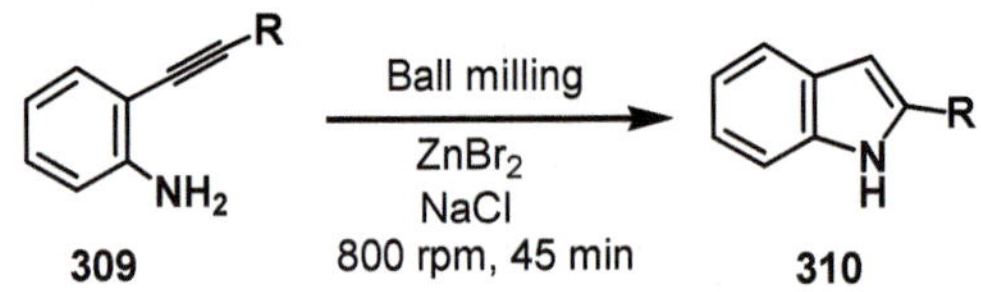

SCHEME 3.85

Synthesis of indoles by intramolecular hydroamination.

Table 3.44 Synthesis of Indoles by Intramolecular Hydroamination[a]

		800 rpm		200 rpm[d]	
Entry	R	Conversion (%)[c]	Yield (%)[b]	Conversion (%)[c]	Yield (%)[b]
1	C_6H_5	>99	72 (91)[e]	>99	82
2	$4\text{-}CH_3C_6H_4$	>99	60	>99	80
3	2-pyridyl	>99	59	>99	71
4	*n*-octyl	>99	69	>99	81
5	TMS	>99	28[c]	30	0

[a]*Fritsch planetary ball mill Pulverisette 7, stainless steel vessel (45 mL), 6 × 15 mm balls, 800 rpm, 45 min, substrate:ZnBr$_2$ 1:1 mol.*
[b]*Isolated products.*
[c]*GC.*
[d]*200 rpm, 450 min.*
[e]*Toluene, reflux 3 h.*
Reproduced from Zille M, Stolle A, Wild A, Schubert US. ZnBr$_2$-mediated synthesis of indoles in a ball mill by intramolecular hydroamination of 2-alkynylanilines. RSC Adv 2014;4:13126–33, with permission from the Royal Society of Chemistry.

3.12.4 ADDITIONS TO IMINE BOND

Several studies report synthesis of nitrogen-containing products in ball mill in which the initial reaction step is formation of an imine bond, followed by addition across the C=N bond in the successive reaction step. Olefinic amines and heterocycles such as thiazolidine and benzimidazole were prepared in this way. Kaupp et al. have carried out several mechanochemical reactions by milling of stoichiometric quantities of (L)-cysteine **311** and paraformaldehyde (in Retsch MM2000 mill, with a 10 mL stainless steel beaker and balls) affording a quantitative yield of the (*R*)-thiazolidine **312** within 1 h (Scheme 3.86) [31]. The crystal water was then removed by drying. The formation of product takes place via an initial imine formation and subsequent cyclization.

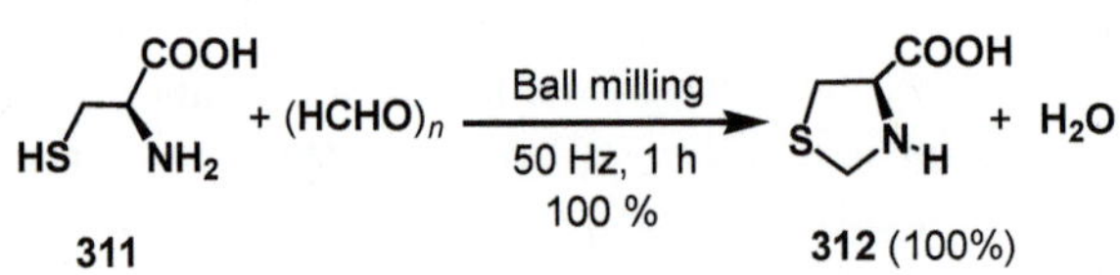

SCHEME 3.86

Synthesis of thiazolidine.

The use of solid paraformaldehyde as the most convenient reagent in a ball mill is an innovative approach in sustainable solid-state chemistry.

Even methylene iminium salts can be formed from ammonium salts and paraformaldehyde in the ball mill [18]. Kaupp found that they are extremely labile but were efficiently trapped by adjacent thiol groups, such as in the reaction of (L)-cysteine hydrochloride monohydrate yielding the (L)-thiazolidine hydrochloride **315**, presumably via an imine intermediate **314** (Scheme 3.87).

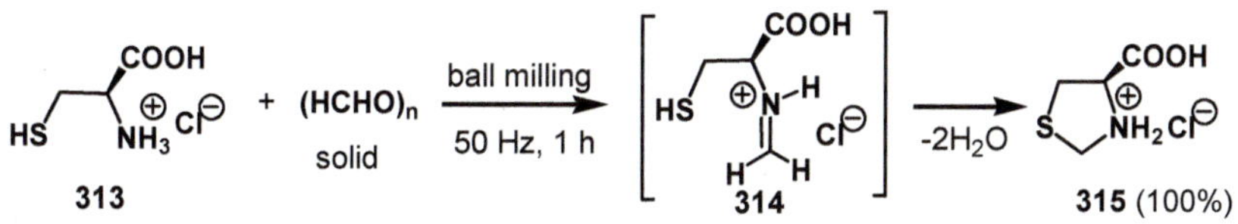

SCHEME 3.87

Synthesis of (L)-thiazolidine hydrochloride.

Imines generated in situ by mechanosynthesis could be used for further chemical transformations in ball mill. For instance, solid-state radical reactions of 1, 3-cyclohexanediones with in situ generated imines mediated by $Mn(OAc)_3{\cdot}2H_2O$ were efficiently carried out under mechanical-milling conditions [56]. Wang obtained good to excellent yields of products by sequential two-step procedure (Scheme 3.88). Milling of 4-methylaniline with aromatic aldehydes afforded quantitatively imines **318**, which were directly used as radical acceptors in the following reaction step with 1,3-cyclohexanediones and 2 equiv. of manganese acetate (Table 3.45). Other metal salts which are traditionally used for the generation of carbon-centered radicals were less efficient (entry 5). In comparison, traditional $Mn(OAc)_2$-mediated radical reactions in solution usually give yields lower than 70%. Another advantage of ball-mill process is that in some cases it is very difficult to obtain the imines quantitatively in the presence of organic solvents, for instance 4-chlorobenzaldehyde/4-methylaniline condensation in CH_3CN at room temperature. Reaction completion was slow and isolation of imine after the second step and 6h in these conditions led to a very complex reaction mixture, containing just a trace of desired product (entry 4).

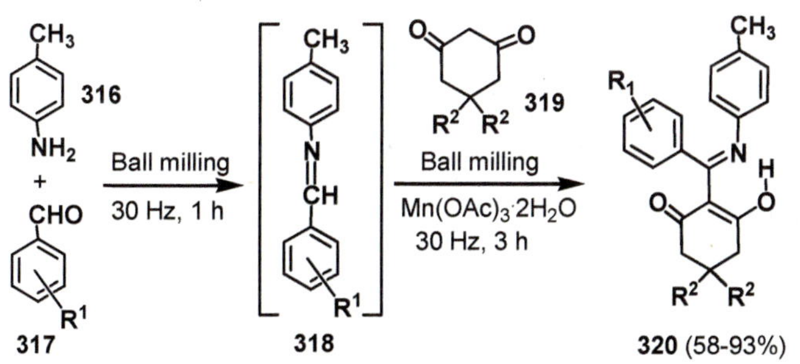

SCHEME 3.88

Solid-state radical reactions of in situ generated imines mediated by $Mn(OAc)_3{\cdot}2H_2O$.

Table 3.45 Radical Reactions of 1,3-Cyclohexanedione and Dimedone With In Situ Generated Imines[a]

Entry	R^1	R^2	Yield (%)[b]
1	4-NO_2	CH_3	84
2	3-NO_2		81
3	4-CN		94
4	4-Cl		89 (Trace)[c]
5	4-Br		93 (36,0)[d]
6	3,4-Cl		91
7	3,4-OCH_2O		84
8	4-NO_2	H	76
9	3-NO_2		73
10	4-CN		78
11	4-Cl		71
12	4-Br		71
13	3,4-Cl		74
14	3,4-OCH_2O		58

[a]Retsch MM200 mixer mill, 30 Hz, 1 h, stainless steel jar (5 mL), one 7 mm ball, aldehyde:aniline 1:1, then 1,3-cyclohexadione:$Mn(OAc)_3{\cdot}2H_2O$ 1:2, 3 h.
[b]Isolated yield.
[c]Acetonitrile 6 h.
[d]CAN and $Cu(OAc)_2$.
Reproduced from Zhang Z, Wang G-W, Miao C-B, Dong Y-W, Shen Y-B. Solid-state radical reactions of 1,3-cyclohexanediones with in situ generated imines mediated by manganese(III) acetate under mechanical milling conditions. Chem Commun 2004;1832–3, with permission from the Royal Society of Chemistry.

Mechanochemical preparation of benzimidazoles is a further example of synthesis of nitrogen heterocycles via initial imine formation. Li et al. have carried efficient solvent-free synthesis of benzimidazoles in ball mill (Scheme 3.89) [57]. Optimized reaction conditions required 2 equiv. of aromatic aldehyde and $FeCl_3{\cdot}6H_2O$ (25 mol%) as Lewis acid catalyst, in conjunction with NaCl as grinding auxiliary. Thus, milling for 2 h provided the corresponding benzimidazoles **325** in high yields, with the exception of 2-nitrobenzaldehyde (Table 3.46). This simple method does not require any oxygen or moisture precautions, giving selectively one product. Limitations were encountered in the case of aliphatic aldehydes, as *n*-butanal and citral both gave a complex reaction mixtures. Further condensations of *o*-phenylenediamine with 1 or 2 equiv. of aldehyde in ball mill without Fe catalyst afforded quantitatively the imine **324** and bisimine **325**, respectively. In the ball-mill conditions, it was difficult for imines to cyclize or aromatize to benzimidazole derivatives without catalyst, and this behavior is in contrast to reported solvent preparations where cyclizations are facilitated by subsequent oxidation steps. Solvent-free experiments indicated that $FeCl_3{\cdot}6H_2O$ was more inclined to act as a Lewis acid rather than a redox reagent in this system, which is at variance to solvents. These results showed that the reaction pathway and chemical selectivity were changed in this solid state and mechanically activated solvent-free system. Possible mechanism starts with the formation of bis-imines, followed by Fe(III) promoted cyclization and subsequent hydride shift.

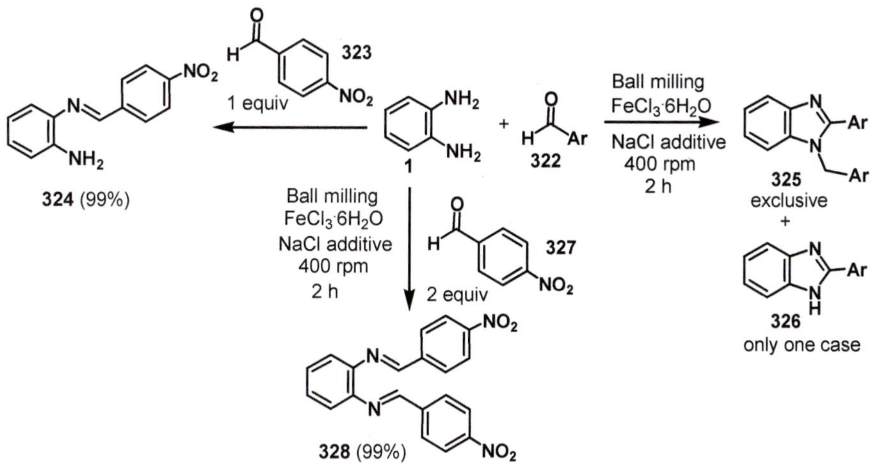

SCHEME 3.89

Synthesis of benzimidazoles.

Table 3.46 Synthesis of Benzimidazoles[a]

Entry	Ar	Yield (%)[b]
1	C_6H_5	96
2	4-HOC_6H_4	85
3	4-$MeOC_6H_4$	93
4	4-$Me_2NC_6H_4$	89
5	4-ClC_6H_4	81
6	4-BrC_6H_4	91
7	4-$NO_2C_6H_4$	92
8	3-$NO_2C_6H_4$	79
9	2-$NO_2C_6H_4$	–
10	2-FC_6H_4	87
11	3-MeO-4-HOC_6H_3	90
12	3,4-$OCH_2OC_6H_3$	91
13	2,4-$Cl_2OC_6H_3$	57
14	2-furyl	91
15	2-thienyl	94

[a]Retsch planetary ball-milling apparatus Pulverisette 7, 400 rpm, 2 h, Teflon jar (50 mL), 5×5 mm agate balls, diamine:aldehyde:catalyst 1:2:0.25 mol, NaCl grinding auxiliary.
[b]Isolated yields.

Analogous reaction starting with imine formation is solvent-free method developed by Jang et al. for the synthesis of benzothiazole, benzimidazole, and benzoxazole derivatives in the presence of ZnO nanoparticles (ZnO-NPs) capped with a ligand **330** as the catalyst (Scheme 3.90) [58]. For synthesis of benzothiazoles (benzimidazole and benzoxazole) **331**, a mixture of 2-aminothiophenol (*o*-phenylenediamine or 2-aminophenol) **329**, appropriate benzaldehyde and 0.5 mol% ZnO-NPs was reacted in a tungsten carbide ball mill. At the end of the reaction, the crude products were simply washed with water and methanol to afford products in high yield (Table 3.47). The scalability of this solvent-free methodology was studied on the reaction of 2-aminothiophenol with salicylaldehyde. When performed on a larger scale, 80% yield of the corresponding benzothiazole derivative was achieved, demonstrating that the reaction is applicable to multi-gram–scale syntheses. The recovered catalyst was successfully recycled five times without any detrimental effect on catalytic efficiency. Mechanistic studies (time-dependent ^{1}H NMR) suggest that the reaction proceeds through an imine intermediate.

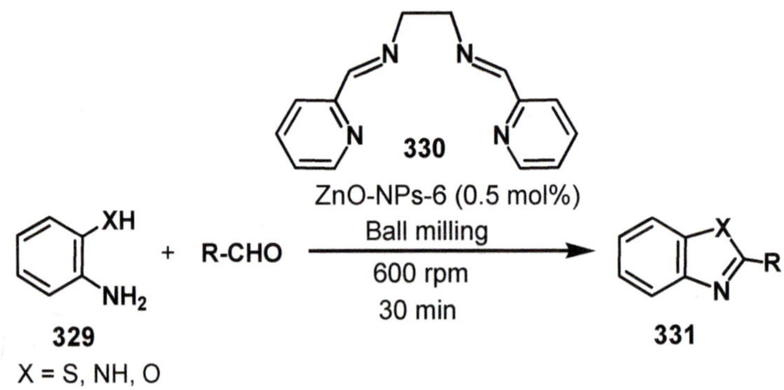

SCHEME 3.90

Synthesis of benzothiazoles, benzimidazoles, and benzoxazoles.

Table 3.47 Synthesis of Benzothiazoles, Benzimidazoles, and Benzoxazoles[a]

R	X	Yield (%)[b]	X	Yield (%)[b]	X	Yield (%)[b]
H	S	81	NH	92	O	87
C_6H_5	S	91	NH	94	O	90
4-OH-C_6H_4	S	87	NH	90	O	92
3-OH-C_6H_4	S	84	NH	91	O	89
2-OH-C_6H_4	S	87	NH	93	O	94
CH_3	S	88	NH	85	O	88
3,4-$(OH)_2$-C_6H_3	S	85				
2,3-$(OH)_2$-C_6H_3	S	87				
2-OH-4-NEt_2-C_6H_3	S	84				
4-OH-2-NO_2-C_6H_3	S	79				

[a]*Fritsch planetary micro mill Pulverisette 7, 600 rpm, 30 min, tungsten carbide vial (80 mL) and 20 × 5 mm balls, substrate:aldehyde: catalyst 1:1:0.05 mol.*
[b]*Isolated yields.*
Reproduced from Sharma H, Singh N, Jang DO. A ball-milling strategy for the synthesis of benzothiazole, benzimidazole and benzoxazole derivatives under solvent-free conditions. Green Chem 2014;16:4922–30, with permission from the Royal Society of Chemistry.

The mechanochemical synthesis of spirocyclic quinazolinones by the liquid–solid and solid–solid condensations of cyclic ketones with anthranilamide **332** was carried out by Fülöp et al. (Scheme 3.91) [59]. Again, the first mechanistic step is formation of imines. The I_2-catalyzed solvent-less ring-closure reactions of various cycloalkanones **333** are complete within 2 h in the ball mill, and products **334** were formed in high yield (Table 3.48). Further mechanochemical treatment was required for less reactive 1-indane (for 1.5 h at 30 Hz) to achieve yields comparable to other ketones. It was found that during the mechanochemical experiments, the temperature of the reaction vessel reached 60–70°C. Higher yields were obtained in mechanochemical reactions in comparison to solution. Yields obtained by ball milling are similar to those obtained for solvent-less reactions (12 h, at RT), with advantageous shorter time for mechanochemical ones. Two ZrO_2 balls were used instead of stainless steel to overcome the problems of iron particles abrasion. The removal of the I_2 catalyst was achieved by a simple aqueous workup procedure with a solution of $Na_2S_2O_5$ and water. The aqueous suspension was milled at 25 Hz for 5 min, followed by product filtration, washing with water and drying.

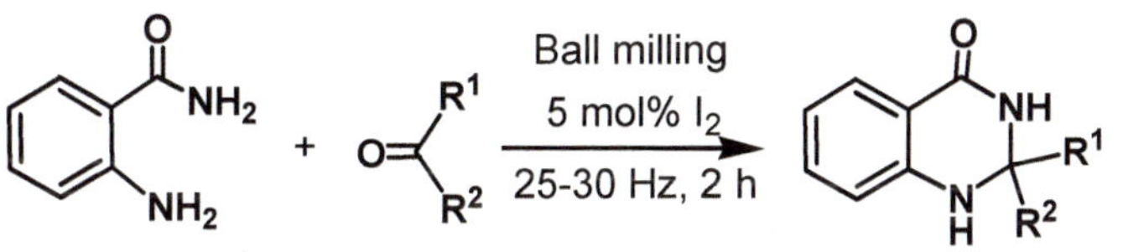

332 333 334

SCHEME 3.91

Synthesis of 2-spiroquinazolinones.

Table 3.48 Mechanochemical Synthesis of 2-Spiroquinazolinones[a]

Entry	R^1 R^2	Time	Yield (%)	Solution[e]
1	$R^1+R^2=(CH_2)_6$	2	98[b] (95)[c]	(88)[f]
2	$R^1+R^2=(CH_2)_7$	2	97 (93)	(67)[g]
3	$R^1+R^2=(CH_2)_{11}$	2	98 (94)	
4	$R^1+R^2=(CH_2)_{14}$	2	98 (91)	
5	$R^1+R^2=$1-indane	2+1.5[d]	97 (94)	
6	$R^1+R^2=2'$-adamantane	2	99 (90)	

[a]*Retsch MM400 mixer mill, 25 Hz, 25°C, stainless steel jar, two ZrO_2 balls 15 mm, amide:ketone:I_2 1:1:0.05.*
[b]*Estimated by ^1H NMR.*
[c]*Isolated yields from aqueous workup.*
[d]*At 30 Hz.*
[e]*In H_2O.*
[f]*100°C, 3 h.*
[g]*78°C, 2 h.*
Reproduced from Miklós F, Hum V, Fülöp F. Eco-friendly syntheses of 2,2-disubstituted- and 2-spiroquinazolinones. ARKIVOC 2014;vi:25–37, with permission from Arkat USA, Inc.

High enantioselectivity and good yields of chiral propargylamines **338** were achieved by Su et al. by addition of acetylene reagent across the in situ formed imine bond [60]. This synthesis was carried out in ball mill by solvent-free three-component asymmetric coupling of aldehydes, alkynes, and amines within 60 min (Scheme 3.92). The optimization of process revealed that $Cu(OTf)_2$/Ph-Pybox catalyst **339** (10 mol%) and silica gel as grinding auxiliary are the most efficient combination (representative examples, Table 3.49). Other copper salts (CuOTf, $CuCl_2$, $CuBr_2$, $Cu(OAc)_2$, CuCl, or CuBr) gave very low yields. Interestingly, the use of copper balls as a source of copper was totally dysfunctional. Among several ligands tested, The Ph-pybox ligand **339** promoted reaction and also afforded best enantiomeric excesses. Recovery of the catalyst was achieved by extraction with dichloromethane, and it was reused in next reactions, maintaining an excellent performance in the five runs.

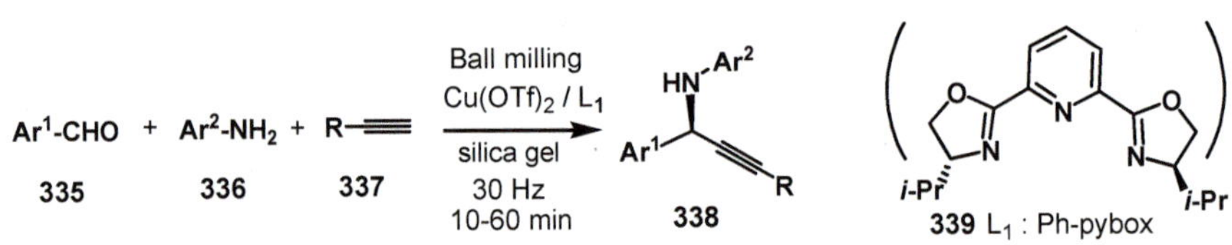

SCHEME 3.92

Synthesis of chiral propargylamines.

Table 3.49 Synthesis of Chiral Propargylamines[a]

Entry	Ar¹	Ar²	R	Time (min)	Yield (%)[b]	ee (%)
1	C_6H_5	C_6H_5	C_6H_5	30	99	99
2	4-F-C_6H_4	C_6H_5	C_6H_5	10	97	98
3	4-Cl-C_6H_4	C_6H_5	C_6H_5	30	97	96
4	4-OCH_3-C_6H_4	C_6H_5	C_6H_5	10	96	85
5	1-naphth	C_6H_5	C_6H_5	30	93	97
6	C_6H_5	4-Cl-C_6H_4	C_6H_5	60	92	99
7	4-Cl-C_6H_4	4-OCH_3-C_6H_4	C_6H_5	30	93	88
8	4-Cl-C_6H_4	4-OCH_3-C_6H_4	n-C_5H_{11}	30	92	88

[a] *Vibrational micro mill, stainless steel jar (50 mL), 2 × 15 mm balls, 25 Hz, aldehyde:amine:alkyne:$Cu(OTf)_2$:L_1 1:1.2:2:0.1:0.1, silica gel.*
[b] *Isolated yields.*
Reproduced from Li Z, Jiang Z, Su W. Fast, solvent-free, highly enantioselective three-component coupling of aldehydes, alkynes, and amines catalysed by the copper(II)pybox complex under high-vibration ball-milling. Green Chem 2015;17:2330–4, with permission from the Royal Society of Chemistry.

3.13 SUBSTITUTION REACTIONS

Studies on dynamic kinetic resolution (DKR) of α-halo carbonyls by S_N2 halogen replacement employing mechanochemistry by Lamaty revealed that DKR are accelerated in comparison to solution [61]. The configurationally labile halogen atom in the α position of the carbonyl can be irreversibly substituted by a nucleophile, and stereo-differentiation in the S_N2 halogen replacement can be controlled by the chiral environment on carbonyl moiety. Optimization of DKR of α-bromo-(*R*)-pantolactone ester **340** with Bn_2NH in ball mill revealed that $NaHCO_3$ is suitable base to replace commonly used Et_3N, with addition of catalytic amount of tetrabutylammonium iodide (Scheme 3.93). Solvent-free reaction afforded after 30 min amino ester **341**, **342** in similar yield (62%) and with the same diastereoselectivity as obtained in classical solvent reaction (dr 98:2), but in shorter time (in THF reaction needs 5 h). Remarkably, LAG conditions (H_2O, $\eta=2$) further shortens milling time to 15 min, while yield increased to 96%. Optimized ball-mill reaction conditions were applied to several DKRs, and comparison of results with solution indicates similar yields and diastereoselectivities, but in shorter time and with the use of less troublesome reagents (Table 3.50). Oxygen-based nucleophiles such as *p*-methoxyphenol could also be used to perform DKR on α-bromo ester, albeit with a decrease of dr.

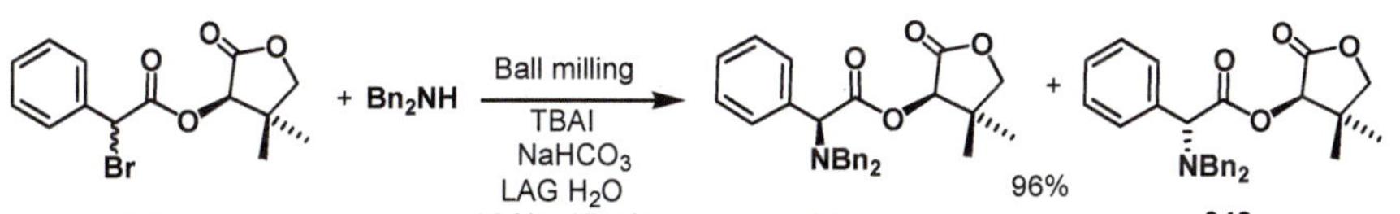

340 **341** **342**

SCHEME 3.93

Dynamic kinetic resolution of α-halo carbonyls.

Table 3.50 Dynamic Kinetic Resolution of α-halo Carbonyls[a]

Entry	Substrate	Product	Base	Time	dr	Yield (%)[b]
1	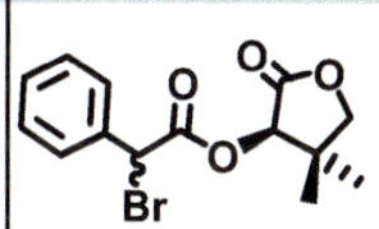	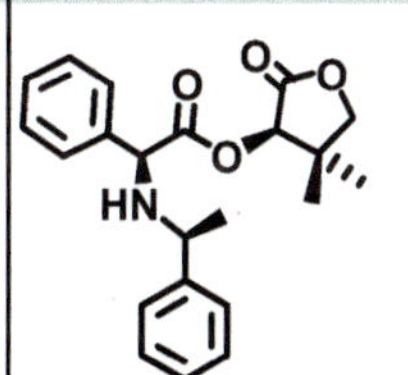	$NaHCO_3$ (Et_3N)	30 min -	>98:2 (>98:2)	80 (84)[c]
2		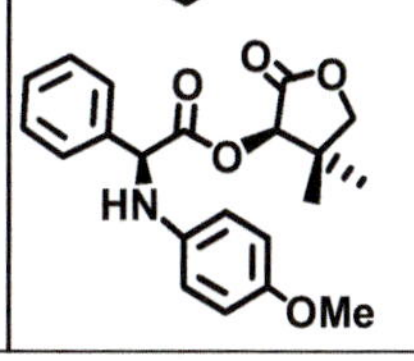	$NaHCO_3$ (Et_3N)	6 h -	88:12 (>98:2)	64 (75)

Continued

Table 3.50 Dynamic Kinetic Resolution of α-halo Carbonyls[a]—cont'd

Entry	Substrate	Product	Base	Time	dr	Yield (%)[b]
3		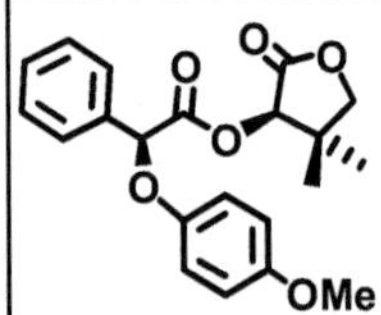	$NaHCO_3$ (NaH)	2h (6h)	72:28 (95:5)	86 (70)
4		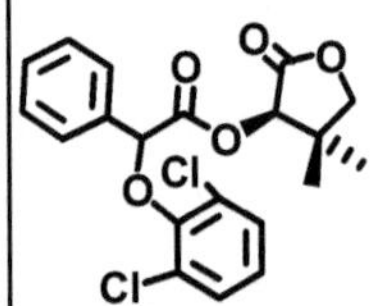	$NaHCO_3$ (NaH)	1.5h (7h)	56:44 (50:50)	77 (78)
5	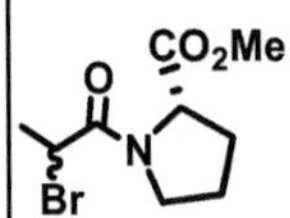	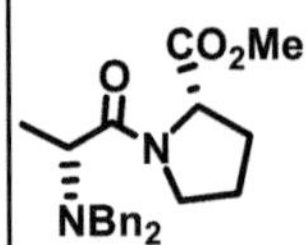	$NaHCO_3$ (Et_3N)	13h (24h)	>98:2 (87:13)	75 (51)
6	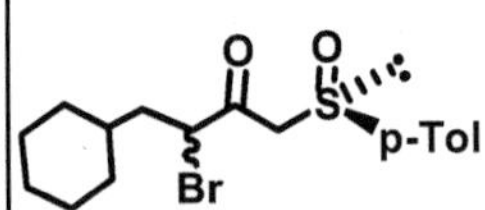	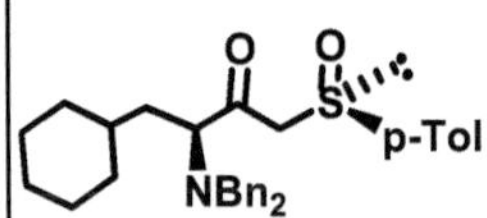	Bn_2NH Bn_2NH	7h (10d)	>98:2 (>95:5)	71 (74)

[a]Retsch MM 200 stainless steel reactor (12mL), 10mm ball, 30Hz, substrate:nucleophile:TBAI:$NaHCO_3$ 1:1.2:0.2:1.2, LAG H_2O (η=3).
[b]Isolated products.
[c]Solution reactions in parentheses.
Reproduced from Métro T-X, Salom-Roig XJ, Reverte M, Martinez J, Lamaty F. Faster and cleaner dynamic kinetic resolution via mechanochemistry. Green Chem 2015;17:204–8, with permission from the Royal Society of Chemistry.

Ball-milling technology has been also applied with success to carbohydrate chemistry. Kartha et al. developed a facile method of solvent-free mechanochemical synthesis of glycosyl azides employing sodium azide and glycosyl halides **343** (glucopyranosyl, galactopyranosyl, lactosyl, and arabinopyranosyl derivatives) (Scheme 3.94) [62]. Several polyfunctional 1,2-*trans*-linked glycosyl azides **344** were prepared by halogen/azide substitution after 8h milling procedure employing a planetary ball mill (Retsch PM-100, 50mL stainless steel vial and balls 10×10mm) in good to excellent yields as illustrated for 2,3,4,6-tetra-*O*-acetyl-*α*-D-glucopyranosyl bromide **343a**, which gives 2,3,4,6-tetra-*O*-acetyl-*β*-D-glucopyranosyl azide **344a**. Results support involvement of the cyclic oxocarbonium ion intermediate and anchimeric assistance rather than direct displacement mechanism. This is safe method for preparation of organic azides, most of which are unstable shock- and heat-sensitive explosives and reaction could be scaled up to 5g size without the incidents.

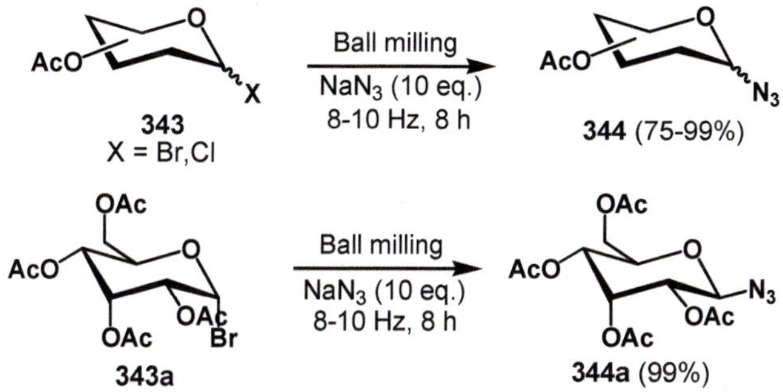

SCHEME 3.94

Synthesis of glycosyl azides.

Similarly, synthesis of glycosylamines promoted by mechanical milling was carried out by Wadouachi et al. (Scheme 3.95) [63]. Preparation of glycosylamines **346** from L-rhamnose **345** and different primary alkyl amines was performed with an excess of amine, whereas for liquid amines grinding auxiliary (SiO_2) was necessary to homogenize the mixture (selected products, Table 3.51). To carry out reaction, amine and sugar were initially mixed in mortar and then transferred to milling vessel. Comparison of the classical method to a solvent-less strategy carried out in the ball mill for the preparation of octylrhamnosylamine showed the superiority of solvent-less method in terms of shorter reaction time, better yield, and no need for heating.

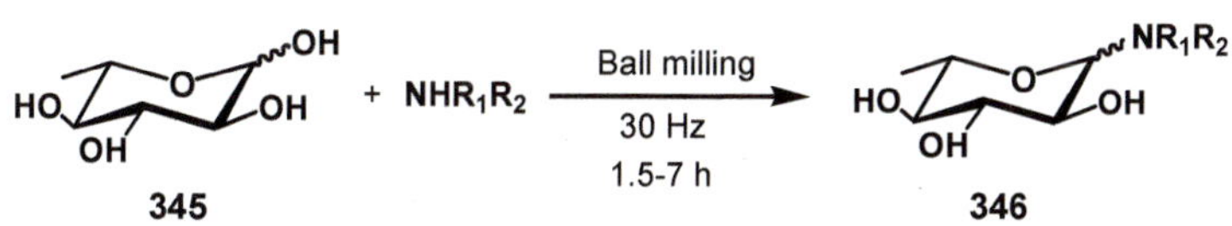

SCHEME 3.95

Glycosylamine synthesis.

Table 3.51 Glycosylamine Synthesis[a]

Entry	Sugar	Time (h)	Grinding Auxiliary	Product	α/β	Yield (%)[b]
1	L-rhamnose	1.5	SiO_2	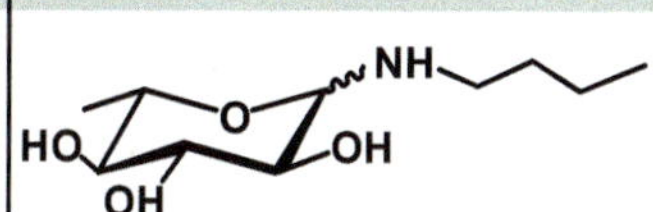	93/7	91
2		1.5	–	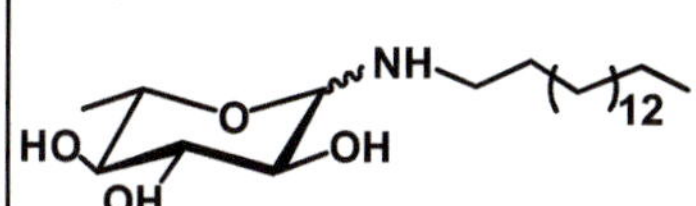	92/8	99
3		7	SiO_2	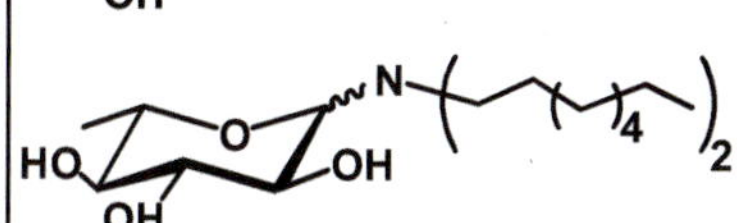	–	<5

Continued

Table 3.51 Glycosylamine Synthesis[a]—cont'd

Entry	Sugar	Time (h)	Grinding Auxiliary	Product	α/β	Yield (%)[b]
4		3	–	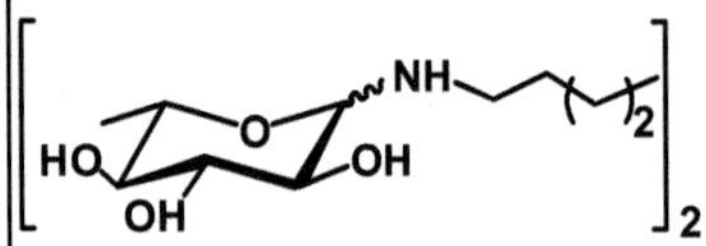	81/16	95
5		1.5	–	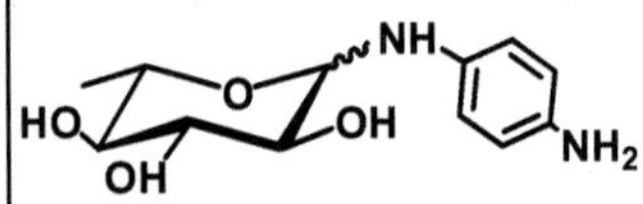	85/15	95
6	D-glucose	1.5	–	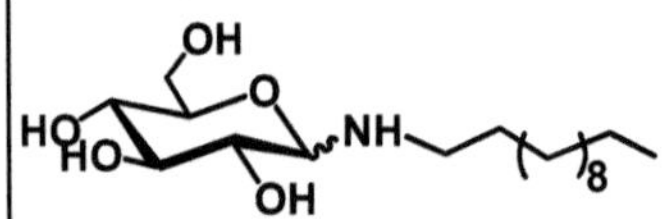		97
7	D-galac-tose	1.5	–	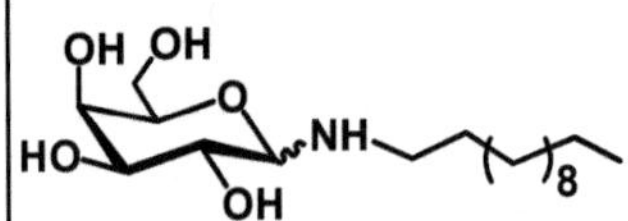		94
8	Maltose	1.5	–	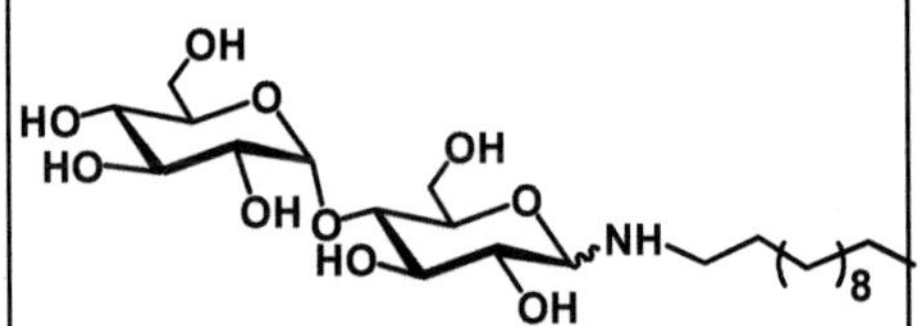		99

[a]Spex 8000M high-speed vibrational ball mill, stainless steel jar (60mL), 13mm balls, 30Hz, sugar:amine 1:1.1 eq.
[b]Isolated yields.
Reproduced from Epoune Lingome C, Pourceau G, Gobert-Deveaux V, Wadouachi A. Efficient synthesis of glycosylamines in solventless conditions promoted by mechanical milling. RSC Adv 2014;4:36350–6, with permission from the Royal Society of Chemistry.

A rapid, solvent-free and general synthesis of short- and long-chain substituted zwitterionic *meta*- and *para*-amino benzoquinones was reported by Cuccia, Friščić et al. (Scheme 3.96) [64]. Yields obtained by ball milling were compared with solution procedure, melt reactions, and mechanosynthesis in a cell lysis mill (Table 3.52). Authors demonstrated that cell lysis mill could be successfully employed to mechanosynthesis of organic compounds, and it is also suitable for coordination-based reactions. In transamination synthesis of *meta*-aminobenzoquinones **350**, an excess of amine was used, whereas *para*-aminobenzoquinones **352** are synthesized by condensation of 1,4-benzoquinone and an alkylamine in a 3:2 ratio. After chromatographic workup, it was determined that the yields were comparable for all four methods employed, while the solution reactions are the slowest.

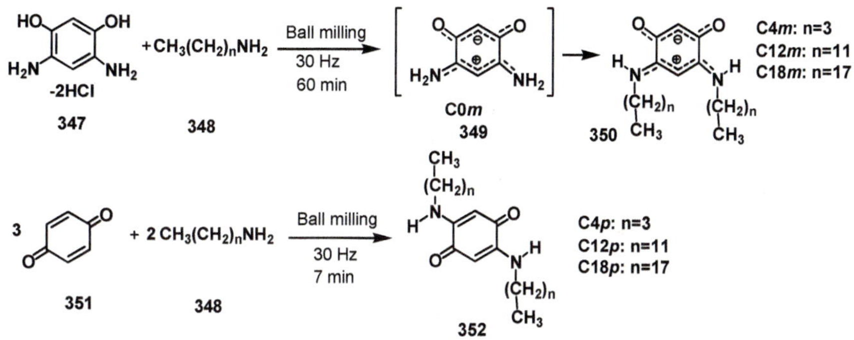

SCHEME 3.96

Synthesis of di(alkylamino) *meta*- and *para*-benzoquinones.

Table 3.52 Synthesis of Di(alkylamino) *meta*- and *para*-Benzoquinones[a]

Entry	*n*	Method	Time (min)	Amine (equiv.)	Yield (%)[b,c]
***m*-quinones[d]**					
1	3	A	120	7	88 (74,80)
2	3	B	10	7	83 (85,90)
3	3	C	60	7	89 (86,87)
4	3	D	10	10	92 (84,69)
***p*-quinones[e]**					
5	3	A	300	0.66	32 (29,22)
6	3	B	5	0.66	27 (30,28)
7	3	C	7	0.66	25 (32,27)
8	3	D	5	0.66	20 (22,21)

[a]*A: In MeOH. B: FastPrep-24; MP-Biomedicals lysis mill. C: Retsch MM200 mixer mill, stainless steel vessel (10 mL), 2×7 mm balls, 30 Hz. D: melt, 70°C.*
[b]*Isolated.*
[c]*In parentheses yields for n = 11 and n = 17.*
[d]*Quinone:amine 1:4–7 equiv.*
[e]*Quinone:amine 3:2 equiv.*
Reproduced from Fang Y, Salamé N, Woo S, Bohle DS, Friščić T, Cuccia LA. Rapid and facile solvent-free mechanosynthesis in a cell lysis mill: preparation and mechanochemical complexation of aminobenzoquinones. Cryst Eng Comm 2014;16:7180–5, with permission from the Royal Society of Chemistry.

Condensation of ninhydrin with (thio)ureas via substitution/addition reaction cascade led after 1–2 h of ball milling to quantitative conversion to the corresponding heterocyclic *bis-N/O*-semiacetals **355** (Scheme 3.97) [2]. These have been previously prepared in solution with inferior yields (Table 3.53).

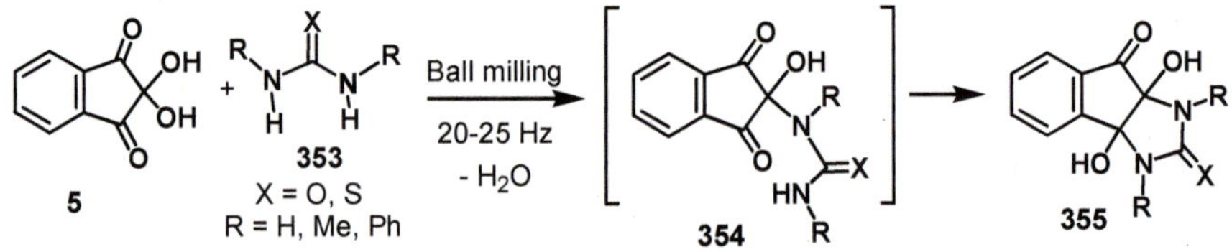

SCHEME 3.97

Condensations of ninhydrin with (thio)ureas.

Table 3.53 Mechanochemical Condensations of Ninhydrin With (Thio)ureas[a]

Entry	X	R	Conditions	Time	T (°C)	Yield (%)[b]	Yield in Solution (%)
1	O	H	Ball milling	1 h	80	100	78
2			Mortar	1 d	80	100	
3	S	H	Ball milling	45 min	RT	100	91
4	O	Me	Ball milling	2 h	RT	100	92
5			Mortar	3 d	RT	100	
6	S	Me	Ball milling	2 h	RT	100	
7			Mortar	3 d	RT	100	
8	S	Ph	Ball milling	1 h	60	100	

[a]Retsch MM 2000 swing mill, 20–25 Hz, stainless steel jar (10 mL) double walled with fittings for circulating coolants, two stainless steel balls 12 mm; (thio)urea:ninhydrin 1:1 mol.
[b]Isolated yield.

The wealth of products which could be obtained in solid-state reactions could be further illustrated by the L-proline reaction in a substitution, elimination, and decarboxylation cascade with ninhydrin in ball-milling conditions to give the versatile azomethine ylide **359**, waste-free, and in quantitative yield (Scheme 3.98) [2]. As the decarboxylation takes place in reaction vial, the Teflon seal was tightened at 15 N m to avoid dangerously high pressure. The same azomethine ylide can be prepared in solution in lower yield (82%). Furthermore, the solid enaminoester (3-methylcrotonate) reacts quantitatively with ninhydrin in stoichiometric ball-mill runs to give the heterocyclic building block with α-hydroxyketone and *N/O*-semiaminal functionality. For comparison, solution reaction gives the product in 82% yield. The three-cascade process consists of vinylogous substitution, cyclization, and 1,3-hydrogen shift.

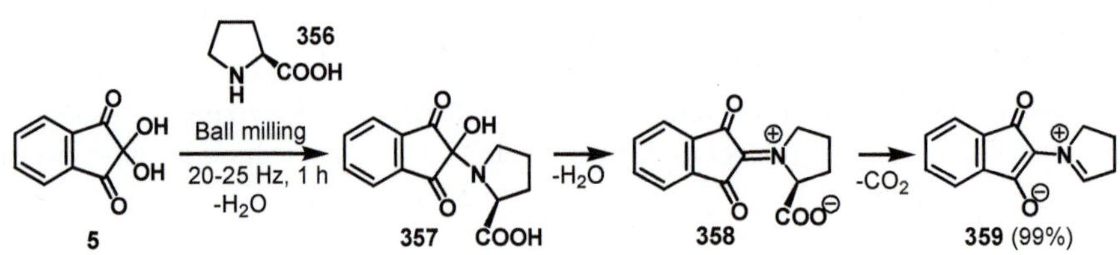

SCHEME 3.98

Solid-state condensations of ninhydrin.

3.14 *N*-ARYLATION OF AMINES

Solvent-free *N*-arylation of amines with arylboronic acids under ball-milling conditions was carried out by Su et al. (Scheme 3.99) [65]. Efficacy of several copper catalyst/base systems was explored showing that $Cu(OAc)_2{\cdot}H_2O$ (1 equiv.)/K_2CO_3 (2.5 equiv.) system is the most satisfactory. Silica gel was found to be the most effective among grinding auxiliaries. A number of aromatic and aliphatic amines were subjected to mechanochemical reaction in planetary mill with arylboronic acids possessing various electron-withdrawing and electron-donating groups on the phenyl ring and moderate to high yields were obtained (selected results, Table 3.54). For this purpose, two different sizes of milling balls were employed. Nitrogen on imidazole ring could also be effectively arylated by this procedure in 55% yield.

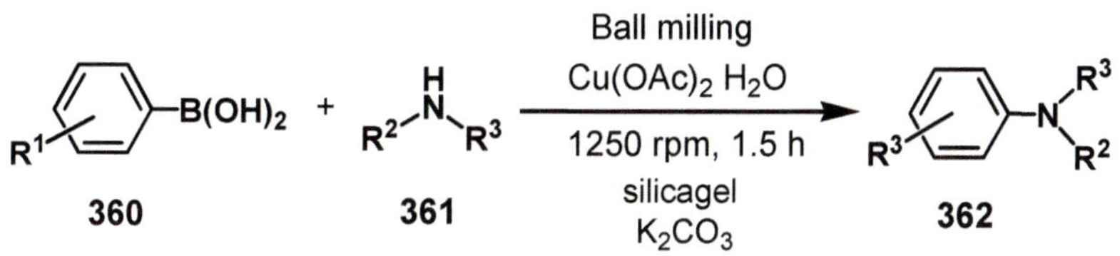

SCHEME 3.99

Copper-promoted coupling of arylboronic acid with amines.

Table 3.54 Copper-Promoted Coupling of Arylboronic Acid With Amines[a]

Entry	R^1	R^2	R^3	Yield (%)[b]
1	H	C_6H_5	H	86
2	H	4-CH_3-C_6H_4	H	83
3	H	4-F-C_6H_4	H	76
4	H	4-CF_3-C_6H_4	H	74
5	H	1-Naphthyl	H	61
6	4-CF_3	H	H	81
7	4-Cl	H	H	69
8	2-CH_3	4-OCH_3-C_6H_4	H	61
9	H	Bn	H	84
10	H	*n*-Pr	H	68

[a]*AGO-2 planetary centrifugal mill, drum (150mL), 50×8.9mm+60×5mm balls (175g), 25Hz, 1.5h, acid:amine:$Cu(OAc)_2$:K_2CO_3 5:7:1:2.5, silica gel.*

[b]*Isolated yields.*

Reproduced from Zhu X, Zhang Q, Su W. Solvent-free N-arylation of amines with arylboronic acids under ball milling conditions. RSC Adv 2014;4:22775–8, with permission from the Royal Society of Chemistry.

High chemoselectivity was obtained in solvent-free Cu^0-catalyzed Ullmann-type *N*-arylation reactions of iodobenzenes by Cravotto et al. (Scheme 3.100). [66]. In planetary ball-milling *N*-arylations reactions, stoichiometric amount of copper powder, an excess of base (KOH) and alumina as grinding auxiliary gave the best results.

High N/O selectivity was demonstrated in reactions with various aminoalcohols, which were added in excess (Table 3.55). The most striking example is site selectivity obtained with *N*-methyl isoserinol (entry 6, Table 3.56). Ball-milling conditions significantly outperform ultrasound activation and conventional stirring in terms of time: while ultrasonic reaction takes 3 h, only partial conversion (35%) was reached after 2 days stirring at room temperature. The success of mechanochemical reaction conditions is explained in terms of the formation of copper-chelating intermediate, making the aromatic position more electrophilic and susceptible for substitution. At the same time, amine was brought closer to the attack position which accelerates the reaction rate. This mechanism is corroborated by the lack of reaction of simple primary amines. Interesting finding is the reactivity change going from solution to solid-state conditions. While in solution reactivity increases by addition of electron-withdrawing groups to aromatic moiety, and limited increase was obtained with electron-donating groups, such as methoxy and amino, in solid-state ball-milling conditions this relative reactivity order differs: 4-iodoanisole and 2-iodoaniline reacted well, while 4-nitro aniline is relatively less reactive (Scheme 3.101,

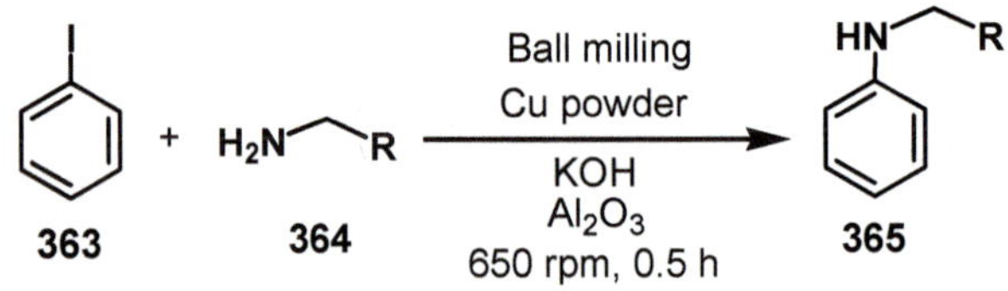

363 364 365

SCHEME 3.100

Copper-promoted *N*-arylation of amines with iodobenzene.

Table 3.55 *N*-arylation of Amines With Iodobenzene[a]

Entry	R	Yield (%)[b]
1	CH_2OH	98 (99)
2	CH_2CH_2OH	92 (95)
3	CH_2NH_2	99 (>99)[c]
4	CH_2OMe	0 (3)
5	$CH_2CH(CH_3)OH$	85 (89)

[a]*Retsch PM 100 planetary mill, stainless steel vessel (50 mL), 1500 × 2 mm + 48 × 5 mm balls, 650 rpm, 0.5 h, PhI:amine:KOH:Cu 1:3:3.4:1, Al_2O_3.*
[b]*GC–MS yields.*
[c]*Amine 10 mmol, 15 min.*
Reproduced from Martina et al. Synlett 2015;26:http://dx.doi.org/10.1055/s-0035-1560173, with permission from Thieme Publishers.

Table 3.56). An explanation was put forward by postulated formation of π-complex of cuprous chelate which coordinates to aryl halide could be affected by nitro group drawing electron density from the benzene moiety.

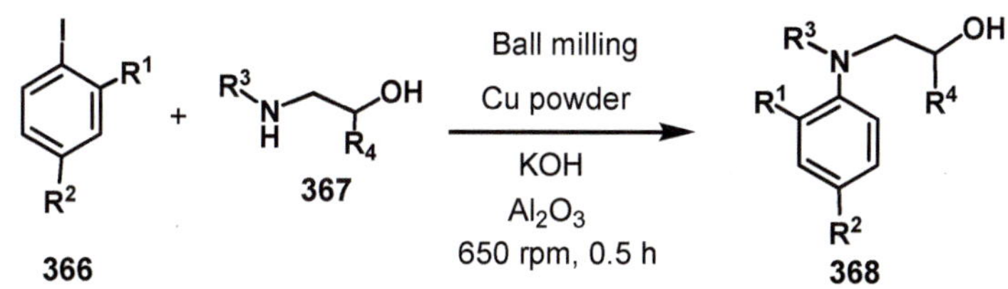

SCHEME 3.101

Copper-promoted *N*-arylation of ethanolamines with iodobenzenes.

Table 3.56 *N*-Arylation of Ethanolamines With Iodobenzenes[a]

Entry	R^1	R^2	R^3	R^4	Yield (%)[b]
1	H	OCH_3	H	H	96 (98)
2	H	NO_2	H	H	81 (85)[c]
3	H	Br	H	H	96 (99)
4	H	CHO	H	H	87 (92)
5	H	COOEt	H	H	91 (95)
6	NH_2	H	H	H	86 (89)
7	H	OCH_3	CH_3	CH_2OH	32 (34)[c]

[a]*Retsch PM 100 planetary mill, stainless steel vessel (50mL), 1500×2mm+48×5mm balls, 650rpm, 0.5h, PhI:amine:KOH:Cu 1:3:3.4:1, Al_2O_3.*
[b]*GC–MS yields.*
[c]*4h, without Al_2O_3 and KOH.*
Reproduced from Martina et al. Synlett 2015;26:2789-94, with permission from Thieme Publishers.

Generality of this mechanochemical protocol was demonstrated in one-pot, two-step synthesis of triazole (Scheme 3.102). The first synthetic step, 1,3-dipolar cycloaddition ("click reaction") of 4-iodobenzyl azide with phenyl acetylene employing copper powder (1 equiv.) was complete in 10min and follows the synthetic procedure described in chapter "Cycloaddition Reactions" [67]. Ullmann *N*-arylation of triazole **371** with ethylenediamine in optimized reaction conditions provided the corresponding *N*-arylated triazole **372** in 56% yield, albeit reaction required 2h of ball milling to complete.

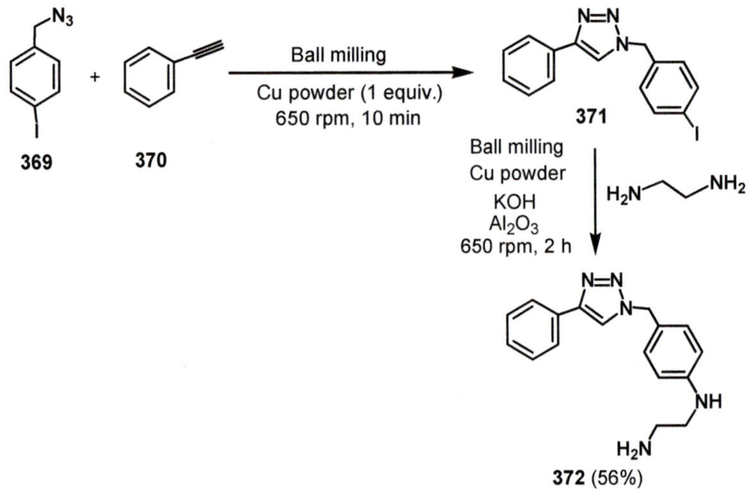

SCHEME 3.102

Two-step synthesis of triazoline **372**.

3.15 BIGINELLI REACTION

One-pot multicomponent Biginelli reaction which consists of the formation of two new C—N bonds and one new C—C bond was carried under catalyst-free conditions by ball-milling procedure (Scheme 3.103) [68]. M'hamed and Alduaij subjected an equimolar amounts of an aromatic aldehyde, malononitrile, and (thio)urea to ball milling for 40 min to afford 2-(thi)oxo-pyrimidine-5-carbonitrile products **375** in pure form without need for purification. Reaction starts with nucleophilic addition by the urea to the aldehyde, resulting in the formation of an imine. The β-ketoester then adds to the imine bond, which is followed by second condensation and the ring closure by the nucleophilic attack by the amine onto the carbonyl group. In the course of the optimization of reaction conditions, authors established a relationship between (balls/reagents weight) ratio: when ratio is equal to 1, the reaction does not

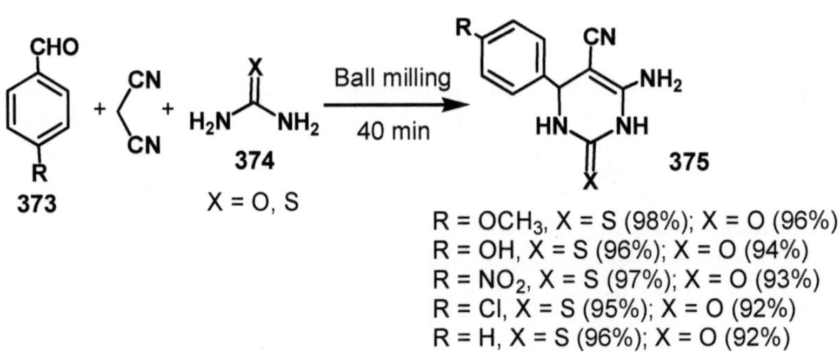

SCHEME 3.103

Multicomponent synthesis of 2-(thi)oxo-pyrimidine-5-carbonitriles.

change even after 4 h of grinding, while with increasing value of the balls' weight, the conversion increases until the optimal value for the ratio balls/reagents is equal to 5 (SPEX 8000 mixer mill, 10 mL stainless steel vial, balls 27.8 g).

3.16 NITROSO BOND FORMATION/DISSOCIATION

Dissociation and dimerization of C-nitroso compounds was studied by Vančik et al. under different solvent-free conditions, including mechanochemical [69]. Reaction rates and kinetic isotope effects were obtained from experiments carried out with ^{15}N labeled C-nitroso compound (*p*-bromonitrosobenzene) as shown for homodimers **376** and **377** (Scheme 3.104). Heterodimers **380** and **381** were prepared by mixing equimolar amounts of *p*-Cl, *p*-NO_2 *p*-I nitrosobenzene derivatives, and ^{15}N isotope labeled *p*-bromonitrosobenzene at room temperature. The obtained mixture was sublimed, melted, or milled (Retsch MM200 grinder mill, 25 Hz, 30 min, 10 mL stainless steel jar, 2×7 mm balls, yield of 50% in both cases). The observed rate constant strongly depends on the individual experimental conditions, while the kinetic isotope effect is much less dependent on the particular experiment. Resemblance of the kinetics of the reactions in mechanical mill with the kinetics of the thermal reactions in sublimed crystals of monomers implies that the mechanisms of two processes are similar.

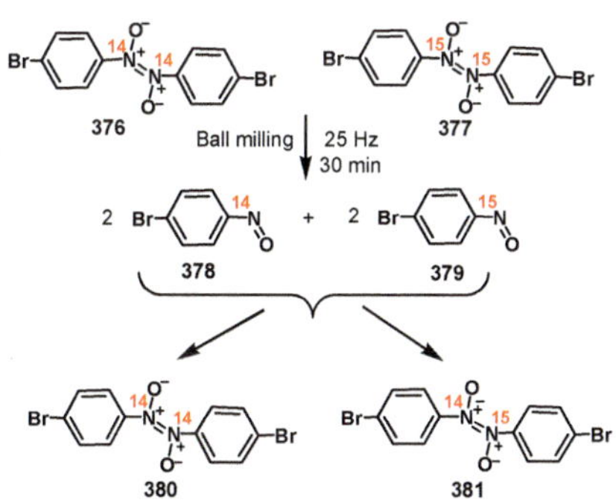

SCHEME 3.104

Dissociation and dimerization of C-nitroso compounds.

REFERENCES

[1] Zefirov NS, Sereda GA, Volkov PV, Tkatchenko SE, Zyk NV. Solid-phase synthesis of 1,2-benzophenazine and some fused imidazole derivatives. Chem Heterocycl Comp 1996;32:577–9. (Khim Geterotsikl Soedin 1996;672–4); ECHET98. http://www.ch.ic.ac.uk/ectoc/echet98/pub/097/index.htm.

[2] Kaupp G, Naimi-Jamal MR, Schmeyers J. Quantitative reaction cascades of ninhydrin in the solid state. Chem Eur J 2002;8:594–600.

[3] Etman HA, Metwally HM, Elkasaby MM, Khalil AM, Metwally MA. Green, two components highly efficient reaction of ninhydrin with aromatic amines, and malononitrile using ball-milling technique. Am J Org Chem 2011;1:10–3.

[4] Kaupp G, Naimi-Jamal MT. Quantitative cascade condensations between *o*-phenylenediamines and 1,2-dicarbonyl compounds without production of wastes. Eur J Org Chem 2002:1368–73.
[5] Carlier L, Baron M, Chamayou A, Couarraze G. Use of co-grinding as a solvent-free solid state method to synthesize dibenzophenazines. Tetrahedron Lett 2011;52:4686–9.
[6] Carlier L, Baron M, Chamayou A, Couarraze G. Greener pharmacy using solvent-free synthesis: investigation of the mechanism in the case of dibenzophenazine. Powder Technol 2013;240:41–7.
[7] Margetić D. Mechanochemical organic reactions without the use of solvent. Kem Ind 2005;54:351–8.
[8] Margetić D, Mann DA, Warrener RN. Ninhydrin as a building block in scaffold-linked dyad construction. ARKIVOC 2014;15:210–24.
[9] Schmeyers J, Toda F, Boy J, Kaupp G. Quantitative solid-solid synthesis of azomethines. J Chem Soc Perkin Trans 1998;2:989–93.
[10] Ferguson M, Giri N, Huang X, Apperley D, James SL. One-pot two-step mechanochemical synthesis: ligand and complex preparation without isolating intermediates. Green Chem 2014;16:1374–82.
[11] Stilinović V, Cinčić D, Zbačnik M, Kaitner B. Controlling solvate formation of a Schiff base by combining mechanochemistry with solution synthesis. Croat Chem Acta 2012;85:485–93.
[12] Chen S, Wang P, Jia C, Lin Q, Yuan W. Spectrochim Acta A 2014;133:223–8.
[13] Naimi-Jamal MR, Mokhtari J, Dekamin MG, Javanshir S. Efficient synthesis and deprotection of semicarbazones under solvent-free conditions. Iran J Chem Chem Eng 2012;2:1–8.
[14] Mokhtari J, Naimi-Jamal MR, Hamzeali H, Dekamin MG, Kaupp G. Kneading ball-milling and stoichiometric melts for the quantitative derivatization of carbonyl compounds with gas-solid recovery. ChemSusChem 2009;2:248–54.
[15] Nun P, Martin C, Martinez J, Lamaty F. Solvent-free synthesis of hydrazones and their subsequent *N*-alkylation in a ball-mill. Tetrahedron 2011;67:8187–94.
[16] Oliveira PFM, Baron M, Chamayou A, André-Barrés C, Guidetti B, Baltas M. Solvent-free mechanochemical route for green synthesis of pharmaceutically attractive phenol-hydrazones. RSC Adv 2014;4:56736–42.
[17] Magyar T, Miklós F, Lázár L, Fülöp F. Application of a ball milling technique for the condensation of anthranilic hydrazides with aromatic aldehydes towards 4-quinazolinone derivatives. Chem Heterocycl Comp 2015;50:1463–9. Khim Geterotsikl Soedin 2014;10:1590–5.
[18] Kaupp G, Schmeyers J, Boy J. Iminium salts in solid-state syntheses giving 100% yield. J Prakt Chem 2000;342:269–80.
[19] Kaupp G, Schmeyers J. Solid-state reactivity of the hydrazine-hydroquinone complex. J Phys Org Chem 2000;13:388–94.
[20] Declerck V, Nun P, Martinez J, Lamaty F. Solvent-free synthesis of peptides. Angew Chem Int Ed 2009;48:9318–21.
[21] Bonnamour J, Métro T-X, Martinez J, Lamaty F. Environmentally benign peptide synthesis using liquid-assisted ball-milling: application to the synthesis of Leu-enkephalin. Green Chem 2013;15:1116–20.
[22] Hernández J, Juaristi E. Green synthesis of α,β- and β,β-dipeptides under solvent-free conditions. J Org Chem 2010;75:7107–11.

[23] Ravalico F, James SL, Vyle JS. Synthesis of nucleoside analogues in a ball mill: fast, chemoselective and high yielding acylation without undesirable solvents. Green Chem 2011;13:1778–83.

[24] Štrukil V, Bartolec B, Portada T, Đilović I, Halasz I, Margetić D. One-pot mechanosynthesis of aromatic amides and dipeptides from carboxylic acids and amines. Chem Commun 2012;48:12100–2.

[25] Métro T-X, Bonnamour J, Reidon T, Sarpoulet J, Martinez J, Lamaty F. Mechanosynthesis of amides in the total absence of organic solvent from reaction to product recovery. Chem Commun 2012;48:11781–3.

[26] Métro T-X, Bonnamour J, Reidon T, Duprez A, Sarpoulet J, Martinez J, Lamaty F. Study of the organic-solvent-free CDI-mediated acylation of various nucleophiles by mechanochemistry. Chem Eur J 2015;21:12787–96.

[27] Kumar V, Giri SK, Venugopalan P, Kartha KPR. Synthesis of cross-linked glycopeptides and ureas by a mechanochemical, solvent-free reaction and determination of their structural properties by TEM and x-ray crystallography. ChemPlusChem 2014;79:1605–13.

[28] Tireli M, Juribašić M, Kulcsá M, Cindro N, Gracin D, Biliškov N, Borovina M, Ćurić M, Halasz I, Užarević K. Mechanochemical reactions studied by in situ Raman spectroscopy: base catalysis in liquid-assisted grinding. Chem Commun 2015;51:8058–61.

[29] Swinburne AN, Steed JW. The mechanochemical synthesis of podand anion receptors. CrystEngComm 2009;11:433–8.

[30] Konnert L, Reneaud B, de Figueiredo RM, Campagne J-M, Lamaty F, Martinez J, Colacino E. Mechanochemical preparation of hydantoins from amino esters: application to the synthesis of the antiepileptic drug phenytoin. J Org Chem 2014;79:10132–42.

[31] Kaupp G, Schmeyers J, Boy J. Quantitative solid-state reactions of amines with carbonyl compounds and isothiocyanates. Tetrahedron 2000;56:6899–911.

[32] Štrukil V, Igrc MD, Eckert-Maksić M, Friščić T. Click mechanochemistry: quantitative synthesis of "ready to use" chiral organocatalysts by efficient two-fold thiourea coupling on vicinal diamines. Chem Eur J 2012;18:8464–73.

[33] Štrukil V, Margetić D, Igrc MD, Eckert-Maksić M, Friščić T. Desymmetrisation of aromatic diamines and synthesis of non-symmetrical thiourea derivatives by click-mechanochemistry. Chem Commun 2012;48:9705–7.

[34] Zhang Z, Wu H-H, Tan Y-J. A simple and straightforward synthesis of phenyl isothiocyanates, symmetrical and unsymmetrical thioureas under ball milling. RSC Adv 2013;3:16940–4.

[35] Štrukil V, Gracin D, Magdysyuk OV, Dinnebier RE, Friščić T. Trapping reactive intermediates by mechanochemistry: elusive aryl *n*-thiocarbamoylbenzotriazoles as bench-stable reagents. Angew Chem Int Ed 2015;54:8440–3.

[36] Tan D, Štrukil V, Mottillo C, Friščić T. Mechanosynthesis of pharmaceutically relevant sulfonyl-(thio)ureas. Chem Commun 2014;50:5248–50.

[37] Giao J, Wang G-W. Direct oxidative amidation of aldehydes with anilines under mechanical milling conditions. J Org Chem 2008;73:2955–8.

[38] Phung C, Ulrich RM, Ibrahim M, Tighe NTG, Lieberman DL, Pinhas AR. The solvent-free and catalyst-free conversion of an aziridine to an oxazolidinone using only carbon dioxide. Green Chem 2011;13:3224–9.

[39] Sikchi SA, Hultin PG. Solventless protocol for efficient bis-*n*-Boc protection of adenosine, cytidine, and guanosine derivatives. J Org Chem 2006;71:5888–91.

[40] Konnert L, Gauliard A, Lamaty F, Martinez J, Colacino E. Solventless synthesis of *n*-protected amino acids in a ball mill. ACS Sustain Chem Eng 2013;1:1186–91.

[41] Konnert L, Lamaty F, Martinez J, Colacino E. Solventless mechanosynthesis of *n*-protected amino esters. J Org Chem 2014;79:4008–17.

[42] Chuev VP, Lyagina LA, Ivanov EY, Boldyrev VV. Mechanochemical phthalazole synthesis. Dokl Akad Nauk SSSR 1989;307:1429–32.

[43] Colacino E, Nun P, Colacino FM, Martinez J, Lamaty F. Solvent-free synthesis of nitrones in a ball-mill. Tetrahedron 2008;64:5569–76.

[44] Thorwirth R, Stolle A. Solvent-free synthesis of enamines from alkyl esters of propiolic or but-2-yne dicarboxylic acid in a Ball mill. Synlett 2011;15:2200–2.

[45] Fan W, Gao J, Jia J, Han L, Sheng W, Li Y. Solvent-free synthesis of β-enamino ketones (esters) under high-speed vibration milling (HSVM) condition. Chin J Org Chem 2010;30:1732–6.

[46] Waddell DC, Thiel I, Bunger A, Nkata D, Maloney A, Clark T, Smith B, Mack J. Investigating the formation of dialkyl carbonates using high speed ball milling. Green Chem 2011;13:3156–61.

[47] Glasovac Z, Trošelj P, Jušinski I, Margetić D, Eckert-Maksić M. Synthesis of highly basic hexasubstituted biguanides by environmentally friendly methods. Synlett 2013;24:2540–4.

[48] Tan D, Mottillo C, Katsenis AD, Štrukil V, Friščić T. Development of C–N coupling using mechanochemistry: catalytic coupling of arylsulfonamides and carbodiimides. Angew Chem Int Ed 2014;53:9321–4.

[49] Li Y-J, Cao Y-W, Xu F-S, Fang W-M, Yu W-B, Jia J-H, Gao J-R. Solvent and catalyst free azo-Michael addition under high-speed vibration milling. Sci China Chem 2012;55:1252–6.

[50] Kaupp G, Schmeyers J, Kuse A, Atfeh A. Cascade reactions in quantitative solid-state syntheses. Angew Chem Int Ed 1999;38:2896–9.

[51] Zhu X, Li Z, Jin L, Xu L, Wu Q, Su W. Mechanically activated synthesis of 1,3,5-triaryl-2-pyrazolines by high speed ball milling. Green Chem 2009;11:163–5.

[52] Zhang Z, Tan Y-J, Wang C-S, Wu H-H. One-pot synthesis of 3,5-diphenyl-1*H*-pyrazoles from chalcones and hydrazine under mechanochemical ball milling. Heterocycles 2014;89:103–12.

[53] Wang G-W, Wu X-L. Mechanochemical aminochlorination of electron-deficient olefins with chloramine-T promoted by (diacetoxyiodo)benzene. Adv Synth Catal 2007;349:1977–82.

[54] Wu X-L, Xia J-J, Wang G-W. Aminobromination of olefins with $TsNH_2$ and NBS as the nitrogen and bromine sources mediated by hypervalent iodine in a ball mill. Org Biomol Chem 2008;6:548–53.

[55] Zille M, Stolle A, Wild A, Schubert US. $ZnBr_2$-mediated synthesis of indoles in a ball mill by intramolecular hydroamination of 2-alkynylanilines. RSC Adv 2014;4:13126–33.

[56] Zhang Z, Wang G-W, Miao C-B, Dong Y-W, Shen Y-B. Solid-state radical reactions of 1,3-cyclohexanediones with in situ generated imines mediated by manganese(III) acetate under mechanical milling conditions. Chem Commun 2004:1832–3.

[57] Jin M, Song G, Li Z, Zhou F, Fan B, Ouyang P. Efficient synthesis of 2-Aryl-1-arylmethyl-1H-benzimidazoles in ball mill without solvent. J Heterocycl Chem 2014;51:1838–43.

[58] Sharma H, Singh N, Jang DO. A ball-milling strategy for the synthesis of benzothiazole, benzimidazole and benzoxazole derivatives under solvent-free conditions. Green Chem 2014;16:4922–30.

[59] Miklós F, Hum V, Fülöp F. Eco-friendly syntheses of 2,2-disubstituted- and 2-spiroquinazolinones. ARKIVOC 2014;vi:25–37.

[60] Li Z, Jiang Z, Su W. Fast, solvent-free, highly enantioselective three-component coupling of aldehydes, alkynes, and amines catalysed by the copper(II)pybox complex under high-vibration ball-milling. Green Chem 2015;17:2330–4.
[61] Métro T-X, Salom-Roig XJ, Reverte M, Martinez J, Lamaty F. Faster and cleaner dynamic kinetic resolution via mechanochemistry. Green Chem 2015;17:204–8.
[62] Mugunthan G, Kartha RKP. Application of ball milling technology to carbohydrate reactions-II. Solvent-free mechanochemical synthesis of glycosyl azides. J Carbohydr Chem 2008;27:294–9.
[63] Epoune Lingome C, Pourceau G, Gobert-Deveaux V, Wadouachi A. Efficient synthesis of glycosylamines in solventless conditions promoted by mechanical milling. RSC Adv 2014;4:36350–6.
[64] Fang Y, Salamé N, Woo S, Bohle DS, Friščić T, Cuccia LA. Rapid and facile solvent-free mechanosynthesis in a cell lysis mill: preparation and mechanochemical complexation of aminobenzoquinones. CrystEngComm 2014;16:7180–5.
[65] Zhu X, Zhang Q, Su W. Solvent-free *N*-arylation of amines with arylboronic acids under ball milling conditions. RSC Adv 2014;4:22775–8.
[66] Martina K, Rinaldi L, Baricco F, Boffa L, Cravotto G. Highly efficient mechanochemical *n*-arylation of amino alcohols and diamines with Cu^{o} powder. Synlett 2015;26:2789–94.
[67] Rinaldi L, Martina K, Baricco F, Rotolo L, Cravotto G. Solvent-free copper-catalyzed azide-alkyne cycloaddition under mechanochemical activation. Molecules 2015;20:2837–49.
[68] M'hamed MO, Alduaij OK. An efficient one-pot synthesis of new 2-thioxo and 2-oxo-pyrimidine-5-carbonitriles in ball-milling under solvent-free and catalyst-free conditions. Phosph Sulfur Silicon 2014;189:235–41.
[69] Maganjić A, Šolić I, Milovac S, Halasz I, Biljan I, Vančik H. Mechanochemically induced crossdimerizations of nitrosobenzenes. Kinetics and solid-state isotope effects. J Phys Org Chem 2014;27:177–82.

CHAPTER

Carbon—Oxygen and Other Bond-Formation Reactions

4

CHAPTER OUTLINE

Mechanochemical Organic Synthesis. http://dx.doi.org/10.1016/B978-0-12-802184-2.00004-2

4.1 C—O BOND FORMATION REACTIONS

4.1.1 SYNTHESIS OF ETHERS

The quantitative solid-state synthesis of the pyrazolyl ethers by ball milling of equimolar amounts of pyrazolone derivatives **1** with phenacyl bromide was achieved by Abdel-Latif and Metwally [1]. In this procedure, it was unnecessary to use base catalysis, solvents, or liquid phases to obtain ether products **2a–c** in 100% yield at room temperature after washing with sodium carbonate solution (Scheme 4.1). *O*-alkylation of equimolar amounts of reagents was carried out in Retsch MM2000 swing mill in stainless steel 10 mL vial and balls (2 × 12 mm).

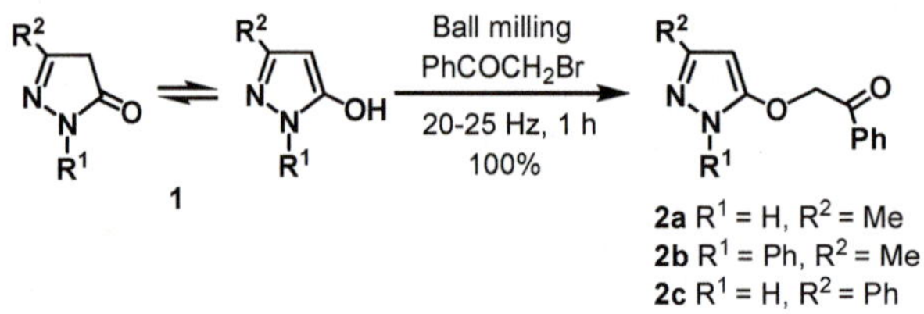

SCHEME 4.1

O-alkylation of pyrazolones.

4.1.1.1 Glycosylation

Kartha et al. developed an efficient solvent-free method for glycosylation of acetobromosugars in a planetary ball mill (Scheme 4.2) [2]. Milling of equimolar amounts of acetobromo-α-D-glucose **3** and alcohols with 1.5 equiv. of carbonates ($CdCO_3$/$ZnCO_3$) gives glycoside products **5** in moderate to high yield (Table 4.1). Employment of other carbonates resulted in lower yields often with side reactions. For less-reactive substrates, longer milling time is required, the longest in the case of 3β-cholesteryl required 15 h, still giving only 35% yield. Optimized reaction conditions were applied to other acetobromosugars (such as acetobromo-α-D-galactopyranose, acetobromopentose, and disaccharide-derived glycosyl halide lactosyl bromide), and the corresponding glycosides **5** were obtained in 78–87% yield after 1.5–8 h milling.

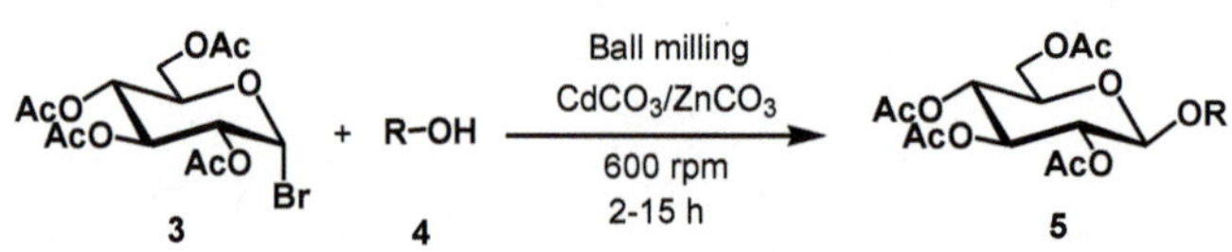

SCHEME 4.2

Glycosylation of acetobromo-α-D-glucose.

Glycosylation of *O*-trityl carbohydrate **6** was also achieved by mechanochemical reaction employing $In(OTf)_3$ as catalyst (Scheme 4.3) [3]. One-pot, two-step planetary ball-milling reaction afforded disaccharide **7** in 64% yield after isolation as a

Table 4.1 Glycosylation of Acetobromo-α-D-Glucose[a]

Entry	R	Time (h)	Yield (%)[b]
1	Propargyl	2	87
2	Allyl	2.5	82
3	4-Pentenyl	3	87
4	2-(Trimethylsilyl)ethyl	3	88
5	*n*-Octyl	3	88
6	*n*-Hexadecyl	3	85
7	Menthyl	9	74
8	1-(3-*O*-Tosyl)glyceryl	8	72
9	3β-Cholesteryl	15	35
10	Bn	4	85

[a]*Retsch PM 100 planetary mill, stainless steel vial 50mL, 10×10mm balls, 600rpm, substrate:alcohol: carbonates 1:1:1.5mol.*
[b]*Isolated yields.*
Reproduced from Tyagi M, Khurana D, Kartha KPR. Solvent-free mechanochemical glycosylation in ball mill. Carbohydr Res 2013;379:55–9, Copyright (2013), with permission from Elsevier.

1:4 mixture of the α- and β-anomers. This suggests the in situ anomerization of the inter-glycosidic linkage in the disaccharide catalyzed by the metal triflate. Prerequisite for this synthesis was development of mechanochemical In(III) triflate-catalyzed detritylation reaction. Methodology was developed for *O*-trityl derivatives of carbohydrates, phenols, and alcohols which are given in Scheme 4.4. Optimized procedure includes milling of tritylated substrate **10** and $In(OTf)_3$ (25mol%) employing planetary ball-mill PM 100, in stainless steel vial (50mL) with 10×10mm balls. Results obtained in ball mill were compared with literature procedures, which all suffer from drawbacks such as low-tolerance to acid-sensitive groups and by-product formation, while the mechanochemical method is free from such limitations. Deprotection in solid state of model substrate per-*O*-acetylated methyl 6-*O*-trityl-glucoside was found to be advantageous to all methods in solution tested and $In(OTf)_3$ was the most satisfactory catalyst.

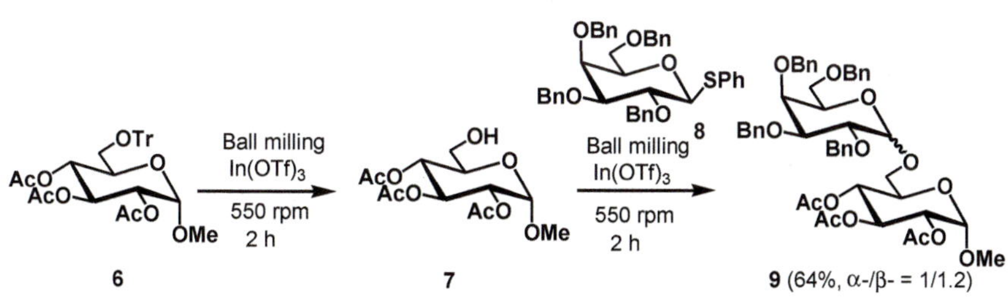

SCHEME 4.3

One-pot detritylation and glycosylation.

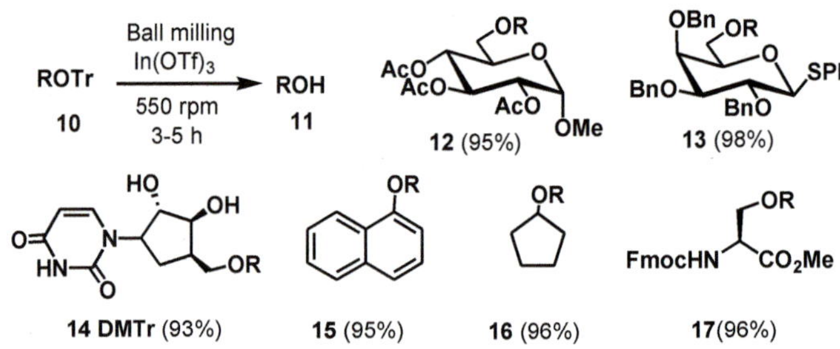

SCHEME 4.4

Mechanochemical solvent-free detritylation.

Finally, grinding in the ball mill of 6-*O*-trityl-protected phenyl thioglucoside **18** in the presence of $In(OTf)_3$ (25 mol%) afforded a library of seven 1,6-linked galacto-oligosaccharides of varying degree of oligomerization in 83% total yield (Scheme 4.5).

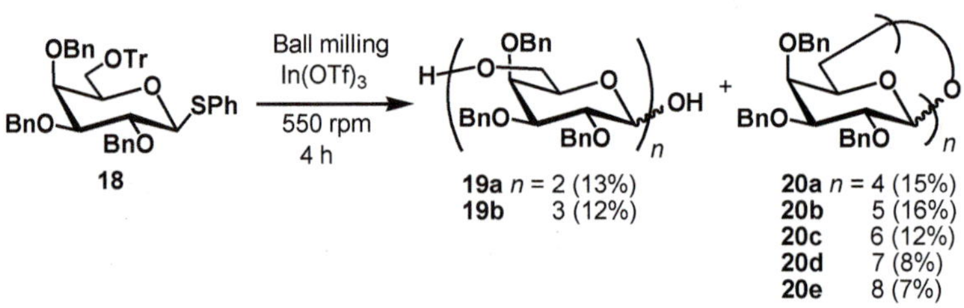

SCHEME 4.5

Indium triflate–mediated synthesis of oligomeric 1,6-linked galacto-oligosaccharides.

4.1.1.2 O-Protection

Chemoselective protection and preparation of monosaccharide 6-*O*-substituted ethers was also performed by Kartha mechanochemically (Scheme 4.6) [4]. The reaction of methyl α-D-glucopyranoside with trityl chloride and 1,4-diazabicyclo[2.2.2]octane (DABCO) under solvent-free conditions was optimized to find that the best results are obtained by 1:2.5:2.5 M ratio of reagents in a planetary ball-mill Retsch PM-100, using stainless steel jar (capacity 50 mL) containing stainless steel balls (10 × 10 mm). Among the different organic and inorganic bases used, DABCO was the most efficient on the small scale, while reactions were significantly more efficient, when carried out in multigram quantities of the substrate. Thus, the 6-*O*-tritylation of methyl α-D-glucopyranoside **21** in the presence of Na_2CO_3 when carried out at a 15 g scale went to completion in 6 h at 500 rpm and in 2 h at 600 rpm 92% and 93% yield, while 1 mmol scale afforded 71% yield of **23** after 15 h of milling. 4-*N,N*-(dimethylamino) pyridine (DMAP) was inactive catalyst, which is due to formation of stable pyridinium salt by reaction with TrCl. In large-scale preparations Na_2CO_3, a greener (and also cheaper) substitute for DABCO, was equally efficient.

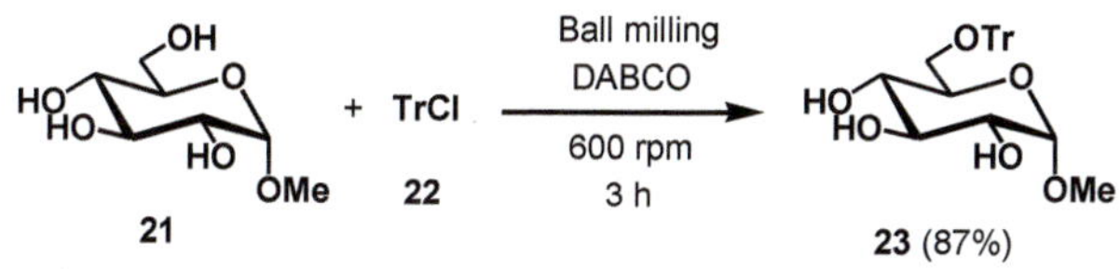

SCHEME 4.6

Tritylation of methyl α-D-glucopyranoside.

Using optimized reaction conditions, other substrates were subjected to tritylation reaction in ball mill (Scheme 4.7). Yields vary from moderate to good, and uridine besides being converted to its 5′-*O*-trityl derivative **27** was also converted to 5′-*O*-dimethoxytrityl derivative **28** by milling with DMTCl (2.5 mol equiv.), respectively, in the presence of DABCO (2.5 mol equiv.).

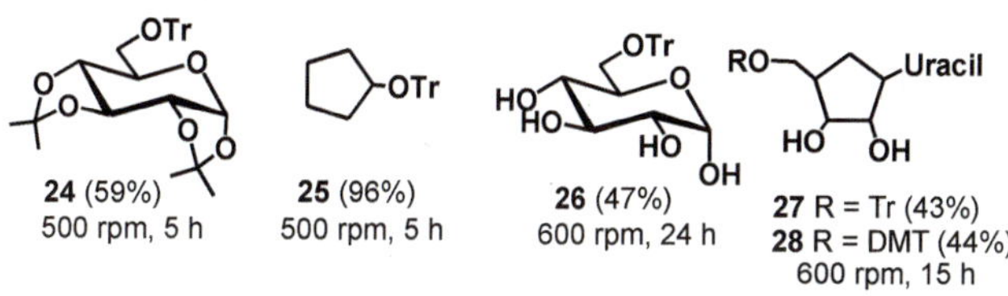

SCHEME 4.7

Tritylation by ball milling.

Chemoselectively protected monosaccharides obtained by ball milling were further acetylated with acetic anhydride (Ac_2O) and one-pot, two-step synthesis was developed (Scheme 4.8). Thus, in the second reaction step, upon completion of the protection reaction, DMAP (1 mol equiv.) and Ac_2O (1.5 mol equiv. for each OH group) were added to the reaction mixture, and further milled for 30 min. Chromatographic separation yielded fully trityl-protected monosaccharide **31**, and this two-step mechanochemical sequence was also applied for *t*-butyldiphenylsilyl and *t*-butyldimethylsilyl chemoselective protection with TBDPSCl and TBDMSCl. The procedure was then successfully extended to other gluco- and galacto-monosaccharide glycosides to obtain the corresponding *O*-acetylated 6-*O*-trityl ethers in 39–82% yields.

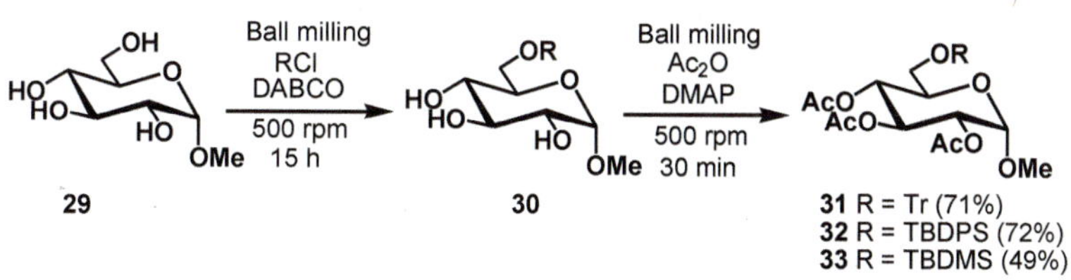

SCHEME 4.8

One-pot two-step synthesis of protected carbohydrates.

4.1.1.3 Cyclodehydration

The conversion of 1-(2-hydroxyphenyl)-3-aryl-1,3-propanediones **34** to flavones **35** was effected by Su et al. by solid-state ball milling (Scheme 4.9) [5]. Cyclodehydration of 1,3-propanediones takes place in the presence of $KHSO_4$, which is an efficient reagent for different functional group transformations under heterogeneous conditions. Planetary centrifugal-mill AGO-2 was employed in stainless steel vial (35 mL), with 5 mm balls (75 g). It was found that 20 mol% is optimal amount of catalyst, and silica gel was used as milling additive. Employment of the optimal reaction conditions to variety of 1-(2-hydroxyphenyl)-3-aryl-1,3-propanediones afforded flavone products **35** in high yields (88–97%), regardless on the electronic nature of the substituents. To prevent "overheating" of the reaction mixture, a milling cycle with a rotational speed of 1290 rpm for 5 min followed by a 5 min pause was employed.

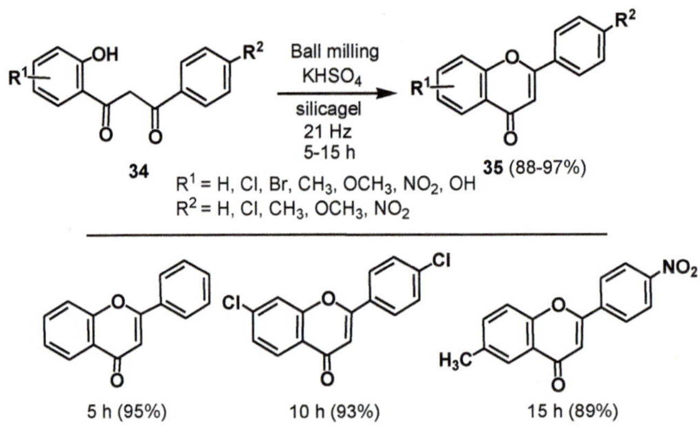

SCHEME 4.9

Synthesis of flavones.

The authors also showed that catalyst could be recycled ($KHSO_4$ with silica gel was collected and regenerated by drying at 120°C for 4 h). The reusability of the catalyst was examined with the 1-(2-hydroxyphenyl)-3-phenyl-1,3-propanedione and resulted in 94%, 92%, 92%, and 88% yields over four 5 min cycles (using substrate:$KHSO_4$ 10:1 mol ratio).

4.1.1.4 Epoxides

Hapiot et al. used ball-milling technique to transform α-, β-, and γ-cyclodextrins (CDs) **36–38** into their epoxides **42–44** (Scheme 4.10). This reaction was achieved with zirconia jar and ball in two reaction steps: *O*-tosylation of vicinal diols and subsequent epoxide formation by internal displacement of tosylate [6]. Authors have found that CDs could be tosylated with *N*-tosylimidazole (TsIm), in the presence of an inorganic base (in stoichiometric proportions). Results were dependent on the inorganic base applied, but for all bases applied in short time (optimal conversion within 6 min), for which KOH gave the highest yields in only 80 s. Mechanochemical formation of mono-2-tosyl-CDs **39–41** takes place on the secondary face of CDs. The most appropriate base was the most water-soluble Rb_2CO_3, which possesses counterion with the ability to rapidly activate the 2-OH group of α-CD. Although reaction is carried out in the solid state, residual water molecules contained in CDs effect the diffusion of the base through the aqueous interstitial channels of the CD network and the activation of the C-2 hydroxyl group. Very fast conversion (>90%) of α-CD to tosylate was achieved within 1 min, without literature precedence for such a rapid process for conversion of CDs into CDs functionalized on their secondary face. The current mechanochemical approach is advantageous (in terms of yields and purity) over synthetic literature procedures for which a protection of the CD primary face was often required. The scaling-up procedure was also successfully carried out. Tosylation experiments carried out at longer period of time (>15 min) indicated slow conversion into mono-(2,3-manno-epoxide)-CDs. Epoxide-formation reaction was completed when CD-tosylates **39–41** were subjected to the second ball milling in the presence of stoichiometric amount of KOH, and mono-epoxide-CDs **42–44** were obtained in high yield.

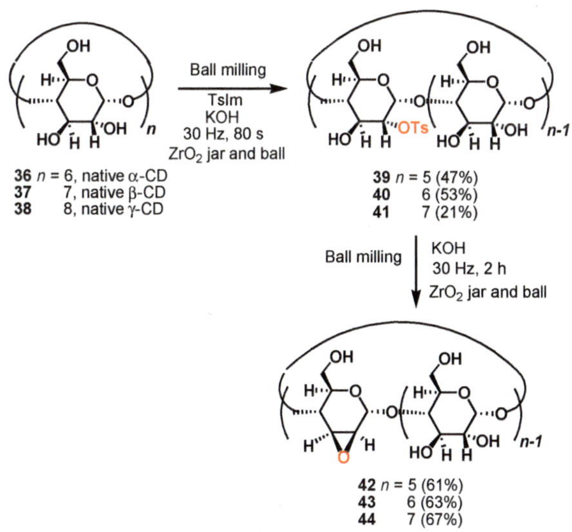

SCHEME 4.10

Epoxide modification of cyclodextrins.

A mechanism for selective tosylation at position 2 has been put forward: formation of supramolecular inclusion complex **46** between CD and TsIm in which the hydrophobic methylphenyl moiety of TsIm is included into the CD cavity (Scheme 4.11). Inclusion is followed by deprotonation of 2-OH and 3-OH groups, where complex selectively orients tosylation reaction on the inward-facing 2-OH group. Deprotonated 3-OH group remains unchanged, as being pointed outside the CD cavity.

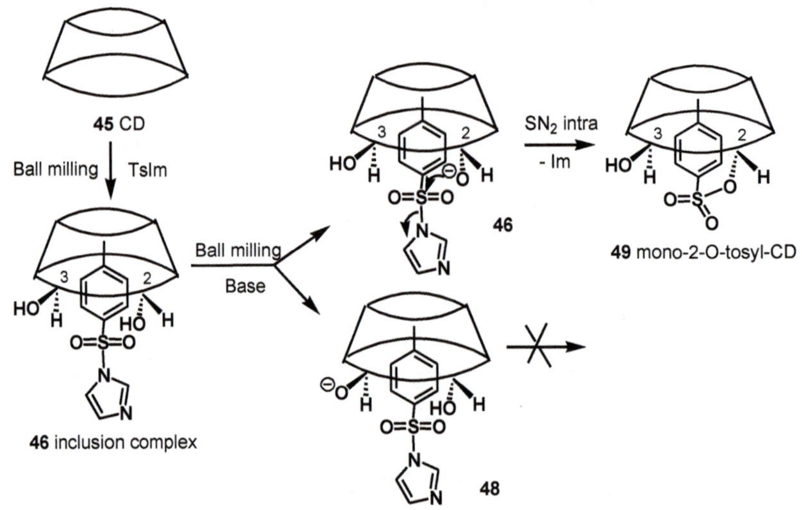

SCHEME 4.11

Reaction selectivity by encapsulation of tosylate in cyclodextrins.

4.1.2 ESTERS

4.1.2.1 Alkylation

Besides acetylation which is described in the previous section, several other reactions were used for synthesis of esters in ball mill. Dialkyl carbonates **52** were formed by alkylation of metal carbonates **50** (environmentally benign, but solubility problematic) with benzyl halides in ball mill by Mack et al. (Scheme 4.12) [7]. It was found that the metal cations that are known to be more tightly bound to oxygen (Li^+ and Na^+) showed no reactivity, while less tightly bound cations (K^+ and Cs^+) allowed for a reaction to occur (Table 4.2). Different methods of increasing the nucleophilicity of oxygen using ball milling were applied. Addition of 18-crown-6 to potassium carbonate not only increased its basicity but also acts to increase nucleophilicity and create dialkyl carbonates in high conversion. Employment of Polysorbate 80 or TweenR80A, a safer alternative to 18-crown-6 to complex the metal ion proved to be an adequate substitute, although less effective and with formation of side products **53** and **54** which arise from the loss of carbon dioxide from reaction intermediate. Utilizing Le Chatelier's principle, addition of dry ice into the reaction mixture quenched this side reaction, and conversion to product was significantly increased, while formation of side products was greatly reduced. Using cesium as the counter-cation of the metal carbonate was another method to increase the nucleophilicity of the metal carbonate's oxygen. Again, addition of dry ice effectively improved conversion, with minimal amounts of side product formed (entry 10).

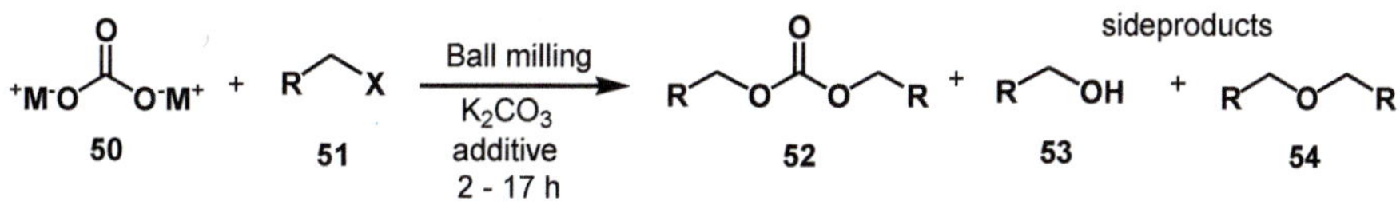

SCHEME 4.12

Synthesis of dialkyl carbonates.

Table 4.2 Synthesis of Dialkyl Carbonates[a]

Entry	R	M	Additive	Substrate:Bromide: M_2CO_3:Additive	Time (h)	Conversion 52 (%)
1	4-BrC_6H_4	K	18-cr-6	1:2:4:2	2	74
2	4-BrC_6H_4	K	no	1:2:4:no	17	2
3	C_6H_5	K	18-cr-6	1:2:4:2	2	67
4	4-Br$C_6H_4CH_2$	K	18-cr-6	1:2:4:2	17	58
5	4-ClC_6H_4	K	18-cr-6	1:2:4:2	17	62
6	4-BrC_6H_4	K	Tween80	1:1:4:0.04	17	22 (53)[b]
7	4-BrC_6H_4	K	Tween80	1:1:4:0.04[c]	17	41 (1)[b]
8	4-BrC_6H_4	K	Tween80	1:1:4:0.05[d]	17	56
9	4-BrC_6H_4	Cs		1:2:4:0	17	18 (13)[b]
10	4-BrC_6H_4	Cs		1:2:4:0[c]	17	37 (2)[b]

[a]*8000M SPEX Certiprep Mixer Mill, 2.0×0.5 in. stainless steel vial, 1/8 in. ball.*
[b]*Side products.*
[c]*Vial full of dry ice.*
[d]*¼ Vial full of dry ice.*
Reproduced from Waddell DC, Thiel I, Bunger A, Nkata D, Maloney A, Clark T, et al. Investigating the formation of dialkyl carbonates using high speed ball milling. Green Chem 2011;13:3156–61, with permission from the Royal Society of Chemistry.

4.1.2.2 Decarboxylative Esterification

The *O*-protection of *N*-protected amino acids **55** was achieved by decarboxylative esterification in solvent-free conditions by Colacino et al. (Scheme 4.13). It was found that the use of planetary ball mill was more effective than vibratory mill [8]. Commonly used *O*-activation reagents (dialkyl dicarbonate (Boc_2O), carbonate (*N,N'*-disuccinimidyl carbonate, DSC), and alkyl chloroformates (ROCOCl, R = Bn, Et, allyl)) were employed in combination with DMAP as base. Reaction parameters had to be optimized for each individual reagent to achieve acceptable yields (selection of results, Table 4.3). Due to high reactivity of benzyloxy chloroformate (Z-Cl), the two-step cycled milling was executed by addition of Z-Cl in 2 equiv. portions, so as to consume the chloroformate and reduce the formation of the undesired byproducts. Acidic workup with 10% aqueous citric acid of ether extracts eliminates DMAP and affords the *N*-protected amino ester derivatives **56** in good yields.

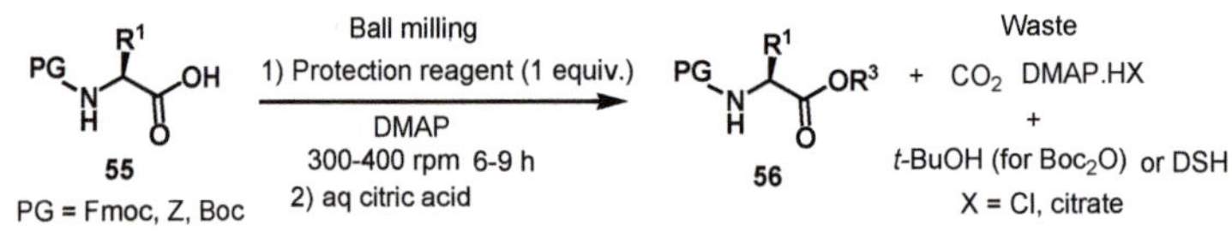

SCHEME 4.13

Decarboxylative esterifications of amino acids.

Table 4.3 *O*-Protection of Amino Acids by Decarboxylative Esterifications[a]

Entry	Substrate	Reagent (equiv.)	DMAP (equiv.)	rpm	Time (h)	Amino Ester	Yield (%)[b]
1	Z-Phe-OH	Boc_2O/1	0.3	300	6	Z-Phe-O^{t}Bu	79
2	Boc-Thr(Bn)-OH					Boc-Thr(Bn)-O^{t}Bu	65
3	Z-Phe-OH	DSC/1	0.3	450	6	Z-Phe-OSu	69
4	Boc-Met-OH					Boc-Met-OSu	35
5	Z-Phe-OH	BnOCOCl/1	1.3	300	6	Z-Phe-OBn	48
6	Boc-Met-OH					Boc-Met-OBn	59
7	Z-Phe-OH	EtOCOCl/1.2	1.5	300	9	Z-Phe-OEt	90
8	Boc-Cys(SBn)-OH					Boc-Cys(SBn)-OEt	91
9	Z-Phe-OH	AllylOCOCl/1.2	1.5	300	9	Z-Phe-OAllyl	90
10	Boc-Met-OH					Boc-Met-OAllyl	82

[a]*Retsch planetary mill PM 200, stainless steel jar (12 mL), 50 × 5 mm balls, 300–400 rpm, substrate:reagent 1:1.*
[b]*Isolated yields.*
Reproduced with permission from Konnert L, Gauliard A, Lamaty F, Martinez J, Colacino E. Solventless synthesis of N-protected amino acids in a ball mill. ACS Sustain Chem Eng 2013;1:1186–91, Copyright 2013, American Chemical Society.

4.1.2.3 Transesterification

Ranu et al. developed simple and efficient solvent-free transesterification in a ball mill over alumina surface (Scheme 4.14) [9]. The use of basic Al_2O_3 was proven to be the most efficient grinding auxiliary for model reaction of benzyl alcohol and allyl acetate. Variation of time and grinding auxiliary showed that acidic alumina and silica gel failed to initiate reaction, while reaction time variations between 1 and 3 h for basic Al_2O_3 revealed that elongation of reaction time from 2 to 3 h does not increase yield. In a typical procedure, alcohol and carboxylic ester **57** (in little excess) adsorbed on the basic alumina were milled for a given time and mixture then filtered through a column of neutral alumina to provide ester product **59**. A wide range of carboxylic esters underwent transesterification with structurally diverse alcohols (selected results, Table 4.4). Reactions are very clean and the products were obtained in high purity. The yields of transesterified products are reasonably good considering the reversible nature of the reaction, with the advantageous mild reaction conditions, which permits the application of ball-milling process to sensitive ester groups.

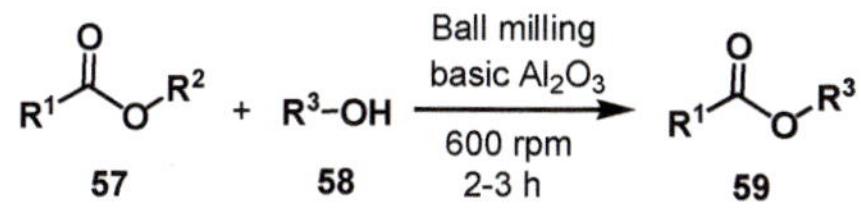

SCHEME 4.14

Transesterification in a ball mill.

Table 4.4 Mechanochemical Transesterifications[a]

Entry	R^1	R^2	R^3	Time (h)	Yield (%)[b]
1	Me	Allyl	Bn	2	76
2	Me	Allyl	3-Cl-C_6H_4-CH_2-	2	75
3	Et	Et	3-Cl-C_6H_4-CH_2-	2	71
4	Me	Allyl	4-F-C_6H_4-CH_2-	2	70
5	n-Pr	Me	4-OCH_3-C_6H_4-CH_2-	2	81
6	Ph	Bn	4-OCH_3-C_6H_4-CH_2-	3	80
7	*n*-Pr	Me	4-NO_2-C_6H_4-CH_2-	2.5	72
8	Vinyl	Me	4-NO_2-C_6H_4-CH_2-	2	70
9	*n*-Pr	Me	Ph-$(CH_2)_2$-	2	71
10	Me	Allyl	Ph-$(CH_2)_3$-	2	73
11	*n*-Pr	Me	Ph-$(CH_2)_4$-	2	70
12	Et	Et	Ph-$(CH_2)_5$-	2	72
13	Me	Allyl	$CH_3CH(Ph)(CH_2)_2$-	2	72
14	*n*-Pr	Me	$CH_3CH(Ph)(CH_2)_2$-	2	70
15	*n*-Pr	Me	2-Thienyl-CH_2-	3	66
16	Et	Et	2-Pyridyl-CH_2-	2.5	71
17	*n*-Pr	Me	2-Furyl-CH_2-	2.5	65
18	*n*-Pr	Me	Ph-CH=CH-CH_2-	2	69
19	Ph	Bn	Ph-CH=CH-CH_2-	3	68
20	Et	Et	*n*-Hex	2	74
21	Me	Allyl	$CH_3(CH_2)_5$-CH(-)C=CH_2	2.5	67
22	Et	Et	Chex	2.5	66
23	Et	Et	*t*-Bu	3	0

[a]*Retsch PM 100 planetary mill, stainless steel vial (25 mL), 6×10 mm balls, 600 rpm, substrate:alcohol 1.2:1 mol.*
[b]*Isolated yields.*
Reproduced from Chatterjee T, Saha D, Ranu BC. Solvent-free transesterification in a ball-mill over alumina surface. Tetrahedron Lett 2012;53:4142–4, Copyright (2012), with permission from Elsevier.

4.1.2.4 Anhydride Ring Opening

Ring opening of cyclic anhydride was used by Bolm to furnish dicarboxylic acid monoesters by ball milling (Scheme 4.15) [10]. Asymmetric anhydride ring opening was achieved in reactions of equimolar amounts of anhydride **60** and benzyl alcohols **61** in combination with small excess of quinidine alkaloid as mediator, in planetary mill with grinding vial and balls, both composed of chemically inert and nonabrasive zirconium oxide. To prevent "overheating" of the reaction mixture, a milling cycle with a rotational speed of 250 rpm for 25 min followed by a 5 min cooling pause was used. Conventionally reaction is performed at low temperatures (−60°C) in organic solvent, to achieve high enantiomeric excess and yield. In mechanochemically carried out reaction, an enantiomeric excess of 61% was achieved for *p*-methylbenzyl alcohol, which compares well with reactions performed in solution at comparable temperatures (Table 4.5). The ball-milling reaction is more atom efficient because the reaction in solution usually requires 3 equiv. of the nucleophile, while the milling requires only 1 equiv. without significantly affecting the yield. As a consequence, extractive workup can be avoided and acid wash suffices to remove the alkaloid and obtain pure hemiester **62**. The asymmetric anhydride opening with *p*-methylbenzyl alcohol **64** also proceeded well with other cyclic anhydrides **63**, affording products **65** in high yield and moderate enantioselectivities (26–57% enantiomeric excess)

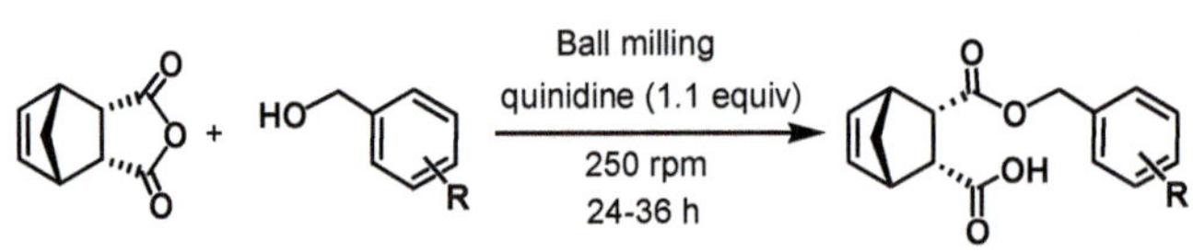

60 **61** **62** (2*R*,3*S*)

SCHEME 4.15

Asymmetric opening of cyclic anhydrides.

Table 4.5 Asymmetric Opening of Cyclic Anhydrides[a]

Entry	R	Yield (%)[b]	ee (%)
1	4-CH_3	91	61
2	4-NO_2	88	40
3	4-CH_3	85	40
4	2-Br	92	64

[a]*Fritsch Planetary Micro Mill Pulverisette 7, ZrO_2 milling bowl (45 mL), 60×5 mm ZrO_2 balls, 250–400 rpm, sequential intervals of 25 min milling followed by a 5 min pause, anhydride:alcohol:quinidine 1:1:1.1 mol.*

[b]*Isolated yields.*

Reproduced with permission from Rantanen T, Schiffers I, Bolm C. Solvent-free asymmetric anhydride opening in a ball mill. Org Process Res Dev 2007;11:592–7, Copyright 2013, American Chemical Society.

(Scheme 4.16, Table 4.6). In the case of synthesis of (2*R*,3*S*)-3-*endo*-*p*-methylbenzyloxycarbonyl-1,2,3,4-tetrahydro-1,4-methano-naphthalene-2-carboxylic acid (entry 1), the use of ball milling was advantageous, since in solution the starting anhydride was unreactive, whereas under ball-milling conditions the asymmetric anhydride opening with *p*-methylbenzyl alcohol proceeded well, affording hemiester with 32% enantiomeric excess in 78% yield.

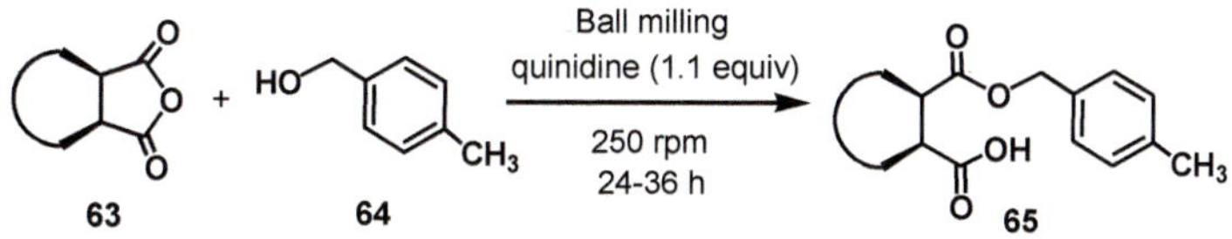

SCHEME 4.16

Asymmetric opening of cyclic anhydrides.

Table 4.6 Asymmetric Opening of Cyclic Anhydrides[a]

Entry	Anhydride	Yield (%)[b]	ee (%)	Product
1	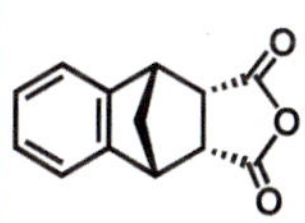	78	32	(2*R*,3*S*)-
2	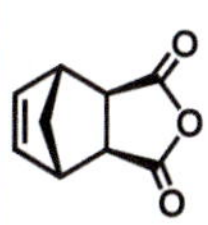	89	57	(2*R*,3*S*)-
3	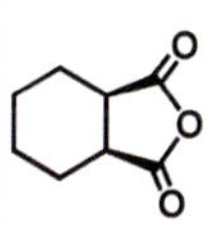	91	26	(1*R*,2*S*)-
4	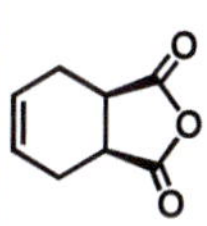	90	55	(1*R*,2*S*)-
5	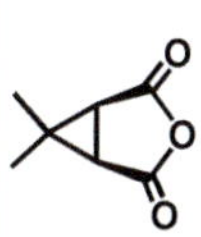	84	46	(1*R*,2*S*)-

[a]*Fritsch Planetary Micro Mill Pulverisette 7, ZrO_2 milling bowl (45 mL), 60 × 5 mm ZrO_2 balls, 250–400 rpm, sequential intervals of 25 min milling followed by a 5 min pause, anhydride:alcohol:quinidine 1:1:1.1 mol.*
[b]*Isolated yields.*
Reproduced with permission from Rantanen T, Schiffers I, Bolm C. Solvent-free asymmetric anhydride opening in a ball mill. Org Process Res Dev 2007;11:592–7, Copyright 2013, American Chemical Society.

4.1.2.5 Tishchenko Reaction

Waddell and Mack have studied Tishchenko reaction (the conversion of aromatic aldehydes **66** to their dimeric esters **67**) in high-speed milling conditions (Scheme 4.17) [11]. Reaction could be effectively carried out within 30 min to afford products in high yields (Table 4.7). Notable reaction acceleration was observed: benzaldehyde was transformed into benzyl benzoate in as little as 30 min in ball-milling conditions compared to 5 h in solution. Interesting results were obtained in the study on the role of the ball material on this reaction, since the Tishchenko reaction is catalyzed by various metals. Stainless steel and brass steel balls were equally effective. Even without metal ball reaction takes place, demonstrating that the movement of container provides enough energy to cause the mixing and reaction. In addition, reactions carried out when reaction vial was immersed in liquid nitrogen at –196°C provided Tishchenko product in 5% yield. This result demonstrates the ability to conduct ball-milling experiments in a low-temperature environment which would allow this process to be implemented with highly exothermic reactions.

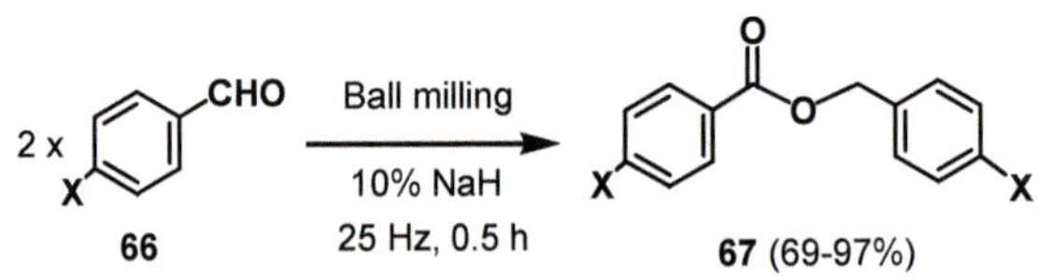

SCHEME 4.17

Tishchenko reaction.

Table 4.7 Tishchenko Reaction of Aryl Aldehydes[a]

Entry	Ar-CHO	Time (h)	Conversion (%)	Yield (%)	TOF
1	4-Br-C_6H_4	2	>99	91	2.5
2	4-Cl-C_6H_4	0.5	>99	92 (5)[b] (>95)[c]	10
3	C_6H_5	0.5	>99	98	10
4	3-Pyridyl	0.5	86	80	10
5	3-CH_3-C_6H_4	2	94	86	2.5
6	4-SCH_3-C_6H_4	2	>99	93	2.5
7	3-F-C_6H_4	0.5	>99	91	10
8	3-OCH_3-C_6H_4	2	>99	97	2.5
9	4-CH_3-C_6H_4	2	80	69	2.5
10	2-Thienyl	2	80	70	2.5
11	2-Naphthyl	16	95	91	1
12	2-Furyl	2	86	69	2.5
13	2-CHO-C_6H_4	2	>99	93	2.5
14	4-OCH_3-C_6H_4	16	>99	97	1

[a]*8000M SPEX Certiprep Mixer Mill, 2.0×0.5 in. stainless steel vial, 0.150 in. ball.*
[b]*SPEX Certiprep freezer mill, –196°C, 30 min; aldehyde:NaH 10:1 mol.*
[c]*Mg metal.*
Reproduced from Waddell DC, Mack J. An environmentally benign solvent-free Tishchenko reaction. Green Chem 2009;11:79–82, with permission from the Royal Society of Chemistry.

Replacement of sodium hydride catalyst with other catalysts that could give the Tishchenko reaction using ball-milling conditions was unsuccessful. The use of catalysts such as lithium bromide and calcium oxide in conjunction with benzaldehyde did not provide any product. Ball milling in the custom-made vials out of nickel and molybdenum showed that these metals did not lead to Tishchenko products. In contrast, ball milling of *p*-chlorobenzaldehyde along with magnesium metal produced >95% yield of dimeric ester. Ball milling of aliphatic and α,β-unsaturated aldehydes was poor, and gave only small amount of products, in conjunction with aldol condensation products.

4.1.2.6 Acylation

Mechanochemical synthesis from carboxylic acids activated by CDI is applicable to O-, S-, and C-nucleophiles. Two-step process involves CDI activation of acid **68** in the first milling step, followed by milling with (thio)alcohol or malononitrile (Scheme 4.18). Métro, Lamaty and coworkers obtained the corresponding esters, thioesters, and alkenol products **70** in high yields (Table 4.8) by simple treatment with water after reaction completion [12].

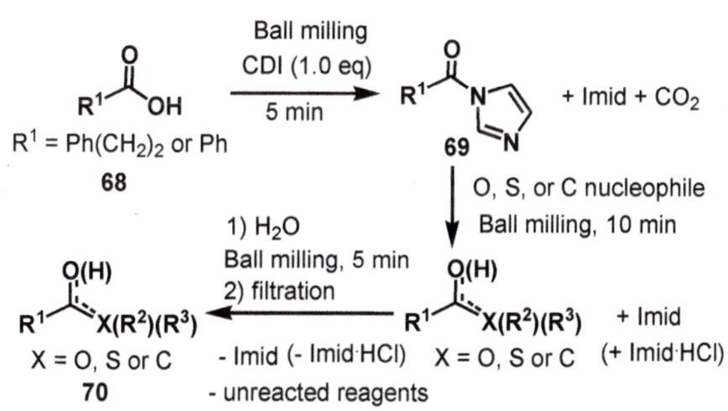

SCHEME 4.18

CDI-mediated acylation of O-, S-, and C-nucleophiles.

The mechanochemical synthesis of piroxicam 4-nitrobenzoate **76** was carried out in two steps including acylation (Scheme 4.19) [13]. Pankrushina et al. carried out the reactions in SPEX 8000 mill at ambient temperature and in a planetary mill AGO-2 with water cooling. The reason for cooling to 0°C is due to the fact that mechanochemical acylation of piroxicam performed in the ball mill resulted in rapid temperature increase due to the exothermic nature of the reaction and friction during milling. Synthesis in planetary mill with cooling proceeded in moderate yield of acylated product (28%). Synthesis started by pretreatment in the mill for 15 min to convert piroxicam **71** into the more-active zwitterionic form **72**. In the second reaction step, zwitterionic piroxicam was milled with 4-nitrobenzoyl chloride (1 equiv.) and aluminum oxide (2 equiv.) as a base, with an addition of triethylamine for increase basicity of the aluminum oxide for 10–20 min to afford the acylated product in 28% yield.

Table 4.8 CDI-Mediated Acylation of O-, S-, and C-Nucleophiles[a]

Products 70 (Yield in %)

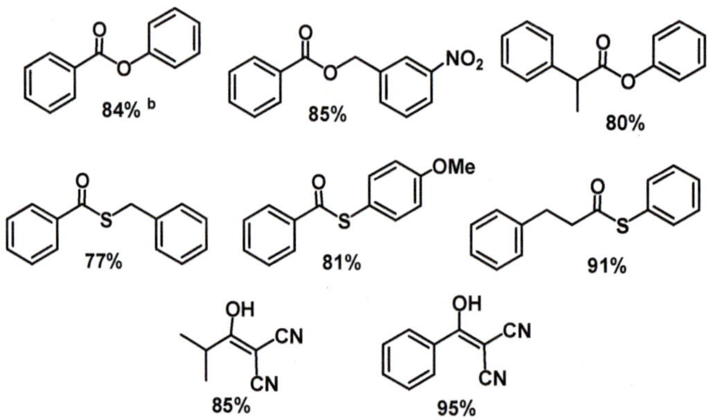

[a]Retsch MM200 mixer mill, stainless steel vessel (10 mL), 2 × 10 mm balls, 25 Hz.
[b]Retsch PM 100 planetary mill, stainless steel grinding bowl (12 mL), 50 × 5 mm balls, 500 rpm,

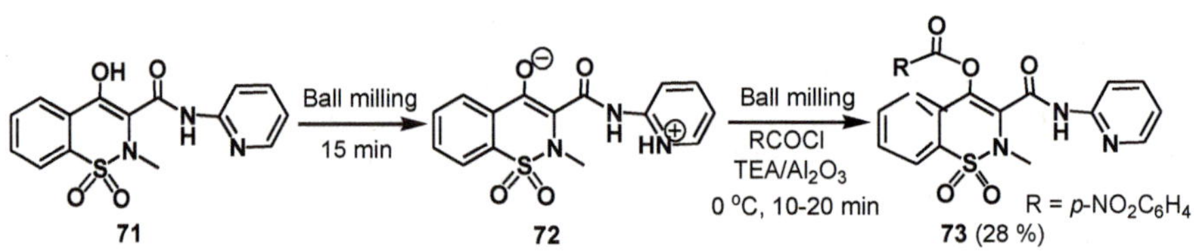

SCHEME 4.19

Synthesis of piroxicam 4-nitrobenzoate.

4.2 SULFUR BOND FORMATION REACTIONS

In addition to acylation of thiols by carboxylic acids activated by CDI which is described in the previous section, other reactions were used for sulfur bond construction in ball-milling conditions. Several reaction cascades which are given in this section start with the formation of sulfur bond.

4.2.1 ALKYLATION

Thiuronium salts **76** and **78** were quantitatively obtained when 2-imidazolidine thione **74** or 2-mercaptobenzimidazole **77** were ball milled with phenacyl bromide by the procedure reported by Kaupp et al. (Scheme 4.20) [14]. For this reaction, equimolar amounts of reagents were used. Similar reactions in solution, on the other hand, are incomplete and generate corrosive wastes.

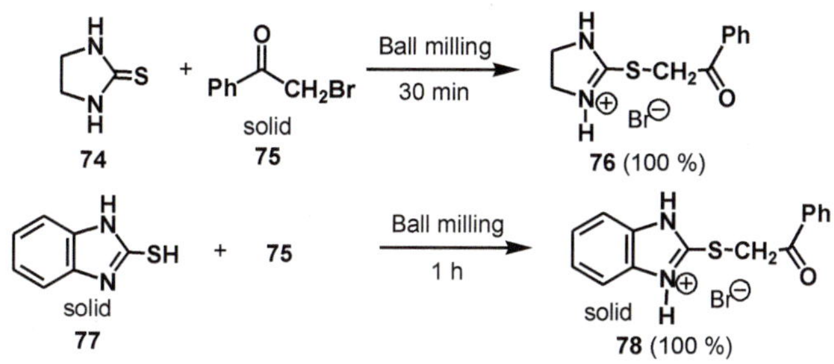

SCHEME 4.20

Synthesis of thiuronium salts.

4.2.2 HANTZSCH REACTION

Kaupp et al. employed ball-milling technique to transform thioureas **79** by reaction with phenacyl bromide to 2-amino-4-phenyl-thiazole-hydrobromides **80** in quantitative yields from stoichiometric mixtures of the reagents at room temperature (Scheme 4.21) [14]. In solid-state conditions, base catalyst was not needed. The water formed in the reaction does not hydrolyze phenacyl bromide under applied mild conditions and was removed by heating at 80°C in vacuo. When the same reaction was performed in a melt at 110°C, partial hydrolysis occurred, which diminishes yield, while yields obtained in solution were lower (80–90%). This Hantzsch thiazole synthesis starts with nucleophilic substitution on sulfur and formation of the carbon–sulfur bond (S-alkylation), followed by further reaction cascade which results in heterocyclic ring.

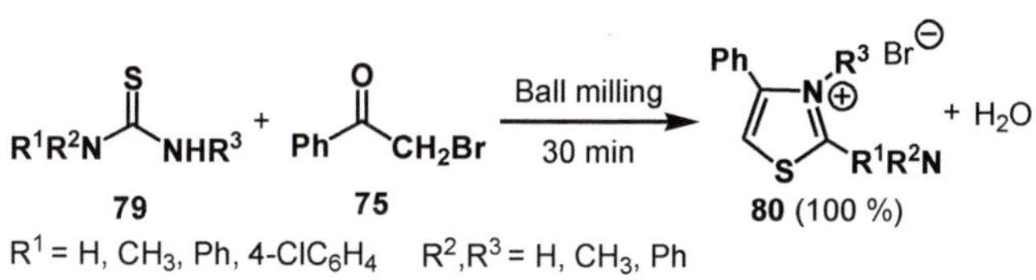

SCHEME 4.21

Synthesis of 2-amino-4-phenyl-thiazoles.

For thiazole synthesis Abdel-Latif and Metwally used arylidenethiosemicarbazones [1]. The stoichiometric ball-milling reactions of α-haloketones **81** with the arylidenethiosemicarbazone derivatives **82** afforded the corresponding iminium salts **83** with quantitative yield without the aid of basic catalysts and solvents (Scheme 4.22). Washing with aqueous Na_2CO_3 was used to liberate the free bases of 4-substituted 2-(arylidenehydrazino) thiazoles **84**. The ball-milling reaction of phenacyl bromide proceeds more easily at room temperature, while the corresponding reaction of 4-(chloroacetyl)antipyrine requires ball milling at higher temperatures (50°C, Retsch MM2000 with double-walled stainless steel beaker with fittings for circulation of coolants) for a quantitative yield.

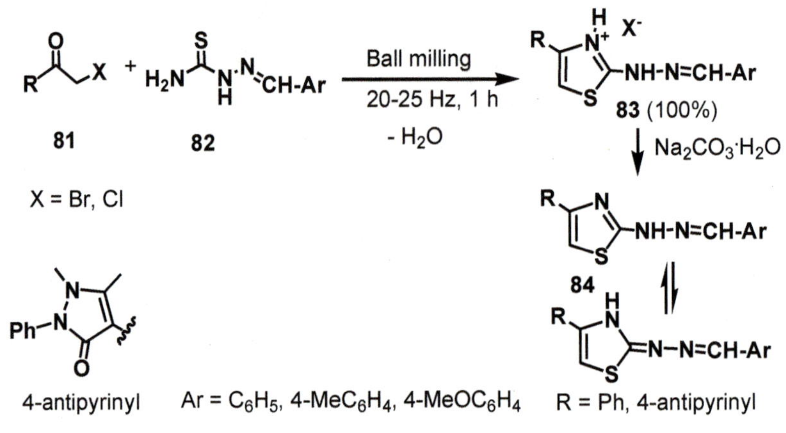

SCHEME 4.22

Synthesis of 4-substituted 2-(arylidenehydrazino) thiazoles.

4.2.3 NINHYDRIN CONDENSATION

In a similar manner as used for 1,2-dione/diamine condensations, Kaupp carried out stoichiometric mechanochemical reactions of *o*-mercaptoaniline **86** with ninhydrin in the three reaction cascade (substitution, cyclization, and elimination), which gave the salt of 10a-hydroxyindeno[2,1-*b*]benz[1,4]thiazin-11(10*aH*)-one **89** (Scheme 4.23) [15]. Washing of product with $NaHCO_3$ was the simple workup in this efficient solvent-free synthetic protocol.

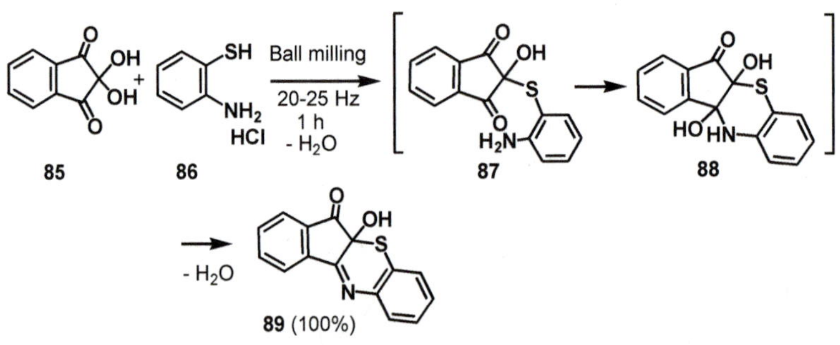

SCHEME 4.23

Condensation of ninhydrin with *o*-mercaptoaniline.

4.2.4 FORMATION OF C—S, C—Se, AND C—Te BONDS FROM DICHALCOGENIDES

Reaction of aryl diazonium tetrafluoroborate **90** and diaryl dichalcogenides (disulfane, diselenide, and ditelluride) **91** under ball milling at ambient temperature was applied in the synthesis of unsymmetrical aryl chalcogenides **92** by Ranu et al.

(Scheme 4.24) [16]. The best reaction conditions employ KOH base (0.75 equiv.) and neutral Al_2O_3 as grinding auxiliary. By mechanosynthesis, unsymmetrical aryl sulfides, selenides, and tellurides were obtained in high yield within 15–30 min (selected examples, Table 4.9). Although the diazonium salts are, in general, susceptible to explosion on heating/grinding, authors did not encounter any such incidence during milling process. This method is also effective for the solvent-free synthesis of *S*-aryl dithiocarbamates **96** by the two-step, one-pot reaction of stoichiometric amounts aryl diazonium fluoroborate **95**, carbon disulfide, and an amine **93** (Scheme 4.25). Reaction of CS_2 and secondary amine takes place within 2 min of grinding, which is followed by the addition of aryl diazonium fluoroborate and continuation of milling, using basic alumina as grinding auxiliary. The employment of this synthetic procedure afforded after 15–20 min a various functionalized dithiocarbamates in 77–88% yields.

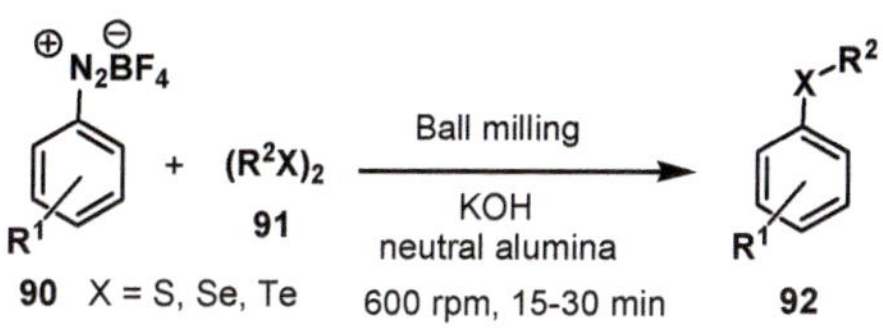

SCHEME 4.24

Synthesis of unsymmetrical sulfides, selenides, and tellurides.

Table 4.9 Synthesis of Unsymmetrical Sulfides, Selenides, and Tellurides[a]

Entry	R^1	R^2	Time (min)	Yield (%)[b]
Sulfides				
1	4-CH_3	4-Cl	15	90
2	3-$COCH_3$	2-Pyridyl	15	78
3	4-CF_3	4-NO_2	20	73
Selenides				
4	4-OCH_3	H	15	78
5	2,5-Di(CH_3)	H	20	73
6	4-CN	H	30	74
Tellurides				
7	4-OCH_3	H	15	76
8	4-$COCH_3$	H	15	78
9	4-CN	H	30	62

[a]*Retsch PM 100 stainless steel vial (25 mL), 6×10 mm balls, 600 rpm, diazonium salt:KOH:($R^2X)_2$ 1:0.75:0.5 equiv.*
[b]*Isolated.*
Reproduced with permission from Mukherjee N, Chatterjee T, Ranu BC. Reaction under ball-milling: solvent-, ligand-, and metal-free synthesis of unsymmetrical diaryl chalcogenides. J Org Chem 2013;78:11110–4, Copyright 2013, American Chemical Society.

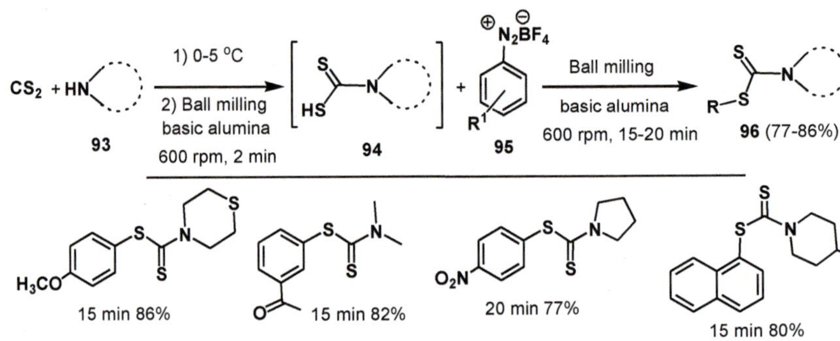

SCHEME 4.25

Synthesis of dithiocarbamates.

4.2.5 FORMATION OF SULFILIMINE BOND

Bolm and collaborators have demonstrated that the sulfoxide to sulfilimine conversion can be performed under solvent-reduced conditions in a ball mill [17]. The reaction of sulfoxide **97** and 2 equiv. of Burgess-type reagent **98** produced the corresponding sulfilimine **99** in 68% yield, after a milling for 80 min (Scheme 4.26). For comparison, in solution (THF, reflux, 2 h) the same product was obtained in slightly higher yield (78%).

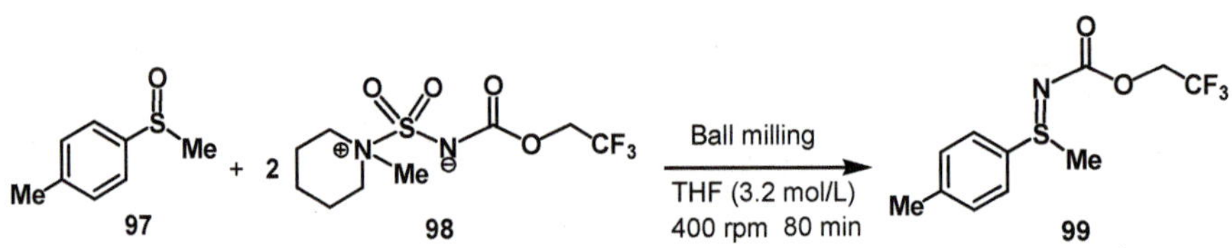

SCHEME 4.26

Synthesis of sulfilimine employing Burgess-type reagent.

4.2.6 S—S BOND METHATHESIS

Sanders et al. carried out mechanochemical interconversions of two aromatic disulfides to heterodimer (Scheme 4.27) [18]. The two homodimers, 2-nitrophenyl disulfide **100** and 4-chlorophenyl disulfide **101** in equimolar amounts, gave upon milling the corresponding heterodimer **102**, using base catalyst 1,8-diazabicyclo[5.4.0]undec-7-ene (2 mol%). Process involves reversible covalent S—S bond breaking and its subsequent reforming. Detailed kinetic study and direct observation of intermediates demonstrated that this solid-state reaction was under thermodynamic control. Two milling conditions applied (solvent free or liquid-assisted grinding (LAG) with MeCN) determine the exclusive formation of two different polymorphs **102** and **103**. By seeding experiments it was shown that the stable polymorphs formed under ball-mill neat or LAG conditions are the metastable polymorphs under respective grinding conditions and rapidly transform to more stable polymorphs. Obtained results confirm thermodynamic control observed in their earlier study in which 4-nitrophenyl and 4-methylphenyl disulfides were also employed [19].

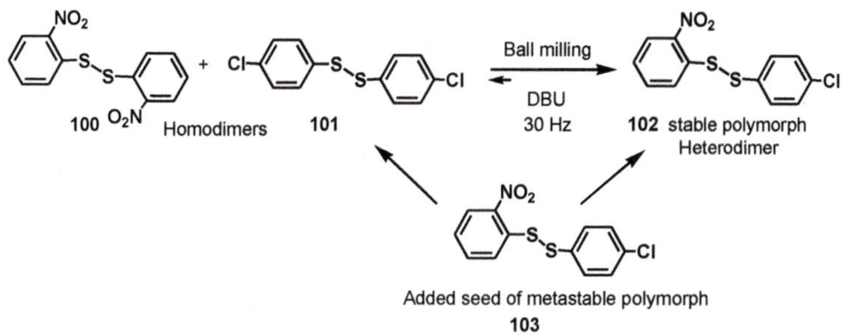

SCHEME 4.27

Base-catalyzed disulfide metathesis.

4.3 HALOGEN BOND FORMATION REACTIONS

Several halogenation reactions were carried out in ball mill, mostly on aromatic substrates, and mechanochemistry was also successfully employed for aromatic dehalogenation reactions.

4.3.1 AROMATIC SUBSTITUTION

Halogenation by aromatic substitution in ball mill was carried out by Stolle et al. by solvent-free method which does not require the direct use of dihalogens, instead they are generated in situ [20]. Reaction conditions were established using bromination of mesitylene (Scheme 4.28). For optimal yield and selectivity toward monobrominated product **105**, quartz sand was the most suitable milling auxiliary. The optimal ratio of reagents was established to be equimolar amounts of mesitylene and NaX (X=Cl or Br) and oxone as oxidant (0.5 equiv.). Solvents for LAG were also employed, but additional solvents (either H_2O or organic) deliver no advantage for solvent-free halogenation with oxone and NaBr.

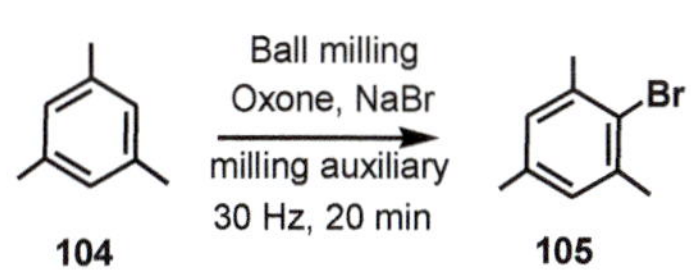

SCHEME 4.28

Synthesis of aryl halides.

After developing the optimal reaction conditions, a number of aromatic substrates were subjected to bromination and chlorination reactions with NaBr and NaCl (Table 4.10). Benzene, *m*-toluidine, benzonitrile, and chlorobenzene did not react in

Table 4.10 Mechanochemical Bromination and Chlorination of Aromatic Compounds[a]

			Hal = Br	Hal = Br	Hal = Cl	Hal = Cl
Entry	**Aromatic Compound**	**Product**	**Conv. (%)[b]**	**Selectivity (%)[b]**	**Conv. (%)[b]**	**Selectivity (%)[b]**
1	Mesitylene	Hal-mesitylene	85	96	85	96
2	Toluene	Hal-toluene	95	99 (*o*/*p* = 1:2)	92	96 (*o*/*p* = 1:2)
3	*o*-Xylene	4-Hal-1,2-dimethylbenzene	90	97	76	93
4	*m*-Xylene	1-Hal-2,4-dimethylbenzene	94	98	93	98
5	*p*-Xylene	2-Hal-1,4-dimethylbenzene	91	94	80	91
6	Ethylbenzene	Hal-ethylbenzene	71	90 (*o*/*p* = 1:3)	75	98 (*o*/*p* = 1:3)
7	Anisole	Hal-anisole	88	97 (*o*/*p* = 1:3)	86	96 (*o*/*p* = 1:3)
8	Phenol	Hal-phenol	81	77 (*o*/*p* = 2:9)	70	97 (*o*/*p* = 2:9)
9	2,3,5-Trimethylphenol	Hal-2,3,5-trimethylphenol	84	95 (*o*/*p* = 1:3)	64	89 (*o*/*p* = 1:3)
10	2,3,6-Trimethylphenol	4-Hal-2,3,6-trimethylphenol	89	96	81	76
11	*p*-OH-benzaldehyde	3-Hal-4-OH-benzaldehyde	65	71	67	88
12	*m*-Toluidine	4-Hal-3-methylaniline	91	71	99	10
13	Naphthalene	1-Hal-naphthalene	78	94	39	81
14	Anthracene	9-Hal-anthracene	36	61	19	46

[a] *Retsch MM400 mixer mill, 30 Hz, 20 min, yttrium-stabilized zirconia (YSZ) jar (35 mL), 10 × 10 mm magnesia-stabilized zirconia (MSZ) balls, Ar:NaHal:oxone 2:2:1.*

[b] *GC yields.*

Reproduced from Schmidt R, Stolle A, Ondruschka B. Aromatic substitution in ball mills: formation of aryl chlorides and bromides using potassium peroxomonosulfate and NaX. Green Chem 2012;14:1673–9, with permission from the Royal Society of Chemistry.

reaction conditions applied (either in bromination or chlorination reactions), while other aromatic substrates showed good conversions and selectivity as high as 99%.

Application of this mechanochemical halogenation methodology to aromatics **106–108** possessing unsaturated functionalities like alkenyl or alkynyl groups led to the halogen addition to the C—C double or triple bond, without formation of electrophilic substitution products (Scheme 4.29).

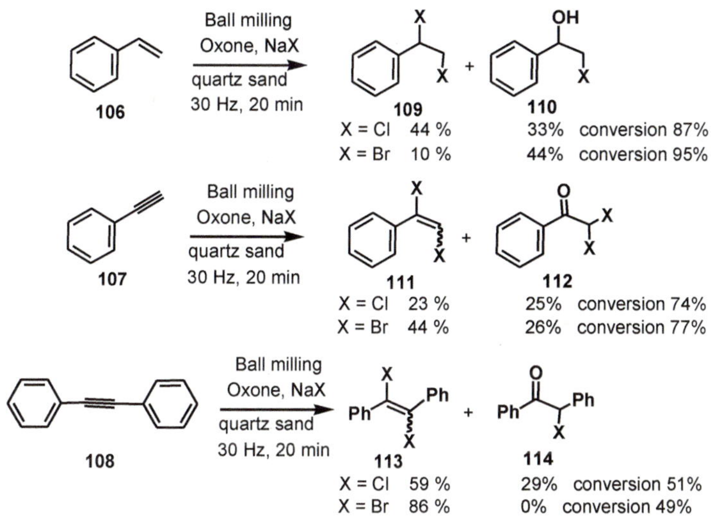

SCHEME 4.29

Mechanochemical halogenation of unsaturated aromatic compounds.

In addition to the described halogenation method, bromination of mesitylene with $CuBr_2$ has been developed, with the best conversion and selectivity (59% and 98%) obtained with Na_2SO_4 as grinding auxiliary. However, the application of copper halides is disadvantageous due to the need for 2 equiv. of $CuBr_2$ for the substitution of 1 equiv. of a C—H function. Another halogenation method was also explored, employing phenylboronic acid in conjunction with $CuBr_2$ and the best conversion was obtained for neutral γ-alumina (28%).

Further aromatic halogenation reactions in ball mill were carried out by Bose and Mal, for which *N*-halosuccinimides were employed to obtain aryl halogenides in high yields and short reaction time (Schemes 4.30–4.32) [21]. Mechanochemical brominations were carried out in much shorter time, at room temperature and without the use of toxic reagents compared to solvent synthesis procedures. In Retsch MM2001 ball mill (10 mL stainless steel jar, one 5 mm ball), substrate and NBS in 1:2 M ratio were grounded upon completion. Products were dissolved in organic solvent and purified, while succinimide was recovered and NBS regenerated. This methodology is found to be chemoselective and no benzylic bromination was observed, with high regioselectivity, and conditions tolerate variety of substituents (Scheme 4.30). Bromination of aniline and anisoles with 1 equiv. of NBS produced regioselectively monobrominated products (4-bromo aniline and 4-bromo anisole) in good yields. Bromination method could be also carried out on larger scale: milling of methoxy-3,5-dimethylbenzene (1.3 g) for 1 h afforded the corresponding product in 87% yield. The bromination methodology was also used for catalyst and additive-free iodination with NIS under ball-milling conditions and electron-rich compounds gave products in chemoselective and regioselective manner (Scheme 4.31). The direct aromatic chlorination was also successful by ball milling with NCS; however, it required the presence of cerric ammonium nitrate (Scheme 4.32). The yields are found to be very good to excellent.

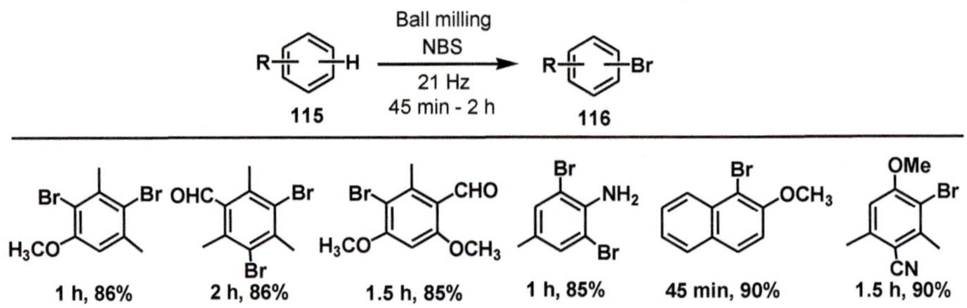

SCHEME 4.30

Solvent-free bromination using *N*-bromosuccinimide.

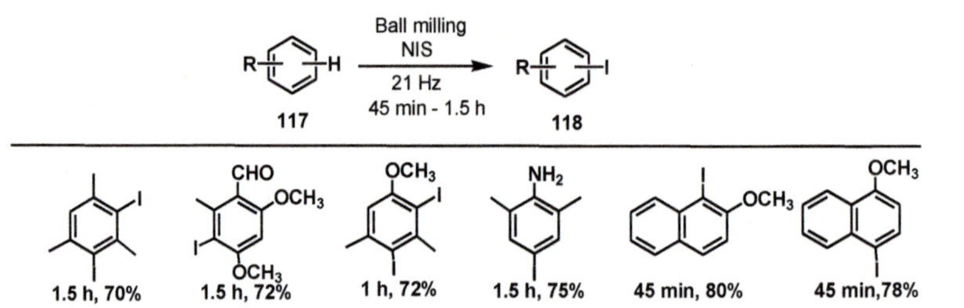

SCHEME 4.31

Solvent-free iodination using *N*-iodosuccinimide.

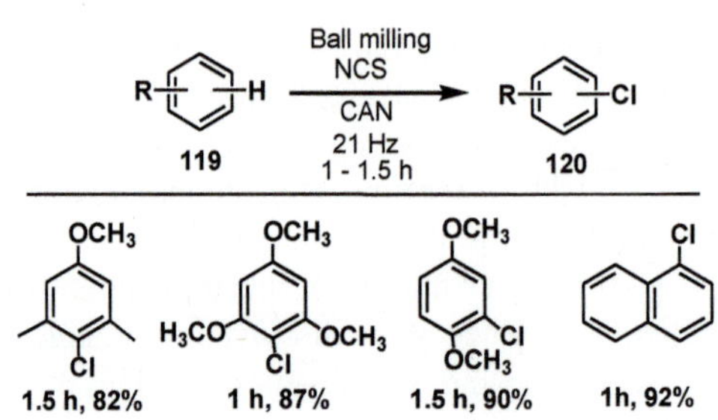

SCHEME 4.32

Solvent-free chlorination using *N*-chlorosuccinimide.

Equally effective iodination of electron-rich aromatic compounds under ball milling was achieved employing I_2-oxone system (Scheme 4.33) [22]. Iodinated products were obtained in very good to excellent yields, in similar reaction times. This solvent-free method required 1 equiv. of oxone, 1 equiv. of I_2 for monoiodination and 2 equiv. of I_2 for diiodination to chemoselectively and regioselectively afford iodinated products. In spite of poor electrophilicity of iodonium ion generated from I_2, the aryl-iodination of various types of aromatic compounds, even iodination of nitrobenzene derivative, was carried out. After reaction completion, products were extracted with ethyl acetate (environmentally friendly solvent) and chromatographed. This methodology is considerably less expensive in comparison with literature procedures in solution. In addition, under ball-milling conditions monoiodinated product of 1,3,5-trimethoxy benzene was isolated in 88% yield using I_2-oxone combination. However, with 0.5 equiv. of iodine, dimerized product **124** was obtained in 48% yield as a single product. This biarylation reaction with I_2 also worked for two more aromatic substrates, and it is thought to proceed via charge-transfer Ar/I_2 complex, followed by single-electron transfer leading to cation radical species, which forms adduct with another aryl compound, and finally goes to biaryl derivative after losing a single electron and two protons.

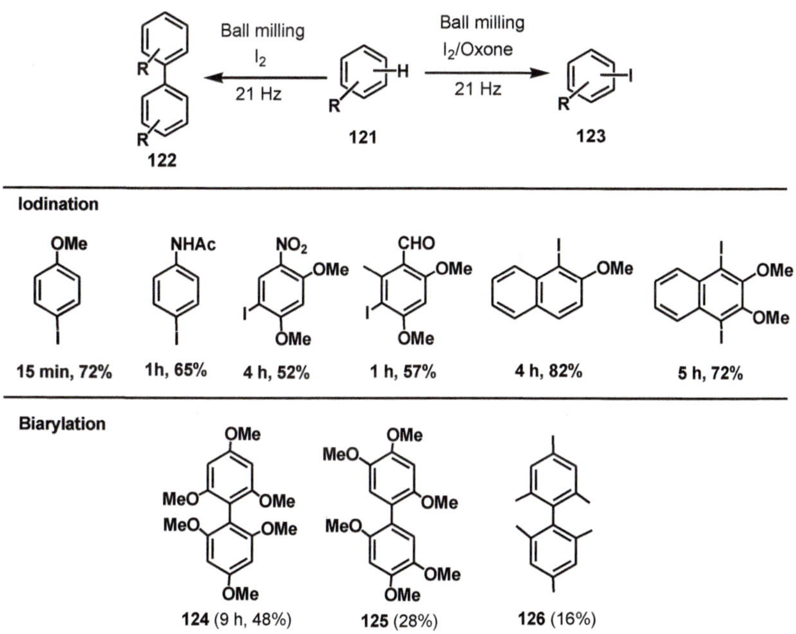

SCHEME 4.33

Aryl-iodination and biarylation with iodine.

Finally, one-pot multistep synthesis by iodination/oxidation process was carried out for the iodination of 3-methoxy benzyl alcohol **127**, to afford the corresponding iodo-aldehyde **129** in 54% yield (Scheme 4.34).

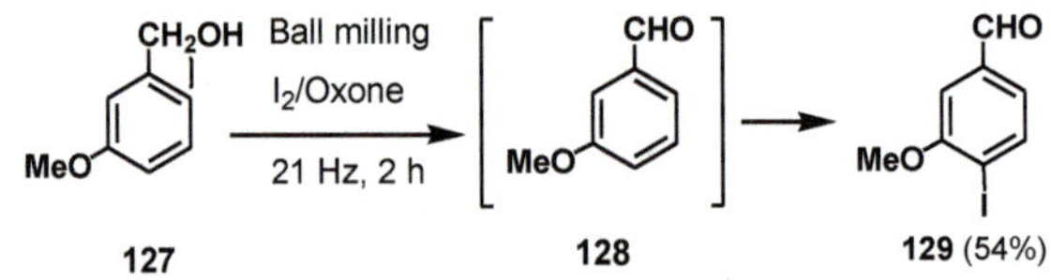

SCHEME 4.34

Aryl-iodination of 3-methoxy benzyl alcohol.

Different reaction conditions were employed for aryl halogenation using NBS and NIS reagents by Bolm [23]. For this purpose, direct synthesis of $[Cp^*RhCl_2]_2$ complex **133** was achieved by mechanochemical reaction of rhodium(III) chloride hydrate with pentamethylcyclopentadiene (Scheme 4.35). LAG (MeOH, $\eta=0.25$) in planetary ball mill, using tungsten carbide–milling vessel and balls conditions were employed for the optimum result. Another rhodacycle **131** (Rh-II) was prepared by reaction of 2-phenylpyridine with $[Cp^*RhCl_2]_2$ in the presence of sodium acetate in a planetary mill by ZrO_2 grinding in high yield. These catalysts were employed in halogenation of 2-phenylpyridine by NBS and NIS in conduction with $AgSbF_6$ to obtain halogenated products **134** in high yield (Table 4.11). With 1.1 equiv. of NBS milling afforded a mixture of the *ortho*-mono- and dibrominated phenylpyridines. This lack of selectivity by the pyridine moiety as a directing group is a common difficulty known from the solution-based methods, which was circumvented by addition of 2.2 equiv. of NBS to obtain dibrominated 2-phenylpyridine as the major product. The long milling time in the planetary mill was reduced along with an increase in yield after conducting the reactions in a mixer mill.

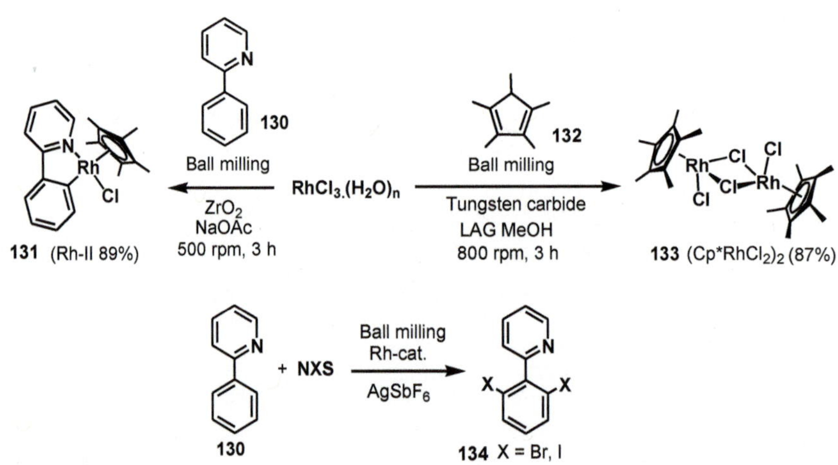

SCHEME 4.35

Halogenation of 2-phenylpyridine.

Table 4.11 Halogenation of 2-Phenylpyridine[a]

Entry	Rh Cat.	Rh Cat. (mol %)	Type of Mill	Milling Media	NXS (equiv.)	Time (h)	Yield (%)[b]
1	$[Cp^*RhCl_2]_2$	5	PM	ZrO_2	NBS (1.1)	4	Dimono products
2		5	PM	ZrO_2	NBS (2.2)	8	60
3		5	PM	ZrO_2	NIS (2.2)	8	78
4		5	MM	SS	NBS (2.2)	3	74
5		5	MM	SS	NIS (2.2)	3	84
6	Rh-II	5	MM	SS	NBS (2.2)	3	71

[a]*Substrate:$AgSbF_6$ 1:0.4; PM: Fritsch Planetary Micro Mill Pulverisette 7, 800 rpm, ZrO_2 vessel (12 mL), ZrO_2 balls (20×5 mm), SS: Retsch MM400 mixer mill, stainless steel jar (10 mL), 1×10 mm ball.*
[b]*Isolated yield.*
Reproduced from Hernández JG, Bolm C. $[Cp^*RhCl_2]_2$: mechanosynthesis and applications in C—H bond functionalisations under ball-milling conditions. Chem Commun 2015;51:12582–4, with permission from the Royal Society of Chemistry.

Mechanochemical chlorination of hydrocarbons with $CuCl_2$ catalyst was performed in high-energy planetary ball mill (Scheme 4.36) [24]. The chlorination of selected hydrocarbons resulted in various chlorinated products, but relatively high amounts of the starting compounds remained unreacted (100% for hexadecane and 26% for biphenyl, Table 4.12). In these conditions, $CuCl_2$ acts as a chlorinating agent yielding the products of classical electrophilic chlorination. It was noted that the inner temperature of the jar reached about 80°C after 5 min of operation and around 170°C after 20 min of milling. Compared to chlorine or other metal chlorides used for halogenation, the action of $CuCl_2$ is much less vigorous, easy to handle, and the reaction stoichiometry can be easily controlled. The process is advantageous over existing ones, which require high temperatures and long reaction time, while the preparation of higher chlorinated naphthalenes was not possible.

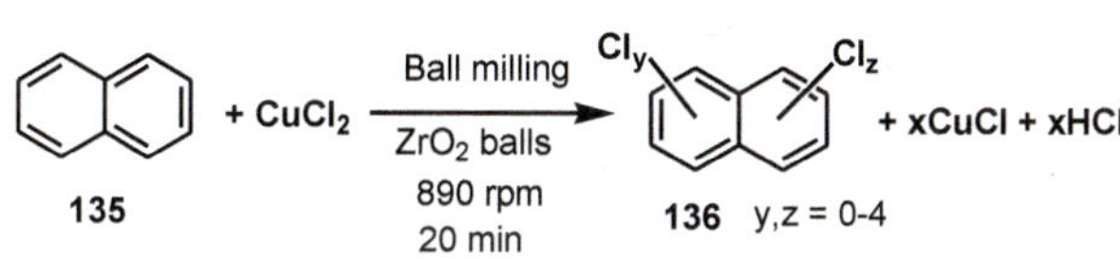

SCHEME 4.36

The chlorination of naphthalene by ball milling.

Mechanochemical bromination method developed by Wang involves the bromide generated *in situ* from the oxidation of bromide ions [25]. This procedure employs a mixture of oxone (1–2 equiv.) and excess of sodium bromide (2.5 equiv.) as bromination reagent. Efficient ball-milling procedure in Retsch MM200 mill (5 mL stainless steel jar, one 7 mm ball) takes from 20 to 40 min to obtain products in high yields. Variety

Table 4.12 The Chlorination of Naphthalene by Ball Milling[a]

			Substrate Number of Cl Atoms in the Molecule									
Entry	**Substrate**	**% Unreacted Substrate**	**1**	**2**	**3**	**4**	**5**	**6**	**7**	**8**	**9**	**10**
1	*n*-Hexadecane	100										
2	Biphenyl	26	22.7	20.1	11.6	7.9	7.3	3.1	1.3	0.3	0.1	0.1
3	Naphthalene	1	20.9	32	19.9	14.1	8	3.2	0.7	0.2		

[a]*High-energy planetary ball mill, TB-1 (Kadaň Ltd., Slovakia) 890 rpm, 20 min, stainless steel jar (0.3 dm^3), 10 mm yttrium-stabilized ZrO_2 ceramic balls, $CuCl_2$:aromatic hydrocarbon mixture weight ratio of 20:1.*

Reproduced from Višňovský J, Billik P, Kubinec R, Podolec P, Hengerics Szabó A, Juriga M, et al. Solvent-free mechanochemical chlorination of hydrocarbons with $CuCl_2$. Tetrahedron Lett 2013;54:7180–2, Copyright 2013, with permission from Elsevier.

of substrates could be brominated: 1,3-dicarbonyl compounds **137** (Scheme 4.37), phenols (Scheme 4.38), α,β-unsaturated carbonyl compounds: chalcones, azachalcones, 4-phenylbut-3-en-2-one, methyl cinnamate (Scheme 4.39), as well as styrene and 1,3-cyclohexadiene (Scheme 4.40). Ball-milling bromination of cyclic 1,3-dicarbonyl compounds afforded α,α-dibromo derivatives, while acyclic 1,3-dicarbonyl compounds selectively produced α-monobromo derivatives. Also, bromination of phenols to multibrominated products was selective to *ortho*- and *para*-positions. Reactions of α,β-unsaturated carbonyl compounds provided *anti*-α,β-dibromo derivatives, and reaction conditions tolerate a number of functional groups (ether, cyano, chloro, ester, nitro, and keto). Notably, bromination of 1,3-cyclohexadiene produced selectively *trans*-3,6-dibromocyclohexene is clearly favorable to solution reaction which produces a mixture of three dibromocyclohexenes. Another advantage of this solvent-free method is the use of environmentally friendly oxone and NaBr reagents.

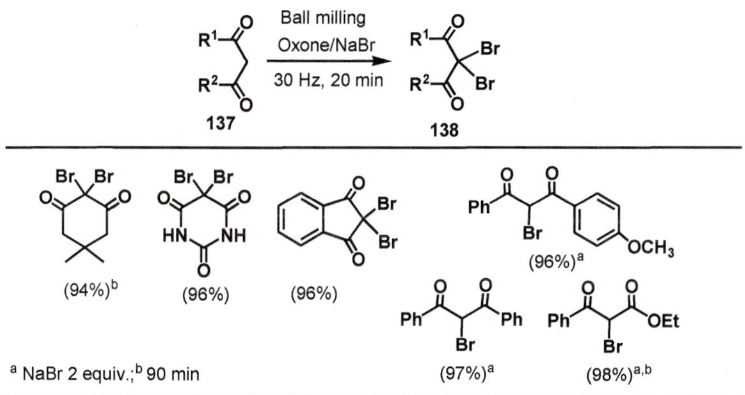

SCHEME 4.37

Bromination of 1,3-dicarbonyl compounds.

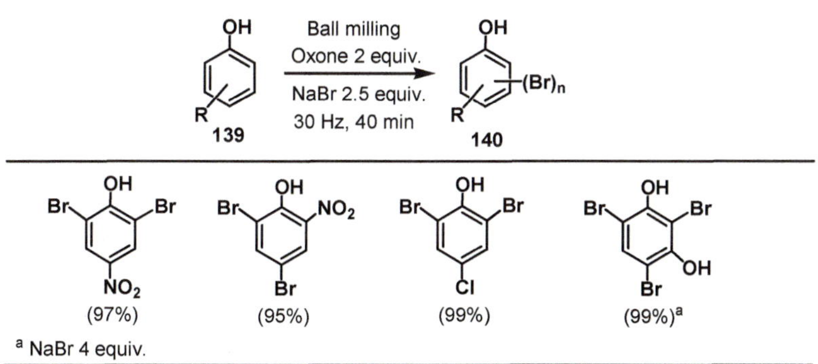

SCHEME 4.38

Bromination of phenols.

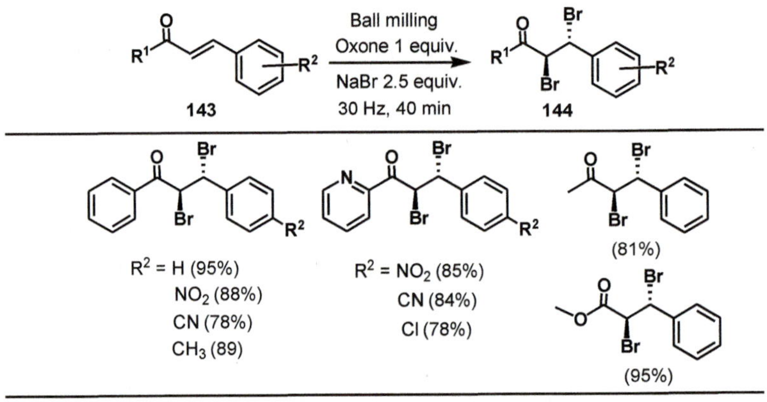

SCHEME 4.39

Bromination of α,β-unsaturated carbonyl compounds.

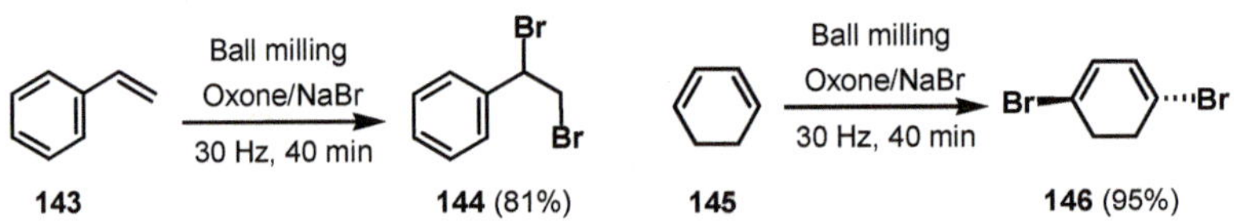

SCHEME 4.40

Bromination of styrene and 1,3-cyclohexadiene.

Dushkin et al. carried out solid-phase mechanochemical synthesis of fluoroaromatic compounds [26]. Synthesis of fluoroaromatic compounds from chloroaromatic predecessors was conducted by milling with a fluorinating agent in a planetary-centrifugal mill AGO-2 (drum 35 mL, 5 mm steel balls 75 g). Alkaline and earth metals fluorides were used as fluorinating agents (molar ratio OXH:fluorinating agent = 1:5), and KF and CsF were the most active in 10 min. Hexachlorobenzene, pentachloropyridine, and octachloronaphthalene were used as substrates. The degree of substitution of chlorine by fluorine kept increasing up to 25° and yield of fluoroderivatives decreased upon further increase in temperature. The highest degree of conversion into F-derivatives at room temperature (10 min) was 48.5%. This procedure has the advantages of lowering energy expenditure and reaction temperature, as well as eliminating the need for solvents.

4.3.2 BENZYLIC BROMINATION: WOHL–ZIEGLER REACTION

Solvent-free benzylic bromination (Wohl–Ziegler reaction) of diquinolines **147** in a ball mill with NBS has been reported by Rahman (Scheme 4.41) [27]. For completion of this reaction, 6 h of milling with *N*-bromosuccinimide (3.5 equiv.) in Retsch MM200 mill was required. Mechanochemical procedure is advantageous over solution reaction, since employment of traditionally used highly toxic and ozone-depleting carbon tetrachloride solvent was avoided, while regio- and stereoselectivity encountered in solution phase is retained.

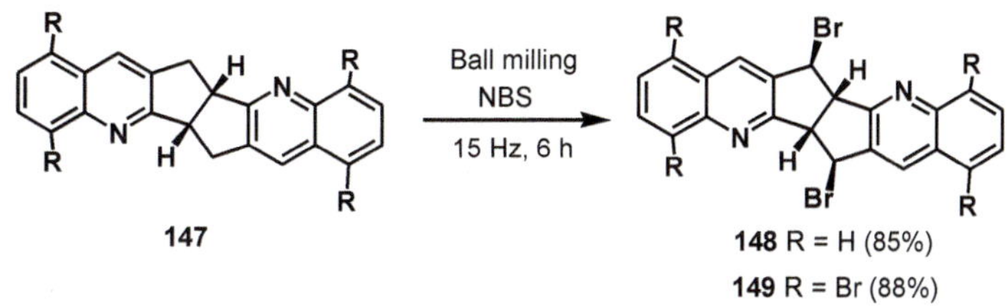

SCHEME 4.41

Benzylic bromination.

4.3.3 NUCLEOPHILIC SUBSTITUTION

Chemoselective nucleophilic substitution reactions of *p*-bromobenzyl bromide in solvent-free, ball-milling conditions were carried out by Mack et al. (Scheme 4.42). Grinding with variety of nucleophile salts (3 equiv.) for 1 h provided nucleophilic substitution products **151** (Table 4.13) [28]. At the conclusion of the reaction, the product was washed with 10% HCl to remove the residual salt and recrystallized in methanol. The increase in the nucleophilicity of the reagents was achieved through the addition of 18-crown-6 (1 mmol) for the potassium salt nucleophiles. These reaction conditions led to an increase in yield and conversion for all of the potassium salt nucleophiles including those which were previously unsuccessful such as fluoride, acetate, and cyanide. It was found that crown ether does not act as the phase-transfer catalysts under ball-millling conditions, but instead greatly increases the reaction rate.

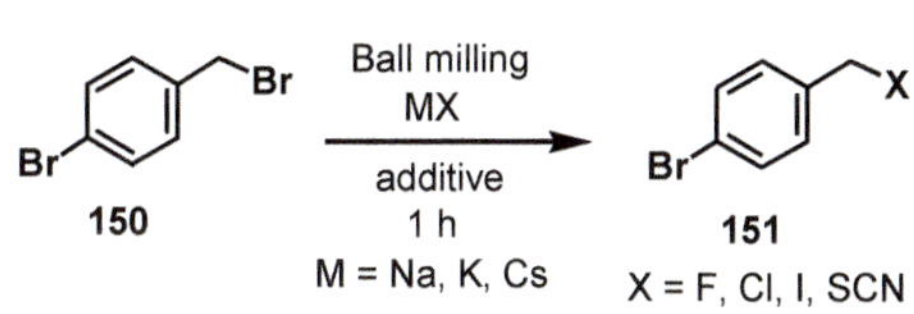

SCHEME 4.42

Nucleophilic substitution reactions of *p*-bromobenzyl bromide.

4.3.4 IODOKETONES IN HANTZSCH PYRROLE SYNTHESIS

The efficient synthesis of highly substituted pyrroles by multicomponent reaction was developed by Menéndez et al. employing the high-speed vibration milling (Scheme 4.43). In generalized Hantzsch pyrrole synthesis, pyrroles **158** were prepared in a one-pot, two-step reaction, with relatively high yields (illustrative examples, Table 4.14) [29]. This reaction involves in the first-step milling of ketones with

Table 4.13 Nucleophilic Substitution Reactions of *p*-Bromobenzyl Bromide[a]

Entry	Metal	Nucleophile	Additive	Conversion (%)	Yield (%)[b]
1	Cs	F		50	50
2	K	Cl		14	14
3	Na	I		>99	94
4	K	I		64	58
5	Na	SCN		>99	97
6	K	SCN		>99	96
7	Na	N_3		75	75
8	K	F	18-cr-6	80	80
9	K	Cl	18-cr-6	90	90
10	K	I	18-cr-6	70	70
11	K	OAc	18-cr-6	>99	90
12	K	CN	18-cr-6	>99	90

[a]*5000M SPEX Certiprep Mixer Mill, 2.0×0.5in. stainless steel vial, 1/8in. ball. substrate:MX:additive 1:3:1.*
[b]*Isolated.*
Reproduced from Vogel P, Figueira S, Muthukrishnan S, Mack J. Environmentally benign nucleophilic substitution reactions. Tetrahedron Lett 2009;50:55–6, Copyright 2009, with permission from Elsevier.

N-iodosuccinimide (1 equiv.) and *p*-toluenesulfonic acid (10 mol%) for 1 h, to form an α-iodoketone **153** in situ. In the second reaction step, to this milling vessel were added primary amine (2 equiv.), β-dicarbonyl compound 155 (1.5 equiv.), cerium(IV) ammonium nitrate (5%), and silver nitrate (1 equiv.) and ball milled for another 1 h. The scope of this pyrrole synthesis was broader than the one described in the literature for previous variations of the Hantzsch reaction, since nitrogen-unsubstituted pyrroles and also double Hantzsch-like reactions based on the use of diamines as starting materials were successfully carried out.

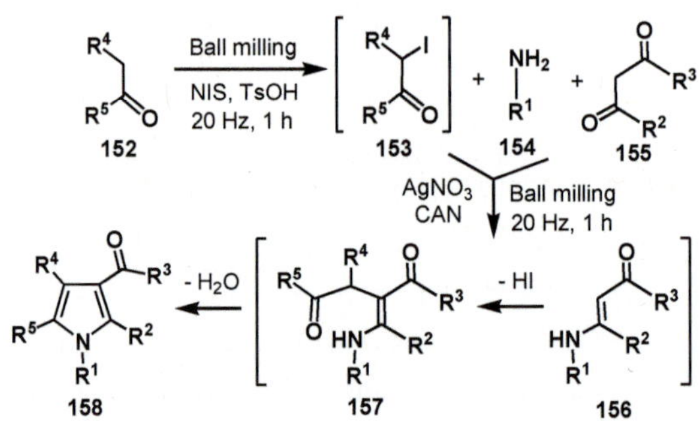

SCHEME 4.43

Synthesis of pyrroles.

Table 4.14 Synthesis of Pyrroles[a]

Entry	R^1	R^2	R^3	R^4	R^5	Yield (%)[b]
1	4-$CH_3C_6H_4$	Me	CO_2Et	H	Ph	85
2	*i*-Bu	Me	CO_2Et	H	Ph	93
3	NMe_2	Me	CO_2Et	H	Ph	60
4	*n*-Hex	Et	CO_2Et	H	Ph	70
5	*n*-Bu	CH_2CO_2Et	CO_2Et	H	Ph	61
6	*n*-Bu	Me	CO_2Et	H	2-Naphthyl	97
7	*n*-Bu	Me	CO_2Et	H	3-Indolyl	75
8	*n*-Bu	Me	CO_2Et	H	2-Thiophenyl	97
9	*n*-Bu	Me	CO_2Et	H	*t*-Bu	67
10	H	Me	CO_2Et	H	Ph	82

[a]*Retsch MM200 mixer mill, ZrO_2 grinding jar (25 mL), 20 mm ZrO_2 ball, 20 Hz.*
[b]*Isolated yield.*
Reproduced from Estévez V, Villacampa M, Menéndez JC. Three-component access to pyrroles promoted by the CAN-silver nitrate system under high-speed vibration milling conditions: a generalization of the hantzsch pyrrole synthesis. Chem Commun 2013;49:591–3, with permission from the Royal Society of Chemistry.

4.3.5 OXIDATIVE HALODECARBOXYLATION

Oxidative halodecarboxylation of β-arylacrylic acids **159** in solvent-free conditions was achieved with the employment of ceric ammonium nitrate/alkali halide system (Scheme 4.44). Kapustina and coworkers ball milled β-arylacrylic acids for 4 h with various alkali halides to obtain β-halostyrenes **160** in moderate to high yield (Table 4.15) [30]. In addition, 2-furyl carboxylic acid was converted to 2-chlorofuran in the same reaction conditions, showing the wider scope of method. Comparison with reactions carried in solution or solvent-free reactions (carried out by simple mixing of reagents which were then allowed standing at room temperature) revealed the shortening of reaction time and increase in yields. Reaction is thought to proceed by initial formation of phenylvinyl radicals.

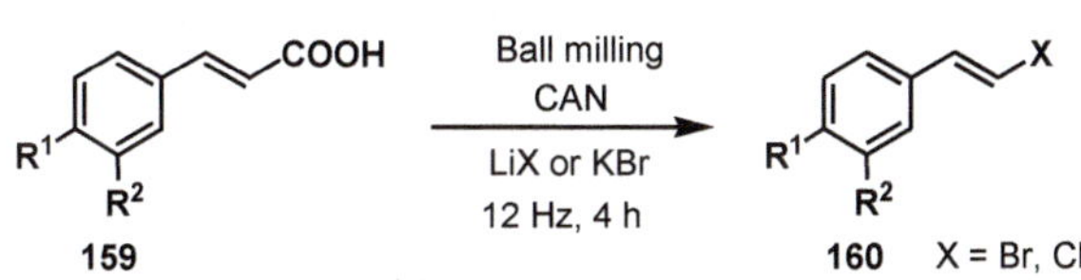

SCHEME 4.44

Oxidative halodecarboxylation of β-arylacrylic acids.

4.3.6 DECHLORINATION

Mechanochemical method was recognized as suitable methodology for reductive dechlorination of aromatics applied by several research groups. For instance,

Table 4.15 Oxidative Halodecarboxylation of β-Arylacrylic Acids[a]

Entry	R^1	R^2	MX	X	Molar Ratio Substrate:CAN:MX	Time	Yield (%)[b]
Halodecarboxylation							
1	H	H	LiBr	Br	1:3:3	4h	80 (56)[c]
2	H	H	KBr	Br	1:2:2	4h	49
3	H	H	LiCl	Cl	1:2:2	4h	59 (45)[c]
4	MeO	H	LiBr	Br	1:2:2	4h	91 (78)[d]
5	MeO	H	KBr	Br	1:2:2	4h	69
6	MeO	H	LiCl	Cl	1:2:2	4h	80
7	MeO	MeO	LiBr	Br	1:2:2	4h	71 (78)[d]
8	**2-FurylCOOH**		LiCl	Cl	1:2:2	4h	43 (39)[e]

[a]*Vibrational mill, stainless steel reactor (80mL), 12.3mm balls (150g), 12Hz.*
[b]*Isolated products.*
[c]*Reagents mixed and allowed to stand for 72h.*
[d]*In MeOH/H_2O, 3h.*
[e]*Standing for 30h.*

simple method for synthesis of quinolines by *N*-deformylation and aromatization via solvent-free, high-speed ball milling (Micro Mill QM-3A, stainless steel jar (35mL), 5mm balls) was carried out by Su et al. (Scheme 4.45) [31]. Known methods for *N*-deformylation include reflux in strongly acidic or basic solutions and treatment with metal-oxidizing agent in methanol solution. Here, 4-chloro-2-phenyl-*N*-formyl-1,2-dihydroquinolines **163** were ball milled with solid base (NaOH) and NaCl as grinding auxiliary. In addition, catalytic amount of polyethylene glycol (PEG 2000) additive was used, as PEG usually plays similar role as 18-crown-6 in organic reactions to improve the yields and they are nontoxic, recoverable, inexpensive media of phase-transfer catalysts and promoters. In these conditions, mechanochemical reactions were greatly promoted and quinolines **162** were obtained in high yield in short time. These results are much better than that obtained in the solvent systems. When reaction was carried out without the PEG 2000 catalyst, mixture of quinoline **164** and 4-chloro quinoline **165** was obtained. The strong electron-withdrawing effect of substituents in the 2-position of dihydroquinoline has a profound effect on the reaction, as shown for *p*-nitrophenyl substituent: these substrates yielded quinoline contaminated with small amounts of the corresponding 2-chloroquinoline. After milling, PEG 2000 and NaCl were recycled, and the regenerated material had the same activity in five consecutive runs/regenerations. Reaction also shows high regioselectivity, and chlorine at position 4 was exclusively removed, while chlorine present on the phenyl substituent remained intact. Similarly, dechlorination conditions are not applicable for the debromination, as shown by milling of 6,8-dibromo substrate. This method is also amenable to larger-scale preparation:

5 g of 4-chloro-2-phenyl-*N*-formyl-1,2-dihydroquinoline was converted into the corresponding quinoline **164** within 10 min in 96% yield.

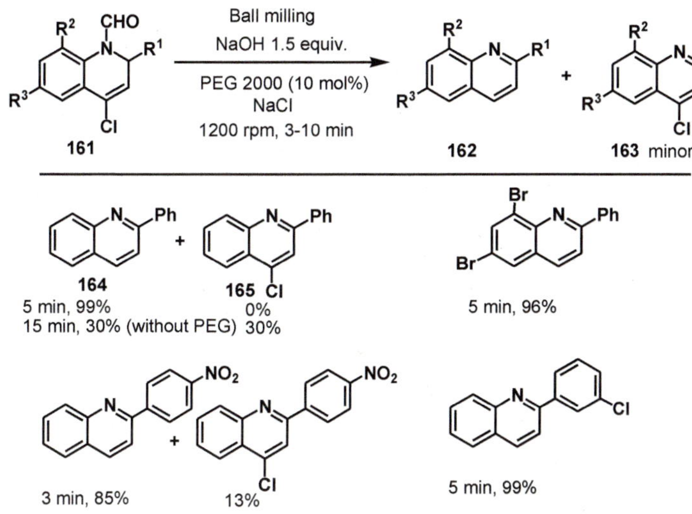

SCHEME 4.45

Dehalogenation, deformylation, and aromatization of 1,2-dihydroquinolines.

Mechanochemical reductive dehalogenation of polyhalogenated pollutants was also reported by Birke et al. [32]. Polychlorinated biphenyls (PCBs) and pentachlorophenol (PCP) were converted to their parent hydrocarbons in high yields, to biphenyl and phenol, respectively, employing magnesium, hydrogen donors and sand as grinding auxiliary (Scheme 4.46) by milling in Retsch S 1 and Retsch S 1000 centrifugal ball mills, in stainless steel grinding jar (50 mL) with 3×20 mm balls.

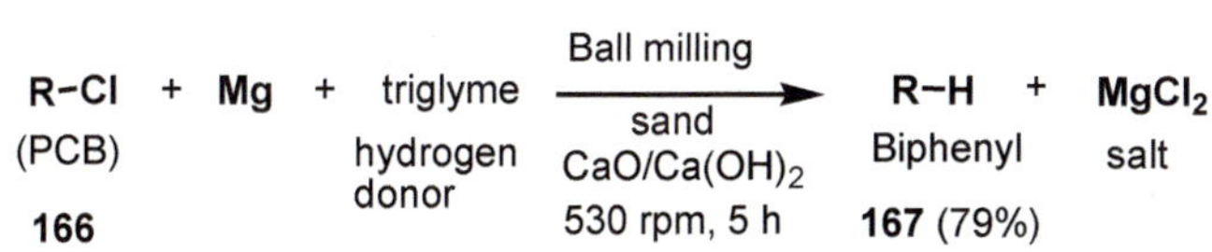

SCHEME 4.46

Reductive dehalogenation of PCB.

Furthermore, Yu et al. developed combined process for mechanochemical destruction of pentachloronitrobenzene (PCNB) in solid waste, coupled with the dechlorination of aqueous 4-chlorophenol (4-CP) using the nonhazardous residue from the solid-phase destruction step [33]. Using the mixture of iron powder, nickel powder, and quartz sand (Fe-Ni-SiO_2) as the additives in a planetary ball mill under the room temperature, complete destruction of PCNB was realized. For this purpose the following conditions were applied: planetary ball mill (QM-3SP2, Nanjing University Instrument Corp.), stainless steel pot, and balls (5.6 mm, total weight 180 g) at room temperature, 550 rpm. Samples which were grounded with Fe-Ni-SiO_2 achieved the highest degradation efficiency of 94.49% after 2 h, and chlorinated organic compounds could be hardly detected after grinding for 4 h (ratio of reagents: Fe:Ni:SiO_2:PCNB 4:0.5:0.3:0.2 g).

Mechanochemical degradation of PCP onto phyllomanganate birnessite (KBi, manganese oxide constituted layers of MnO_6 octahedra, with K cations in the interlayers) highly reactive surface was studied by Di Leo and coworkers [34]. PCP was mixed with KBi and ground in a planetary ball-mill Fritsch Pulverisette-7 in zirconia pots (45 mL) and 14 × 10 mm balls. The planetary ball mill was operated at rotation speed of 700 rpm for different milling times (four steps of 15 min followed by 15 min cooling periods to prevent excess of heating). The synthesized KBi is able to remove PCP and after 60 min the pollutant is almost totally removed (~99%). The surface reaction mechanism is likely a consequence of a redox reaction in which first surface complex between one PCP molecule and the Mn(IV) from the birnessite layer forms, which involves the phenolic group of the pollutant through an oxygen bridge. Then, the electron transfer from the organic molecule to Mn forms a phenoxyl radical, which may be sorbed onto birnessite and then released into water or methanol. Simultaneously, structural changes are induced in the MnO_6 layers: the reduction of Mn(IV) implies the formation of Mn(III) and further Mn(II). The observed dechlorination of PCP molecules once adsorbed is likely the consequence of an oxidative coupling between the newly formed radicals, which leads to the formation of quinonic, dimeric, and/or polymeric molecules. The departure of Cl^- ions from the PCP molecules induces the formation of derivative molecules exhibiting lower toxicity than the parent molecules.

Dechlorination of 2,4,6-trichlorophenol **168** in ball mill was carried out by Lu et al. [35]. TCP with CaO and SiO_2 as additives SiO_2 in the mass ratio of 158:717:125 (equivalent to molar ratio: TCP:CaO = 1:16) were grinded using a ball mill and the obtained degradation rate was 99.0% after 6 h milling. In this system, SiO_2 plays a significant role of a grinding aid, not reactant, it is chemically stable against TCP under these grinding conditions. Planetary ball–milling system at 400 rpm was used, in stainless steel grinding jars (85 mL) filled with 7 g mixture of 24 large stainless steel balls (9.6 mm, 3.6 g) and 99 small balls (5.6 mm, 0.7 g). Mechanism of degradation process of 2,4,6-TCP during ball milling is thought to be realized in four reaction pathways: dechlorination, degradation, carbonation, and carbonization (Scheme 4.47). It was proposed that CaO plays an important role through induction of free radicals by grinding on CaO surface and subsequently the charge transfer on the organic TCP.

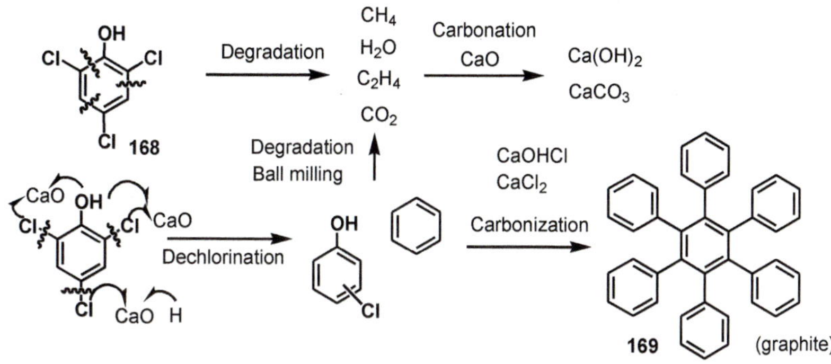

SCHEME 4.47

Mechanism of degradation ways of 2,4,6-TCP.

4.4 PHOSPHORUS BOND FORMATION REACTIONS

Hardacre et al. have shown that hydrolytic sensitivity of reagents can be overcome by combining ionic liquids with mechanochemistry [36]. Phosphitylation reaction of nucleosides and 2-deoxynucleosides **171** using a range of chlorophosphoramidite reagents **170** was carried out in ball mill, with the presence of very small amounts of ionic liquids ([C_4*mim*][FAP] or [C_6*mim*][FAP], 1.5:1 mol ratio IL:reagent) (Scheme 4.48). Ionic liquids used as additives act as stabilizers of sensitive chlorophosphoramidite reagents, which in combination with efficient mass transfer induced by ball milling enabled excellent yields (Table 4.16). Product purification was achieved by concentration of solvent, then filtration of the crude residue through silica gel and washing with ethyl acetate.

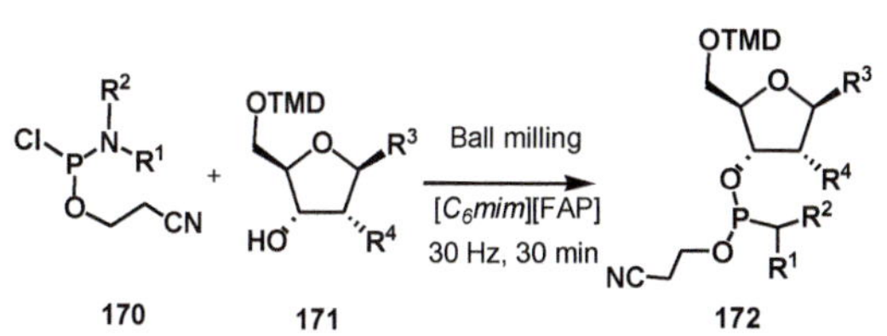

SCHEME 4.48

Phosphitylation of nucleosides.

Table 4.16 Mechanochemical Nucleoside Phosphitylation Reactions[a]

Entry	Protected Nucleoside R¹	R²	Amine R¹R²	Yield (%)[b] Milling	Yield (%)[b] Solution[c]
1	2-Deoxy adenosine	H	-N^iPr_2	80	92
2			-NEt_2	82	15
3			-NEtMe	79	0
4	2-Deoxy cytidine	H	-N^iPr_2	80	84
5			-NEt_2	91	9

Continued

Table 4.16 Mechanochemical Nucleoside Phosphitylation Reactions[a]—cont'd

Entry	Protected Nucleoside R[1]	R[2]	Amine R[1]R[2]	Yield (%)[b] Milling	Yield (%)[b] Solution[c]
6			-NEtMe	74	5
7	Guanosine	OTBDMS	-N^{i}Pr$_2$	89	46
8			-NEt$_2$	82	8
9			-NEtMe	73	5

[a]*Retsch MM400 mixer mill, 30 Hz, 0.5 h, stainless steel jar (1.5 mL), one 5 mm ball, nucleoside:Hünig's base:1:1 [C$_6$mim][FAP] 1:4:1.5.*
[b]*Isolated yields.*
[c]*DCM, Hünig's base, 24 h.*
Reproduced from Hardacre C, Huang H, James SL, Migaud ME, Norman SE, Pitner WR. Overcoming hydrolytic sensitivity and low solubility of phosphitylation reagents by combining ionic liquids with mechanochemistry. Chem Commun 2011;47:5846–8, with permission from the Royal Society of Chemistry.

4.5 BORON BOND FORMATION REACTIONS

Schnürch and Stanetty have shown that synthesis of boronic acid esters could be effectively carried out in ball mill (Scheme 4.49) [37]. Equimolar quantities of pinacol or 2,2-dimethylpropan-1,3-diol milled with boronic acids for 1 h and washed up with organic solvent afford boronic esters **174** and **175** in excellent purity and high yields (Table 4.17). This ball-milling method is atom efficient and waste free, requiring only a small amounts of nontoxic solvents (ether, ethanol) for isolation. Reaction is much faster (1 h) than the classical methods (several hours or days) which also need removal of water by azeotropic distillation. The scale-up by factor 20 was performed on a Fritsch Analysette 3 PRO ball mill, with the same efficiency as in small-scale batches.

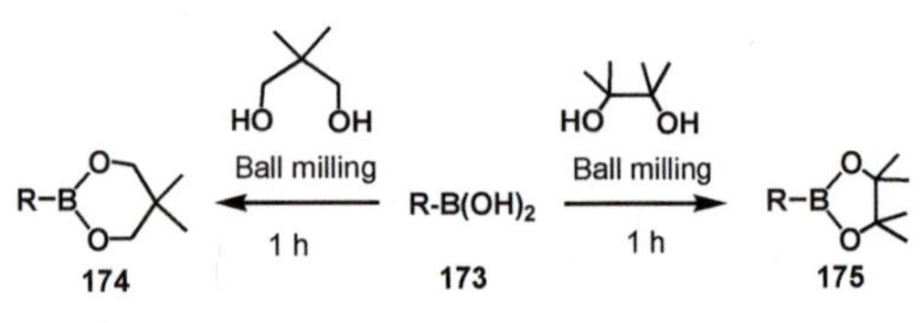

SCHEME 4.49

Mechanochemical synthesis of boronic esters.

Severin et al. have demonstrated that borasiloxanes are also accessible by ball-milling synthesis. Furthermore, they showed that the mechanochemical syntheses of borasiloxane macrocycles can be achieved with high yields, outperforming more classical solution-based methods [38]. Stepwise and one-pot synthetic routes to macrocycles were investigated. In stepwise synthesis di-*tert*-butylsilanediol and 4-formylbenzeneboronic acid were condensed using an MM200 Retsch mill (5 mL jar) to obtain **178** in 75% yield (in solution reflux for 12 h afforded 65%) (Scheme 4.50). Then, products **179a** and **179b** were prepared in solution from **178** in moderate yields (33% and 50%, respectively). Macrocycles **179a** and **179b** were also obtained as one-pot reaction by ball milling of t-Bu$_2$Si(OH)$_2$, 4-formylbenzeneboronic acid, and the corresponding diamines.

Table 4.17 Mechanochemical Synthesis of Boronic Esters[a]

Entry	R	174 Yield (%)[b]	175 Yield (%)[b]
1	C_6H_5	96	99
2	4-F-C_6H_4	86	85
3	3-OCH_3-C_6H_4	86	91
4	2-CHO-C_6H_4	88	82
5	3-CHO-C_6H_4	96	76
6	2,5-$(CH_3)_2C_6H_3$	60	97
7	2-F-4-pyridyl	89	90
8	2-CHO-3-thienyl	83	92
9	3-CHO-2-thienyl	80	84
10	Phenethyl	95	95

[a]*Swing mill, 1 h at 70% of maximum capacity, boronic acid:diol 1:1.*
[b]*Isolated yields.*
Reproduced from Schnürch M, Holzweber M, Mihovilovic MD, Stanetty P. A facile and green synthetic route to boronic acid esters utilizing mechanochemistry. Green Chem 2007;9:139–45, with permission from the Royal Society of Chemistry.

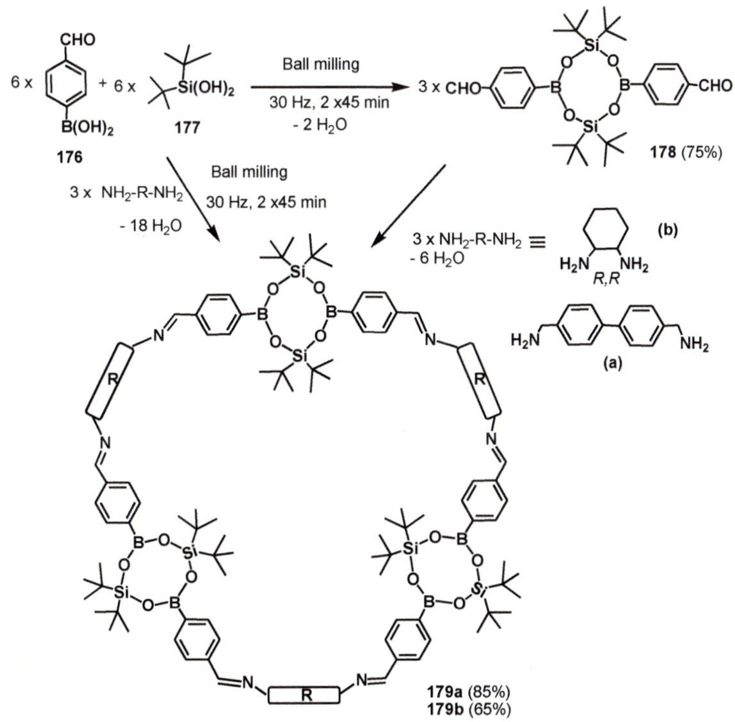

SCHEME 4.50

Macrocycle synthesis.

Ball milling of the multicomponent mixture of the boronic acid, the silane diol, and the 4,4′-bis(aminomethyl) biphenyl for 2×45 min gave after washing with diethyl ether product **179a** in an isolated yield of 85%. Macrocycle **179b** was prepared in a similar fashion by reaction employing (1*R*,2*R*)-1,2-diaminocyclohexane and recrystallized in 65% yield.

Employing similar synthetic methodology in ball mill (Retsch MM400, 10 mL jar), Severin and coworkers prepared three-dimensional cage structures featuring boronate ester units using 1,3,5-trisaminomethyl-2,4,6-triethylbenzene **182** [39]. This one-pot [6+3+2] condensation reaction of the multicomponent mixture of trisamine **182**, 4-formylphenylboronic acid and pentaerythritol in 1 h of milling afforded cage product **183** in 94% yield (Scheme 4.51). Replacement of 4-formylphenylboronic acid with a longer spacer, 4-(4-formylphenyl)phenylboronic acid produced a larger cage structure **184** in 71% yield. After completion of milling, products were simply heated under vacuum at 80°C to remove water and extracted with organic solvent. This ball-milling procedure is superior to the solution-based methods which gave products **183** and **184** in 56% and 40% yield, respectively. Regardless of conditions used for solution reactions, the product was always contaminated with significant amount of an incomplete condensation product.

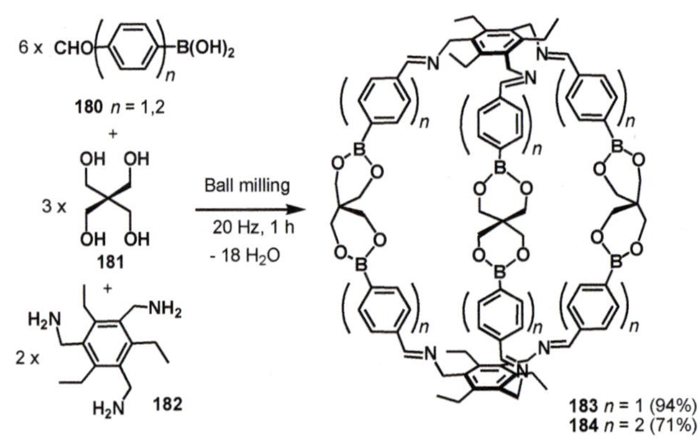

SCHEME 4.51

Cage macrocycle synthesis.

Kaupp et al. have shown that protection of diamines, dialcohols, polyalcohols, and anthranilic acid could be conveniently carried out in ball mill using phenylboronic acid by formation of new N—B and N—O bonds [40]. Grinding of substrates with phenyl boronic acid **185** for 1 h in ball mill (Retsch MM2000 swing mill with stainless steel double-walled beaker with fittings for circulating coolants) afforded products in quantitative yields and without the need for purification, just drying to remove water formed in reaction (Scheme 4.52). Depending on the stability of substrates and products, reaction temperature varied from cooling at 0°C to heating up to 95°C. Employment of this synthetic protocol on 1,8-diaminonaphthalene provides the six-membered diazaborinine **187**, anthranilic acid gave oxazaborinin-4-one **193**, while pyrocatechol, pinacol, [3.3.3]heteropropellanes, mannitol, and *myo*-inositol give dioxaboroles **189**, **191**, **195**, **197**, and **199**, respectively. Yields were higher and reaction times shorter than the methods used previously to obtain these products.

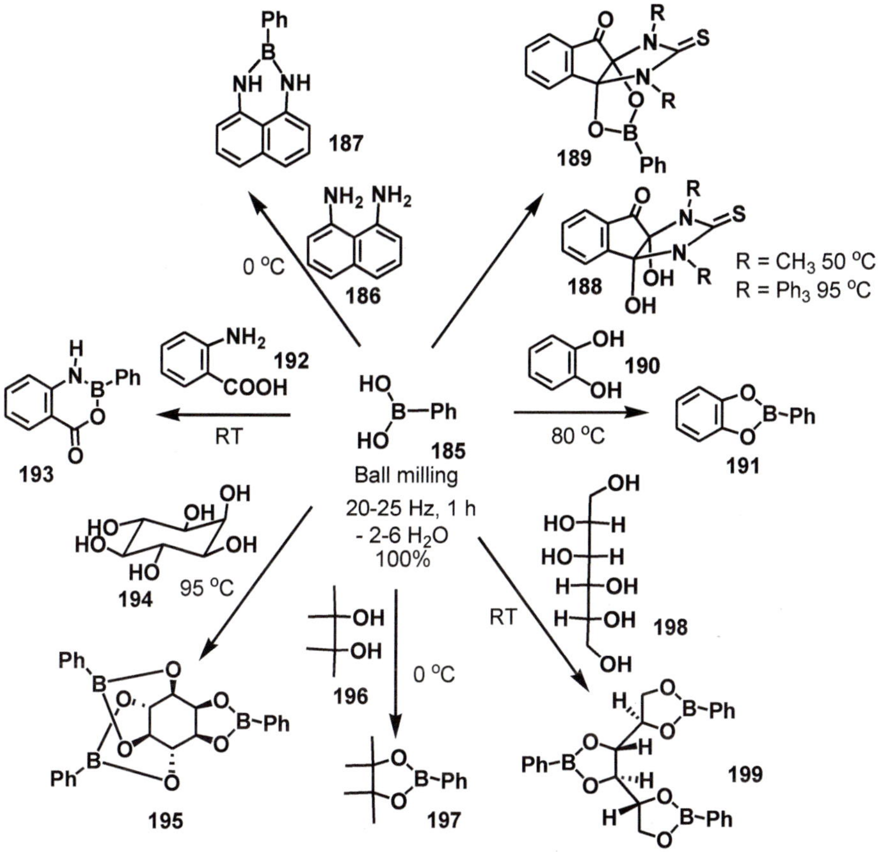

SCHEME 4.52

Protection of amines and alcohols with phenylboronic acid.

4.6 SILICON BOND FORMATION REACTIONS

James et al. used ball milling for hydroxyl group protection with TBDMS and formation of the corresponding silyl ethers [41]. Phenolic hydroxyls were effectively silylated under solvent-free conditions with 1.1 equiv. of *t*-butyldimethylsilylchloride (TBDMSCl) in the ball mill in the presence of 2 equiv. of imidazole (Scheme 4.53). Yields were quantitative and by-products readily removed by extraction, after washing the mixture from the reactor with a small amount of methanol (Table 4.18). These reaction conditions were applied on the series of ribonucleosides **202a–e** using 4 equiv. of TBDMSCl reagent to achieve persilylation in high yields. Instead of overnight procedure typically used for the solution-based processes, mechanosynthesis was complete within 1–3 h. The employment of the solvent-free ball milling in nucleoside chemistry is particularly interesting because the solubility properties of these compounds often require solvents such as DMF and pyridine, which are carcinogenic or highly toxic. It is notable that inosine **202d** which is only sparingly soluble even in DMF could be efficiently derivatized in this fashion. Despite the presence of exocyclic amino groups in cytidine and guanosine, no silylation was observed at these sites. Only with adenosine *N*-silylation was observed, and when larger reagent excess was used, which is reflected in the slightly lower yield. Furthermore, the trityl-protecting group is also compatible with these reaction conditions in the ball mill, despite its thermal and acid sensitivity. Finally, authors performed a one-pot persilylation and N^4-benzoylation of cytidine to yield **206** quantitatively. The reaction was conducted simply by adding the benzoic anhydride and a catalytic amount of DMAP directly to the crude persilylated material in the ball mill. Under classical solvent-based conditions two steps are used, involving dry DMF or pyridine. This alternative synthesis avoids the use of such solvents and the intermediate isolation, thus offering a more efficient synthetic procedure.

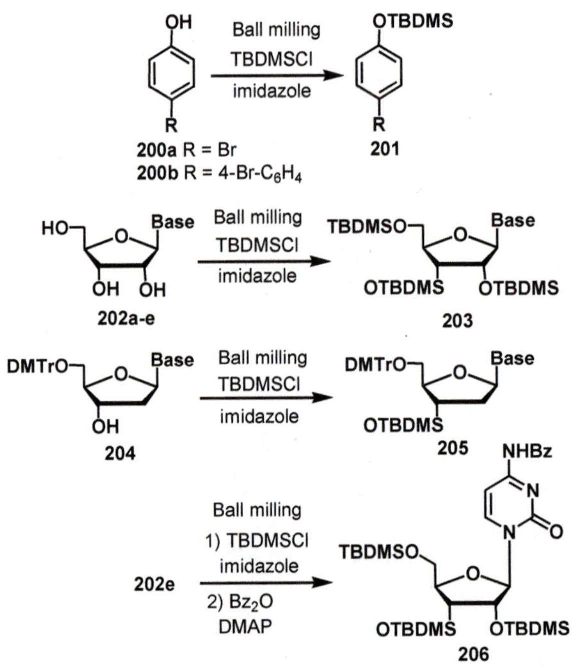

SCHEME 4.53

Mechanochemical silylation reactions.

Table 4.18 Mechanochemical Hydroxyl Silylation Reactions[a]

Entry	Substrate	Base	TBDMSCl (equiv.)	Imidazole (equiv.)	Frequency (Hz)	Time (min)	Yield (%)[b]
1	**200a**		1.1	2	25	45	98
2	**200b**		11	2	25	45	95
3	**202a**	Cytosine	4	8	30	60	95
4	**202b**	Guanine	4	8	30	180	99
5	**202c**	Hypoxanthine	4	8	30	180	94
6	**202d**	Uracil	4	8	30	180	96
7	**202e**	Adenine	5	10	30	90	87
8	**204**	thymine	1.3	8	30	90	99
9	**202a**	Cytosine	4	8	30	60/120	90[c]

[a]*Retsch MM200 mixer mill, stainless steel jar (20mL), one ball.*
[b]*Isolated yields.*
[c]*One-pot persilylation-N*^{4}*-benzoylation (60min persilylation followed by 120min with 2equiv. Bz_2O/0.2equiv. DMAP).*
Reproduced from Giri N, Bowen C, Vyle JS, James SL. Fast, quantitative nucleoside protection under solvent-free conditions. Green Chem 2008;10:627–8, with permission from the Royal Society of Chemistry.

4.7 BISMUTH BOND FORMATION REACTIONS

Suzuki and coworkers developed an alternative synthetic method for the preparation of triarylbismuthanes in ball-milling conditions (Scheme 4.54) [42]. Aryl halides are inert toward bismuth metal on either simple mixing or heating at elevated temperatures. However, ball milling of aryl iodides **207** activated with an electron-withdrawing substituent at the *ortho*-position with bismuth shots and calcite grains in the presence of Cu powder and CuI, affords corresponding *ortho*-functionalized triarylbismuthanes **208** in moderate to good yields (Table 4.19). The nature and relative position of substituent groups on the aromatic ring strongly influence the yield of the products, and a variety of substituents are compatible with the reaction conditions employed. The novel procedure avoids the use of aryl Grignard reagents and circumvents the problem of substituents incompatibility.

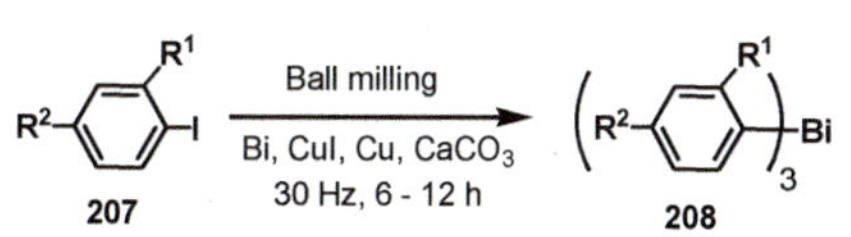

SCHEME 4.54

Synthesis of triarylbismuthanes.

Table 4.19 Synthesis of Triarylbismuthanes[a]

			Reaction Time and Yield (%)[b]	
Entry	**R¹**	**R²**	**6h**	**12h**
1	H	H	16	–
2	H	F	41	66
3	F	H	52	81
4	F	F	73	88
5	Cl	H	30	86
6	Br	H	47[c]	83[c]
7	CF_3	H	25	71
8	H	CF_3	29	62
9	—N=CHC_6H_5	H	25[c]	48[c]
10	NO_2	H	16[d]	–
11	OCH_3	H	30[c]	61[c]
12	$CO_2C_2H_5$	H	27[c]	30[c]
13	3-Pyridyl	H	0	–
14	CN	H	0	–

[a]Retsch mixer mill MM200, 30Hz, stainless steel vial (5mL), 2×7mm balls; aryl iodide:Bi shots:calcite:Cu:CuI 1:14:2:0.7:1.4.
[b]Isolated yield.
[c]NMR yield.
[d]3h, Prolonged reaction led to extensive decomposition, giving 2,2′-dinitrobiphenyl and unidentified products.
Reproduced from Urano M, Wada S, Suzuki H. A novel dry route to ortho-functionalized triarylbismuthanes that are difficult to access by conventional wet routes. Chem Commun 2003;1202–03, with permission from the Royal Society of Chemistry.

A combination of bismuth shots and Cu or CuI is required for the reaction, and the combined use of Cu powder and CuI gave better results than their separate employment. The application of bismuth powder was less effective, indicating the crucial role of the nascent bismuth surface in the formation of organobismuth compounds and advantageous in situ generation of the highly active bismuth surface in ball mill. Addition of 0.5–0.7M equiv. of calcite helped to modulate the milling rate of bismuth shots. In the case when aryl iodides with a keto or an ester group in the *ortho*-position were milled, the first and second arylation of bismuth occurred smoothly, but the third arylation became slow and diaryliodobismuthane **210** were obtained as the major product (Scheme 4.55). Further ball milling after addition of another aryl iodide led to the preparation of unsymmetrical triarylbismuthane **212**.

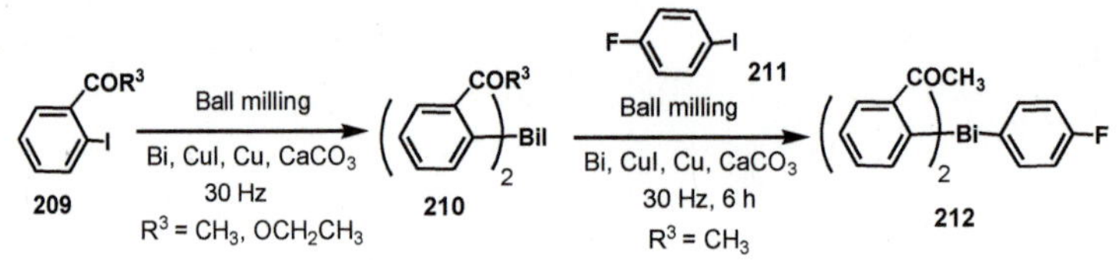

SCHEME 4.55

Synthesis of unsymmetrical triarylbismuthanes.

4.8 C—H BOND FORMATION: HYDROGENATION

Sternhell and coworkers discovered that aromatic hydrocarbons treated mechanically with apparently nonreactive matrices (alumina and silica) were converted to carbon. The reaction in the alumina matrix generated significant amounts of products arising from partial reduction of aromatic substrates [43]. Milling of naphthalene, biphenyl, anthracene, and phenanthrene converts them largely to graphite and partly hydrogenated derivatives in short time (Scheme 4.56). Reactions were carried out in SPEX 8000 Mixer Mill in steel beaker (50 mL) with chromium steel balls (4 × 12 mm) for 24 h. Silica gel was found to be more effective for the hydrocarbon destruction than alumina, while the presence of oxygen slows down the reaction leading to the disappearance of hydrocarbons. Milling with deuterated naphthalene showed that hydrocarbons are the source of hydrogen, with less likely alternative pathway by hydrogen transfer from moisture.

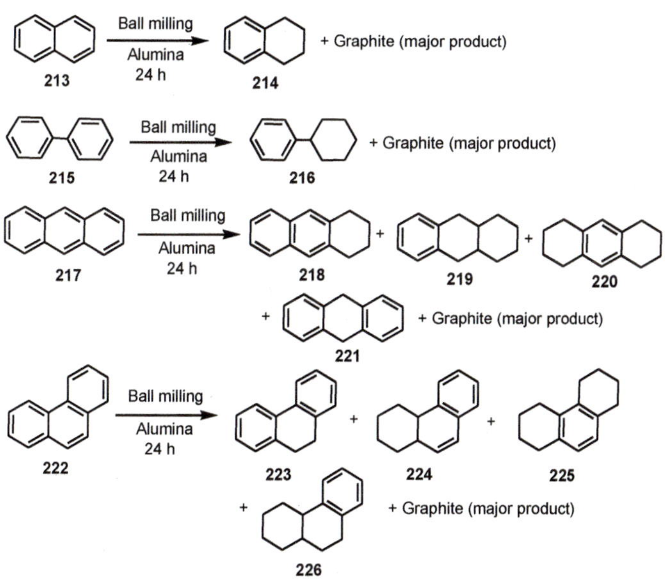

SCHEME 4.56

Ball milling of aromatic hydrocarbons.

4.9 MULTISTEP MECHANOCHEMICAL REACTION

Ability of performing multistep mechanochemical reactions is demonstrated by one-pot, four-step preparation of β-hydroxy triazoles from acetobromo-α-D-glucose **227** (Scheme 4.57) [2]. Optimized synthetic protocol for *O*-alkylation of glycosides described earlier in this chapter was successfully used in the first reaction step, and *n*-pentenyl glycoside **228** was obtained by 2 h milling in 83% yield. Glycoside was epoxidized under solvent free–milling conditions by *m*-chloroperbenzoic acid (1.1 equiv.) within 2 h in a planetary ball mill. To the same milling vessel, NaN_3 (1.1 equiv.) was added and product subjected to epoxide ring opening by ball milling in the presence of PEG-400. Finally, 1,3-dipolar cycloaddition reaction of the in situ formed azido alcohol **230** with 4-ethynyltoluene **231**, $CuSO_4$ (0.4 equiv.) and sodium ascorbate (0.8 equiv.) was also performed in the same ball-milling vessel. The final product **232** was isolated by chromatography in 85% yield (overall of three steps from *n*-pentenyl glycoside **228**).

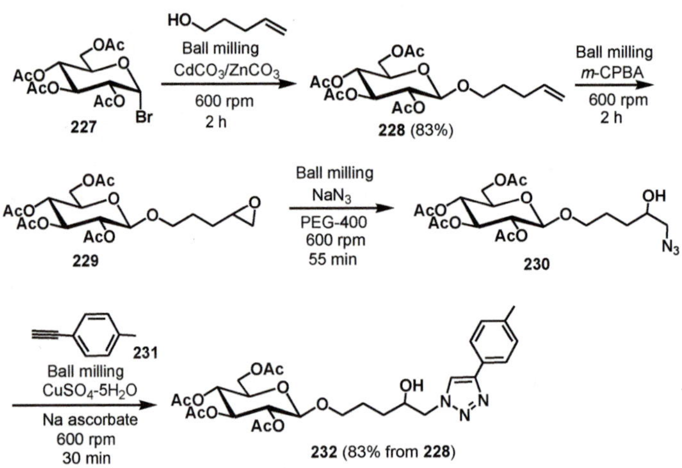

SCHEME 4.57

One-pot preparation of β-hydroxy triazoles from *n*-pentenyl glycosides.

REFERENCES

[1] Abdel-Latif E, Metwally MA. Waste-free solid-state organic syntheses: solvent-free alkylation. Monatsh Chem 2007;138:771–6.

[2] Tyagi M, Khurana D, Kartha KPR. Solvent-free mechanochemical glycosylation in ball mill. Carbohydr Res 2013;379:55–9.

[3] Kumar V, Yadav N, Kartha KPR. In(III) triflate-catalyzed detritylation and glycosylation by solvent-free ball milling. Carbohydr Res 2014;397:18–26.

[4] Patil RR, Kartha KPR. Application of ball milling technology to carbohydrate reactions: I. regioselective primary hydroxyl protection of hexosides and nucleoside by planetary ball milling. J Carbohydr Chem 2008;27:279–93.

[5] Zhu X, Li Z, Shu Q, Zhou C, Su W. Mechanically activated solid-state synthesis of flavones by high-speed ball milling. Synth Commun 2009;39:4199–211.

[6] Menuel S, Doumert B, Saitzek S, Ponchel A, Delevoye L, Monflier E, Hapiot F. Selective secondary face modification of cyclodextrins by mechanosynthesis. J Org Chem 2015;80:6259–66.

[7] Waddell DC, Thiel I, Bunger A, Nkata D, Maloney A, Clark T, Smith B, Mack J. Investigating the formation of dialkyl carbonates using high speed ball milling. Green Chem 2011;13:3156–61.

[8] Konnert L, Gauliard A, Lamaty F, Martinez J, Colacino E. Solventless synthesis of *N*-protected amino acids in a ball mill. ACS Sustain Chem Eng 2013;1:1186–91.

[9] Chatterjee T, Saha D, Ranu BC. Solvent-free transesterification in a ball-mill over alumina surface. Tetrahedron Lett 2012;53:4142–4.

[10] Rantanen T, Schiffers I, Bolm C. Solvent-free asymmetric anhydride opening in a ball mill. Org Process Res Dev 2007;11:592–7.

[11] Waddell DC, Mack J. An environmentally benign solvent-free Tishchenko reaction. Green Chem 2009;11:79–82.

[12] Métro T-X, Bonnamour J, Reidon T, Duprez A, Sarpoulet J, Martinez J, Lamaty F. Study of the organic-solvent-free CDI-mediated acylation of various nucleophiles by mechanochemistry. Chem Eur J 2015;21:12787–96.

[13] Pankrushina N, Nikitina I, Chernjak E, Myz C, Shakhtshneider T, Boldyrev V. Solvent-free mechanochemical modification of lappaconitine and piroxicam. Mater Manufact Proc 2008;23:561–5.

[14] Kaupp G, Schmeyers J, Boy J. Iminium salts in solid-state syntheses giving 100% yield. J Prakt Chem 2000;342:269–80.

[15] Kaupp G, Naimi-Jamal MR, Schmeyers J. Quantitative reaction cascades of ninhydrin in the solid state. Chem Eur J 2002;8:594–600.

[16] Mukherjee N, Chatterjee T, Ranu BC. Reaction under ball-milling: solvent-, ligand-, and metal-free synthesis of unsymmetrical diaryl chalcogenides. J Org Chem 2013;78:11110–4.

[17] Hendriks CMM, Lamers P, Engel J, Bolm C. Sulfoxide-to-sulfilimine conversions: use of modified Burgess-type reagents. Adv Synth Catal 2013;355:3363–8.

[18] Belenguer AM, Lampronti GI, David J, Wales DJ, Sanders JKM. Direct observation of intermediates in a thermodynamically controlled solid-state dynamic covalent reaction. J Am Chem Soc 2014;136:16156–66.

[19] Belenguer AM, Friščić T, Day GM, Sanders JKM. Solid-state dynamic combinatorial chemistry: reversibility and thermodynamic product selection in covalent mechanosynthesis. Chem Sci 2011;2:696–700.

[20] Schmidt R, Stolle A, Ondruschka B. Aromatic substitution in ball mills: formation of aryl chlorides and bromides using potassium peroxomonosulfate and NaX. Green Chem 2012;14:1673–9.

[21] Bose A, Mal P. Electrophilic aryl-halogenation using *N*-halosuccinimides under ball-milling. Tetrahedron Lett 2014;55:2154–6.

[22] Maiti S, Mal P. Electron-rich aromatics under ball milling: oxidative aryl-iodinaton using I_2-oxone and biarylation with I_2. Synth Commun 2014;44:3461–9.

[23] Hernández JG, Bolm C. $[Cp^*RhCl_2]_2$: mechanosynthesis and applications in C-H bond functionalisations under ball-milling conditions. Chem Commun 2015;51:12582–4.

[24] Višňovský J, Billik P, Kubinec R, Podolec P, Hengerics Szabó A, Juriga M, Čabala R, Kubincová J, Blaško J. Solvent-free mechanochemical chlorination of hydrocarbons with $CuCl_2$. Tetrahedron Lett 2013;54:7180–2.

[25] Wang G-W, Gao J. Solvent-free bromination reactions with sodium bromide and oxone promoted by mechanical milling. Green Chem 2012;14:1125–31.

[26] Dushkin AV, Karnatovskaia LM, Chabueva EN, Pavlov SV, Kobrin VS, Boldyrev VV, et al. Solid-phase mechanochemical synthesis of fluoroaromatic compounds. Synth Commun 2001;31:1041–5.

[27] Rahman ANMM, Bishop R, Tan R, Shan N. Solid-state regio- and stereo-selective benzylic bromination of diquinoline compounds using *N*-bromosuccinimide. Green Chem 2005;7:207–9.
[28] Vogel P, Figueira S, Muthukrishnan S, Mack J. Environmentally benign nucleophilic substitution reactions. Tetrahedron Lett 2009;50:55–6.
[29] Estévez V, Villacampa M, Menéndez JC. Three-component access to pyrroles promoted by the CAN-silver nitrate system under high-speed vibration milling conditions: a generalization of the hantzsch pyrrole synthesis. Chem Commun 2013;49:591–3.
[30] Nikishin GI, Sokova LL, Makhaev VD, Kapustina NI. Solid-phase oxidative halodecarboxylation of β-arylacrylic acids with the ceric ammonium nitrate-alkali halide system. Russ Chem Bull Int Ed 2008;57:118–23.
[31] Yu J, Li Z, Su W. Synthesis of quinolines by *N*-deformylation and aromatization via solvent-free, high-speed ball milling. Synth Commun 2013;43:361–74.
[32] Birke V, Mattik J, Runne D. Mechanochemical reductive dehalogenation of hazardous polyhalogenated contaminants. J Mat Sci 2004;39:5111–6.
[33] Zhang T, Huang J, Zhang W, Yu Y, Deng S, Wang B, Yu G. Coupling the dechlorination of aqueous 4-CP with the mechanochemical destruction of solid PCNB using Fe-Ni-SiO_2. J Haz Mat 2013;250–251:175–80.
[34] Di Leo P, Pizzigallo MDR, Ancona V, Di Benedetto F, Mesto E, Schingaro E, Ventruti G. Mechanochemical degradation of pentachlorophenol onto birnessite. J Haz Mat 2013;244–245:303–10.
[35] Lu S, Huang J, Peng Z, Li X, Yan J. Ball milling 2,4,6-trichlorophenol with calcium oxide: dechlorination experiment and mechanism considerations. Chem Eng J 2012;195–196:62–8.
[36] Hardacre C, Huang H, James SL, Migaud ME, Norman SE, Pitner WR. Overcoming hydrolytic sensitivity and low solubility of phosphitylation reagents by combining ionic liquids with mechanochemistry. Chem Commun 2011;47:5846–8.
[37] Schnürch M, Holzweber M, Mihovilovic MD, Stanetty P. A facile and green synthetic route to boronic acid esters utilizing mechanochemistry. Green Chem 2007;9:139–45.
[38] Pascu M, Ruggi A, Scopelliti R, Severin K. Synthesis of borasiloxane-based macrocycles by multicomponent condensation reactions in solution or in a ball mill. Chem Commun 2013;49:45–7.
[39] Içli B, Christinat N, Tönnemann J, Schüttler C, Scopelliti R, Severin K. Synthesis of molecular nanostructures by multicomponent condensation reactions in a ball mill. J Am Chem Soc 2009;131:3154–5.
[40] Kaupp G, Naimi-Jamal MR, Stepanenko V. Waste-free and facile solid-state protection of diamines, anthranilic acid, diols, and polyols with phenylboronic acid. Chem Eur J 2003;9:4156–60.
[41] Giri N, Bowen C, Vyle JS, James SL. Fast, quantitative nucleoside protection under solvent-free conditions. Green Chem 2008;10:627–8.
[42] Urano M, Wada S, Suzuki H. A novel dry route to *ortho*-functionalized triarylbismuthanes that are difficult to access by conventional wet routes. Chem Commun 2003:1202–3.
[43] Field LD, Sternhell S, Wilton HV. Mechanochemistry of some hydrocarbons. Tetrahedron 1997;53:4051–62.

CHAPTER

Cycloaddition Reactions

5

CHAPTER OUTLINE

5.1 DIELS–ALDER REACTION

Ball milling of stoichiometric amounts of two solid reagents, Diels–Alder partners led to synthesis of silicon-containing heterocyclic product. Cyclic diene 1,1-dimethyl-2,3,4,5-tetraphenylsilole **1** reacts with *N*-methyl maleimide in solid-state milling to afford the *endo*-Diels–Alder cycloadduct **3** stereospecifically (Scheme 5.1) [1]. Notably, transfer of reaction conditions from solution chemistry to solvent-free procedure does not have any effect on stereochemical outcome [2].

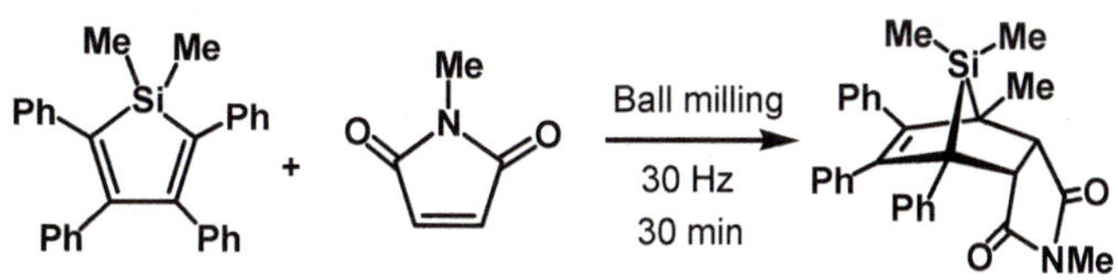

SCHEME 5.1

Diels–Alder cycloaddition of silole.

These results correlate well with the findings from Zhang's group [3]. Under mechanochemical milling conditions, Diels–Alder cycloaddition of cyclopentadiene (CPD) with maleic anhydride and maleimide derivatives **5** proceeded very smoothly, affording *endo*-norbornenes **6** exclusively in quantitative yield (Scheme 5.2). An excess of 4% molar equivalents of CPD is enough to drive reaction to completion. When reactions were carried in THF/hexane at room temperature overnight (1.04 equivalents of CPD) lower yields were obtained (65–77%), and an excess of 3.6 equivalents of CPD was required to obtain comparable results (78–90%). Besides higher yields, solvent-free conditions, smaller amounts of CPD, and significantly shorter reaction times, the other advantage of ball-milling procedure is the *endo*-selectivity. For instance, Diels–Alder reaction of *N*-4-tosylmaleimide in mechanochemical

Mechanochemical Organic Synthesis. http://dx.doi.org/10.1016/B978-0-12-802184-2.00005-4

conditions is *endo*-selective, while solution chemistry leads to formation of mixture of the *endo*- and *exo*-adducts. These results show that the employment of liquid reagent (CPD) in ball mill does not alter *endo*-selectivity.

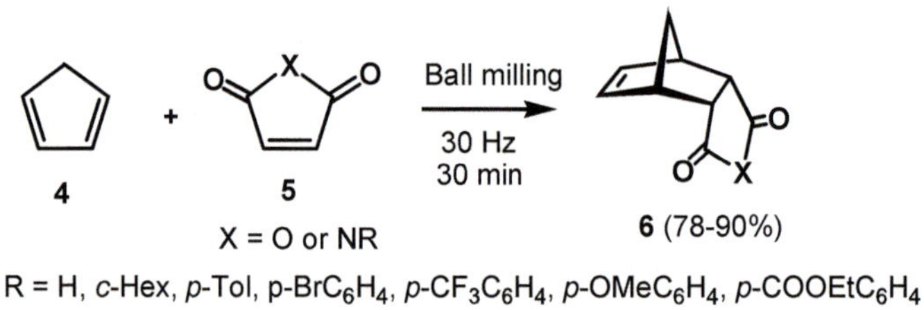

SCHEME 5.2

Diels–Alder cycloadditions of cyclopentadiene.

Solid-state Diels–Alder reactions of anthracenes **7** and benzoquinone **8** were carried out in a vibration mill by Watanabe and Senna (Scheme 5.3) [4]. Only anthracene derivatives with electron-donating functional groups gave rise to Diels–Alder adducts (Table 5.1). However, yields of products **9** obtained under bar-mill conditions are relatively low as compared to the solution reaction. An acceleration of this solid-state Diels–Alder reaction was observed by addition of catalytic amounts of additives which form the charge transfer complex with strong hydrogen bonds. Thus, addition of 2-naphthol (NP) and (*rac*)-1,10-bis-2-naphthol (BN) to solid-state Diels–Alder reaction of dimethylantracene and benzoquinone afforded cycloadducts in quantitative yields (an equivalent amount of dienophile and catalyst OH groups: 4:2:1 and 2:2:1 M ratio, respectively). Less significant catalytic effects of the additives were obtained in the solution compared to those in a solid state (56%).

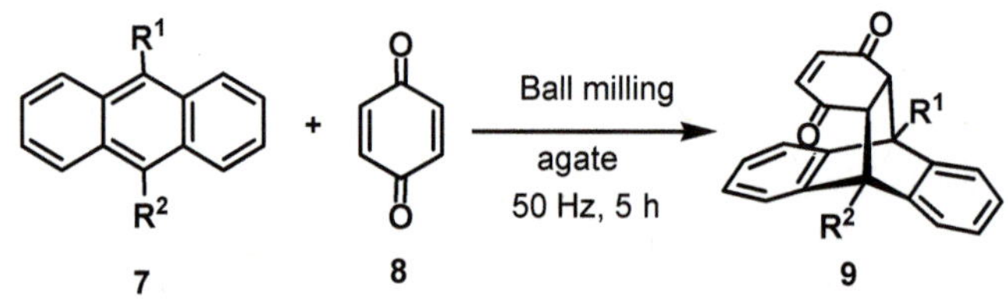

SCHEME 5.3

Diels–Alder reaction between anthracenes and *p*-benzoquinone.

In subsequent paper, Senna et al. employed thymol as an efficient additive and showed that formation of eutectic complex facilitates Diels–Alder reaction in mechanical mill (Scheme 5.4) [5]. It was reasoned that formation of eutectic phase via charge-transfer complex leads to better fusion of two reactants and hence increases reaction rate. An equimolar mixture of 9,10-dimethylanthracene **10** and *p*-benzoquinone **8** with catalytic amount (10 mol%) of thymol **11** were mechanically milled in a closed agate mortar with a single vibrating agate ball to obtain Diels–Alder product **12** in high yield.

Table 5.1 Diels–Alder Reaction Between Anthracenes and *p*-benzoquinone[a]

Entry	R^1	R^2	Condition	Additive: Anthracene:8	Time (h)	Yield (%)[b]
1	Br	Br	Agate	0:1:1	5	0
2	CH_2Cl	CH_2Cl	Agate	0:1:1	5	0
3	H	H	Agate	0:1:1	5	0
4	CH_3	H	Agate	0:1:1	5	8
5	CH_3	CH_3	Agate	0:1:1	5	26 (40)[d]
6	CH_3	CH_3	Mixing[c]	0:1:1	48	Trace
7	CH_3	CH_3	Agate	NP 4:2:1	5	>99
8	CH_3	CH_3	Agate	BN 2:2:1	5	>99 (56)[d,e]
9	CH_3	CH_3	Agate	NP 0.1:2:1	5	75
10	CH_3	CH_3	Agate	BN 0.05:2:1	5	85

[a]*Fritsch Pulverisette 0 vibrational mill, 50 Hz, anthracene:**8** 1:1 equiv.*
[b]*NMR.*
[c]*In rotary evaporator.*
[d]*In chloroform.*
[e]*NP:anthracene:**8** 2:1:1 equiv.*
Reproduced from Watanabe H, Senna M. Acceleration of solid state Diels–Alder reactions by incorporating the reactants into crystalline charge transfer complexes. Tetrahedron Lett 2005;46:6815–8. Copyright (2005), with permission from Elsevier.

Eutectic complex

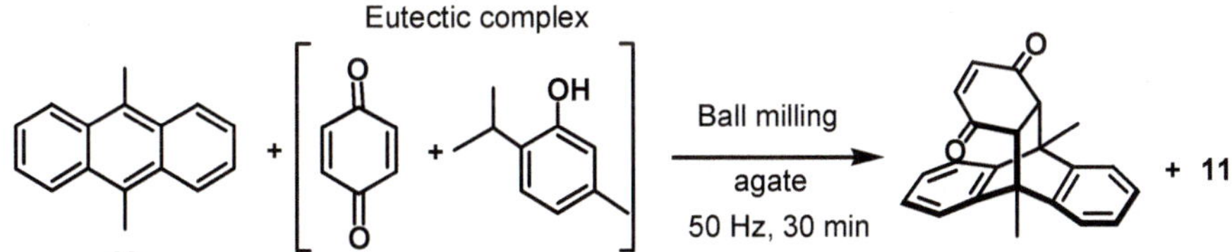

10 **8** **11** **12** (95%)

SCHEME 5.4

Diels–Alder cycloaddition of 9,10-dimethylanthracene.

Examples of hetero-Diels–Alder reactions were also reported. Mechanochemical milling was found to be sufficient to promote solvent-free imino Diels–Alder reaction catalyzed by $FeCl_3$ (Scheme 5.5) [6]. Reaction is diastereoselective, and Li et al. obtained exclusively *cis*-2,4-diaryl-1,2,3,4-tetrahydroquinolines **17**. This two-step one-pot synthetic method converts in a first step a series of substituted anilines and benzaldehydes into corresponding imines **15** which were used without isolation. The addition of styrene and catalytic amounts of $FeCl_3$ (25 M%) in the second reaction step completes the synthetic sequence. Cycloadditions were effectively complete within 90 min of ball milling, which were faster than most conventional solution reactions. These reactions work well with substituted anilines and benzaldehydes bearing either electron-donating or electron-withdrawing groups on the benzene ring (representative examples, Table 5.2). The obtained solvent-free reactivity is in contrast to similar reactions with anilines bearing electron-donating substituents, which exhibited relatively lower reactivity when performed in organic solvents. Good to excellent isolated yield of **17** was provided by simple workup, including simple washing of the reaction mixture with water and recrystallization.

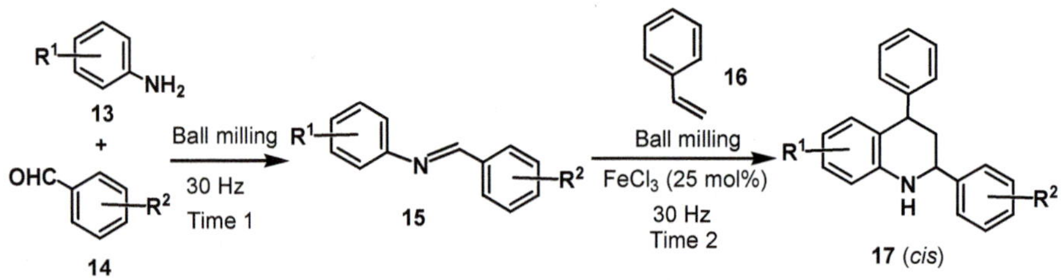

SCHEME 5.5

Synthesis of *cis*-2,4-diaryl-1,2,3,4-tetrahydroquinolines.

Table 5.2 Synthesis of *cis*-2,4-diaryl-1,2,3,4-tetrahydroquinolines[a]

Entry	R^1	R^2	Time 1 (min)	Time 2 (min)	Yield (%)[b]
1	H	H	90	90	71
2	H	4-NO_2	75	90	75
3	4-CH_3	4-Cl	60	90	82
4	4-OCH_3	3-NO_2	30	90	77
5	4-Cl	4-Cl	60	90	91
6	3-Cl	4-NO_2	50	90	87

[a]*Retsch MM400 stainless steel vial (25 mL), 7 mm ball, 30 Hz. Step 1: aniline:aldehyde 1:1 equiv. Step 2: $FeCl_3$ 0.25 equiv.*
[b]*Isolated.*
Reproduced from Tan Y-J, Zhang Z, Wang F-J, Wu H-H, Li Q-H. Mechanochemical milling promoted solvent-free imino Diels–Alder reaction catalyzed by $FeCl_3$: diastereoselective synthesis of cis-2,4-diphenyl-1,2,3,4-tetrahydroquinolines. RSC Adv 2014;4:35635–35638, with permission from the Royal Society of Chemistry.

5.2 1,3-DIPOLAR CYCLOADDITIONS

Another synthetically important cycloaddition reaction is 1,3-dipolar reaction of copper-catalyzed alkynes with azides ("click," Huisgen reaction). Stolle and coworkers applied ball milling to "click" reaction of series or aromatic and aliphatic alkynes and azides to obtain 1,2,3-triazoles **20** in high yield (Scheme 5.6) [7]. Milling of 1.1:1 mixture of **18** and **19** with $Cu(OAc)_2$ catalyst (5 mol%) was carried out with ZrO_2 beaker and milling balls. Remarkably, it was found that the use of standard additive sodium ascorbate is not required, and reactions proceed in 10 min, whereas it was useful additive only in the case of less-reactive substrates. Competing homocoupling reaction of **12** (Glaser reaction) has not been observed. In all experiments, fused quartz sand (SiO_2) was used as inert milling auxiliary.

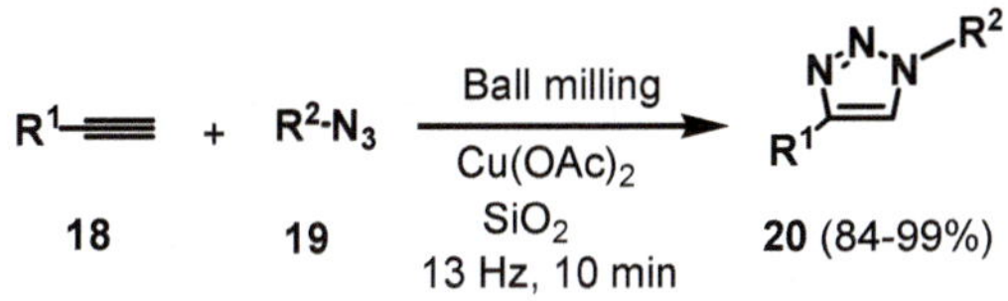

SCHEME 5.6

1,3-Dipolar cycloadditions of azides and alkynes.

Kartha et al. employed ball milling to prepare a series of polyvalent triazoles by azide alkyne click reaction using CuI ($CuSO_4$–sodium ascorbate) (Scheme 5.7) [8]. The synthesis of various sugar-linked triazole compounds **21–25** was carried out by co-grinding of azides with the propargyl glycosides in planetary ball mill at 500 rpm under the solvent-free mechanochemical click–reaction conditions to obtain products in excellent yields (85–92%). The usual aqueous solution–phase CuAAC reaction was not chosen for this purpose, because of the insolubility of the lipophilic substrates in the reaction medium. Considering the potential safety hazard of azide compounds they were always prepared fresh in quantities as required and were never stored for further use.

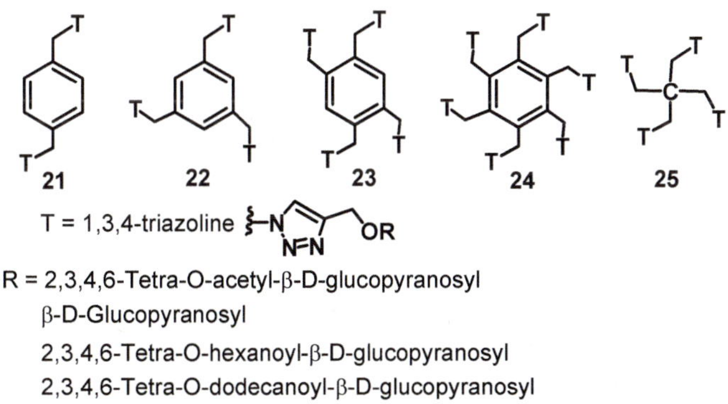

SCHEME 5.7

Various sugar-linked triazole derivatives.

Mechanochemical procedure for 1,3-dipolar reaction developed by Ranu et al. offers advantages in comparison to a related earlier procedures by avoiding handling of hazardous and toxic azides, as well as by the use of environmentally friendly ethanol and recyclability of catalyst [9]. This procedure uses copper(II) sulfate supported on alumina (Cu/Al_2O_3) to catalyze reaction which includes in situ preparation of alkyl azides (Scheme 5.8). Synthesis of 1,2,3-triazole derivatives **28** was achieved by a three-component coupling of equimolar amounts of alkyl halides, NaN_3 and terminal alkynes using 10 mol% of Cu/Al_2O_3 catalyst in ball mill. Triazoles were afforded in excellent yields within 1 h of grinding and isolated by simple washing with ethanol (Table 5.3). Moreover, the catalyst could be recycled and used for subsequent reactions, showing a minimal loss of activity in up to eight runs.

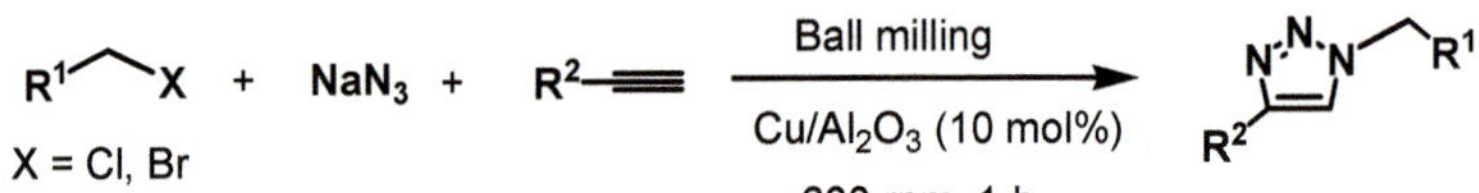

26 27 28

SCHEME 5.8

Cu/Al_2O_3-catalyzed one-pot azide preparation and "click" reaction.

Table 5.3 Cu/Al_2O_3-Catalyzed One-Pot Azide Preparation and "Click" Reaction[a]

Entry	Alkyl Halide	Alkyne	Yield (%)[b]
1	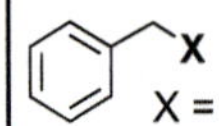 X = Cl X = Br	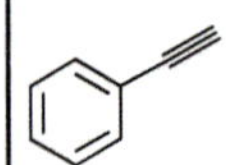	92 96
2	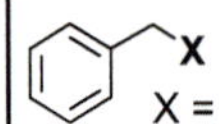 X = Cl X = Br	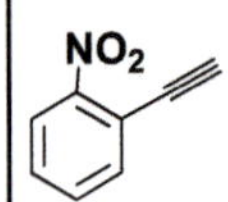	91 94
3	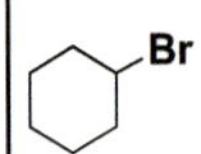	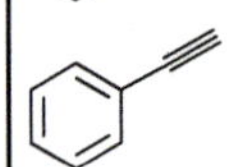	78
4	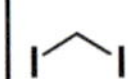	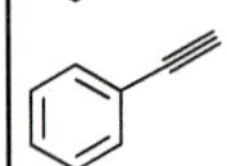	80
5	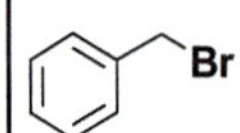	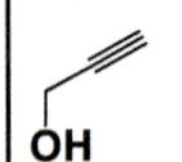	85

[a]Retsch MM400 stainless steel vial (25 mL), 6×10 mm balls, 600 rpm, 1 h, halide:alkyne:NaN_3:Cu/Al_2O_3 1:1:1:0.1 equiv.
[b]Isolated.
Reproduced from Mukherjee N, Ahammed S, Bhadraa S, Ranu BC. Solvent-free one-pot synthesis of 1,2,3-triazole derivatives by the 'Click' reaction of alkyl halides or aryl boronic acids, sodium azide and terminal alkynes over a Cu/Al_2O_3 surface under ball-milling. Green Chem 2013;15:389–97 with permission from the Royal Society of Chemistry.

This synthetic procedure employed to aryl halides failed to produce triazoles, and the method was modified by using aryl boronic acids **29** in place of aryl halides. One-pot, two-step mechanochemical procedure comprises grinding of stoichiometric amounts of aryl halide, NaN_3, K_2CO_3, and catalytic amounts of Cu/Al_2O_3 (10 mol%) in the first milling step (Scheme 5.9). The obtained aryl azides were used in the second step and ball milled with alkynes to obtain the corresponding triazoles **31** in high yields (Table 5.4).

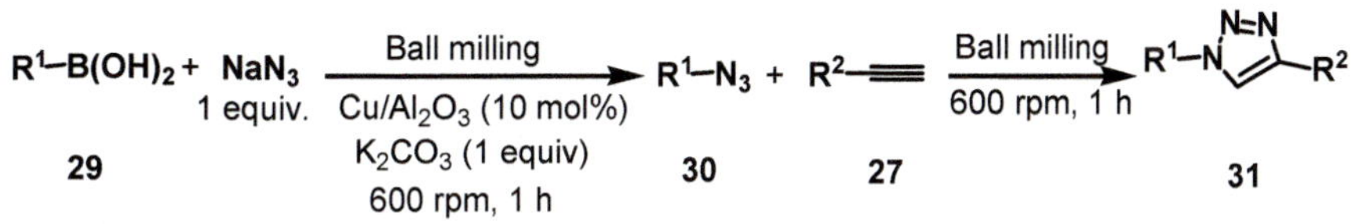

SCHEME 5.9

Cu/Al_2O_3-catalyzed one-pot aryl azide preparation from aryl boronic acids and "click" reaction.

Table 5.4 Cu/Al_2O_3-Catalyzed One-Pot Aryl Azide Preparation from aryl boronic acids and "Click" Reaction[a]

Entry	Boronic Acid[a]	Alkyne	Yield (%)[b]
1	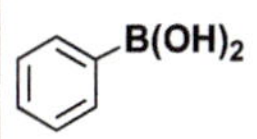	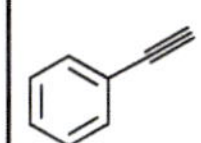	91
2	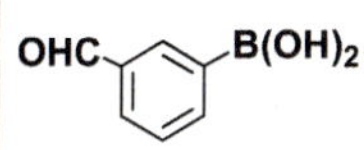	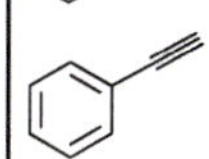	88
3	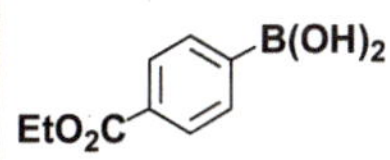	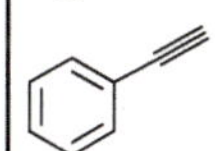	88
4	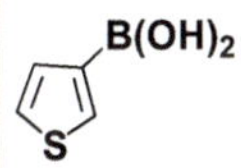	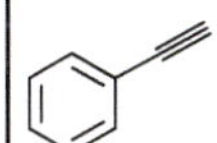	87
5	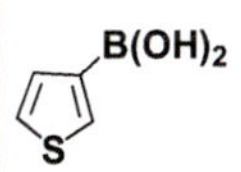	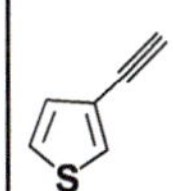	83

[a]*Retsch MM400 stainless steel vial (25 mL), 6 × 10 mm balls, 600 rpm, 1 h, boronic:NaN_3:K_2CO_3:Cu/Al_2O_3 1:1:1:0.1 equiv. second step 1 h, alkyne 1 equiv.*

[b]*Isolated.*

Reproduced from Mukherjee N, Ahammed S, Bhadraa S, Ranu BC. Solvent-free one-pot synthesis of 1,2,3-triazole derivatives by the 'Click' reaction of alkyl halides or aryl boronic acids, sodium azide and terminal alkynes over a Cu/Al_2O_3 surface under ball-milling. Green Chem 2013;15:389–97 with permission from the Royal Society of Chemistry.

Cravotto et al. have shown that copper powder could activate 1,3-dipolar cycloaddition reaction of azides and alkynes in ball-mill solid-state conditions (Scheme 5.10) [10]. In optimized reaction conditions, equimolar amounts of an azide, alkyne, and copper powder were ball milled for short time (up to 30 min). Simple filtration is enough to isolate variously substituted 1,2,3-triazoles **34** in high yield (Table 5.5). This facile method could be easily scaled up to 10 g quantities.

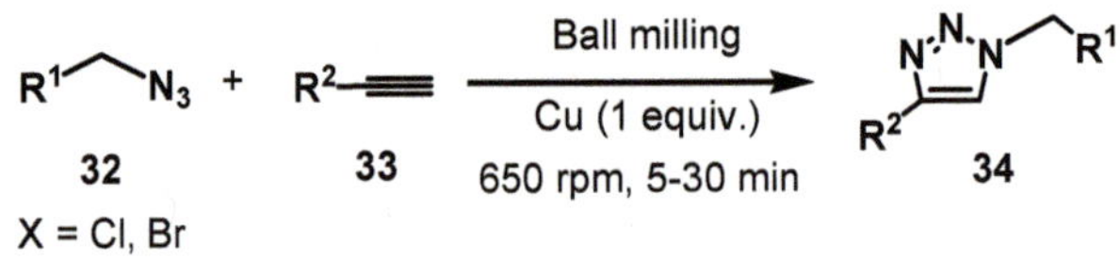

SCHEME 5.10

Copper powder-catalyzed "click" reaction.

Table 5.5 Copper Powder-Catalyzed "Click" Reaction [a]

Entry	Azide	Alkyne	Alkyne:azide:Cu	Time (min)	Yield (%)[b]
1	$CH_3(CH_2)_7N_3$	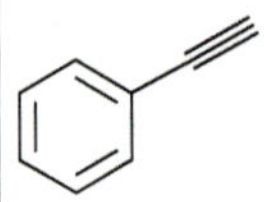	1:1:1	5	99
2	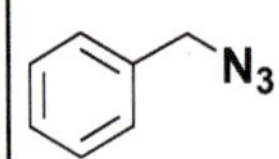	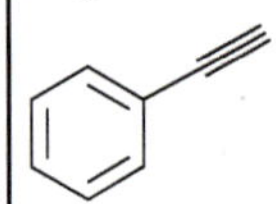	1:1:1	10	98
3	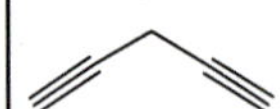	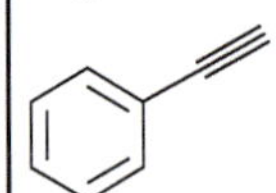	1:2:1	5	98
4	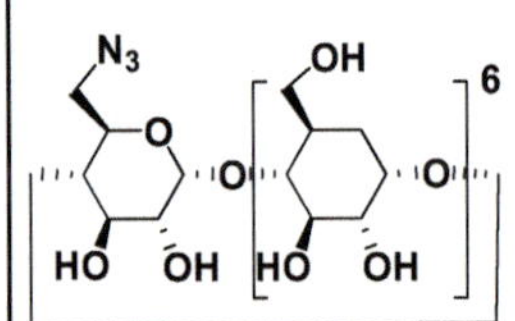	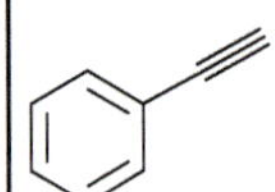	1:0.2:0.2	30	81

[a]*Retsch PM100 stainless steel vial (50mL), 1500 × 2mm + 48 × 5mm balls, 650rpm, azide:alkyne:Cu 1:1:1 equiv.*

[b]*Isolated.*

Reproduced from Rinaldi L, Martina K, Baricco F, Rotolo L, Cravotto G. Solvent-free copper-catalyzed azide-alkyne cycloaddition under mechanochemical activation. Molecules 2015;20:2837–49 with permission from MDPI.

The above-described method which employs copper powder to activate 1,3-dipolar cycloaddition reaction of azides and alkynes in ball mill is related to work by Mack et al. [11]. Here, copper catalyst was replaced by Cu vial and Cu milling balls as a source of metal catalyst. The efficacy and simplicity of the method is clearly shown by reaction of phenylacetylene and benzylazide in Spex Certiprep 8000M mixer mill (Scheme 5.11). Minute amounts of copper peeled off the balls were sufficient to afford 1,2,3-triazole **38** in excellent yield, without the need for purification of product. The ICP-MS analysis determined very low level of copper metal present in product after isolation (4.61 mg/g of copper in the product). Multicomponent variant of this reaction includes the in situ preparation of azide from benzyl bromide and sodium azide in the presence of phenyl acetylene. These conditions work equally well; however, reaction takes much longer time to complete (16h). This one-step multicomponent reaction is applicable on various alkyne compounds providing very high yields of the desired products after 24h of milling. Copper vial was used for over 100 reactions, without noticeable change in yields or reaction rate.

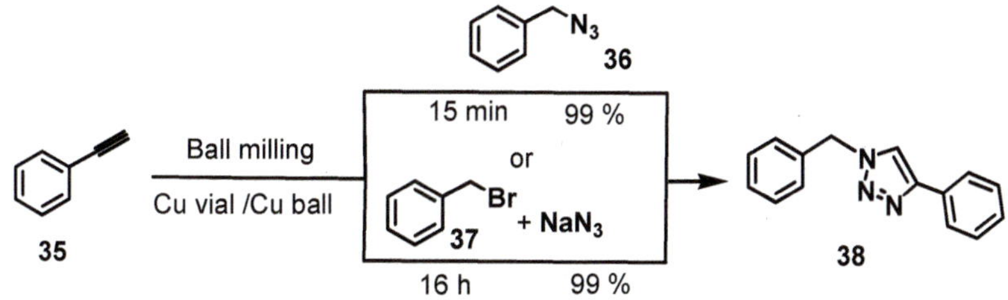

SCHEME 5.11

Copper vial and ball-catalyzed "click" reaction.

Additional examples of cycloadditions carried out in ball mill could be found in other chapters, in particular in chapter "Applications of Ball Milling in Nanocarbon Material Synthesis" (fullerenes).

REFERENCES

[1] Margetić D. Mechanochemical organic reactions without the use of solvent. Kem Ind 2005;54:351–8.

[2] Margetić D, Murata Y, Komatsu K, Eckert-Maksić M. Synthesis, X-ray, and DFT study of the double-bond pyramidalization in 1,7,8,9-Tetraphenyl-4,10,10-trimethyl-4-aza-10-silatricyclo[5.2.1.0.^{2,6}]deca-8-ene-3,5-dione and its Germanium analogue. Organometallics 2006;25:111–7.

[3] Zhang Z, Peng Z-W, Hao M-F, Gao J-G. Mechanochemical Diels–Alder cycloaddition reactions for straightforward synthesis of *endo*-norbornene derivatives. Synlett 2010;19:2895–8.

[4] Watanabe H, Senna M. Acceleration of solid state Diels–Alder reactions by incorporating the reactants into crystalline charge transfer complexes. Tetrahedron Lett 2005;46:6815–8.

[5] Watanabe H, Hiraoka R, Senna M. Diels–Alder reaction catalyzed by eutectic complexes autogenously formed from solid state phenols and quinones. Tetrahedron Lett 2006;47:4481–4.

[6] Tan Y-J, Zhang Z, Wang F-J, Wu H-H, Li Q-H. Mechanochemical milling promoted solvent-free imino Diels–Alder reaction catalyzed by $FeCl_3$: diastereoselective synthesis of *cis*-2,4-diphenyl-1,2,3,4-tetrahydroquinolines. RSC Adv 2014;4:35635–8.

[7] Thorwith R, Stolle A, Ondruschka B, Wild A, Schubert US. Fast, ligand- and solvent-free copper-catalyzed *click* reactions in a ball mill. Chem Commun 2011;47:4370–2.

[8] Tyagi M, Taxak N, Bharatam PV, Nandanwar H, Kartha KPR. Mechanochemical click reaction as a tool for making carbohydrate-based triazole-linked self-assembling materials (CTSAMs). Carbohydr Res 2015;407:137–47.

[9] Mukherjee N, Ahammed S, Bhadraa S, Ranu BC. Solvent-free one-pot synthesis of 1,2,3-triazole derivatives by the 'Click' reaction of alkyl halides or aryl boronic acids, sodium azide and terminal alkynes over a Cu/Al_2O_3 surface under ball-milling. Green Chem 2013;15:389–97.

[10] Rinaldi L, Martina K, Baricco F, Rotolo L, Cravotto G. Solvent-free copper-catalyzed azide-alkyne cycloaddition under mechanochemical activation. Molecules 2015; 20:2837–49.

[11] Cook TL, Walker Jr JA, Mack J. Scratching the catalytic surface of mechanochemistry: a multi-component CuAAC reaction using a copper reaction vial. Green Chem 2013;15:617–9.

CHAPTER

Oxidations and Reductions

6

CHAPTER OUTLINE

6.1 OXIDATIONS

6.1.1 OXIDATION OF β-PINENE

Stolle et al. have shown that potassium permanganate could be used for solvent-free oxidation of β-pinene **1** to nopinone **2** (Scheme 6.1) [1]. Reactions with $KMnO_4$ as oxidant are normally performed in aqueous conditions, and the main product is the sodium salt of nopinic acid; however, mixtures of up to eight products are usually obtained [2]. Several mechanical and chemical parameters were optimized. The reaction was performed with $KMnO_4$ under solvent-free conditions in a planetary ball mill in the presence of grinding auxiliary, which is necessary for milling processes that employ liquid substrates.

Mechanochemical Organic Synthesis. http://dx.doi.org/10.1016/B978-0-12-802184-2.00006-6

Quartz sand was found to be superior to alumina and all other grinding auxiliaries tested (Table 6.1). Addition of small amount of water, a large excess of $KMnO_4$ (6 equiv.) and high rotation frequency (800 rpm = 13.3 Hz) were necessary for better yields. After only 10 min of milling using ZrO_2 vessels and balls, **2** was obtained in 89% yield. For a comparison, mortar and pestle grinding for 10 min furnished **2** in only 27% yield (and 73% selectivity). The optimal number of milling balls was six and no significant difference was observed when different materials for grinding vessels and balls were used. Investigation on the selection of oxidants showed that $KMnO_4$ is the most suitable oxidant for the reaction and gave the highest conversions, while oxidants such as Oxone [3] and $NaIO_4$ gave total conversion, but very low amounts of desired products were obtained.

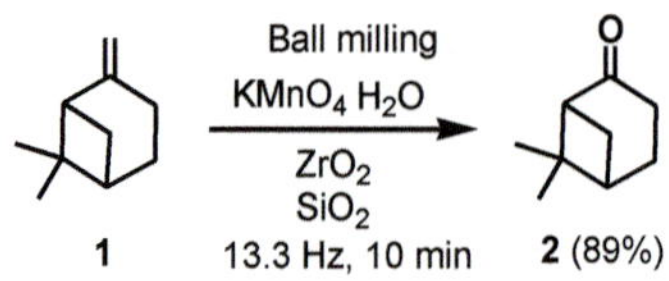

SCHEME 6.1

Oxidation of β-pinene.

Table 6.1 Oxidation of Pinene With $KMnO_4$ Yields (%)[a]

$KMnO_4$ equiv.	ν_{rot} (min)$^{-1}$	Al_2O_3	SiO_2
0.5	400	8 (67)[b]	2 (50)
1	400	14 (61)	6 (46)
2	400	25 (67)	8 (53)
4	400	49 (69)	36 (68)
6	400	44 (73)	15 (65)
6	600	72 (90)	60 (73)
6	800	79 (95)	89 (91)
6[c]	–	27 (73)	5 (33)

[a]*Fritsch Pulverisette planetary mill, ZrO_2 beaker (45 mL), ZrO_2 balls (6 × 15 mm), 800 rpm, 10 min, pinene:oxidant 1:3 equiv., water 0.1 g.*
[b]*Selectivity for nopinone is given in brackets.*
[c]*Mortar and pestle.*

Addition of small amount of water has a positive effect on $KMnO_4$ oxidations (0.1 g H_2O/2 mmol of substrate **1**). Authors proposed a mechanism of its action in which the presence of water is necessary to generate metallaoxetane **4** (Scheme 6.2). It is assumed that the water and the MnO_4^- ion are added to the surface of grinding auxiliary. The MnO_4^- ion is now activated and coordinates to C=C bond of **1**, thus making a π-complex **3**. In the following steps, **3** shifts via **4** to the cyclic manganate(V) ester **5**. Finally, the cyclic ester **5** rearranges to **2**, formaldehyde and the manganate(IV) intermediate **6**, which on hydrolysis affords MnO_2.

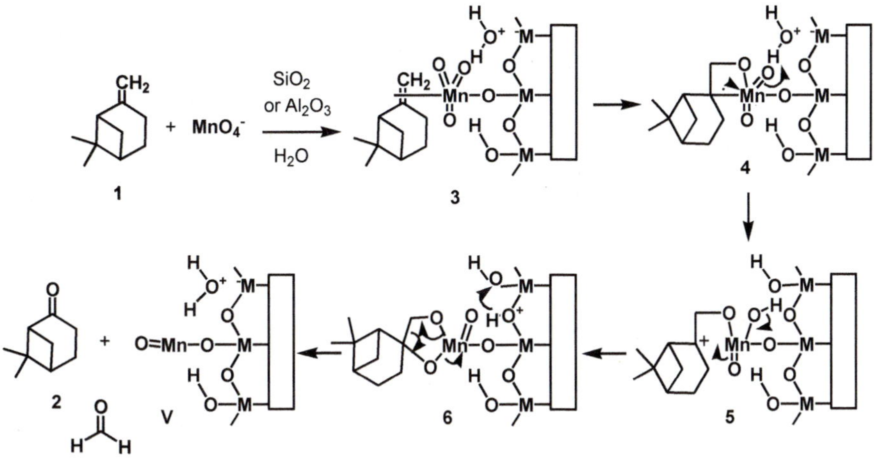

SCHEME 6.2

Mechanism of mechanochemical oxidation of β-pinene by $KMnO_4$.

6.1.2 OXIDATION TO KETONES

Mechanochemical oxidation of organic compounds using potassium permanganate was reported by Nüchter et al. [4]. Employment of planetary ball mill using potassium permanganate as an oxidant and Al_2O_3 as grinding auxiliary gives ketones in low yield, while addition of water to the milling jar greatly improves yields (Scheme 6.3). Optimized reaction conditions were applied to various substrates, and again, the presence of water is crucial for increase of reactivity of oxidizing agent (Table 6.2). Two explanations on the role of water were devised: water on the surface of the substrate forms a thin film in which permanganate is dissolved helping to pass through surface. Secondly, water adsorbed on the alumina surface reacts with the surface, which acts as an acid and lowers the pH locally. Since the substrates are only sparingly soluble in water the first role of water seems to be of less importance.

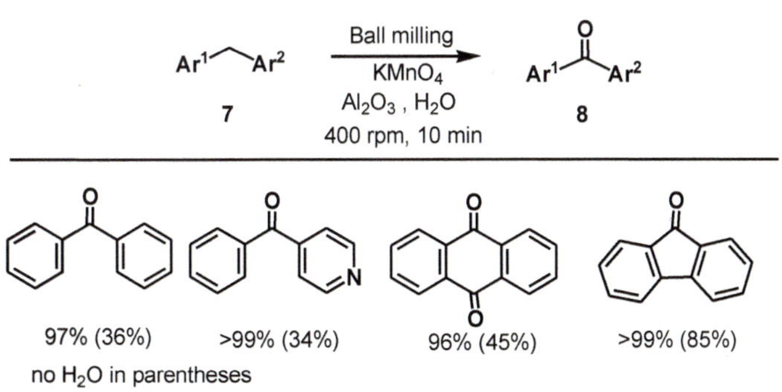

SCHEME 6.3

Oxidations of benzyl aromatics with $KMnO_4$.

Table 6.2 Oxidations of Organic Substrates With $KMnO_4$[a]

Entry	Substrate	Product	Yield (%)[b]
1	Cyclohexane	Cyclohexanol	18 (–)[c]
2	Toluene	Benzoic acid	22 (17)
3	Cyclohexanol	Adipinic acid, cyclohexanone	49 (25), 7 (2)
4	1-Phenylethanol	Benzoic acid	78 (68)
5	Cycloheptanone	1,2-Cycloheptadione, pimelic acid	15 (15), 59 (42)
6	Acetophenone	Benzoic acid	77 (73)

[a]*Fritsch Planetary Ball Mill Pulverisette 7, stainless steel jar (10 mL), agate balls (7 × 10 mm), 400 rpm, 10 min, substrate:$KMnO_4$:Al_2O_3:H_2O 1:3:19:6.*
[b]*Isolated.*
[c]*Reactions without water.*
Nüchter M, Ondruschka B, Trotzki R. Mechanochemical oxidation of organic model compounds by means of potassium permanganate. J Prakt Chem 2000;342:720–4, Copyright Wiley-VCH Verlag GmbH & Co. KGaA, Reproduced with permission.

6.1.3 OXIDATIVE DEHYDROGENATION OF γ-TERPINENE

Reaction parameters for oxidative dehydrogenation of γ-terpinene **9** to *p*-cymene **10** were studied in great detail by Stolle et al. (Scheme 6.4) [5]. It was found that potassium permanganate is the most efficient oxidant in mechanochemical reaction. Alumina was used as milling auxiliary in planetary ball mill Pulverisette, with ZrO_2 milling beakers (V = 45 mL) and 5 × 15 mm balls. In analogy to previous account, addition of water or the presence on the grinding auxiliary surface is beneficial for solvent-free transformation in ball mill. A great excess of $KMnO_4$ is necessary to guarantee quantitative yields (3 equiv.). A further increase of the amount of oxidant afforded a decrease in selectivity due to formation of *p*-cymenene as side product.

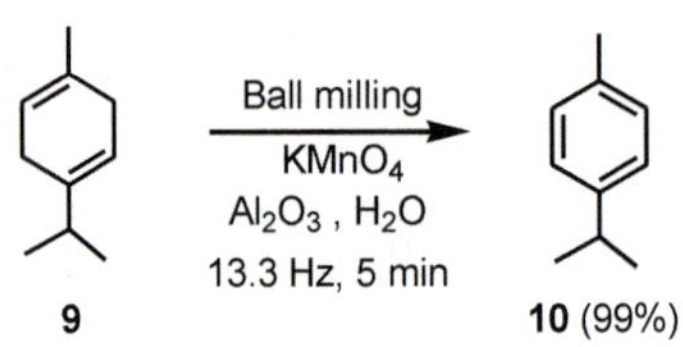

SCHEME 6.4

Oxidative dehydrogenation of γ-terpinene.

6.1.4 OXIDATION OF *p*-TOLUIDINE

A detailed investigation on the influence of the type of oxidant and grinding auxiliary in the oxidation reactions of *p*-toluidine was carried out by Stolle et al. (Scheme 6.5) [6]. Notable differences between the oxidants, as well as significant differences for the same oxidant in combination with different grinding auxiliaries, were observed. This finding allowed the reaction to be directed toward all three major oxidation products (azo, azoxy, and nitro) by choosing the right combination of oxidant and grinding auxiliary (Table 6.3). The azo product **12** was chemoselectively generated with $KMnO_4/Al_2O_3$, while the application of Oxone/urea or sodium percarbonate with Al_2O_3 was the best choice for azoxy product **12**. Finally, oxidations with Oxone and SiO_2 drive the reaction to the nitro product **14**. All reactions were carried in agate milling beakers and agate milling balls.

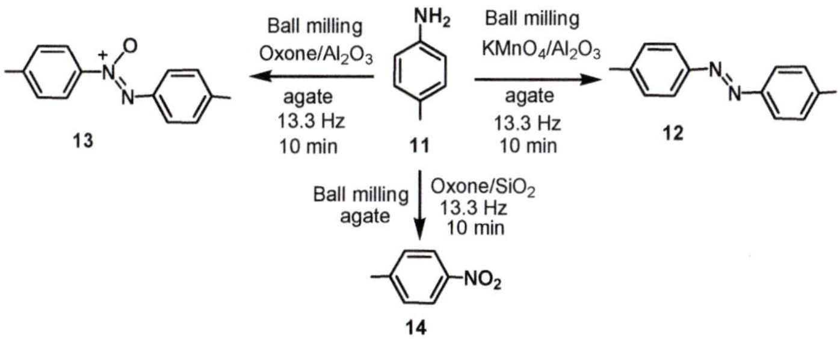

SCHEME 6.5

Oxidation of anilines.

Table 6.3 Mechanochemical Oxidation of *p*-Toluidine[a]

Conversion (%)			12 (%)		13 (%)		14 (%)	
Oxidant	Al_2O_3	SiO_2	Al_2O_3	SiO_2	Al_2O_3	SiO_2	Al_2O_3	SiO_2
$KMnO_4$	96	95	95	88	2	3	2	–
K_2MnO_4	71	49	98	94	–	3	–	–
$NaIO_4$	48	97	74	91	2	–	–	–
Oxone	94	99	3	2	88	10	6	85
H_2O_2/urea	96	7	1	20	92	45	4	–
$Na_2S_2O_8$	29	88	83	97	6	–	1	–
Na percarbonate	84	2	4	99	84	–	11	–
$K_2Cr_2O_7$	5	3	67	50	–	–	–	–
KIO_3	3	2	50	–	–	–	–	–
$KMnO_2$	3	3	50	50	–	–	–	–
$NaBO_3$	6	3	50	50	–	–	–	–

[a]*Fritsch planetary ball mill Pulverisette 7, agate jar (45 mL), balls (6 × 15 mm), 800 rpm, 10 min, 0.1 g water, p-toluidine:oxidant 2:4 equiv., grinding auxilliary (4 g). Copyright (2010), Wiley. Used with permission from Thorwirth R, Bernhardt F, Stolle A, Ondruschka B, Asghari J. Switchable selectivity during oxidation of anilines in a ball mill. Chem Eur J 2010;16:13236–42.*

Investigation of the influence of the grinding auxiliary with $KMnO_4$ as oxidant indicated that Al_2O_3 (γ-modification) and montmorillonite K-10 gave the highest conversion and selectivity for formation of the azo product **12** (>95%).

A number of *p*-, *m*-, and *o*-substituted anilines were transformed into **12** and **13** using $KMnO_4$ and Oxone as oxidants (Table 6.4). Comparison of the performance

Table 6.4 Mechanochemical Oxidation of Anilines Using $KMnO_4$ and Oxone[a]

Oxidant	$KMnO_4$					Oxone				
		12 (%)		13 (%)			12 (%)		13 (%)	
	Conversion (%)	Agate	ZrO_2	Agate	ZrO_2	Conversion (%)	Agate	ZrO_2	Agate	ZrO_2
Ph	89 (97)[b]	95	95	2	3	83 (98)[b]	3	4	92	89
m-MeO	66 (80)	94	91	4	3	81 (96)	3	5	93	87
o-Me	81 (92)	84	80	2	3	79 (97)	3	3	89	88
m-Me	80 (93)	91	89	3	3	81 (97)	3	4	93	88
p-Me	96 (98)	95	95	2	2	94 (99)	3	4	88	87
p-Ac	29 (32)	67	75	17	15	52 (82)	14	17	81	70
o-Cl	61 (79)	82	85	7	5	70 (94)	4	3	84	89
m-Cl	79 (92)	92	86	6	7	69 (92)	6	8	89	87
p-Cl	97 (99)	93	92	4	5	76 (96)	5	8	86	83
o-Br	61 (76)	82	78	9	10	79 (95)	3	4	87	87
m-Br	78 (89)	86	84	6	6	67 (92)	3	6	86	81
p-Br	92 (98)	91	88	6	6	73 (95)	7	8	85	76
o-I	48 (70)	84	80	9	7	60 (82)	13	17	73	58
p-I	77 (88)	92	8	3	1	61 (88)	16	23	69	41

[a]*Fritsch planetary ball mill Pulverisette 7, agate jar (45 mL), balls (6 × 15 mm), 800 rpm, 10 min, 0.1 g water, aniline:oxidant 2:4 equiv., grinding auxilliary (4 g).*
[b]*γ-Al_2O_3 activity 90, neutral (ZrO_2 in parentheses). Copyright (2010), Wiley.*
Used with permission from Thorwirth R, Bernhardt F, Stolle A, Ondruschka B, Asghari J. Switchable selectivity during oxidation of anilines in a ball mill. Chem Eur J 2010;16:13236–42.

of agate and ZrO_2 as material for milling beakers and balls showed that the conversions could be increased by changing the milling material from agate (SiO_2; $\rho = 2.65$ g cm^{-3}) to more dense ZrO_2 ($\rho = 6.5$ g cm^{-3}), while keeping the other reaction conditions constant.

Further study showed that mass loss of milling balls is influenced by the grinding auxiliary. The deterioration of the milling balls caused by friction and impact processes in milling with SiO_2 as the auxiliary is more than threefold larger compared to the use of Al_2O_3, in which the mass lost did not affect the reproducibility of the results. The variation of conversion when using completely new milling balls compared with those obtained with older balls that had been applied more than 100 times was less than 2%.

Comparison of solvent-free mechanosynthesis with solution reactions indicate that milling is more effective than the oxidation performed in solution, and the reaction time is longer for solution reactions. Additional advantage of the ball-mill procedure is the easy workup (extraction of product from powdered reaction mixture with an organic solvent). Microwave-conducted oxidations afforded lower yields, which is ascribed to the less-effective mixing. Finally, energy consumption was measured, and molar energy conversion for vibrational mill was found to be superior to solvent-based methods.

6.1.5 OXIDATION OF *p*-ANILINES TO NITROSOBENZENES

A series of *p*-substituted nitrosobenzenes **18** was prepared in high yield by oxidations of the corresponding anilines by Oxone in ball mill in solvent-free conditions (Scheme 6.6) [7]. Reaction takes place via hydroxylamine intermediate **16**, which was not detected; however, reaction mixture was contaminated with azoxybenzene side products, obtained by coupling reaction of hydroxylamines with nitrosobenzenes. Purification was achieved by sublimation of nitrosobenzenes, thus achieving a synthesis and purification of products without the use of solvent. Comparison of solid-state route with the solution approach revealed that *p*-Br nitrosobenzene was obtained in higher yield and with the comparable purity (Table 6.5). Formation of *p*-NO_2 nitrosobenzene was slowest, and the yield could be increased by longer grinding, although at the expense of purity. Purity and yield could be substantially improved upon addition of equimolar amount of $NaHCO_3$ (Table 6.5). It was proposed that $NaHCO_3$ may act as a base to neutralize acidic species formed during oxidation, and prevents the acid-catalyzed reactions between nitrosobenzene and hydroxylamine or amine. When *p*-aminophenol and *p*-aminobenzoic acid were subjected to reaction conditions, noncharacterized tars were produced, which may be due to the more acidic environment. The liquid-assisted grinding reduced purity and the yield of nitrosobenzenes, in which aprotic and nonpolar solvents were more successful.

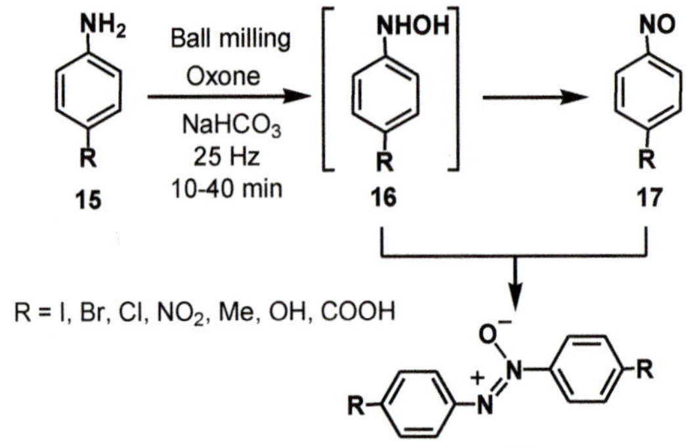

SCHEME 6.6

Oxidation of anilines to nitrosobenzenes.

Table 6.5 Nitrosobenzene Syntheses by Ball Milling[a]

Substituent	I	Br	Cl	NO_2	Me
Yield (%)	85 (~12)[b]	80 (~15) 70[c]	70 (~8)	(26)	52
Purity (%)	92 (~80)	90 (~60) 95[c]	80 (~65)	(90)	69
Time (min)	20	20	20	30	20

[a]*Retsch MM 200 ball mill 25 Hz, 10 min.*
[b]*In parentheses without* $NaHCO_3$ *additive.*
[c]*Solution synthesis.*
Reproduced from Huskić I, Halasz I, Friščić T, Vančik H. Mechanosynthesis of nitrosobenzenes: a proof-of-principle study in combining solvent-free synthesis with solvent-free separations. Green Chem 2012;14:1597–1600, with permission from the Royal Society of Chemistry.

6.1.6 OXIDATION OF 1,2,3-TRIMETHOXYBENZENE TO *P*-QUINONE

Oxidation of 1,2,3-trimethoxybenzene **19** to *p*-quinone **20** using Oxone was carried out in aqueous solution and in rotary rock tumbler (Scheme 6.7) [8]. This report by Crabtree et al. provides a remarkable example of a complete change of selectivity in mechanochemical activation versus conventional solution.

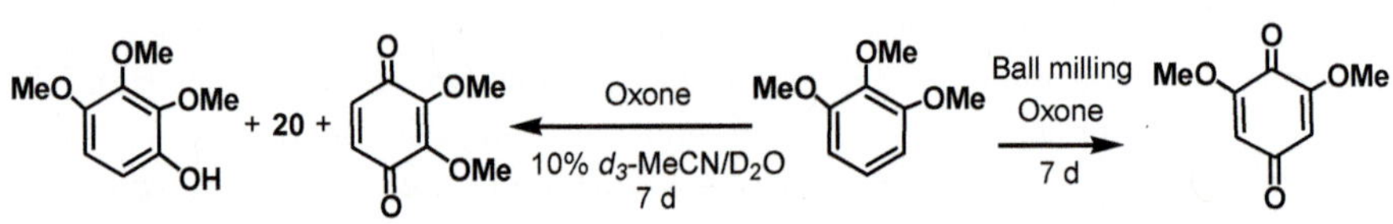

SCHEME 6.7

Oxone oxidation of 1,2,3-trimethoxybenzene.

Mechanochemical oxidation by manual grinding of **19** with 3 equiv. of Oxone was effected in a mortar and pestle by periodical grinding for a total of 1 h per day for 14 days. The starting material was selectively converted into 2,6-dimethoxybenzoquinone **20** which (73% yield). In contrast, the solution-based reaction gave **22** as major product: Oxone, 2 equiv., in 10% d_3-MeCN/D_2O, 5 h, room temperature, gives mixture of **22**:**20**:**21** in 60:7:6 and in 40% yield. Using an automated mortar and pestle mixer to grind the solids, the only product after 7 h was **20**, in low overall isolated yield (13–21%), as the result of partial volatilization of the organics due to the intense localized heating associated with grinding. Mass loss was limited by a ball mill with a closed capsule and full conversion of **19** was observed after 2 h, most likely because the higher localized energy associated with ball milling compared to mortar grinding (**20** and **21** were formed in 19% and 8% yield, respectively). The low yields are probably due to the intense heat generated in the ball mill (which could not be cooled) and the best results were achieved by limiting the milling time to 15 min, allowing the capsule to cool to room temperature before resuming milling (in total of 2 h). The shortcomings of these procedures were avoided by using rotary rock tumbler/polisher which improved reproducibility and maximized product recovery. This approach was selective for **20** (Table 6.6), with isolated yields as high as 82%. As in solution, the yield depended on the amount of Oxone present with the best results obtained using 4 or 6 equiv. In the mechanochemical reactions the presence of excess Oxone does not over-oxidize the product and in the best case the reaction could be run to full conversion without affecting yield of **20**. The reaction outcomes were identical under air or N_2. Isolation of products in mechanochemical reactions was effected by extraction of reaction mixture using chloroform and column chromatography.

Table 6.6 Mechanochemical Oxidation of **19** Using Oxone[a]

Entry	Oxone equiv.	20 Yield (%)
1	1	15
2	2	15
3	4	81
4	6	82 (79)[b]

[a]*Rotary rock tumbler glass vials, stainless steel balls (2 × ~3.5 g), room temperature, 7 days.*
[b]*Inert (N_2) atmosphere.*
Reproduced from Collom SL, Anastas PT, Beach ES, Crabtree RH, Hazari N, Sommer TJ. Differing selectivities in mechanochemical versus conventional solution oxidation using Oxone. Tetrahedron Lett 2013;54:2344–7, Copyright (2013), with permission from Elsevier.

6.1.7 OXIDATION OF ALCOHOLS

Lead tetraacetate/alkali halide system was proven to be an effective oxidant for primary and secondary aliphatic alcohols in mechanochemical conditions (Scheme 6.8). Kapustina and coworkers used this oxidation system to prepare a variety of esters **24** (from primary alcohols **23**) and ketones **26** (from secondary alcohols **25**) by ball milling for 4h in high yields (Table 6.7) [9]. The most efficient halide was found to be KBr, and stark difference in reaction outcome was obtained when LiCl was employed in ball mill and in solution. Unlike transformations of primary and secondary aliphatic alcohols induced by $Pb(OAc)_4$/MX system in the presence of a solvent, under mechanical activation, δ-haloalkanols were not formed. Hence, the oxidation follows different mechanism in ball mill. In solvent the one-electron process with the generation of alkoxyl radicals occurs in the first slop. Under mechanical activation in the absence of a solvent, the reaction follows the two-electron heterolytic mechanism to afford aldehydes and ketones. While ketones are final products and are not involved in further transformations, aldehydes are further oxidized to form esters.

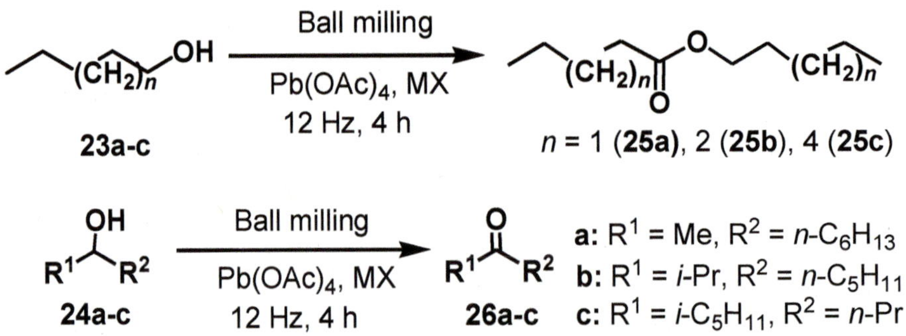

SCHEME 6.8

Oxidation of alcohols with lead tetraacetate.

Table 6.7 Oxidation of Alcohols With Lead Tetraacetate[a]

Entry	Alcohol/Product	MX	Molar Ratio Substrate: $Pb(OAc)_4$:MX	Conversion (%)	Yield (%)[b]
Primary Alkanols					
1	**23a/25a**	LiCl	1:1.2:5	55	38
2	**23a/25a**	LiCl	1:1.2:5	60	40[c]
3	**23a/25a**	LiCl	1:2:5	66	42
4	**23a**/4-chloropentanol	LiCl	1:2:5	67	65[d]
5	**23a/25a**	LiBr	1:1.2:5	65	41
6	**23a/25a**	LiBr	1:1.2:5	84	65[c]
7	**23a/25a**	KBr	1:1.2:5	86	67
8	**23a/25a**	KBr	1:2:5	83	71
9	**1b/3b**	LiCl	1:2:5	72	49
10	**23b**/5-chlorohexan-1-ol	LiCl	1:2:5	70	67[d]
11	**23c/25c**	KBr	1:2:5	85	64

Table 6.7 Oxidation of Alcohols With Lead Tetraacetate[a]—cont'd

Entry	Alcohol/Product	MX	Molar Ratio Substrate: $Pb(OAc)_4$:MX	Conversion (%)	Yield (%)[b]
Secondary Alkanols					
12	**24a/26a**	LiCl	1:2:5	54	42
13	**24a/26a**	KBr	1:2:5	78	63
14	**24a**/5-chlorooctan-2-ol	LiCl	1:2:5	54	42
15	**24b/26b**	LiCl	1:2:5	72	51
15	**24b/26b**	KBr	1:2:5	95	79
16	**24b**/*n*-hexanal	LiCl	1: 2:5	63	60
17	**24c/26c**	LiCl	1:2:5	60	43
18	**24c/26c**	KBr	1:2:5	98	88
19	**24c**/7-methyl-7-chlorooctan-4-ol	LiCl	1:2:5	27	23

[a]*Vibrational mill, stainless steel reactor (80mL), 12.3mm balls (150g), 12Hz.*
[b]*Isolated product.*
[c]*19Hz.*
[d]*Benzene, 80°C.*
Reproduced from Kapustina NI, Sokova LL, Makhaev VD, Borisov AP, Nikishin GI. Oxidation of aliphatic alcohols with the lead tetraacetate-metal halide system under mechanical activation. Russ Chem Bull Int Ed 2000;49:1842–5 [Izv Akad Nauk Ser Khim 2000;49:1870–3], with permission of Springer.

IBX (2-iodoxybenzoic acid) was used as an oxidant in several transformations by Mal et al. [10]. A number of aromatic and primary alcohols **27** were successfully oxidized to the corresponding aldehydes and ketones **28** within 3h of ball milling (Scheme 6.9, Tables 6.8 and 6.9). Using IBX reaction times were shortened compared to traditional methods; for instance, IBX/CH_3CN/AcOH procedure requires 8h, while the reaction is complete in 45min by ball milling (Table 6.9, entry 5). Furthermore, an 11.5h procedure in MeCN/H_2O was shortened to 1h for entry 3 (Table 6.9). Comparatively better yields were obtained by ball milling for entry 13 (77% vs <5% by cucurbit[8]uril-catalyzed oxidation in aqueous solvent). Ball-milling reactions were carried out in ZrO_2 jar using balls of the same material. Isolation of product consists of dissolving in dichloromethane, followed by separation from waste IBA by filtration through a paper filter and evaporation of filtrate.

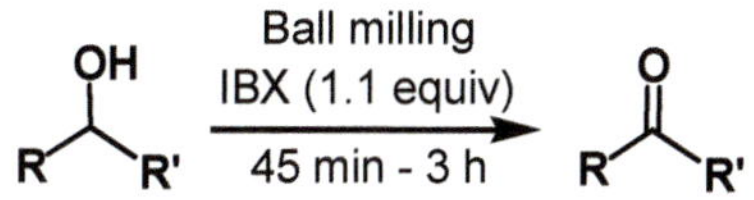

27 28

SCHEME 6.9

IBX oxidations of alcohols.

Table 6.8 IBX oxidations of Primary Alcohols to Aldehydes[a]

Entry	Substrate		Time	Yield (%)
1	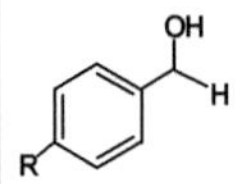	R = CH_3	45 min	87
2		R = Br	45 min	93
3		R = NO_2	2.5 h	99
4	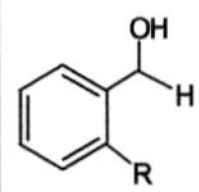	R = CH_3	1.5 h	89
5		R = Br	45 min	86
6		R = CHO	2 h	86
7	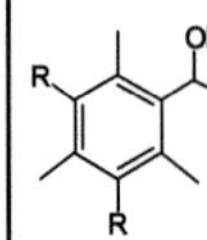	R = H	45 min	94
		R = Br	2 h	88
8	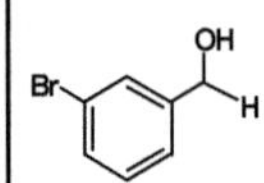		45 min	89
9	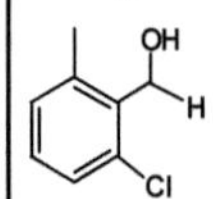		1 h	94
10	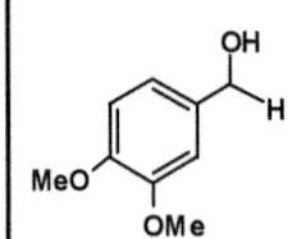		45 min	97
11	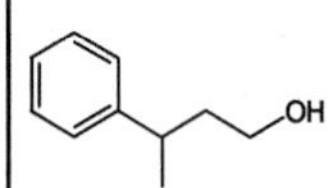		2 h	81
12	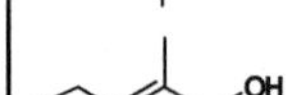		1.5 h	79
13	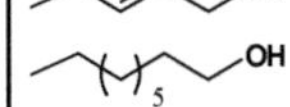		1.5 h	77
14	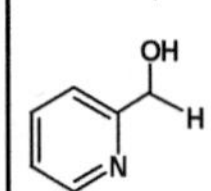		2.5 h	88
14	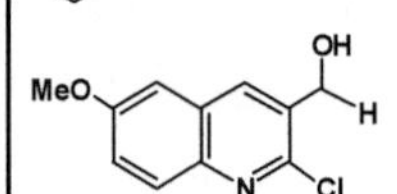		2.5 h	90
14	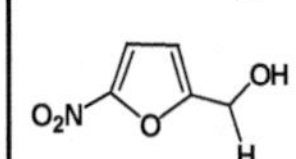		2 h	89

[a]IBX (1.1 equiv.), ZrO_2 ball (d = 15 mm), 10 mL ZrO_2 jar, Retsch MM 200, 21 Hz.
Reproduced from Achar TK, Maiti S, Mal P. IBX works efficiently under solvent free conditions in ball milling. RSC Adv 2014;4:12834–9, with permission from the Royal Society of Chemistry.

Table 6.9 IBX Oxidations of Secondary Alcohols to Ketones[a]

Entry	Substrate		Time	Yield (%)
1	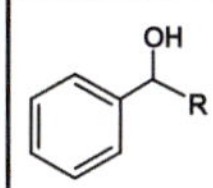	R = Et	45 min	95
2		R = *n*-Pr	1 h	93
3		R = *n*-Bu	1 h	85
4		R = Ph	1 h	94
5	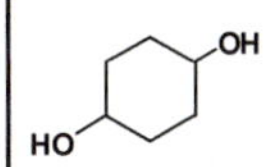		1 h	97

[a]*IBX (1.1 equiv.), ZrO_2 ball (d = 15 mm), 10 mL ZrO_2 jar, Retsch MM 200, 21 Hz.*

Reproduced from Achar TK, Maiti S, Mal P. IBX works efficiently under solvent free conditions in ball milling. RSC Adv 2014;4:12834–9, with permission from the Royal Society of Chemistry.

Oxidations by IBX could be scaled up to gram scale, which was demonstrated for (2-bromophenyl)methanol (2.5 g, 13.4 mmol) and IBX (4.12 g, 14.7 mmol) in 25 mL ZrO_2 jar containing one 15 mm diameter ZrO_2-grinding ball. After 1 h milling, the product 2-bromobenzaldehyde was isolated in 85% yield (2.10 g). The reaction is advantageous over traditional oxidation in solution (dimethyl sulfoxide, DMSO) in terms of total cost of chemicals and materials (estimated saving cost of 48%), and additional cost-effectiveness (31%) was achieved by IBA **32** to IBX waste recycling in one cycle by Oxone (Scheme 6.10).

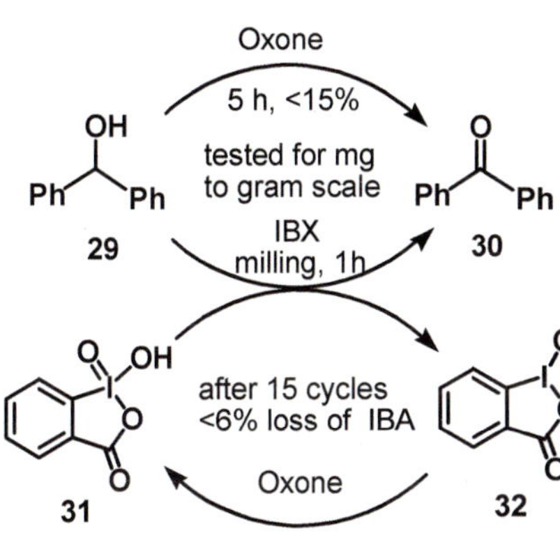

SCHEME 6.10

Efficiency of ball milling and recycling of IBX for benzhydrol oxidation to benzophenone.

The synthetic advantages of IBX oxidation procedure were employed in the one-pot synthesis of benzimidazoles **36** (Scheme 6.11), in which in situ generated aldehydes **34** were further reacted with *o*-phenylenediamine **35**. Products were obtained in high yield within 2–3 h of milling. For this procedure, timing of the diamine addition to the reaction mixture was not given (prior to the milling, or after initial aldehyde formation).

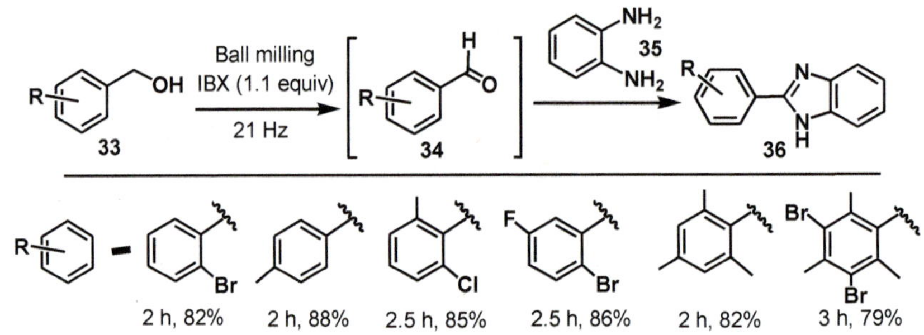

SCHEME 6.11

One-pot synthesis of benzimidazoles.

6.1.8 OXIDATION OF ALDEHYDES

Oxone was employed by Mack for the chemoselective purification of α,β-unsaturated esters **37** obtained from a solvent-free Horner–Wadsworth–Emmons (HWE) reaction in ball mill (Scheme 6.12, Table 6.10) [11]. Substrates with electron-withdrawing group reacted faster, while electron-donating substituents decrease the reactivity; however, after allowing the reaction to proceed overnight (12 h), all of the excess anisaldehyde had been oxidized. Benzaldehyde derivatives **38** under the high-speed ball-milling conditions were oxidized to the corresponding benzoic acids **39** and separated from the esters by an acid–base extraction.

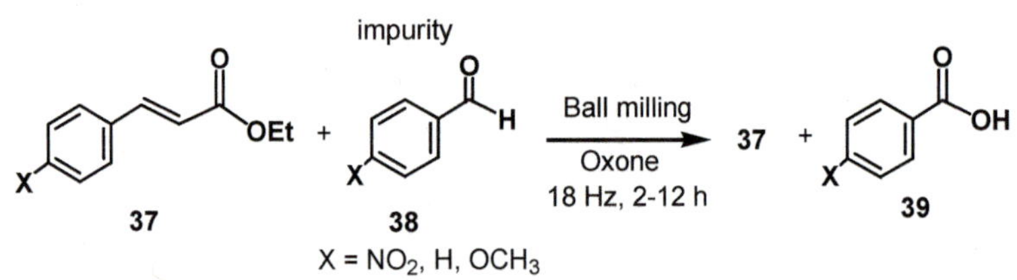

SCHEME 6.12

Oxidation of aldehydes by Oxone.

6.1.9 OXIDATION OF SULFIDES AND THIOPHENES TO SULFONES

Cravotto and coworkers have developed solvent-less selective oxidation of several organosulfur compounds to sulfones, under mechanical-milling conditions (Scheme 6.13) [12]. Oxone was selected by preliminary screening of several solid oxidants (sodium persulphate, potassium superoxide, and sodium tungstate). Using Oxone as oxidant, all substrates, apart from nitrothiophene, gave high conversions and yields (Table 6.11). Mechanochemical synthesis is far superior to conventional conditions (heating at 80°C for 90 min, neat or in toluene solution under stirring) in which only starting material was detected. Advantageously, excess of Oxone could be recycled and easily reused when it is present in stoichiometric excess.

Table 6.10 Mechanochemical Oxidation of Benzaldehydes Using Oxone[a]

Entry	X	K_2CO_3 (equiv.)	Oxone (equiv.)	Time (h)	Aldehyde Remaining (%)
1	H	2	2	2	19
2	H	1	2	2	10
3	H	1	3	1	26
4	NO_2	1	2	2	9
5	NO_2	1	3	2	<5
6	OMe	1	4	2	33
7	OMe	1	4	12	<5

[a]Spex 8000M mixer mill, stainless steel vial and ball (5 mm), 18 Hz, stainless steel ball (5 mm). Reproduced from Shearouse WC, Korte CM, Mack J. A two-step ball milling method synthesizes and purifies α,β-unsaturated esters. Green Chem 2011;13:598–601, with permission from the Royal Society of Chemistry.

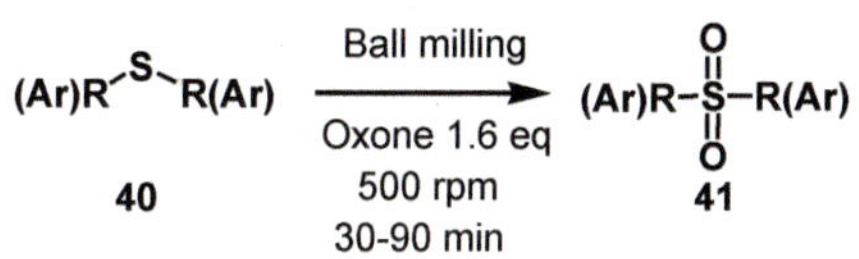

SCHEME 6.13

Chemoselective oxidation of sulfides to sulfones.

Table 6.11 Sulfur Compound Oxidation Using Oxone[a]

Substrate	Time (min)	Conversion (%)	Yield (%)[b]
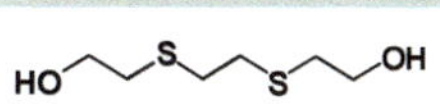	30 90	9 100	8 99
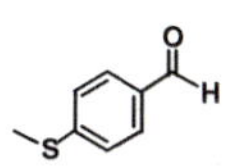	30 90	13 100	11 97
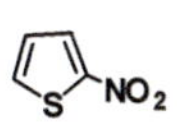	30 90	0 0	– –
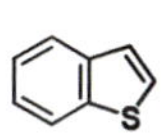	30 90	35 100	34 98
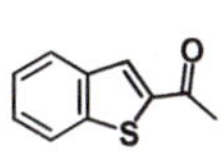	30 90	31 100	27 97
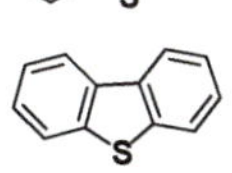	30 90	20 100	19 99
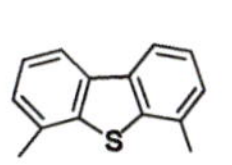	30 90	27 100	25 99

[a]Retsch PM 100 planetary mill, stainless steel beaker (12 × 5 mL), balls (25 × 10 mm), substrate:Oxone 1:1.6 mol, 500 rpm.
[b]Isolated products.
Reproduced from Cravotto G, Garella D, Carnaroglio D, Calcio Gaudino E, Rosati O. Solvent-free chemoselective oxidation of thioethers and thiophenes by mechanical milling. Chem Commun 2012;48:11632–4, with permission from the Royal Society of Chemistry.

Establishing the optimal reaction conditions, oxidation procedure was applied to paraffin-oxidative desulfurization. Two types of paraffins, P1 and P2, that differ in their sulfur content (330 and 3000 $mg\,kg^{-1}$, respectively) were ball milled with Oxone. After oxidative treatment in the ball mill, the sulfur content of paraffins was cut down to 13 $mg\,kg^{-1}$ for P1 and 155 $mg\,kg^{-1}$ for P2, indicating reaction potential for industrial applications.

An alternative oxidation reagent, IBX was used by Mal et al. [10] to convert sulfides to sulfones by solvent-free method (Scheme 6.14). Ball milling of sulfides with IBX (1.1 equiv), using one ZrO_2 ball (d = 15 mm) and ZrO_2 jar (10 mL) in Retsch MM 200 mill afforded sulfones in high yield.

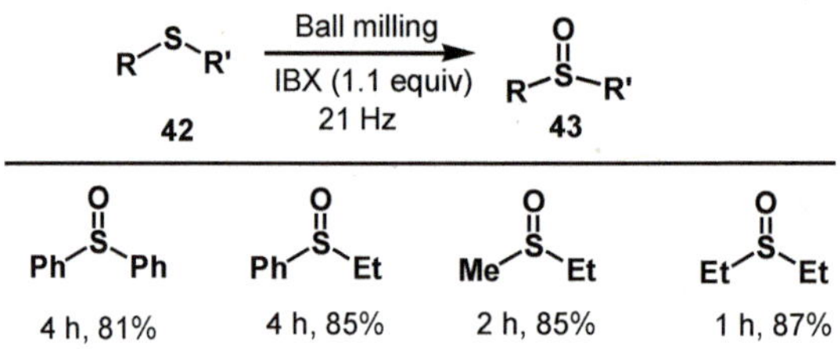

SCHEME 6.14

IBX oxidations of sulfides to sulfoxides.

6.1.10 OXIDATION OF THIOLS TO DISULFIDES

Ranu and Chatterjee have developed ball-milling procedure of aerial oxidation of thiols to disulfides (Scheme 6.15) [13]. Process is chemoselective, and carried out in the absence of any catalyst, co-oxidant, solvent, or base employing neutral aluminum oxide as grinding auxiliary. The influence of grinding auxiliary on the reaction outcome was studied using thiophenol as a model compound and the best results were obtained with neutral Al_2O_3 as grinding auxiliary in planetary mill, and reaction was complete in short time (15 min). When ball milling was carried out in an inert atmosphere of argon, reaction was virtually stopped, demonstrating the necessity of air. The optimal reaction conditions were found to be superior to reaction in solvent (oxygen bubbled through a solution of thiophenol in ethanol).

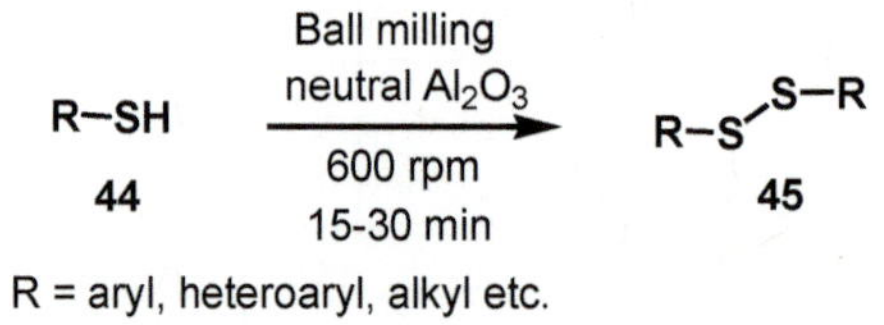

SCHEME 6.15

Chemoselective oxidation of thiols to disulfides.

A variety of substituted thiols underwent aerobic oxidation by this optimized procedure to form the corresponding disulfides (Table 6.12). Aromatic thiols possessing both electron-withdrawing and electron-releasing groups (such as Cl, Br, OMe, OH, NH_2, CH_2OH, COOH, NO_2, F, and BOCNH) as well as sterically hindered thiols readily participate in the reaction. Chemoselectivity was demonstrated in the case of NBoc-lysine and thiols containing CH_2OH group (entries 23, 10, and 21). Fast conversion to high-purity disulfides was obtained by milling within 15–30 min without the use of any catalyst. Products were isolated by simple washing of reaction mixture with ethanol followed by the evaporation of solvent, and generally products were considerably pure (>95%). If needed, further purification was performed either by recrystallization or by column chromatography.

Table 6.12 Oxidation of Thiols to Disulfides[a]

Entry	Substrate		Time (min)	Yield (%)[b]
1	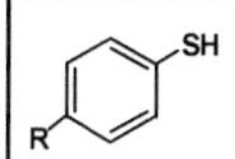	R=H	15	99
2		R=Cl	20	98
3		R=Br	20	98
4		R=Me	15	98
5		R=OMe	15	98
6		R=OH	20	97
7		R=NO_2	25	95
8	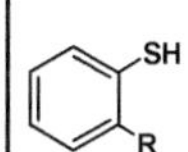	R=Br	20	96
9		R=NH_2	15	97
10		R=CH_2OH	20	96
11		R=COOH	30	95
12	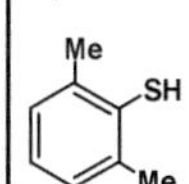		30	95
13	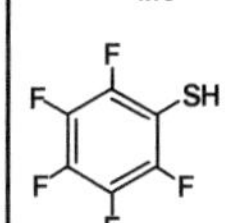		25	96
14	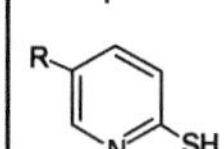	R=H	20	96
15		R=Br	20	98
16	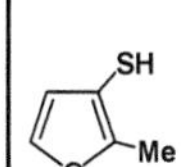		25	97
17	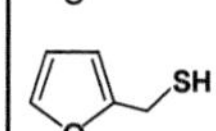		20	96

Continued

Table 6.12 Oxidation of Thiols to Disulfides[a]—cont'd

Entry	Substrate		Time (min)	Yield (%)[b]
18	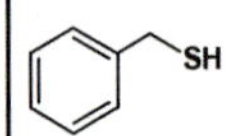		15	98
19		$n=3$	20	98
20		$n=16$	20	95
21	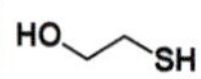		20	96
22	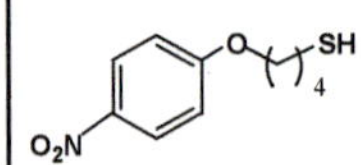		20	97
23	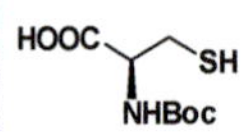		30	96

[a]*Retsch PM 100, stainless steel vessel (25 mL), balls (6×10 mm), 600 rpm, thiol (1 mmol), neutral alumina (3 g).*
[b]*Isolated products.*
Reproduced from Chatterjee T, Ranu BC. Aerobic oxidation of thiols to disulfides under ball-milling in the absence of any catalyst, solvent, or base. RSC Adv 2013;3:10680–6, with permission from the Royal Society of Chemistry.

6.1.11 DITHIANE DEPROTECTION

Another procedure involving IBX and ball milling could be effected, namely deprotection of dithianes **46** to prepare aldehydes (Scheme 6.16). Both aromatic and aliphatic aldehydes were prepared by Mal [10] employing this reaction protocol in high yield within 45 min to 1.5 h, using the oxidation procedure described for IBX oxidation of alcohols to aldehydes (ZrO_2 jar and ball, in Retsch MM 200 mill).

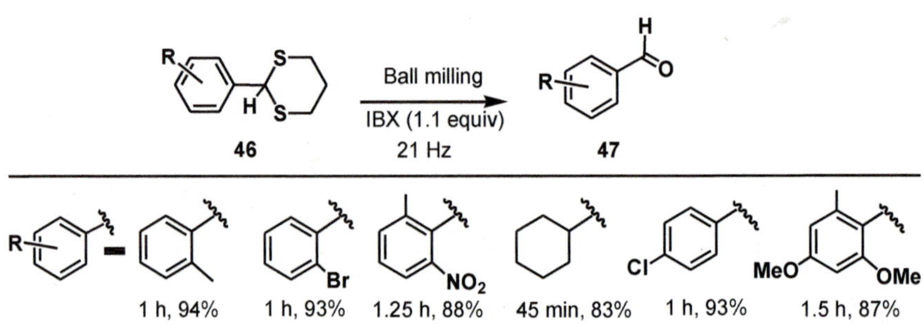

SCHEME 6.16

IBX deprotection of dithiane.

6.1.12 OXIDATION OF AMINES TO IMINES

Mal also employed IBX to oxidize amines to imines (Scheme 6.17) [10]. A short ball milling of benzyl amines with IBX (1.1 equiv.), using ZrO_2 ball (15 mm) and ZrO_2 jar (10 mL) in Retsch MM 200 mill afforded imines in almost quantitative yield.

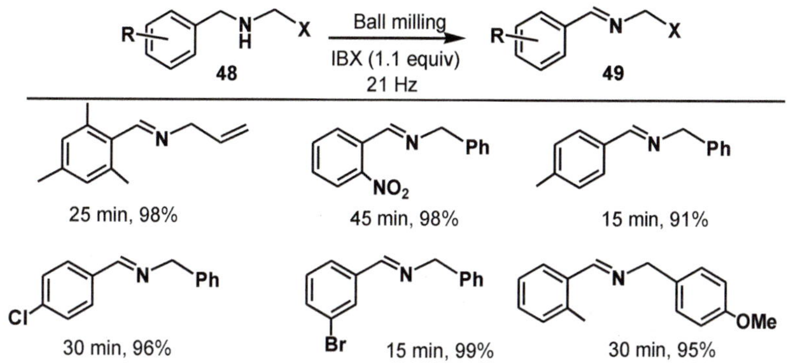

SCHEME 6.17

IBX oxidations of amines to imines.

6.1.13 TRANSFORMATION OF OLEFINS TO α-HALOKETONES

Mechanochemical transformation of olefins to α-haloketones was developed by Mal in which aromatic and aliphatic olefins were ball milled with succinimides and IBX (Scheme 6.18) [10]. The mechanism of this transformation is thought to involve initial halogenation with succinimides, followed by attack of IBX leading to alkoxy periodinanes, which collapse to α-halocarbonyl compounds. Reported reaction yields were high using the mechanochemical oxidation procedure described earlier in the chapter for Mal's IBX oxidation of alcohols to aldehydes. Although no experimental details were given, presumably simultaneous milling with NXS and IBX was carried out, in an analogy to the classical procedure in DMSO [14]. Reaction times and yields of mechanochemical reactions for styrene, indene, and cyclohexene are comparable to the literature results obtained in DMSO.

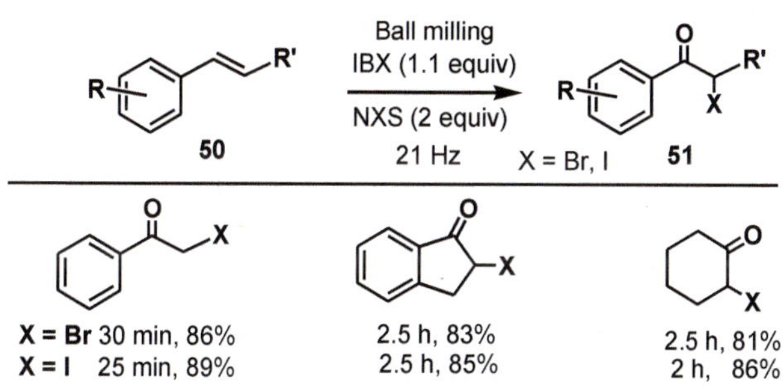

SCHEME 6.18

IBX transformation of olefins to α-haloketones.

6.1.14 OXIDATION OF PORPHYRINS

Simple two-step mechanochemical synthesis of porphyrins was developed by Hamilton and coworkers [15]. The second reaction step in this procedure is oxidation in ball mill, although DDQ oxidation reaction is inferior to solution conditions (see chapter: Carbon–Carbon Bond- Forming Reactions).

6.1.15 ACHMATOWICZ REARRANGEMENT

Achmatowicz oxidative rearrangement of furfuryl alcohols **52** in planetray ball mill provided 6-hydroxy-2*H*-pyrane-3(6*H*)-ones **53**, via mechanism starting with transient furan epoxide formation by Prilezhaev reaction. König et al. established that *m*CPBA was the most efficient among variety of oxidants, when added in excess (Scheme 6.19) [16]. Quartz sand was employed as solid support and reactants milled in planetary mill in agate beaker and balls for 15–30 min to achieve high reaction yields. The conditions for the solid starting material **52d** were modified, and simultaneous use of choline chloride and quartz sand reached 63% yield. Upon the commencement of the grinding, the reaction mixtures were simply diluted with water and filtered, thus entirely organic solvent-free synthesis was achieved. For comparison, the best yield obtained by *m*CPBA reaction with furfuryl alcohol in classical solvent reactions was 88%.

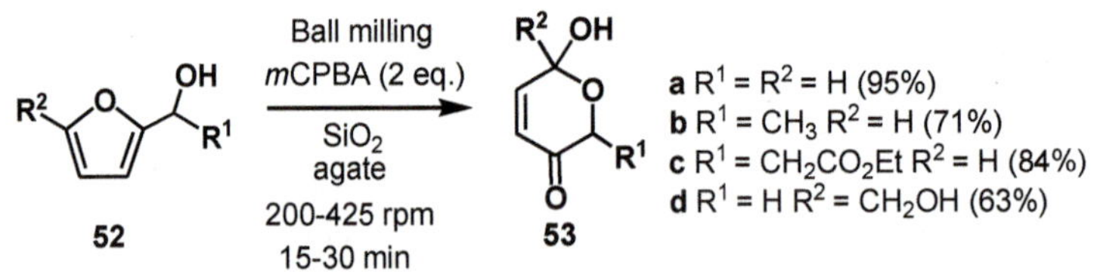

SCHEME 6.19

Achmatowicz rearrangement of furfuryl alcohols.

Some of substrates required for this study were prepared in mechanochemical conditions: ethyl 3-(furan-2-yl)-3-hydroxypropanoate was prepared by Reformatsky reaction (chapter: Carbon–Carbon Bond- Forming Reactions), while reduction of 2-acetylbenzofuran **57** was carried out with $NaBH_4$ (Scheme 6.20). Reduction of aldehyde **54** in planetary ball mill (agate beaker and balls) afforded alcohol **57** in high yield; while the presence of choline chloride gives better yield, reduction of 5-hydroxymethylfurfural **56** without choline chloride was sluggish.

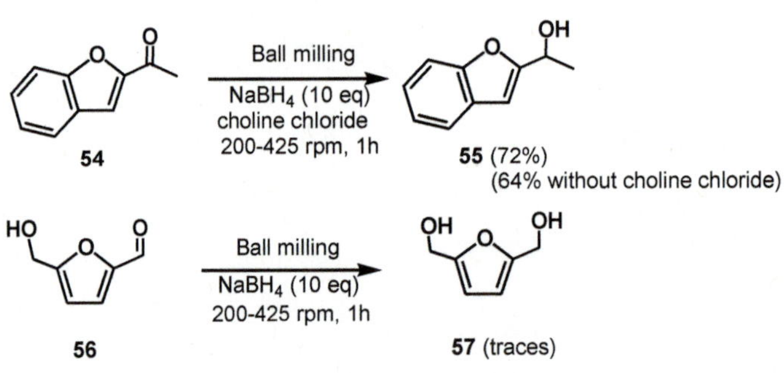

SCHEME 6.20

Synthesis of furfuryl alcohols.

6.1.16 OXIDATION OF LAPPACONITINE

The oxidation of lappaconitine **58** was achieved by the direct reaction with *m*-chloroperbenzoic acid (*m*-CPBA) in solvent-free conditions employing CertiPrep SPEX-8000 vibratory mill (Scheme 6.21). Pankrushina et al. used an excess of *m*CPBA (1.5 equiv.) and aluminum oxide as grinding auxiliary to produce lappaconitine *N*(20)-oxide **59** in 64% yield within 30min of grinding [17]. Authors prevented the abrasive impurities coming from vial walls and balls during the milling process by carrying out a lining of them, by pre-milling of small amounts of a solid hydroperoxide–urea complex for 10min before performing the reaction. Further functionalization of lappaconitine, *N*(20)-deethylation was achieved by the milling with *N*-bromosuccinimide (NBS, 1.1 equiv.). In this reaction, the lining was carried out by pre-milling of NBS before the mechanochemical reaction. The *N*(20)-deethyllappaconitine **60** was obtained in 47% yield by 40min of milling employing Al_2O_3 as grinding auxiliary.

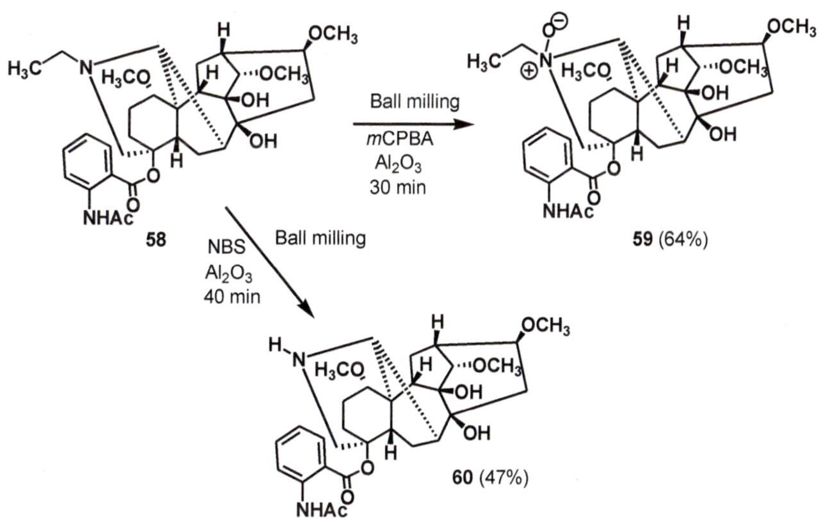

SCHEME 6.21

Oxidation of lappaconitine.

6.1.17 OXIDATION OF β-ARYLACRYLIC ACIDS

Kapustina and coworkers have found that when β-arylacrylic acids **61** were ball milled for 4h with ceric ammonium nitrate, β-arylacrylic acids were converted to corresponding aromatic aldehydes **62** by the cleavage of the C=C bond (Scheme 6.22) [18]. Yields and reaction time for this transformation were comparable to classical solution conditions (Table 6.13). Proposed mechanism includes nitration of olefinic bond, followed by formation of radicals and β-cleavage at the O—N bond.

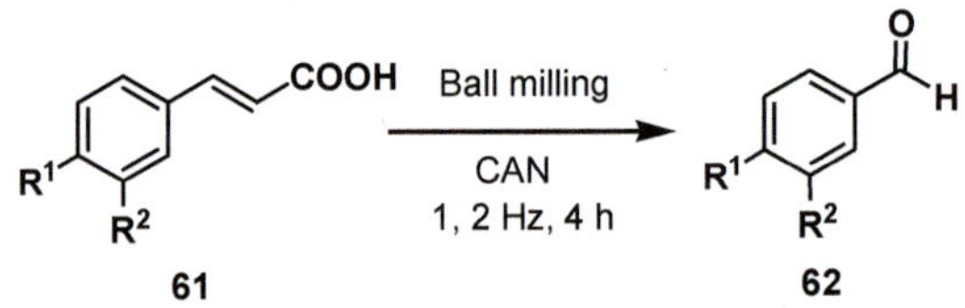

SCHEME 6.22

Oxidation of β-arylacrylic acids.

Table 6.13 Oxidation of β-Arylacrylic Acids[a]

Entry	R^1	R^2	Time	Yield (%)[b]
1	H	H	4h	20 (10)[c]
2	MeO	H	4h	59 (65)[d]
3	MeO	MeO	4h	60 (55)[e]

[a]*Vibrational mill, stainless steel reactor (80 mL), balls (12.3 mm, 150 g), 12 Hz, substrate:CAN 1:2.*
[b]*Isolated products.*
[c]*In MeOH/H_2O, 6 h.*
[d]*In MeOH/H_2O, 3 h.*
[e]*In MeOH/H_2O, 5 h.*
Reproduced from Nikishin GI, Sokova LL, Makhaev VD, Kapustina NI. Solid-phase oxidative halodecarboxylation of β-arylacrylic acids with the ceric ammonium nitrate-alkali halide system. Russ Chem Bull Int Ed 2008;57:118–23, with permission of Springer.

6.2 REDUCTIONS

6.2.1 REDUCTION OF *P*-SUBSTITUTED ALDEHYDES, KETONES, AND ESTERS

Mack et al. have found that *p*-substituted aromatic aldehydes, ketones, and esters **63** can be easily reduced using solvent-free ball-milling conditions (Scheme 6.23) [19]. Sodium borohydride was used to reduce aldehydes and ketones in high yields within 6 h (Table 6.14). These results are superior over previously published solvent-free reduction of ketones using a mortar and pestle which required 5 days of grinding, an inert atmosphere, and a large excess of sodium borohydride [20]. The first solvent-free method for the reduction of esters employs lithium borohydride, a powerful reducing agent which was generated in situ from sodium borohydride and lithium chloride (5:1 ratio). This reagent is expensive, cumbersome to prepare, and it has been known to readily combust in humid air. Through the use of high-speed ball milling, ester functionalities were reduced to respective alcohols in moderate to high yields (Table 6.14), in experimental conditions in open atmosphere with no special precautions taken against moisture. After the conclusion of the reaction the products were washed with diluted HCl and solid products were collected by filtration.

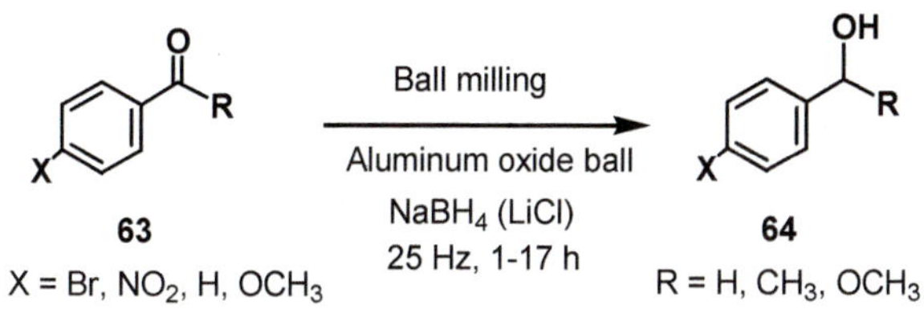

SCHEME 6.23

Reduction of carbonyl group.

Table 6.14 Mechanochemical Reduction of Carbonyl Group Yields (%)[a]

Substituent	Reagent	Time (h)	NO_2	Br	H	OCH_3
CHO	$NaBH_4$	1	–	70 (>95)[b]	70 (>95)	65 (>95)
$COCH_3$	$NaBH_4$	6	70 (>95)	73 (>95)	70 (>95)	65 (>95)
CO_2CH_3	$NaBH_4$/ LiCl	17	69 (>90)	85 (>90)	67 (>90)	32 (38)

[a]*Spex Certiprep 8000M mixer mill, stainless steel vial (0.5×2.0in.), 0.250 in. aluminum oxide ball, substrate. $NaBH_4$ 1:1.*

[b]*Conversions are given in brackets.*

Reproduced from Mack J, Fulmer D, Stofel S, Santos N. The first solvent-free method for the reduction of esters. Green Chem 2007;1041–3, with permission from the Royal Society of Chemistry.

6.2.2 REDUCTION OF ALDEHYDES AND KETONES

Naimi-Jamal, Kaupp and coworkers investigated solvent-free reductions of large series of aldehydes and ketones **65**: liquid, solid, or polymer (Scheme 6.24) [21]. Since $NaBH_4$ did not readily dissolve in them, three types of milling procedures with substrates in a 4:1 ratio were used: (1) solvent-free kneading ball milling, (2) solid-state molecular ball milling, and (3) mechanochemical ball milling (covalent bonds of paraformaldehyde were broken by mechanical action to provide free HCHO). All reactions proceeded quantitatively at the temperatures given, independent of the milling type (Table 6.15). The aldehydes reacted rapidly at 25°C in less than 15 min, while some ketones required higher temperatures, but for the solid ketones, this temperature was well below their melting points. This methodology is also applicable to acetophenone (entry 23) and benzophenone (entry 28), both compounds were practically unreactive when the reduction was attempted in solution or after 2 h grinding with an excess amount of $NaBH_4$. Isolation of product includes hydrolysis of intermediate tetraalkoxyborates with water. These moisture-sensitive intermediate solid tetraalkoxyborates were for the first time collected, handled, and spectroscopically characterized in a moisture-free glove box.

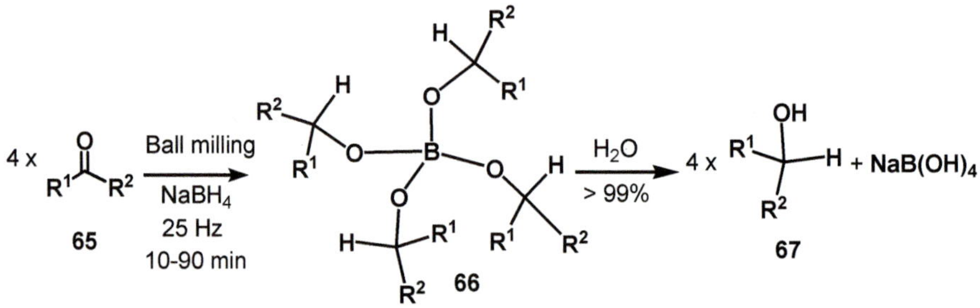

SCHEME 6.24

Reduction of aldehydes and ketones.

Table 6.15 Mechanochemical Reduction of Carbonyl Group[a]

Entry	R^1	R^2	*T* (°C)	*T* (min)
1	H	4-ClC_6H_4	25	10
2	H	4-BrC_6H_4	25	10
3	H	4-$NO_2C_6H_4$	25	10
4	H	3-$NO_2C_6H_4$	25	10
5	H	2-$NO_2C_6H_4$	25	10
6	H	4-CNC_6H_4	25	15
7	H	Ph	25	10
8	H	4-MeC_6H_4	25	15
9	H	4-$OMeC_6H_4$	25	15
10	H	2-$OMeC_6H_4$	25	15
11	H	4-OHC_6H_4	25	15
12	H	2-OHC_6H_4	25	15
13	Vanillin		25	15
14	H	4-$(Me)_2NC_6H_4$	25	20
15	H	2-Furanyl	25	15
16	H	2-Thiophenyl	25	15
17	H	2-Naphthyl	25	10
18	H	$CH_3(CH_2)_6$	25	15
19	Paraformaldehyde		25	20
20	H	$CH_3CH{=}CH$	25	15
21	H	$PhCH{=}CH_2$	25	10
22	H	5-Norbornen-2-Yl	25	20
23	Me	Ph	25	30
24	Me	4-$NO_2C_6H_4$	25	30
25	Me	4-OHC_6H_4	60	60
26	Me	3-OHC_6H_4	60	30
27	Me	$CH_3(CH_2)_4$	25	30
28	Ph	Ph	90[b]	90

Table 6.15 Mechanochemical Reduction of Carbonyl Group[a]—cont'd

Entry	R^1	R^2	*T* (°C)	*T* (min)
29	Ph	CH_2Br	25	20
30	Cyclohexanone		25	20
31	Cyclooctanone		60	30
32	Ph	PhC=O	70	90

[a]*Retsch MM 200 swing mill, stainless steel beaker (10mL) and balls (2×12mm), 20–25Hz, substrate:NaBH$_4$ 4:1, the yield of alcohols was always >99% after workup.*
[b]*In solution reduction gives 20–50% conversion. Copyright (2010), Wiley.*
Used with permission from Naimi-Jamal MR, Mokhtari J, Dekamin MG, Kaupp G. Sodium tetraalkoxyborates: intermediates for the quantitative reduction of aldehydes and ketones to alcohols through ball milling with NaBH$_4$. Eur J Org Chem 2009;3567–72.

Results of reductions collected in Table 6.15 show great regiospecificity and regioselectivity, as there are several substrates in which side reactions with $NaBH_4$ could occur, such as Br elimination, or reduction of C=C bonds (Scheme 6.25).

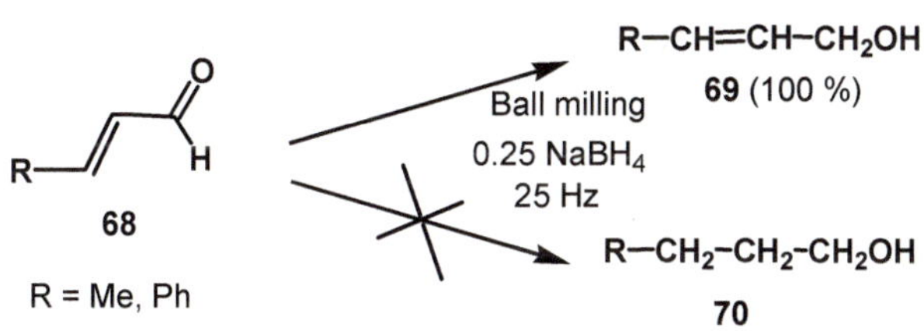

SCHEME 6.25

Regiospecific solvent-free reduction of α,β-unsaturated aldehydes with $NaBH_4$.

Under mechanochemical conditions, bifunctional benzil **71** is specifically reduced by $NaBH_4$ when used in a 4:1 stoichiometric ratio to quantitatively give racemic benzoin **72** (Scheme 6.26). Such result has never been described in solution reactions of these reagents. Furthermore, both carbonyl groups of **71** were quantitatively reduced to dihydrobenzoin (**17/18**), if a 2:1 ratio of **71** and $NaBH_4$ was applied under the otherwise identical conditions of Table 6.17. This stoichiometric synthesis provides *meso*-**73** and *rac*-**74** in 80% and 20% yields, respectively. This stereoselectivity compares with the reported 100:0 ratio in methanol (2h at 25°C) and the 85:15 ratio of **73/74** in ethanol (overnight).

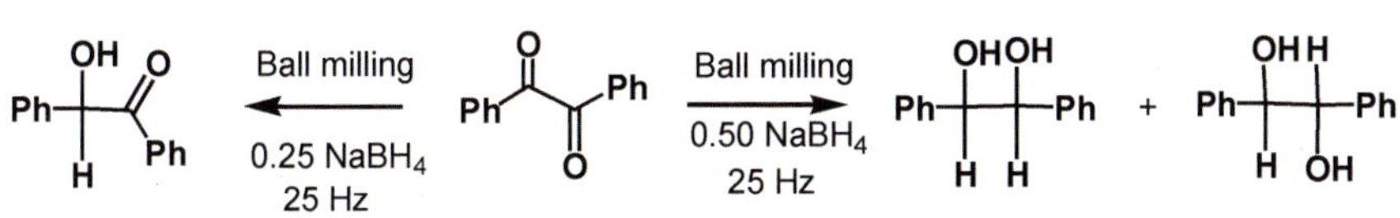

72 (100%) **71** **73** (80%) **74** (20%)

SCHEME 6.26

Specific and stereoselective solvent-free reduction of benzil with $NaBH_4$.

6.2.3 REDUCTION OF NITROARENES

One-step reduction of nitroarenes to azoxy- and azoarenes by the ball-milling technique was carried out by Suzuki [22]. Milling of nitroarenes with bismuth balls (shots) gave as the major products azoxyarenes **76**, while under N_2 azoarenes **77** were formed predominantly (Scheme 6.27, Table 6.16). Addition of one drop of an inert organic solvent (hexane or benzene) led to complete formation of azoxyarenes.

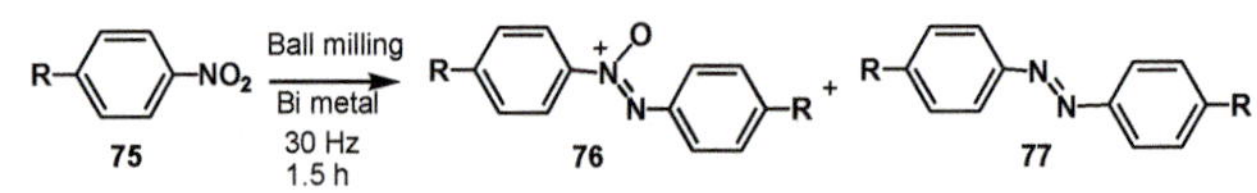

SCHEME 6.27

Reduction of nitroarenes.

Table 6.16 Mechanochemical Reduction of *p*-Nitroanisole With Bismuth[a]

Entry	Conditions	Product Ratio 77/76	Conversion (%)	Yield (%)[b]
1	Under air	78/22	~100	~100
2	Under air[c]	100/0	7[d]	~100
3	Under N_2	17/83	~100	~100
4	Under O_2	100/0	~100	~100
5	Hexane 1 drop	100/0	91[d]	~100
6	Benzene 1 drop	100/0	~100	~100

[a]*Retsch mixer mill, MM200 stainless steel cylinder (5 mL), two stainless steel balls + bismuth balls (1 mm), 30 Hz, substrate:Bi shots 1:8.*
[b]*Based on conversion.*
[c]*Bismuth powder (~100 mesh).*
[d]*Isolated yield.*
Adapted with permission from Wada S, Urano M, Suzuki H. The newborn surface of dull metals in organic synthesis bismuth-mediated solvent-free one-step conversion of nitroarenes to azoxy- and azoarenes. J Org Chem 2002;67:8254–7, Copyright 2002 American Chemical Society.

The reductive dimerization of nitroarenes is very dependent on the substitution in aromatic ring and electron-withdrawing groups (*p*-OCOMe, *p*-CO_2Me, and *p*-NO_2) suppress the reaction (Table 6.17). Reaction was tested with several other metals (tin, manganese, copper, aluminum, antimony, and lead), and metal activity correlates with oxidation–reduction potentials of these metals in the order Al > Mn > Sn > Pb > H > Sb > Bi > Cu (Table 6.18). The pathway for the deoxygenative dimerization of nitroarenes on the activated bismuth surface proceeds via the stepwise reduction processes from nitroarene to nitrosoarene and *N*-arylhydroxylamine, followed by dehydrative coupling.

Table 6.17 Mechanochemical Reduction of Nitroarenes With Bismuth[a]

Nitroarene R	Product Ratio 77/76	Conversion (%)[b]	Yield (%)[c]
p-OMe	100/0	81	~100
p-Me	100/0	96	~100
o-Me	100/0	~100	~100
p-Et	100/0	89	~100
p-Br	100/0	98	~100
m-Br	100/0	43	~100
p-Cl	100/0	~100	~100
m-Cl	100/0	92	~100
p-F	100/0	63	~100
H	0/100	~100	~100

[a]Retsch mixer mill MM200 stainless steel cylinder (5 mL), two stainless steel balls + bismuth balls (1 mm), 30 Hz, substrate:Bi shots 1:8, hexane (1 drop).
[b]Based on conversion.
[c]Isolated yield.
Adapted with permission from Wada S, Urano M, Suzuki H. The newborn surface of dull metals in organic synthesis bismuth-mediated solvent-free one-step conversion of nitroarenes to azoxy- and azoarenes. J Org Chem 2002;67:8254–7, Copyright 2002, American Chemical Society.

Table 6.18 Mechanochemical Reduction of Nitroarenes With Bismuth[a]

Entry	Nitroarene R	Metal	Additive	Product Ratio 77/76[b]	Yield (%)[c]
1	*p*-OMe	Bi	Hexane	100/0	91
2	*p*-OMe	Pb	Hexane	60/40	89
3	*p*-OMe	Pb		0/100	96
4	*p*-F	Bi	Hexane	95/5	~100
5	*p*-F	Pb	Hexane	0/100	71
6	*p*-F	Pb			94
7	H	Sb			0
8	*p*-OMe	Sn			0
9	*p*-OMe	Mn			0
10	*p*-OMe	Al			0

[a]Retsch mixer mill MM200 stainless steel cylinder (5 mL), two stainless steel balls + bismuth balls (1 mm), 30 Hz, substrate:Bi shots 1:8; hexane (1 drop).
[b]Based on NMR analysis.
[c]Isolated yield.
Adapted with permission from Wada S, Urano M, Suzuki H. The newborn surface of dull metals in organic synthesis bismuth-mediated solvent-free one-step conversion of nitroarenes to azoxy- and azoarenes. J Org Chem 2002;67:8254–7, Copyright 2002, American Chemical Society.

The intramolecular reaction of 2,2′-dinitrobiphenyl **77** with bismuth shots was slow and incomplete, giving benzo[*c*]-cinnoline *N*-oxide **78** in 90% yield. However, by replacing bismuth with lead balls, 2,2′-dinitrobiphenyl readily underwent reductive cyclization to form benzo[*c*]cinnoline **79** in 95% isolated yield (Scheme 6.28).

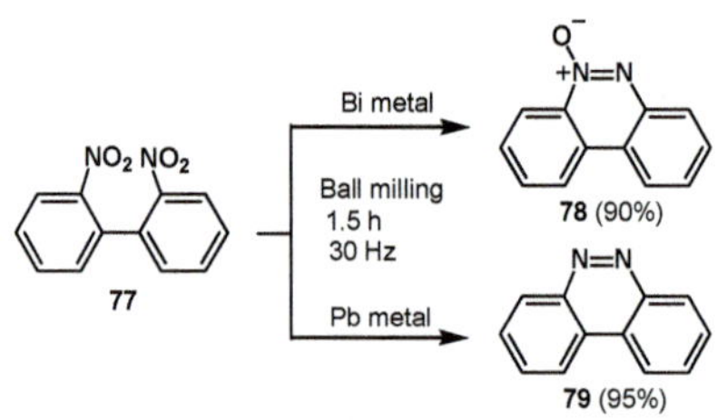

SCHEME 6.28

Reduction of nitroarenes.

REFERENCES

[1] Szuppa T, Stolle A, Ondruschka B, Hopfe W. An alternative solvent-free synthesis of nopinone under ball-milling conditions: investigation of reaction parameters. ChemSusChem 2010;3:1181–91.

[2] Marsaioli AJ, Nurnberg V, Sarragiotto MH, Castellano EE. Envisaging an old reaction from a new point of view. J Org Chem 1989;54:5838–9.

[3] Lang PT, Harned AM, Wissinger JE. Oxidation of borneol to camphor using Oxone and catalytic sodium chloride: a green experiment for the undergraduate organic chemistry laboratory. J Chem Educ 2011;88:652–6.

[4] Nüchter M, Ondruschka B, Trotzki R. Mechanochemical oxidation of organic model compounds by means of potassium permanganate. J Prakt Chem 2000;342:720–4.

[5] Szuppa T, Stolle A, Ondruschka B, Hopfe W. Solvent-free dehydrogenation of γ-terpinene in a ball mill: investigation of reaction parameters. Green Chem 2010;12:1288–94.

[6] Thorwirth R, Bernhardt F, Stolle A, Ondruschka B, Asghari J. Switchable selectivity during oxidation of anilines in a ball mill. Chem Eur J 2010;16:13236–42.

[7] Huskić I, Halasz I, Friščić T, Vančik H. Mechanosynthesis of nitrosobenzenes: a proof-of-principle study in combining solvent-free synthesis with solvent-free separations. Green Chem 2012;14:1597–600.

[8] Collom SL, Anastas PT, Beach ES, Crabtree RH, Hazari N, Sommer TJ. Differing selectivities in mechanochemical versus conventional solution oxidation using Oxone. Tetrahedron Lett 2013;54:2344–7.

[9] Kapustina NI, Sokova LL, Makhaev VD, Borisov AP, Nikishin GI. Oxidation of aliphatic alcohols with the lead tetraacetate-metal halide system under mechanical activation. Russ Chem Bull Int Ed 2000;49:1842–5. [Izv Akad Nauk Ser Khim 2000; 49: 1870–1873].

[10] Achar TK, Maiti S, Mal P. IBX works efficiently under solvent free conditions in ball milling. RSC Adv 2014;4:12834–9.

[11] Shearouse WC, Korte CM, Mack J. A two-step ball milling method synthesizes and purifies α,β-unsaturated esters. Green Chem 2011;13:598–601.

[12] Cravotto G, Garella D, Carnaroglio D, Calcio Gaudino E, Rosati O. Solvent-free chemoselective oxidation of thioethers and thiophenes by mechanical milling. Chem Commun 2012;48:11632–4.

[13] Chatterjee T, Ranu BC. Aerobic oxidation of thiols to disulfides under ball-milling in the absence of any catalyst, solvent, or base. RSC Adv 2013;3:10680–6.

[14] Moorthy JN, Senapati K, Singhal N. An expedient protocol for conversion of olefins to α-bromo/iodoketones using IBX and NBS/NIS. Tetrahedron Lett 2009;50:2493–6.

[15] Shy T, Mackin P, Orvieto AS, Gharbharan D, Peterson GR, Bampos N, Hamilton TD. The two-step mechanochemical synthesis of porphyrins. Faraday Discuss 2014;170:59–69.

[16] Falenczyk C, Pölloth B, Hilgers P, König B. Mechanochemically initiated Achmatowicz rearrangement. Synth Commun 2014;45:348–54.

[17] Pankrushina N, Nikitina I, Chernjak E, Myz C, Shakhtshneider T, Boldyrev V. Solvent-free mechanochemical modification of lappaconitine and piroxicam. Mater Manufact Proc 2008;23:561–5.

[18] Nikishin GI, Sokova LL, Makhaev VD, Kapustina NI. Solid-phase oxidative halodecarboxylation of β-arylacrylic acids with the ceric ammonium nitrate-alkali halide system. Russ Chem Bull Int Ed 2008;57:118–23.

[19] Mack J, Fulmer D, Stofel S, Santos N. The first solvent-free method for the reduction of esters. Green Chem 2007:1041–3.

[20] Toda F, Kiyoshige K, Yagi M. $NaBH_4$ reduction of ketones in the solid state. Angew Chem Int Ed Engl 1989;28:320–1.

[21] [a] Naimi-Jamal MR, Mokhtari J, Dekamin MG, Kaupp G. Sodium tetraalkoxyborates: Intermediates for the quantitative reduction of aldehydes and ketones to alcohols through ball milling with $NaBH_4$. Eur J Org Chem 2009:3567–72.
[b] Naimi-Jamal MR, Mokhtari J, Dekamin MG. Fast and efficient method for quantitative reduction of carbonyl compounds by $NaBH_4$ under solvent-free conditions. In: 11th international electronic conference on synthetic organic chemistry (ECSOC-11). November 2011. p. 1–30. http://wwwmdpiorg/ecsoc-11/.

[22] Wada S, Urano M, Suzuki H. The newborn surface of dull metals in organic synthesis bismuth-mediated solvent-free one-step conversion of nitroarenes to azoxy- and azoarenes. J Org Chem 2002;67:8254–7.

CHAPTER

Applications of Ball Milling in Nanocarbon Material Synthesis

7

CHAPTER OUTLINE

7.1 FUNCTIONALIZATION OF FULLERENES

Several excellent reviews cover the field of ball-milling functionalizations of fullerenes [1–3], but less information is given on the reactions of single-walled carbon nanotubes (SWCNTs), nanohorns, and graphenes.

7.1.1 CYCLOADDITION REACTIONS

Diels–Alder cycloaddition of anthracene to fullerene which was carried by Komatsu and Murata [4] nicely illustrates development of the ball-milling technique for fullerene functionalizations and transfer from standard reaction conditions in solution to mechanochemical solvent-free conditions (Scheme 7.1) [5]. Anthracene and fullerene react in various solvents to obtain 1:1 adduct **3** and equally good yields were obtained in the microwave-assisted reaction, however in significantly shorter time (Table 7.1). Conducting of reaction in solvent-free conditions has two benefits in comparison with solution reactions: fullerene Diels–Alder adducts are thermally reversible, thus room temperature milling is beneficial. In addition, fullerene and its functional derivatives are poorly soluble in common organic solvents [6], and this problem could be readily circumvented by solvent-free methodology. Thus, manual grinding of two reactants with mortar and pestle without the presence of solvent gave small amount of product (2.3%). When the same reaction was carried in grinding mill, efficiency of reaction was increased. This mill was available in the laboratory and routinely used for grinding potassium bromide for IR spectroscopy, but previously not considered for use in organic synthesis. Finally, the use of a special home-built high-speed vibrational mill (HSVM) led to much higher yields and also to formation of 1:2 adduct **4**.

Mechanochemical Organic Synthesis. http://dx.doi.org/10.1016/B978-0-12-802184-2.00007-8

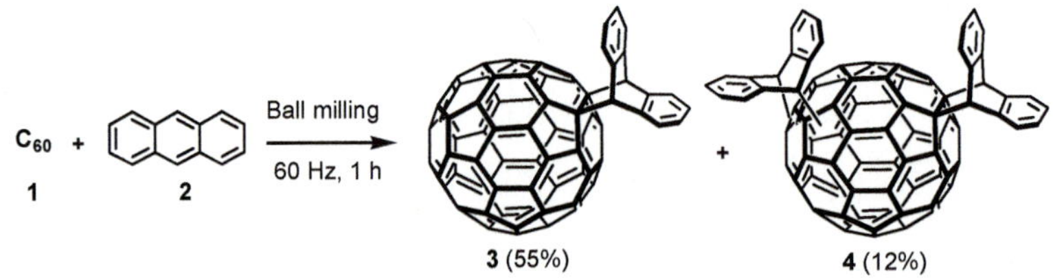

SCHEME 7.1

Reaction of fullerene and anthracene.

Table 7.1 Cycloaddition Reaction of Fullerene and Anthracene[a]

Solvent	Reaction Conditions	Yield of 3 (%)[b]
Naphthalene	2 days, 200°C	39
Toluene	72 h, 115°C	13
Benzene	12 h, 80°C	25
Toluene	15 min, 115°C, MW	35
Manual grinding	5 min	2.3
BM	1 h	55 (+**4**, 19%)

[a]*Home-built ball mill, stainless steel capsule and ball 60 Hz, C_{60}:anthracene 1:1.2*
[b]*isolated yield.*

Wang and Komatsu also studied mechanochemical preparation of [60] fullerene adducts with 9-substituted anthracenes (Scheme 7.2) [7]. Solid-state Diels–Alder reaction of C_{60} with 9-hydroxymethylanthracene **5a** or 9-methoxymethylanthracene (1.2 equiv.) gave cycloadducts **6a** and **6b** in 70% and 74% yields, respectively. In addition, bis(9-anthrylmethyl) adipate **7** (0.5 equiv.) in similar reaction conditions undergoes a double cycloaddition with C_{60} to preferentially afford bisfullerene compound **8** in 34% yield. Solution-phase reactions with same anthracenes in toluene for 24 h at room temperature afforded products **6a** and **6b** in 27% and 28% yield, while after 3 days 4% of the adduct **8** was obtained.

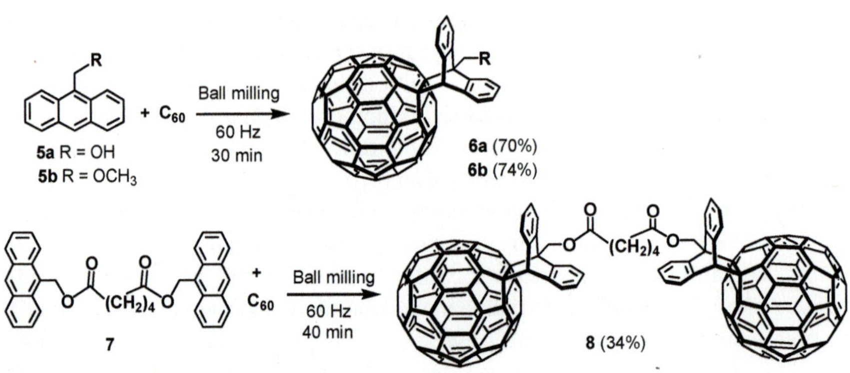

SCHEME 7.2

Diels–Alder reactions of fullerene with 9-substituted anthracenes.

Mechanochemical addition of pentacene to fullerene affords unprecedented mixture of products **10–12** (Scheme 7.3) [4]. Alongside 6,13-adduct **10**, two other products were identified: corresponding *bis*-adduct (two pentacenes added to one C_{60}) **11**, and 2:1 adduct **12**, which if formed via initial formation of 5,14-adduct **13**. Product **13** was not isolated but reacts further with another equivalent of fullerene to afford final adduct **12**.

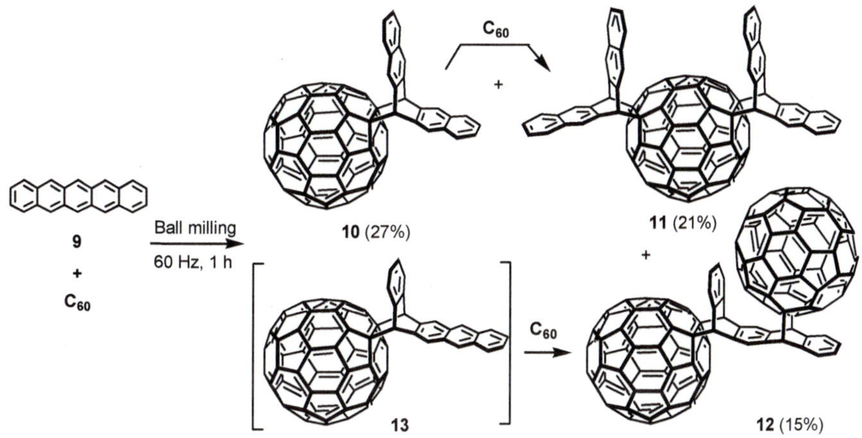

SCHEME 7.3

Reaction of fullerene and pentacene.

In similar reaction conditions, cycloaddition of 2-pyridyl-*s*-tetrazine **13** with C_{60} gives symmetrical 1:1 adduct in more than 90% yield, which in subsequent solid-state reactions affords 2:1 cycloadduct **15**. Novel molecule possess the 2,3-diazabicyclo-[2.2.2] bridge between two fullerene cores (Scheme 7.4) [8]. This reaction outcome is different from *s*-tetrazine coupling described in literature, whereas it was not possible to isolate symmetrical 1:1 adducts [9]. In solution these norbornene adducts undergo irreversible 1,3-prototropic shift yielding the corresponding unsymmetrical 1,4-dienes, while cycloadduct **14** during chromatography on silica undergoes to addition of water and rearrangement [10]. It is evident that by carrying out reaction in solvent-free conditions, this unwanted rearrangement could be avoided and even this intermediate could be isolated.

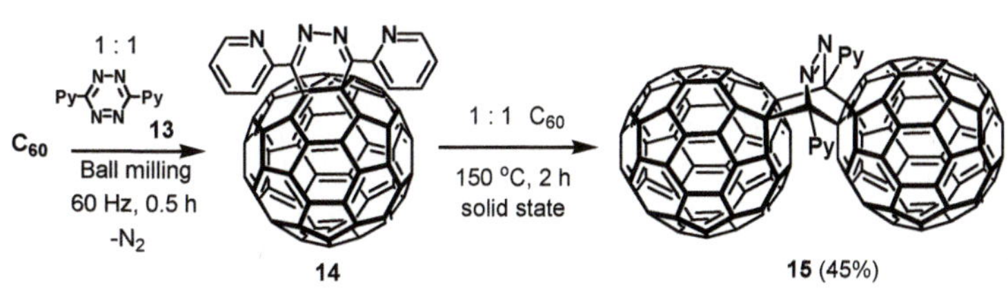

SCHEME 7.4

Reaction of fullerene and 2-pyridyl-*s*-tetrazine.

Another example of chemical reaction which gives different products in ball mill than solution chemistry is phthalazine addition to fullerene [11]. In the ball-milling conditions, intermolecular [4+2] cycloaddition of phthalazine **16** takes place, followed by spontaneous nitrogen elimination from 1:1 adduct **17**. This adduct further in solid state undergoes intermolecular cycloaddition and formation of the corresponding dimer **20** (Scheme 7.5). On the other hand, when reaction is carried out in solution, an intramolecular [4+4] addition takes place with formation of product **20**, followed by retro- [2+2+2] addition and formation of product **21**. This reaction sequence is characteristic for fullerene chemistry, which is not observed in analogous cycloaddition reactions of phthalazines with norbornenes [12].

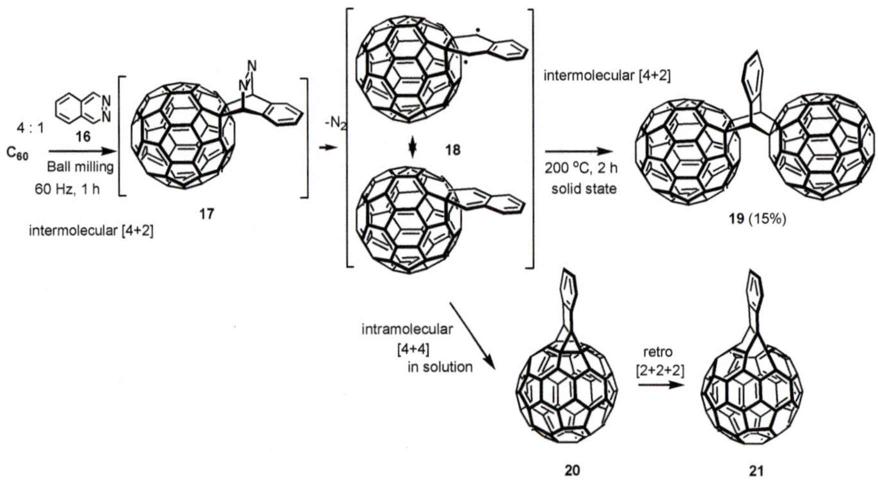

SCHEME 7.5

Reaction of fullerene and phthalazine.

1,3-Dipolar cycloadditions of fullerenes gave expected products. In cycloadditions with azides triazolines **25** were formed, while with diazo compounds such as 9-diazofluorene **22**, spiro cyclopropane products **24** were directly prepared. In this case initially formed pyrazoline intermediate **23** was not isolated (Scheme 7.6) [13].

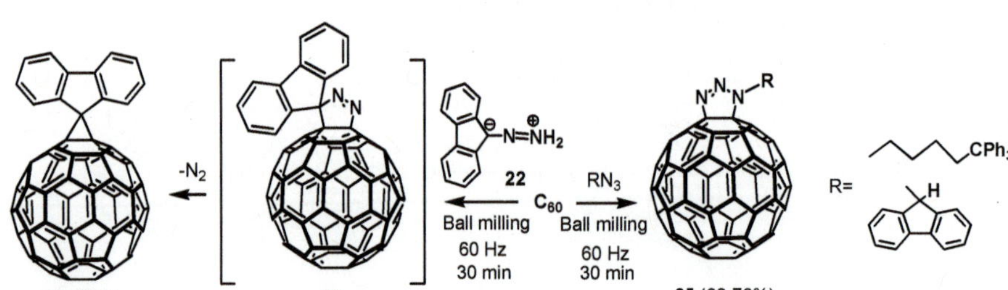

SCHEME 7.6

1,3-Dipolar cycloadditions of fullerene.

The solvent-free mechanochemical reaction of C_{60} with ethyl 2-diazopropionate was investigated by Wang and coworkers [14]. For this purpose, ethyl 2-diazopropionate was prepared in situ from alanine ethyl ester hydrochloride and reacted with fullerene. In the course of reaction, three products, methanofullerene **27** and fulleroids **28** and **29** were obtained in 1.2:1:4.6 ratio (Scheme 7.7). When milling was carried out with preformed ethyl 2-diazopropionate, products **27–29** were obtained in 3.6:1:3.6 ratio and 27% yield. For comparison, reaction carried out in toluene for 16h at room temperature afforded 46% of compounds **27–29** in a ratio of 4:1:10. Although the pyrazoline intermediate was not observed reaction is presumed to take place by initial 1,3-dipolar cycloaddition and subsequent nitrogen elimination from unstable pyrazoline adduct.

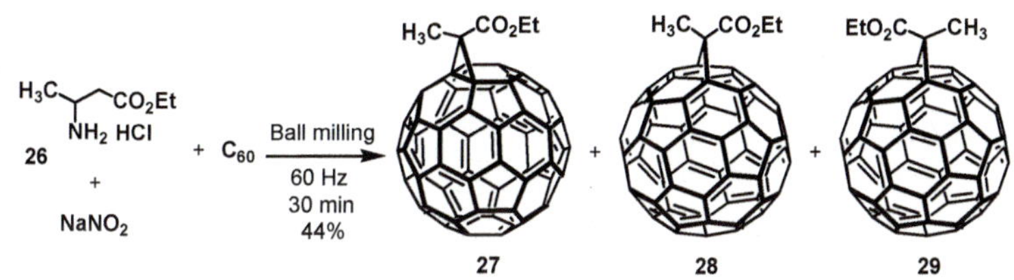

SCHEME 7.7

Fullerene: ethyl 2-diazopropionate reaction.

Question on stability of the pyrazoline intermediates formed in 1,3-dipolar cycloadditions of alkyl diazoacetates with fullerenes was addressed in subsequent publication [15]. The reaction of glycine ethyl ester hydrochloride (or glycine octyl ester hydrochloride) with sodium nitrite was utilized to prepare ethyl diazoacetate in situ under the HSVM conditions. After 30min of vigorous milling at room temperature, 2-pyrazolines **32a** and **32b** were formed in 48% and 49% yield, respectively (Scheme 7.8). These pyrazolines were formed via isomerizations of 1-pyrazolines **31a,b** which are obtained directly by 1,3-dipolar cycloadditions.

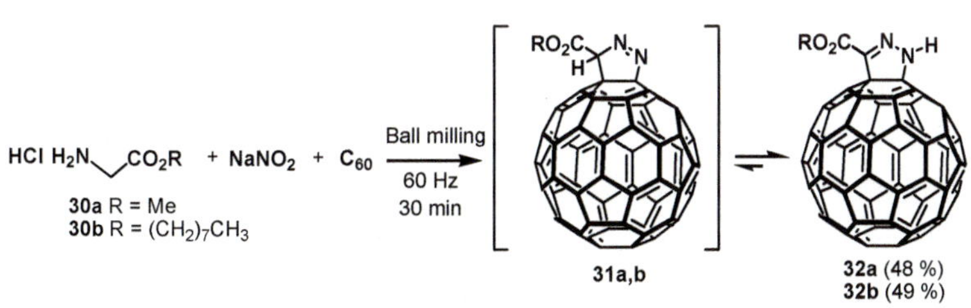

SCHEME 7.8

Cycloaddition of alkyl diazoacetates with fullerenes.

Prato reaction is very characteristic functionalization reaction in fullerene chemistry. A mixture of aldehyde, amino acid, and fullerene provides upon reaction-substituted pyrrolidine ring on fullerene **38** (Scheme 7.9) [16]. Intermediate azomethine ylides **36** are not isolated, but react readily with C_{60} via 1,3-dipolar cycloaddition. Interestingly, product **38a** was also isolated as minor product in reactions with **34b–e**. By appropriate choice of reagents, substitution on nitrogen and α-position could be modified to various fullerene functional derivatives. Formation of **38a** directly from C_{60} and *N*-methylglycine was demonstrated in ball-milling experiment, however, in lower yield and with the formation of the fullerene dimer C_{120} **41**. Similar result was obtained with *N*-ethylglycine. However, when formaldehyde was added in typical Prato conditions, *N*-ethyl pyrrolidine product **40** was obtained in better yield.

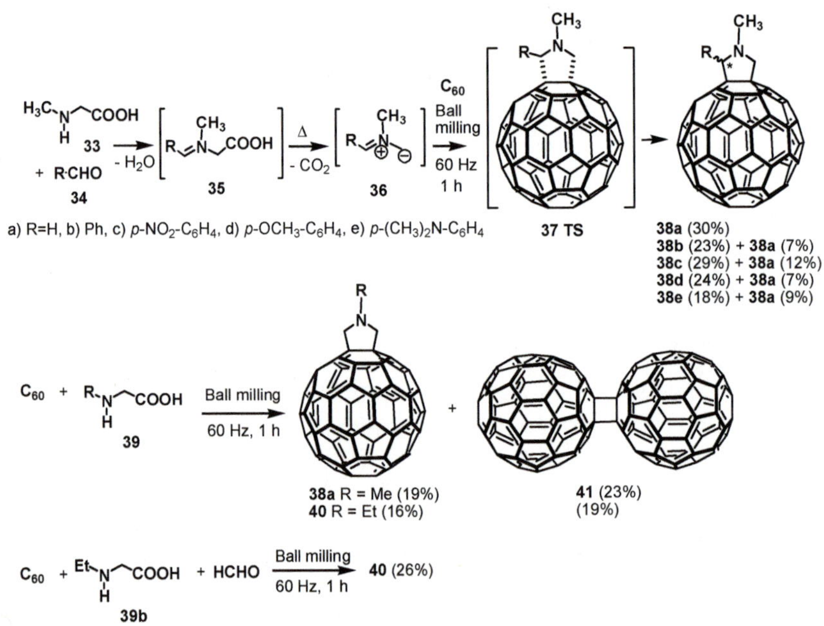

SCHEME 7.9

Prato reaction of fullerene.

Diels–Alder cycloaddition of equimolar amounts of fullerene and exocyclic diene **39** incorporated in norbornene framework gave after 30 min of milling in custom-built ball-mill cycloadduct **40** in 14% yield (Scheme 7.10) [17]. Control experiments carried out in *o*-dichlorobenzene solution at 140°C gave **40** in 40% yield, while at 70°C adduct **40** was formed in 92% yield. Although these results favor solution chemistry, it is an important information that this reaction could be also carried in solvent-free conditions, without the use of highly toxic chlorinated aromatic solvents (*o*-dichlorobenzene).

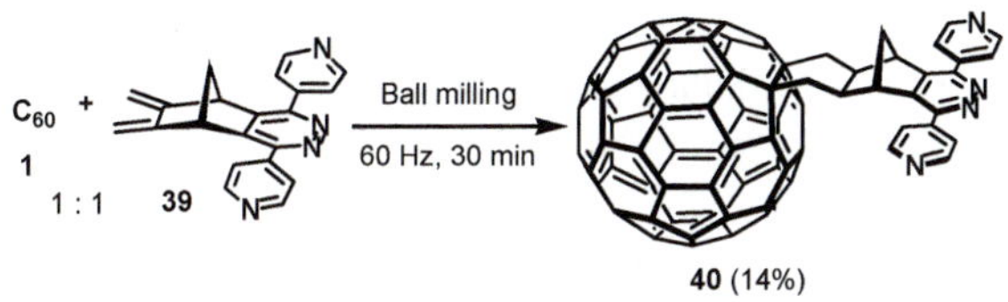

SCHEME 7.10

Diels–Alder cycloaddition of fullerene.

7.1.2 OTHER FULLERENE REACTIONS

Formation of very interesting C_{60} dimer **41** and angular trimer **44** was achieved for the first time by the use of solid-state milling-reaction conditions (Scheme 7.11) [18]. In this reaction potassium and sodium cyanide were used in electrophilic addition, and formation of dimer C_{120} **44** is explained by electron transfer from starting adduct and recombination of radical intermediates [19]. When reaction is conducted in solution, entirely different product of type **43** was obtained. Similarly, the ball milling of C_{60} with 4-aminopyridine as a catalyst for 30 min afforded mixture of C_{120} dimer **41** and C_{180} trimer **44** (4%) [20,21]. Decomposition of the C_{120} product in an association–dissociation equilibrium was noted in high-speed vibrational milling. When C_{120} was ball milled with KCN, dissociation occurred giving a 7:2.5 mixture of C_{60} and C_{120}, which is similar ratio to final product ratio obtained by ball-milling reactions of C_{60} with KCN, suggesting establishment of chemical equilibrium [22].

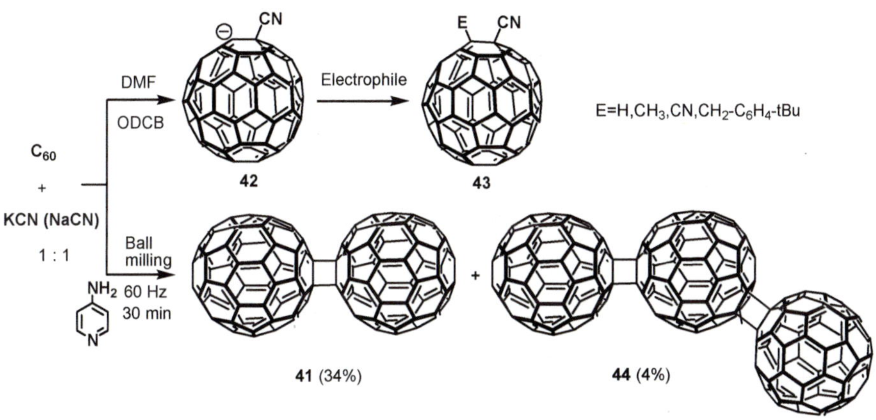

SCHEME 7.11

Formation of fullerene dimer C_{120} and trimer C_{180}.

When equimolar amounts of C_{60} and C_{70} were ball milled together with 4-aminopyridine (2 equiv.), Komatsu and coworkers obtained mixture of two products

(Scheme 7.12) [23]. The main product was C_{120} dimer **41** (10% yield), and as a minor product, formal [2+2] cycloaddition product, cross-dimer C_{130} **45** (3% yield), with a C_{60} cage attached at a 6-6 bond to a C_{70} cage at a 1,2-junction bond in a [2+2] fashion.

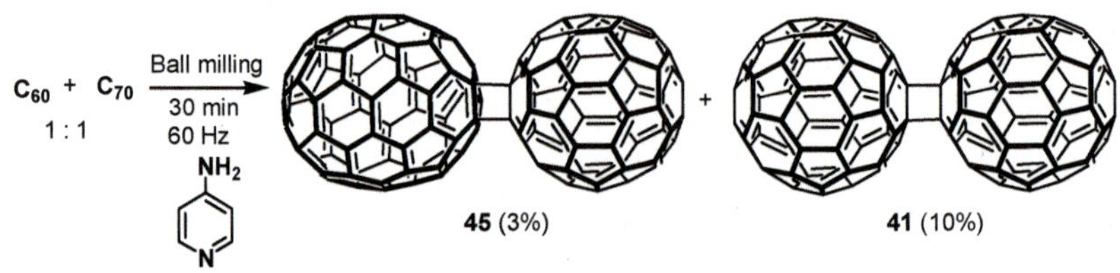

SCHEME 7.12

Formation of fullerene cross-dimer C_{130}.

Two fullerenes were also joined in different fashion by mechanochemical reactions of fullerene with diphenyldichlorosilane and diphenyldichlorogermane (4.2 equiv.), in the presence of lithium (5.8 equiv.), in milling vessel filled with argon which gives novel bisfullerene products **46** and **47** possessing silicon (or germanium) bridges (Scheme 7.13) [24,25].

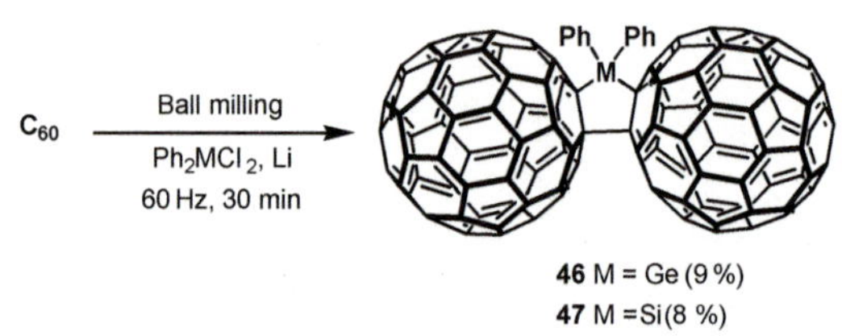

SCHEME 7.13

Reaction of fullerene with diphenyldichlorosilane and diphenyldichlorgermane.

Nucleophilic addition of organozinc reagents is crucial step in Reformatsky reaction which was employed for fullerene derivatization. Ethylbromoacetate reagent (5 equiv.) in conjunction with zinc dust (20 equiv.) reacts with fullerene to obtain product **48** in high yield. In this process, known cyclopropane molecule **49** was formed as a side product, as well as two products unknown in the literature, **50** and **51** (Scheme 7.14) [26]. Formation of these products is rationalized by mechanism which starts with formation of organozinc reagent, which is followed by nucleophilic addition to fullerene.

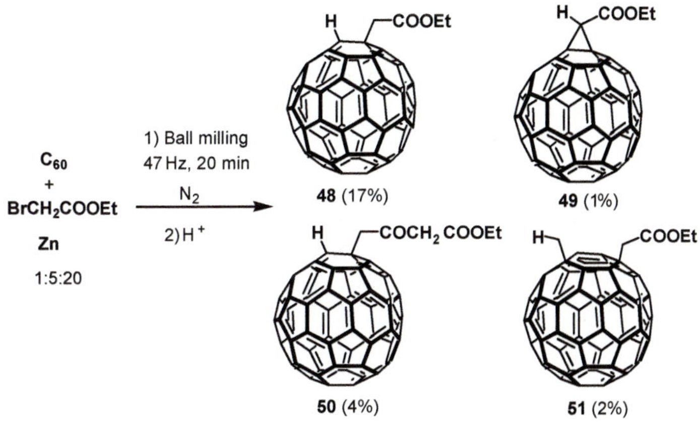

SCHEME 7.14

Reaction of fullerene and organozinc reagent.

Reaction of fullerenes and compounds possessing an active methylene group (Bingel reaction) is usually catalyzed by bases (piperidine, DBU, triethylamine, and Na_2CO_3). Komatsu, Murata, and Wang have shown that Bingel reaction could be carried in solid-state conditions (Table 7.2) [27]. Sodium carbonate performs the best in high-speed vibration-milling conditions with bromomalonate, affording product **52** in 78% yield [28]. Entirely different product **53** was obtained in base-catalyzed reaction of fullerene with diethyl malonate in 19% yield. This product arises via different reaction mechanism, a nucleophilic addition of carbanion, followed by oxidation by air. Furthermore, reaction with ethylacetoacetate in the presence of cerium ammonium nitrate gives an interesting dihydrofuran derivative **54** (manganese acetate was less successful oxidant) [29]. Finally, less-activated aryl methyl ketones, such as 2-acetyl pyrazine gave unexpected products, methanofullerenes such as **55** [30].

Similar to synthesis of **52**, Bingel reaction of C_{70} with diethyl bromomalonate carried out in solvent-free reaction conditions is shown by Komatsu et al. to be dependent on the inorganic base applied. Thus, ball milling with sodium acetate provided cyclopropanated monoadduct **57** in 74% yield, while application of potassium carbonate, alongside the monoadduct **57** also afforded three bisadduct isomers **58a–c** (Scheme 7.15) [28]. This result was also obtained with C_{60} when ball milled with diethyl bromomalonate and K_2CO_3 (1 equiv.) for 30 min at 60 Hz: 29% of mono **57** and 35% of *bis*-adducts were obtained, while in the presence of sodium acetate monoadduct **57** was formed exclusively. It was also shown that the use of organic base DBU, which is the standard reagent in solution-phase Bingel reaction, in solid-state ball-milling conditions no reaction took place.

Table 7.2 Bingel Reactions of Fullerenes[a]

Reactant	Conditions	Product	Yield (%)[b]
$BrCH(COOEt)_2$	Na_2CO_3 30 min	 **52**	78
$CH_2(COOEt)_2$	Na_2CO_3 60 min	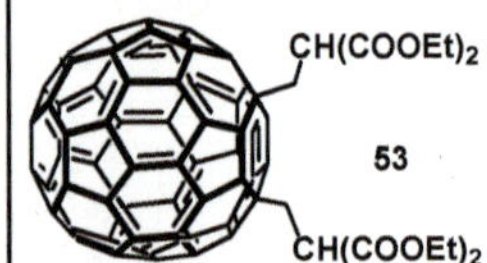 **53**	19
CH_3COCH_2COOEt	$(NH_4)_2Ce(NO_3)_6$ 30 min	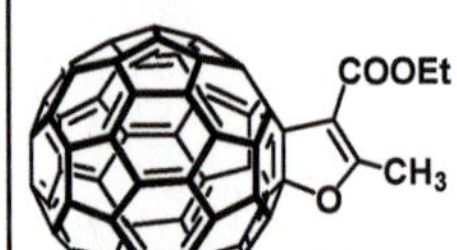 **54**	27
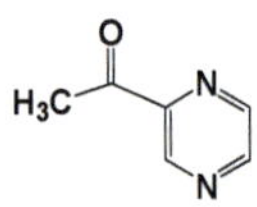	Piperidine 30 min	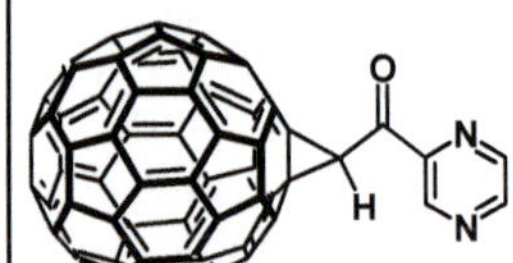 **55**	26

[a]*Custom-made high-speed vibration mill, stainless steel, 60 Hz,* C_{60}*:reactant:base 1:1.5:1.5.*
[b]*Isolated yields.*

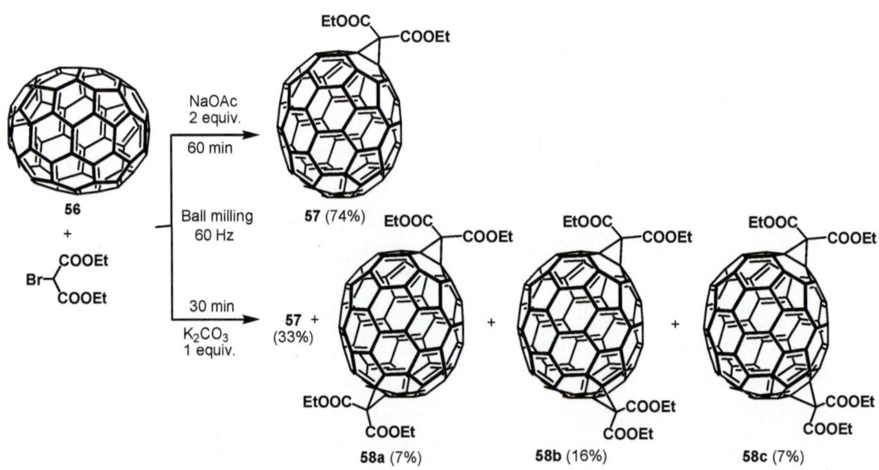

SCHEME 7.15

Bingel reactions of C_{70}.

Wang and coworkers have found that solvent-free mechanochemical reaction of [60]fullerene with phenylhydrazine hydrochlorides **59a–e** in the presence of sodium carbonate under high-speed vibration-milling conditions afforded 1-aryl-1,2-dihydro[60]fullerenes **60a–e** (Scheme 7.16) [31]. Noteworthy difference in comparison to solution conditions is 1-(4-nitrophenyl)-1,2-dihydro[60]fullerene **60e** which could not be obtained in the liquid-phase reaction. When excess sodium nitrite was employed to replace sodium carbonate, fullerotriazoline **61** was formed.

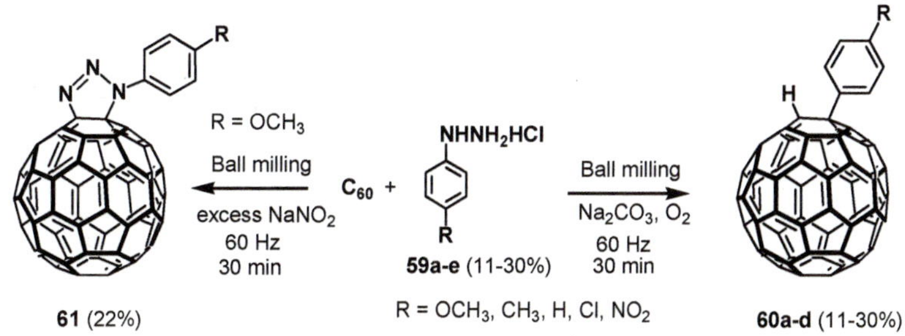

SCHEME 7.16

Reaction of fullerene and phenylhydrazine hydrochlorides.

7.1.2.1 Oxygenation of Fullerene

Senna et al. carried out mechanochemical oxygenation of fullerene in agate or a stainless steel vessel of the vibrational mill (Fritsch, Pulverisette 0) with one ball made of either agate (170 g) or stainless steel (507 g), at 1 atm oxygen pressure. The reaction conversion and the average number of polyoxidized fullerene were increased with increasing the milling time (1–5 h) and ball material. The best conversion and yield (68% and 1.5%) was obtained using agate material after 5 h milling (Scheme 7.17) [32]. Grinding in stainless steel reaction vessel led to polymerization of fullerene epoxide. Polymerization also took place when $C_{60}O$ was milled under argon atmosphere for 5 h.

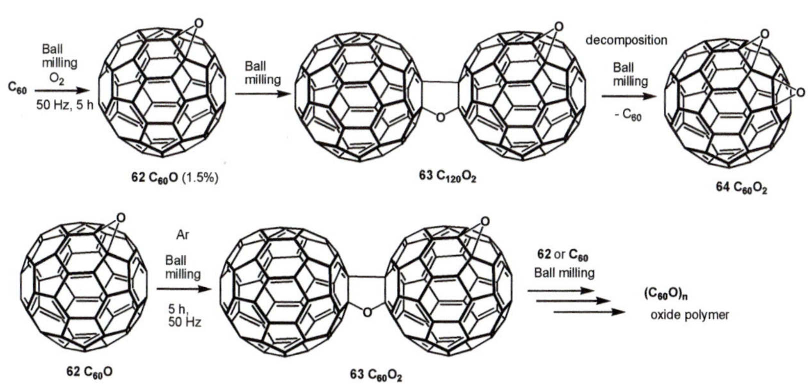

SCHEME 7.17

Mechanochemical oxygenation of fullerene.

7.2 FUNCTIONALIZATION OF SINGLE-WALLED CARBON NANOTUBES

Qin et al. used ball-milling technique to functionalize single-walled carbon nanotubes (SWCNTs) (Scheme 7.18) [33]. Attachment of alkyl chains was carried out by vigorous shaking of mixture of alkyl or aryl chloride, single-walled nanotubes (SWNTs) and potassium in nitrogen atmosphere for 40 min using Wig-L-Bug (Bratt 3110-B, stainless steel capsule and one ball). Mechanistically, electrons were transferred from potassium to SWNTs and the resulting anions **65** were then alkylated to produce derivatives **66**.

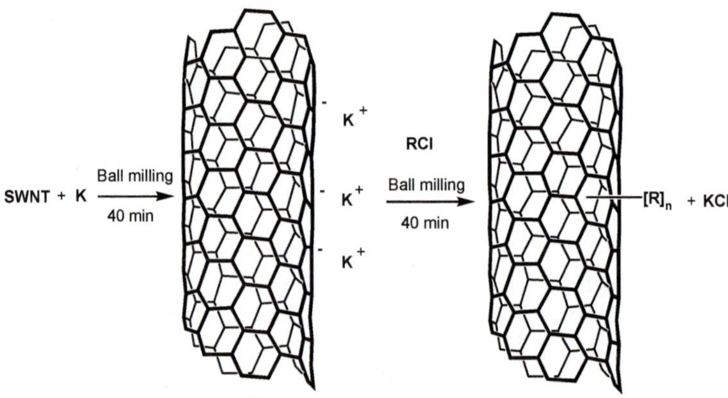

65

66a R = $CH_2(CH_2)_{12}CH_3$
66b R = C_6H_5

SCHEME 7.18

Mechanochemical functionalization of nanotubes.

Gentile et al. prepared multiwalled carbon nanotubes (MWCNTs) functionalized with maleated poly(propylene) (MAPP) by mechanochemical process (Scheme 7.19) [34]. Novel approach based on the use of a ball-milling technique has been optimized to graft-maleated polypropylene on the surface of MWCNTs to prepare CNT/polypropylene composites. Results suggest that oxidation phenomena occurred on the nanotube surface during prolonged ball milling in presence of MAPP, inducing the formation of carbonyl and carboxyl groups that can eventually interact with maleic anhydride groups via hydrogen bonds or other dipole–dipole interactions. Maleic anhydride groups contribute to the increase of the oxygen content by inducing oxidation on the nanotube surface during the ball-milling process.

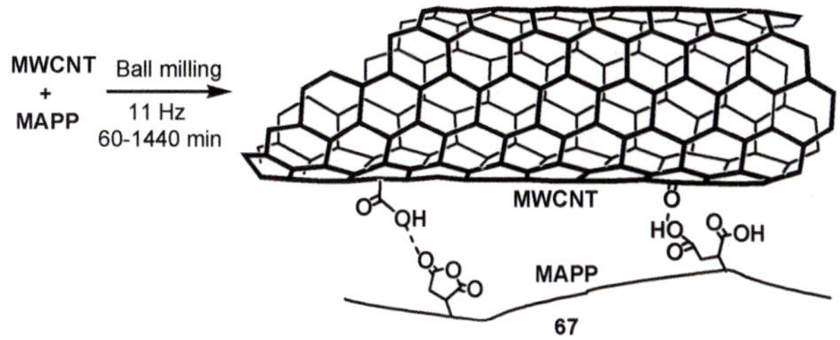

SCHEME 7.19

Interactions between maleic anhydride groups of MAPP MWCNTs.

Ball-milling method was also employed by other researchers for functionalization of SWNTs: functional groups such as thiol, amine, amide, carbonyl, and chlorine were introduced by milling in reactive atmospheres by Konya [35], alkyl halides were used by Barthos [36], whereas Qin employed KOH to prepare SWNTubols [37] and C_{60} to obtain C_{60}-modified SWNTs [38].

A high-speed vibration-milling method was also applied to synthesize water-dispersible single-walled carbon nanohorns (SWNHs) **68** [39–41]. Dorn et al. reported that highly reactive-free radicals ($HOOCCH_2CH_2$•) produced from an acyl peroxide under ball-milling conditions react with hydrophobic SWNHs to produce a highly water-dispersible derivative, a functionalized SWNH-(-$CH_2CH_2COOH)_y$, (f-SWNHs) **69** (Scheme 7.20), which has been characterized in detail by spectroscopic and microscopic techniques. The carboxylic acid–functionalized, water-dispersible SWNHs materials are versatile precursors that have potential applications in the biomedical area.

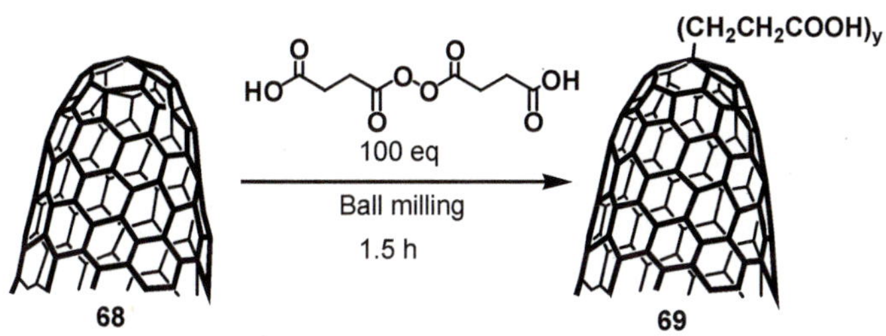

SCHEME 7.20

Functionalization of nanohorns.

7.3 FUNCTIONALIZATION OF GRAPHENES

It was recognized that due to the versatility of mechanochemical reactions driven by ball milling, various functional groups could be introduced at the broken edges of graphite in the ball mill. This methodology is a low-cost, high-yield, versatile approach to the mass production of graphene nanoplatelets (GnPs) with various desired functional groups.

Baek and coworkers prepared maleic anhydride (MA-GnPs), and maleimide-functionalized graphene nanoplatelets (MI-GnPs) were prepared by [4+2] cycloaddition via ball milling of graphite in the presence of MA or MI dienophiles (Scheme 7.21) [42]. Graphite flakes were crushed into graphene nanoplatelets by the kinetic energy of high-speed steel balls which cracks the graphitic C—C bonds and generates reactive carbon species, mostly radicals due to homolytic cleavages and ions (cations and anions) due to heterolytic cleavages along the broken edges, which may lead to promotion of [4+2] cycloaddition. Cycloaddition is expected to occur more along armchair edges rather than zigzag edges and the graphitic basal plane. The rest of the remnant active carbon species at the armchair and zigzag edges could be terminated by exposure to air and moisture. As a result, some oxygenated groups, such as hydroxyl and carboxylic groups, should also be introduced at the cracked edges of the resultant edge-selective functionalized GnPs (EFGnPs). Graphene functionalization was carried out in stainless steel container with balls (500 g, 5 mm) in planetary ball mill (Pulverisette 6, Fritsch). The resultant products were stirred with aqueous HCl to completely acidify the residual active species and to remove residual metallic impurities. Soxhlet extraction with water hydrolyzed the anhydride moieties at the edges of MA-GnPs into carboxylic acids.

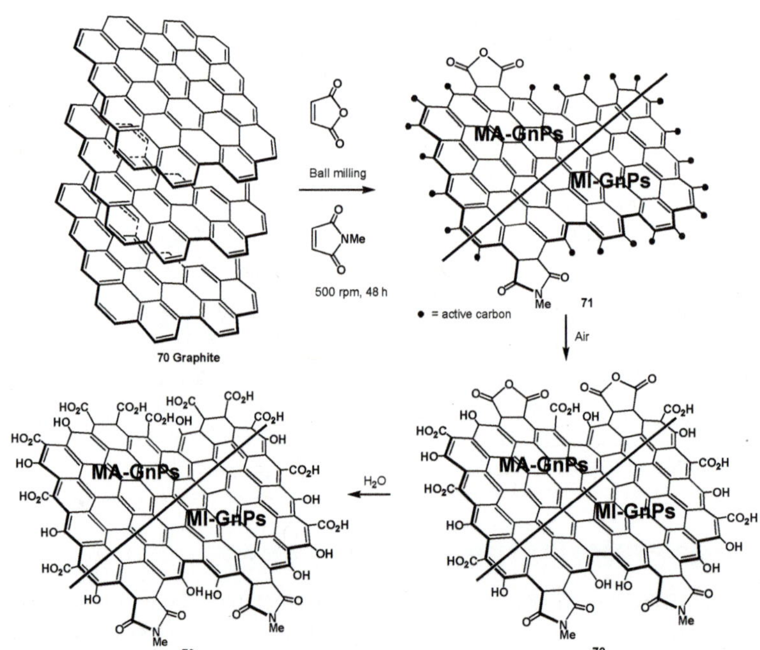

SCHEME 7.21

Diels–Alder reactions of graphene.

Baek et al. [43] prepared a series of EFGnPs **75** by ball-milling graphene in the presence of gases: hydrogen, dry ice [44], sulfur trioxide, or a dry ice/sulfur trioxide mixture (Scheme 7.22). With this process, a large amounts (typically 5.0g for each batch) of hydrogen- (HGnP), carboxylic acid- (CGnP), sulfonic acid (SGnP), and carboxylic acid/sulfonic acid functionalized (CSGnP) GnPs were prepared. The edge-selective functionalization during the ball-milling process involves the reaction between reactive carbon species (radicals and ions) generated by a mechanochemical cleavage of graphitic C—C bonds and gases introduced into a sealed ball mill (Scheme 7.23). The dormant active carbon species unreacted in the mill could be terminated by subsequent exposure to air moisture to introduce oxygenated groups, such as hydroxyl and carboxylic acid at the broken edges of the preformed EFGnPs.

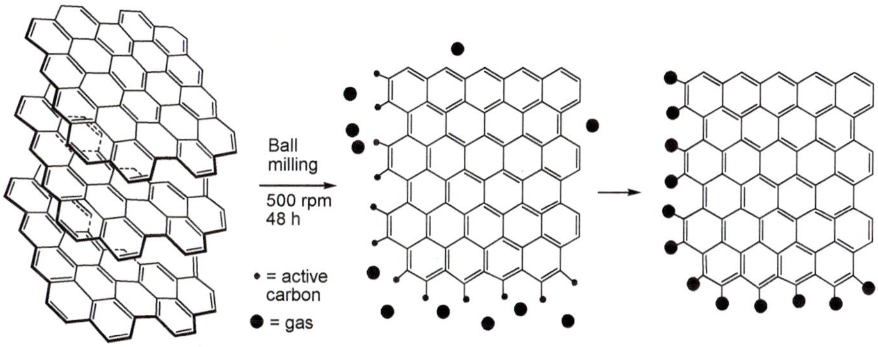

SCHEME 7.22

Reactions of gases with graphene.

The ball milling of graphite was carried out in the presence of corresponding gas in a planetary ball mill using stainless steel capsule containing balls (5 mm). Capsules were charged with graphite (5 g) and hydrogen (10 bar), dry ice (100 g), sulfur trioxide (7 g), or dry ice/sulfur trioxide mixture. After milling, the built-up internal pressure was released through a gas outlet and container lid opened in air for the hydration by air moisture. The resultant products were extracted with aqueous HCl to completely acidify the residual active species and to remove metallic impurities.

This process was used by Mülhaupt to functionalize graphene with carbon dioxide [45]. Milling of graphite under argon and carbon dioxide pressure (13 bar) using a planetary ball mill and the milling chamber with 50 ZrO_2 balls (yttrium stabilized, 10 mm) at 250 rpm for the duration of 48 h. Similarly, dry-ball milling of graphite under nitrogen atmosphere enables the one-step synthesis of edge-functionalized nitrogen-doped graphene **86** (Scheme 7.24) [46]. The physical cracking of graphite flake in a ball-mill crusher (500 mL) containing stainless steel

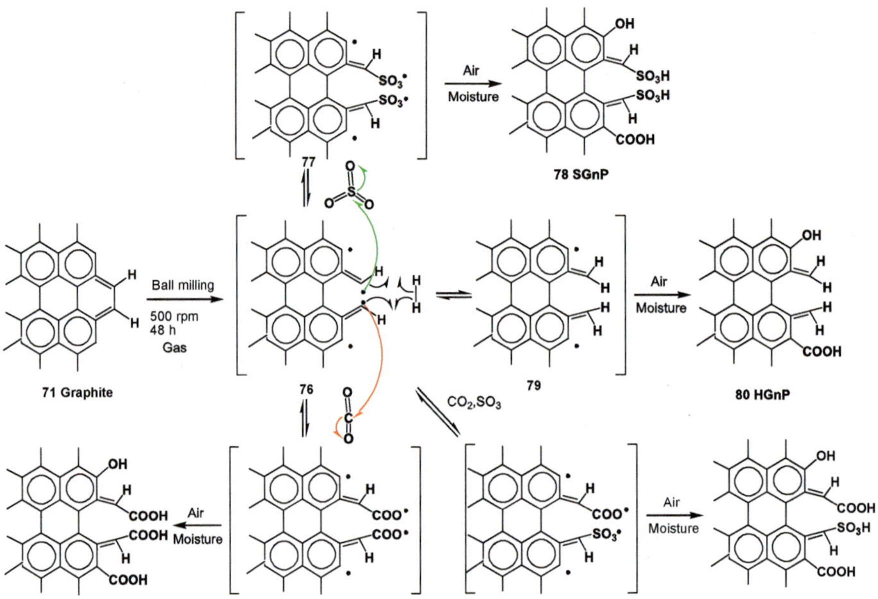

SCHEME 7.23

Mechanism of reactions of gases with graphene.

balls (500 g, 5 mm) in the presence of nitrogen and subsequent exposure to air moisture produced NGnPs **86**. The ring-formation mechanisms depend upon the cracking patterns of unzipped edges: the formation of five-membered pyrazole ring after the reaction between the active zigzag-edge carbon atoms and nitrogen; the formation six-membered pyridazine ring after the reaction between the active armchair-edge carbon atoms and nitrogen.

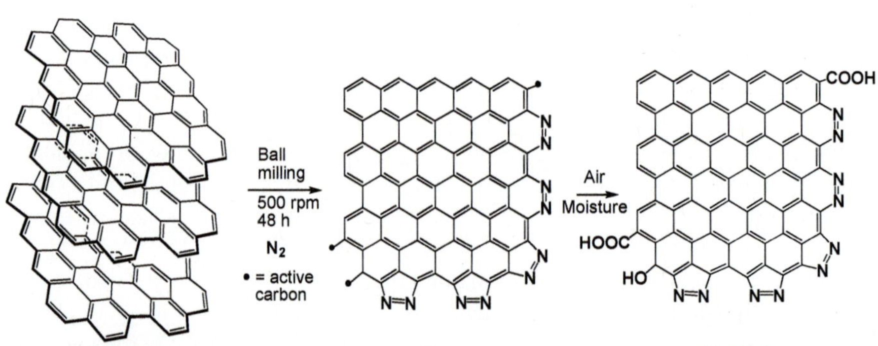

SCHEME 7.24

Graphene functionalization by nitrogen.

Baek's research group also applied ball milling of graphene with gaseous reagent for mechanochemical reduction of graphene oxide (GO) **87** by hydrogen gas (Scheme 7.25) [47].

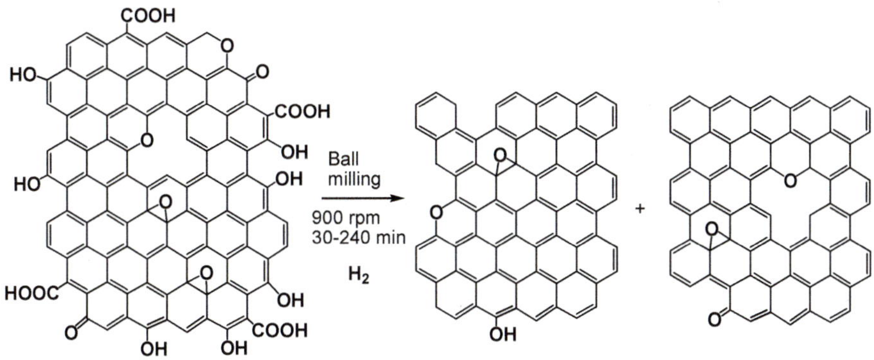

87 GO

88 BMRGO

SCHEME 7.25

Mechanochemical reduction of graphene oxide.

The ball milling of GO to produce reduced GO (BMRGO) **88** was carried out in a planetary micro-ball mill: 2 g of GO and hydrogen gas (10 bar) were charged into a stainless steel capsule containing balls (5 mm) and milled for operation time to obtain ball-milled reduced GO (ie, BMRGO) **88**, including BMRGO30, BMRGO60, BMRGO120, BMRGO180, and BMRGO240 (the number refers to 30, 60, 120, 180, and 240 min, respectively). The resultant powders were purified by extraction with diluted HCl to remove metallic impurities.

Solvent-free mechanochemical oxidation of graphite under comparatively mild conditions was used for the preparation of GO by Posudievsky et al. [48]. Milling of graphite and solid oxidant $KMnO_4$ or $(NH_4)_2S_2O_8$ in molar ratio 1:3 or 1:2, respectively, was effected in an agate grinding bowl using planetary ball mill Pulverisette 6 (Fritsch) for 3 h at a rotation rate of 500 rpm. Prepared GO have different oxidation degrees than those prepared by literature procedure.

REFERENCES

[1] Komatsu K. Organic solid state reactions. In: Toda F, editor. Topics in current chemistry, vol. 254. Berlin: Springer; 2005. p. 185–206. Wang G-W. in Encyclopedia of Nanoscience and Nanotechnology, vol. 3, (Nalwa HS ed.), American Scientific, Stevenson Ranch, 2004, pp. 557–65.

[2] Zhu S-E, Li F, Wang G-W. Mechanochemistry of fullerenes and related materials. Chem Soc Rev 2013;42:7535–70.

[3] Komatsu K, Murata Y. Synthesis of fullerene derivatives with novel structures - liquid-phase versus solid-state reactions. J Synth Org Chem Jpn 2004;72:1138–47.

[4] Murata Y, Kato N, Fujiwara K, Komatsu K. Solid-state [4+2] cycloaddition of fullerene C(60) with condensed aromatics using a high-speed vibration milling technique. J Org Chem 1999;64:3483–8.

[5] Komatsu K, Murata Y, Sugita N, Takeuchi K, Wan TSM. Use of naphthalene as a solvent for selective formation of the 1:1 Diels-Alder adduct of C_{60} with anthracene. Tetrahedron Lett 1993;52:8473–6.

[6] Ruoff RS, Tse DS, Malhotra R, Lorents DC. Solubility of C_{60} in a variety of solvents. J Phys Chem 1993;97:3379–83.

[7] Wang GW, Chen Z-X, Murata Y, Komatsu K. [60]Fullerene adducts with 9-substituted anthracenes: mechanochemical preparation and retro Diels-Alder reaction. Tetrahedron 2005;61:4851–6.

[8] Komatsu K, Murata Y, Fujiwara K, Suzuki M, Murata M. Synthesis of new fullerene dimers and open-cage fullerenes by solid-state and liquid-phase reaction of C_{60} with N-containing aromatics and with organosilicon compounds. Electrochem Soc Proc 2002;12:291–7.

[9] Warrener RN, Margetić D, Russell RA. The preparation of rigid alicyclic molecules bearing effector groups from alkene BLOCKs using *s*-tetrazines and 1,3,4-triazines as stereoselective coupling agents. Synlett 1998:585–7.

[10] Murata Y, Rubin Y, Komatsu K. Structure of the hydration product of the C_{60}-Di(2-pyridyl)-1,2,4,5-tetrazine adduct. Bull Chem Soc Jpn 2003;76:1669–72.

[11] Murata Y, Kato N, Komatsu K. The Reaction of fullerene C_{60} with phthalazine: the mechanochemical solid-state reaction yielding a new C_{60} dimer versus the liquid-phase reaction affording an open-cage fullerene. J Org Chem 2001;66:7235–9.

[12] Margetić D, Murata Y, Komatsu K, Marinić Ž. Rigid alicyclic molecules from bicyclo[2.2.1]hept-2-enes (=8,9,10-trinorbornenes) and 1,4-dipyridin-2-ylphthalazines as stereoselective coupling agents. Helv Chim Acta 2009;92:298–312.

[13] Komatsu K, Murata Y, Wang G-W, Tanaka T, Kato N, Fujiwara K. The solid-state mechanochemical reaction of fullerene C_{60}. Fuller Sci Technol 1999;7:609–20.

[14] Li Y-J, Peng R-F, Liu Y-C, Wang G-W. Solvent-free mechanochemical and liquid-phase reaction of [60]fullerene with ethyl 2-diazopropionate. Chin Chem Lett 2004;15:1265–8.

[15] Wang G-W, Li Y-J, Peng R-F, Liang Z-H, Liu Y-C. Are the pyrazolines formed from the reaction of [60]fullerene with alkyl diazoacetates unstable? Tetrahedron 2004;60:3921–5.

[16] Wang G-W, Zhang T-H, Hao E-H, Jiao L-J, Murata Y, Komatsu K. Solvent-free reactions of fullerenes and *N*-alkylglycines with and without aldehydes under high-speed vibration milling. Tetrahedron 2003;59:55–60.

[17] Margetić D. Mechanochemical organic reactions without the use of solvent. Kem Ind 2005;54:351–8.

[18] Komatsu K, Fujiwara K, Tanaka T, Murata Y. The Fullerene dimer C_{120} and related carbon allotropes. Carbon 2000;38:1529–34.

[19] Wang G-W, Komatsu K, Murata Y, Shiro M. Synthesis and X-ray structure of dumb-bell-shaped C_{120}. Nature 1997;387:583–6.

[20] Komatsu K, Fujiwara K, Murata Y. The mechanochemical synthesis and properties of the fullerene trimer C_{180}. Chem Lett 2000;29:1016–7.

[21] Kunitake M, Uemura S, Ito O, Fujiwara K, Murata Y, Komatsu K. Structural analysis of C_{60} trimers by direct observation with scanning tunneling microscopy. Angew Chem Int Ed 2002;41:969–72.

[22] Komatsu K, Wang G-W, Murata Y, Tanaka T, Fujiwara K. Mechanochemical synthesis and characterization of the fullerene dimer C_{120}. J Org Chem 1998;63:9358–66.

[23] Komatsu K, Fujiwara K, Murata Y. The Fullerene cross-dimer C_{130}: synthesis and properties. Chem Commun 2000:1583–4.

[24] Fujiwara K, Komatsu K. Mechanochemical synthesis of a novel C(60) dimer connected by a silicon bridge and a single bond. Org Lett 2002;4:1039–41.

[25] Murata Y, Han A, Komatsu K. Mechanochemical synthesis of a novel C_{60} dimer connected by a germanium bridge and a single bond. Tetrahedron Lett 2003;44:8199–201.

[26] Wang G-W, Komatsu K, Murata Y, Wan TSM. The solid-phase reaction of [60]fullerene: novel addition of organozinc reagents. Chem Commun 1996:2059–60.

[27] Zhang T-H, Wang G-W, Lu P, Li Y-L, Peng R-F, Liu Y, Murata Y, Komatsu K. Solvent-free reactions of C_{60} with active methylene compounds, either with or without carbon tetrabromide, in the presence of bases under high-speed vibration milling conditions. Org Biomol Chem 2004;2:1698–702.

[28] Peng R-F, Wang G-W, Shen Y-B, Li Y-J, Zhang T-H, Liu Y-C, Murata Y, Komatsu K. Solvent-free reactions of fullerenes with diethyl bromomalonate in the presence of inorganic bases under high-speed vibration milling conditions. Synth Commun 2004;34:2117–26.

[29] Cheng X, Wang G-W, Murata Y, Komatsu K. Solvent-free synthesis of dihydrofuran-fused [60]fullerene derivatives by high-speed vibration milling. Chin Chem Lett 2005;16:1327–9.

[30] Deng F, Wang G-W, Zhang T-H, Jiao L-J, Chen S. Ligand effects on the electrochemical and spectroscopic behaviors of methano[60]fullerene derivatives. Chem Commun 2004:1118–9.

[31] Chen Z-X, Zhu B, Wang G-W. Solvent-free mechanochemical reaction of [60]fullerene with phenylhydrazine hydrochlorides. Lett Org Chem 2008;5:65–8.

[32] Watanabe H, Matsui E, Ishiyama Y, Senna M. Solvent free mechanochemical oxygenation of fullerene under oxygen atmosphere. Tetrahedron Lett 2007;48:8132–7.

[33] Li X, Shi J, Qin Y, Wang Q, Luo H, Guo Z-X, Woo H-S, Park D-K. Alkylation and arylation of single-walled carbon nanotubes by mechanochemical method. Chem Phys Lett 2007;444:258–62.

[34] Ambrogi V, Gentile G, Ducati C, Oliva MC, Carfagna C. Multiwalled carbon nanotubes functionalized with maleated poly(propylene) by a dry mechano-chemical process. Polymer 2012;53:291–9.

[35] Kónya Z, Vesselenyi I, Niesz K, Kukovecz A, Demortier A, Fonseca A, Delhalle J, Mekhalif Z, Nagy JB, Koós AA, Osváth Z, Kocsonya A, Biró LP, Kiricsi I. Large scale production of short functionalized carbon nanotubes. Chem Phys Lett 2002;360:429–35.

[36] Barthos R, Méhn D, Demortier A, Pierard N, Morciaux Y, Demortier G, Fonseca A, Nagy JB. Functionalization of single-walled carbon nanotubes by using alkyl-halides. Carbon 2005;45:321–5.

[37] Pan H, Liu L, Guo Z-X, Dai L, Zhang F, Zhu D, Czerw R, Carroll DL. Carbon nanotubols from mechanochemical reaction. Nano Lett 2003;3:29–32.

[38] Li X, Liu L, Qin Y, Wu W, Guo Z-X, Dai L, Zhu D. C_{60} modified single-walled carbon nanotubes. Chem Phys Lett 2003;377:32–6.

[39] Shu C, Zhang J, Ge J, Sim JH, Burke BG, Williams KA, Rylander NM, Campbell T, Puretzky AA, Rouleau CM, Geohegan DB, More K, Esker AR, Gibson HW, Dorn HC. A Facile high-speed vibration milling method to water-disperse single-walled carbon nanohorns. Chem Mater 2010;22:347–51.

[40] Zhang J, Ge J, Schultz MD, Singh G, Shu C, Fatouros PP, Henderson SC, Corwin FD, Geohegan DB, Puretzky AA, Rouleau CM, More K, Rylander C, Rylander NM, Gibson HW, Dorn HC. In vitro and in vivo studies of single-walled carbon nanohorns with encapsulated metallofullerenes and exohedrally functionalized quantum dots. Nano Lett 2010;10:2843–8.

[41] Huang W, Zhang J, Dorn HC, Geohegan D, Zhang C. Assembly of single-walled carbon nanohorn supported liposome particles. Bioconjugate Chem 2011;22:1012–6.

[42] Seo J-M, Jeon I-Y, Baek J-B. Mechanochemically driven solid-state Diels-Alder reaction of graphite into graphene nanoplatelets. Chem Sci 2013;4:4273–7.

[43] Jeon I-Y, Choi H-J, Jung S-M, Seo J-M, Kim M-J, Dai L, Baek J-B. Large-Scale Production of edge-selectively functionalized graphene nanoplatelets via ball milling and their use as metal-free electrocatalysts for oxygen reduction reaction. J Am Chem Soc 2013;135:1386–93.

[44] Jeon I-Y, Shin Y-R, Sohn G-J, Choi H-J, Bae S-Y, Mahmood J, Jung S-M, Seo J-M, Kim M-J, Wook Chang D, Dai L, Baek J-B. Edge-carboxylated graphene nanosheets via ball milling. Proc Natl Acad Sci 2012;109:5588–93.

[45] Beckert F, Trenkle S, Thomann R, Mülhaupt R. Mechanochemical route to functionalized graphene and carbon nanofillers for graphene/SBR nanocomposites. Macromol Mater Eng 2014;12:1513–20.

[46] Jeon I-Y, Choi H-J, Ju MJ, Choi IT, Lim K, Ko J, Kim HK, Kim JC, Lee J-J, Shin D, Jung S-M, Seo J-M, Kim M-J, Park N, Dai L, Baek J-B. Direct nitrogen fixation at the edges of graphene nanoplatelets as efficient electrocatalysts for energy conversion. Sci Rep 2013;3. Article No. 2260.

[47] Chang DW, Choi H-J, Jeon I-Y, Seo J-M, Dai L, Baek J-B. Solvent-free mechanochemical reduction of graphene oxide. Carbon 2014;77:501–7.

[48] Posudievsky OY, Khazieieva OA, Koshechko VG, Pokhodenko VD. Preparation of graphene oxide by solvent-free mechanochemical oxidation of graphite. J Mater Chem 2012;22:12465–7.

CHAPTER

8 Applications of Ball Milling in Supramolecular Chemistry

CHAPTER OUTLINE

Weak interactions which are characteristic for supramolecular chemistry have been found to also operate in ball-milling conditions. It was demonstrated that supramolecular structures held by weak noncovalent interactions such as co-crystals, cages, rotaxanes, open metal–organic frameworks, and noncovalent organic inclusion complexes could be built by self-assembly processes in the solid state [1].

Several examples of employment of weak interactions in solid-state organic synthesis were given in previous chapters. Their importance in supramolecular catalysis is shown by Juaristi for asymmetric aldol reaction, where an improved stereoinduction under solvent-free conditions was obtained (see chapter: Carbon–Carbon Bond Forming Reactions) [2]. Results were rationalized as an effect of transition state stabilization between catalyst and aldehyde by hydrogen bonding and π–π stacking interactions. In solution these interactions are weakened by solvation and competition of solvent and aldehyde substrate. Similar enhancement of hydrogen-bonding interaction effects in solid-state reactions were observed in asymmetric Michael addition reaction using chiral squaramide catalyst by Xu [3] and chiral thiourea catalysts by Bolm [4] and Šebesta [5]. Additional example is formation of eutectic phase via charge-transfer complexes of benzoquinone and catalytic amounts of additives which form a strong hydrogen bonds (2-naphthol, (*rac*)-1,10-bis-2-naphthol [6], and thymol [7]), being crucial to accelerate solid-state Diels–Alder reactions (see chapter: Cycloaddition Reactions). Senna reported that this complex formation led to better fusion of two reactants.

Supramolecular interactions were also utilized in synthesis of carbon nanotube/polypropylene composites (see chapter: Applications of Ball Milling in Nanocarbon Material Synthesis). Solid-state grafting of maleated polypropylene on the surface of multiwalled carbon nanotubes (MWCNTs) employs interactions of carbonyl and carboxyl groups with maleic anhydride groups via hydrogen bonds or other dipole–dipole interactions [8].

Supramolecular host–guest systems could be formed in ball-milling conditions by self-assembly as well. For instance, Hapiot rationalized a face-selective tosylation of cyclodextrins by the formation of supramolecular inclusion complex between CD and the reagent (see chapter: Carbon–Oxygen and Other Bond Formation Reactions) [9].

Mechanochemical Organic Synthesis. http://dx.doi.org/10.1016/B978-0-12-802184-2.00008-X

8.1 ROTAXANES

The work published by Chiu et al. who employed mechanochemical solvent-free conditions for preparation of [2]rotaxane (Scheme 8.1) demonstrates that the noncovalent interactions stabilizing the recognition of the macrocycle by the dumbbell-shaped salt (N—H·O hydrogen bonding and C—H·π interactions) also operate in ball-milling conditions [10]. General synthetic methodology was established, at first, starting components were mixed in solution to prepare [2]pseudorotaxane complex **3** and solvent removed. This complex was then subjected to the solvent-free ball-milling reaction with 2 equivalents of 1,8-diaminonaphthalene **4** to obtain the [2]rotaxane **5** within 1 h in 80% yield, in which resulting reaction of aldehyde and diamine **4** provided dihydropyrimidine stopper units. Important advantage of ball milling for rotaxane formation is that the solution reaction of three components in acetonitrile did not proceed as efficiently as it did through ball milling: traces of [2]pseudorotaxane complex remained detectable after 24 h. When three components were mixed without first generating the complex **3**, the identical ball-milling conditions afforded a mixture of the [2]rotaxane **5** and the salt **6**. This process is by far the most efficient synthesis of a rotaxane by direct grinding of the macrocyclic, thread-like, and stoppering components as independent solids. This remarkable discovery that the pseudorotaxane complexes are stable in ball-milling conditions was employed in effective synthesis of [4]rotaxanes. In all of [2]rotaxane syntheses described, Retsch MM 200 mill (stainless steel vials 5 mL and 2×7 mm balls) was used.

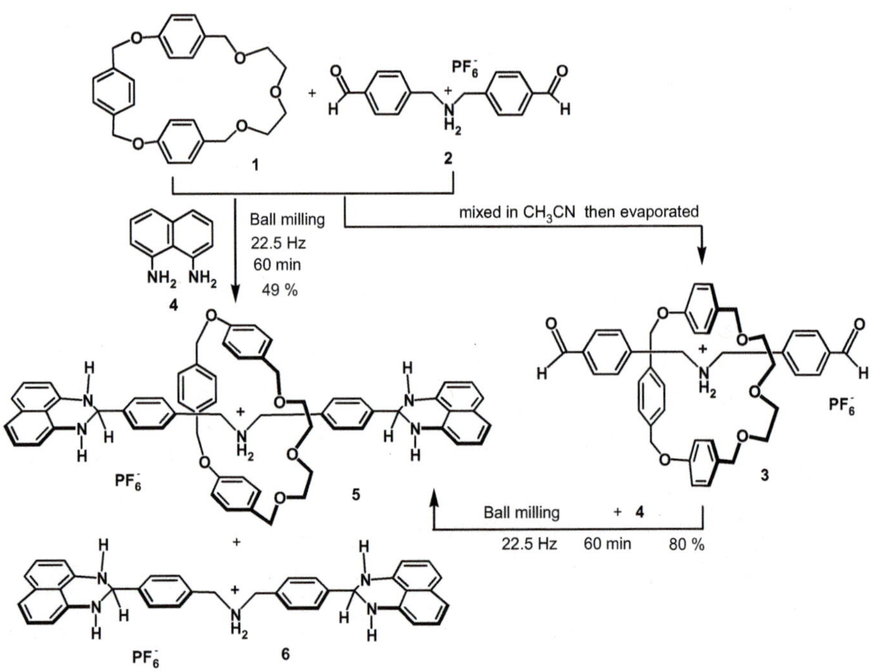

SCHEME 8.1

Synthesis of [2]rotaxanes.

The same research group reported on synthesis of [2]rotaxane employing Diels–Alder reaction of terminal alkyne with 1,2,4,5-tetrazine to provide stopper units (Scheme 8.2) [11]. Synthetic concept is identical as in the previous example. Two components in acetonitrile after evaporation gave pseudorotaxane **9**, which was subjected to ball milling with *sym*-tetrazine **10**. After 9h, [2]rotaxane **11** was obtained in high yield. Silica gel was added to facilitate transformation in solid state. Under mechanochemical solvent-free conditions, small, but sufficiently bulky pyridazine rings were used for stoppering the pseudorotaxanes.

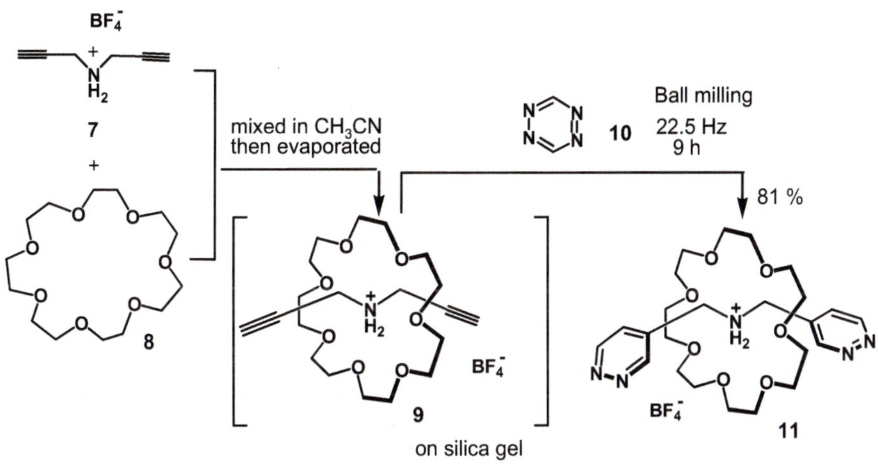

SCHEME 8.2

Synthesis of [2]rotaxanes by Diels–Alder reaction.

Poor regioselectivity was observed in ball milling of unsymmetrical hetero[2]pseudorotaxane possessing terminal alkynes (Scheme 8.3) [12]. Grinding of an equimolar mixture of this solid [2]pseudorotaxane **12** and solid tetrazine **10** for 6h gave a mixture of the [2]rotaxanes **13** and **14** as well as the ammonium salt **15**, suggesting that both the passage of 2-crown-7 onto the threadlike salt and the Diels–Alder reactions at its termini proceeded efficiently through this sequential treatment process, albeit with poor reaction selectivity of tetrazine toward the two terminal alkyne units. Grinding reaction of diphenyltetrazine **16** and the solid [2]pseudorotaxane did not proceed, even after 6h of ball milling; after heating this ground solid mixture for 72h, ammonium salt **17** was obtained as the predominant product. This result suggested that the phenylacetylene terminus of the [2]pseudorotaxane was more reactive toward diphenyltetrazine than was the propargyl terminus during this solid-to-solid heating process.

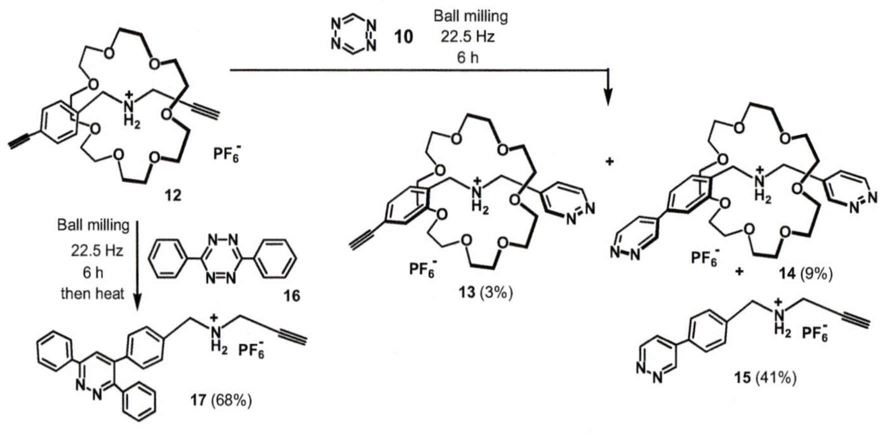

SCHEME 8.3

Synthesis of hetero[2]rotaxanes.

Diphenyltetrazine was successfully employed as a larger-size stopper in mechanosynthesis of [2]rotaxanes **20a,b** (Scheme 8.4) [13]. Its lower reactivity in comparison to 1,2,4,5-tetrazine did not produce any [2]rotaxane after ball milling for 9 h and Diels–Alder reaction was effected by heating of well-ground solid mixture of [2]pseudorotaxane **19** and 2 equivalents of tetrazine **18**. Silica gel was used as a milling auxiliary to convert sticky liquid to a solid. Similar reactions carried out in solution (for 3 days) did not proceed as efficiently and in the solid state, leading to a complicated mixture of products. The generality of methodology was demonstrated by synthesis of [2]rotaxanes **22a,b** starting with dibenzo[24]crown-8 (Scheme 8.5).

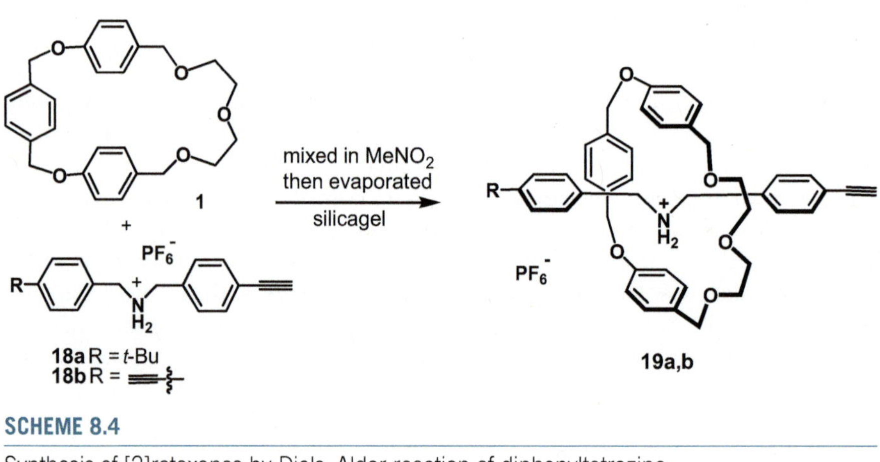

SCHEME 8.4

Synthesis of [2]rotaxanes by Diels–Alder reaction of diphenyltetrazine.

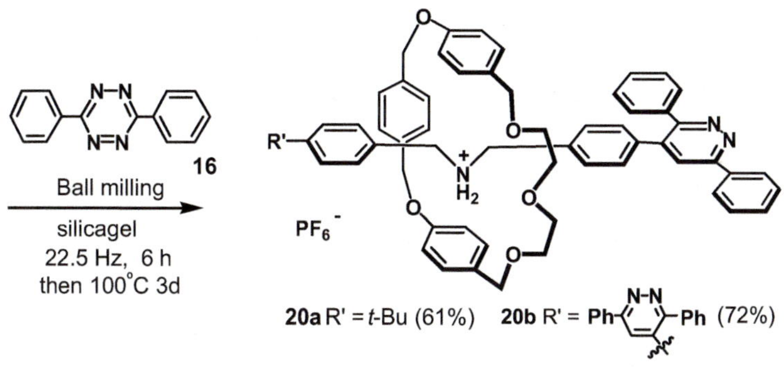

SCHEME 8.4 Continued.

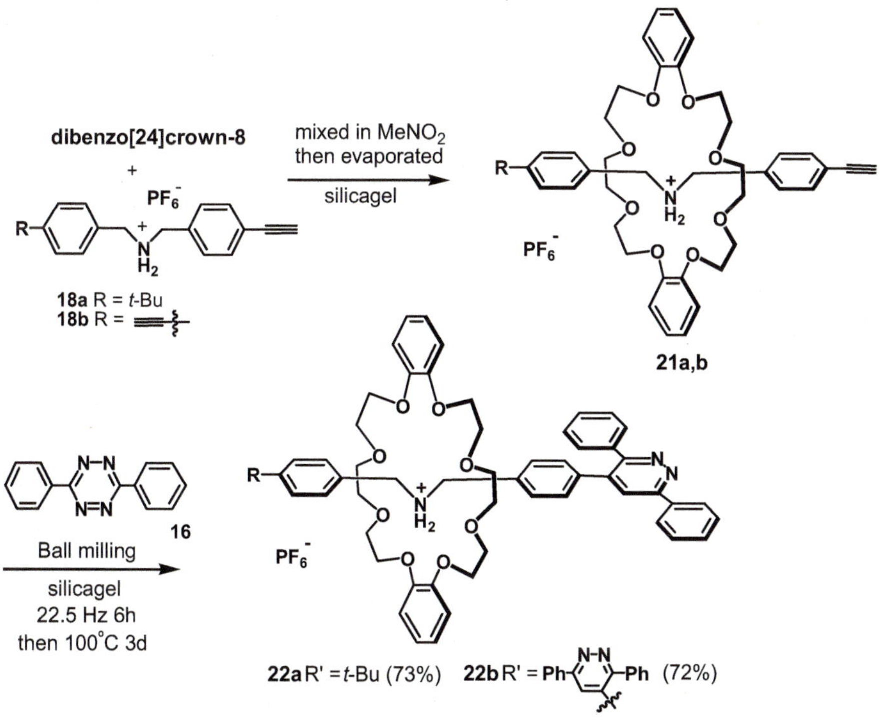

SCHEME 8.5

Synthesis of [2]rotaxanes by Diels–Alder reaction of diphenyltetrazine.

8.2 COMPLEXATION OF FULLERENES

Complexation of fullerene C_{60} with γ-cyclodextrin was attained by employment of high-speed vibrational milling (HSVM) by Komatsu et al. (Scheme 8.6) [14]. Fullerene C_{60} was complexed within 10 min milling with fourfold excess of cyclodextrin, forming a 1:2 complex **25** with a bicapped structure. The solid-state mechanochemical technique was found to be more effective for complexation in contrast to conventional solution phase (reflux, 24 h). Complex **25** exhibits the highest solubility in water than the complex prepared by other methods. It was found that shorter (5 min) or longer (20 min) milling afford complex which is soluble in water to lesser extent. This high-energy milling was more effective than complexation by classical ball milling which was carried out for about 20 h (Fritsch Pulverisette 701, agate vial 50 mL, ball 15 mm) [15].

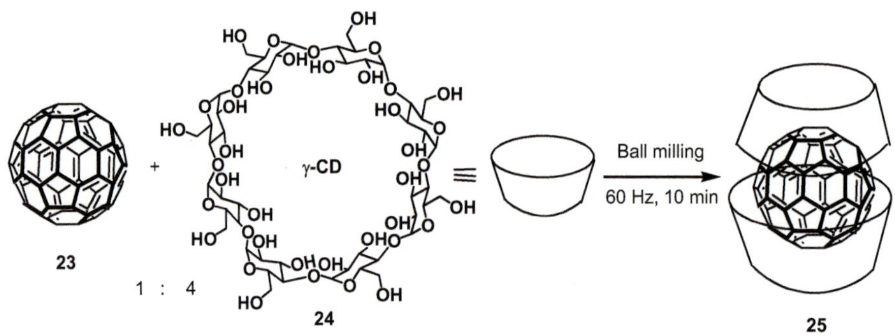

SCHEME 8.6

Fullerene complexation with cyclodextrin.

Equally effective was the complexation of C_{70} by HSVM (10 min, C_{70}:γ-cyclodextrin 1:8), while the other functionalized C_{60} derivatives afforded water-soluble 1:1 complexes, since functional groups prevent the formation of the bicapped 1:2 complexes. By mechanochemical treatment, fullerene C_{60} also complexes with sulfocalix[8]arene (10 min, equimolar amounts), and the obvious advantage of the solid-state complexation is illustrated by the complexation of sulfocalix [8]arene with fullerene dimer (also prepared by mechanosynthesis, see chapter: Applications of Ball Milling in Nanocarbon Material Synthesis). Dimer is hardly soluble in most common organic solvents, but the mechanochemical treatment of an equimolar mixture of sulfocalix[8]arene and fullerene dimer for 10 min afforded bicapped complex, which is about twice more soluble in water than in commonly used ODCB.

Besides γ-cyclodextrin and sulfocalix[8]arene, Geckeler has found that C_{60} forms a complex with cucurbit[7]uril in mechanochemical conditions (Scheme 8.7) [16]. Retsch MM200 mill was employed, in stainless steel capsule (inner diameter 18 mm) and ball (12 mm) at 20 Hz. Irrespective of the molarity of two components, a supramolecular 1:1 complex **27** was formed through noncovalent binding after 4 h in 77% yield, which was more effective than solution preparation (96 h, 55%).

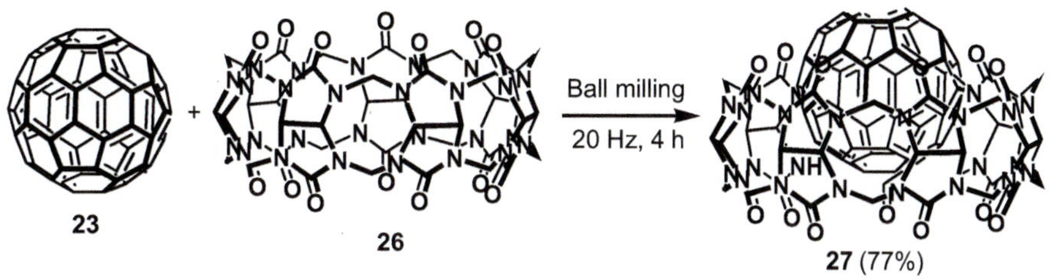

SCHEME 8.7

Fullerene complexation with cucurbit[7]uril.

REFERENCES

[1] Friščić T. Supramolecular concepts and new techniques in mechanochemistry: cocrystals, cages, rotaxanes, open metal-organic frameworks. Chem Soc Rev 2012;41:3493–510.

[2] Hernández J, Juaristi E. Asymmetric aldol reaction organocatalyzed by (*S*)-proline-containing dipeptides: improved stereoinduction under solvent-free conditions. J Org Chem 2011;76:1464–7.

[3] Wang Y-F, Chen R-X, Wang K, Zhang B-B, Li Z-B, Xu D-Q. Fast, solvent-free and hydrogen-bonding-mediated asymmetric Michael addition in a ball mill. Green Chem 2012;14:893–5.

[4] Jörres M, Mersmann S, Raabe G, Bolm C. Organocatalytic solvent-free hydrogen bonding mediated asymmetric Michael additions under ball milling conditions. Green Chem 2013;15:612–6.

[5] Hestericovà M, Šebesta R. Higher enantioselectivities in thiourea-catalyzed Michael additions under solvent-free conditions. Tetrahedron 2014;70:901–5.

[6] Watanabe H, Senna M. Acceleration of solid state Diels–Alder reactions by incorporating the reactants into crystalline charge transfer complexes. Tetrahedron Lett 2005;46:6815–8.

[7] Watanabe H, Hiraoka R, Senna M. Diels–Alder reaction catalyzed by eutectic complexes autogenously formed from solid state phenols and quinones. Tetrahedron Lett 2006;47:4481–4.

[8] Ambrogi V, Gentile G, Ducati C, Oliva MC, Carfagna C. Multiwalled carbon nanotubes functionalized with maleated poly(propylene) by a dry mechano-chemical process. Polymer 2012;53:291–9.

[9] Menuel S, Doumert B, Saitzek S, Ponchel A, Delevoye L, Monflier E, Hapiot F. Selective secondary face modification of cyclodextrins by mechanosynthesis. J Org Chem 2015;80:6259–66.

[10] Hsueh S-Y, Cheng K-W, Lai C-C, Chiu S-H. Efficient solvent-free syntheses of [2]- and [4]rotaxanes. Angew Chem Int Ed 2008;47:4436–9.

[11] Hsu C-C, Chen N-C, Lai C-C, Liu Y-H, Peng S-M, Chiu S-H. Solvent-free synthesis of the smallest rotaxane prepared to date. Angew Chem Int Ed 2008;47:7475–8.

[12] Chen P-N, Lai C-C, Chiu S-H. Self-sorting under solvent-free conditions: one-pot synthesis of a hetero[3]rotaxane. Org Lett 2011;13:4660–3.

[13] Hsu C-C, Lai C-C, Chiu S-H. Using Diels–Alder reactions to synthesize [2]rotaxanes under solvent-free conditions. Tetrahedron 2009;65:2824–9.
[14] Komatsu K, Fujiwara K, Murata Y, Braun T. Aqueous solubilization of crystalline fullerenes by supramolecular complexation with γ-cyclodextrin and sulfocalix[8] arene under mechanochemical high-speed vibration milling. J Chem Soc Perkin Trans 1999;1:2963–6.
[15] Braun T, Buvári-Barcza Á, Barcza L, Konkoly-Thege I, Fodor M, Migali B. Mechanochemistry: a novel approach to the synthesis of fullerene compounds. Water soluble buckminsterfullerene-γ-cyclodextrin inclusion complexes via a solid-solid reaction. Solid State Ionics 1994;74:47–51.
[16] Constabel F, Geckeler KE. Solvent-free self-assembly of C_{60} and cucurbit[7]uril using high-speed vibration milling. Tetrahedron Lett 2004;45:2071–3.

CHAPTER

Experiments for Introduction of Mechanochemistry in the Undergraduate Curriculum

9

CHAPTER OUTLINE

9.1 GREEN CHEMISTRY AT THE UNIVERSITY LEVEL

Education of organic chemists at the university level requires the corresponding curricula to be regularly updated with the latest developments in the field of synthetic methodologies. During 2000, many universities throughout the world have implemented the concept of green chemistry in their curricula to increase the awareness among chemistry students about the importance of changing approach toward

Mechanochemical Organic Synthesis. http://dx.doi.org/10.1016/B978-0-12-802184-2.00009-1

synthesis of new compounds and the design of chemical processes in general [1]. In this respect, green chemistry experiments based on the concepts of solvent-free reactions, and also of reactions in aqueous media in contrast to traditional synthesis in organic solvents have become an integral part of teaching courses on experimental organic chemistry [2]. Typical examples of organic reactions initiated and carried out by manual grinding in a mortar with pestle already incorporated in the undergraduate curricula are aldol condensation, Michael addition [3], and Wittig reaction [4]. However, when it comes to introduction of automated ball-milling techniques at the undergraduate level, there is a noticeable lack of such experiments in organic chemistry laboratory programs. This can be due to the fact that several years back, these techniques were only sporadically employed in organic synthesis. An enormous progress made over the years and the growing interest in alternative environmentally friendly methodologies have brought instrumental mechanochemistry into the focus of contemporary research in preparative organic chemistry. In parallel, the milling instruments have become widely available, both in research and teaching laboratories. For that reason, the main goal of this chapter is to demonstrate the application of automated ball milling in organic synthesis as a "green" means to conduct reactions under conditions of controlled energy input.

9.2 SELECTED EXPERIMENTS

In the following sections, a selection of several mechanochemical syntheses of different organic compounds is proposed to introduce practical organic mechanochemistry in the undergraduate teaching courses. The intention is to familiarize undergraduate students with the potential of mechanochemical milling as an environmentally friendly alternative to conventional solution-based synthesis of organic molecules. The reactions were selected based on the type (carbon–carbon coupling, C—H functionalization, carbon–nitrogen coupling, metal–ligand complexation, and cycloaddition reaction), mode (one-pot and multistep synthesis), time (10 min–3 h), high yields (typically ≥90%), simple workup (scraping the product as synthesized, aqueous workup, and filtration or flash chromatography if necessary), and different milling equipment (planetary and mixer mills, stainless steel and tungsten carbide jars and balls). These parameters should make the proposed experiments fit into the average time frame of an organic chemistry laboratory (4–6 h). All necessary safety precautions required while working in an organic chemistry laboratory need to be strictly followed and chemicals handled in accordance with the directions stated in the material safety data sheet (MSDS).

9.2.1 SUZUKI COUPLING [5]

Suzuki coupling is a prototypical metal-catalyzed carbon–carbon forming reaction normally conducted in an organic solvent under anaerobic conditions. In this experiment, the ligand-free version of the Suzuki reaction is carried out in air by completely avoiding solvents in the synthesis step, while the reaction time is drastically reduced to only 10 min. Also, the students are introduced with the concept of adding an inert solid auxiliary such as sodium chloride to adsorb liquid reagents and facilitate the mass transfer during milling.

9.2.1.1 4-Phenyltoluene (3)

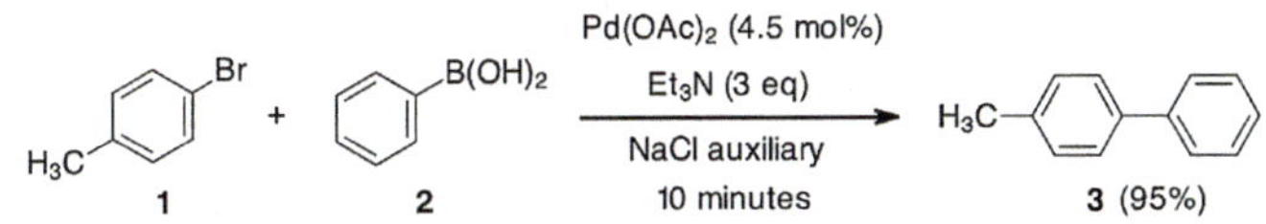

SCHEME 9.1

Suzuki coupling.

Procedure: 4-Bromotoluene (**1**) (171 mg, 123 μL, 1 mmol), phenylboronic acid (**2**) (122 mg, 1 mmol), $Pd(OAc)_2$ (10 mg, 0.045 mmol, 4.5 mol%), and triethylamine (410 μL, 3 mmol) were placed in a 10 mL stainless steel grinding jar. Then c. 2 g of sodium chloride was slowly added to the reaction mixture with agitation until a damp paste was formed. The consistency of the paste was such that no liquid remained in the jar. A 10 mm grinding ball was then added and the reaction mixture ground with a Retsch MM200 mill operating at 30 Hz. After 10 min of milling, the crude product was scraped into a flask and the jar rinsed with a few milliliters of water and then diethylether. More water (30 mL) and ether (30 mL) were poured to the flask and the entire contents transferred to a separating funnel. Then 2 M HCl (40 mL) was added and after agitation, the organic layer was separated, dried over anhydrous $MgSO_4$, filtered and concentrated by rotary evaporation. Flash chromatography on silica gel using hexane as eluent afforded pure biphenyl product **3** in 95% yield (Scheme 9.1).

^{1}H NMR (400 MHz, $CDCl_3$): $\delta = 7.46$ (d, 2H, $J = 7.6$ Hz), 7.38 (d, 2H, $J = 8.0$ Hz), 7.32–7.29 (m, 2H), 7.20 (t, 1H, $J = 7.6$ Hz), 7.13 (d, 2H, $J = 7.6$ Hz), 2.28 (s, 3H); ^{13}C NMR (100 MHz, $CDCl_3$): $\delta = 141.1$, 138.3, 137.0, 129.5, 128.7, 127.0, 126.9, 21.1.

9.2.2 RH-CATALYZED C—H FUNCTIONALIZATION [6]

This experiment illustrates the mechanochemical synthesis of a rhodium catalyst under high-energy milling conditions using tungsten carbide–grinding jars and balls, which is then employed in the mechanochemical metal-catalyzed C—H functionalization. The students will learn about the application of ball milling in organometallic chemistry and the possibility to efficiently conduct metal-catalyzed reactions. In case of short time, the experiment can be divided into two parts: synthesis of the rhodium catalyst and functionalization of phenylpyridine substrate.

9.2.2.1 Di-μ-chloro-dichlorobis(η5-pentamethylcyclopentadienyl) dirhodium(III) (5)

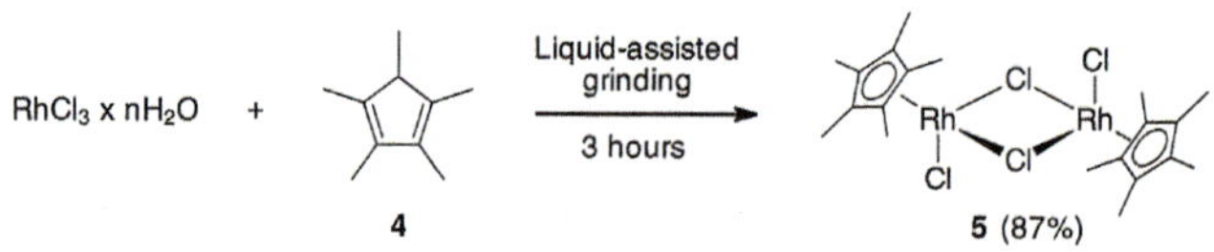

SCHEME 9.2

Mechanosynthesis of Rh complex.

Procedure: A mixture of $RhCl_3{\cdot}(H_2O)_n$ (0.15 g, 0.576 mmol) and 1,2,3,4,5-pentamethyl- cyclopentadiene (**4**) (0.236 g, 1.736 mmol, 3.0 equiv.) was milled under LAG conditions (η = 0.25 μL mg^{-1} of methanol) in a 12 mL tungsten carbide milling jar with 20 tungsten carbide milling balls (5 mm in diameter) in a Fritsch Pulverisette 7 micro planetary mill at 800 rpm for 3 h. After the milling was completed, Rh dimer **5** was isolated by column chromatography using silica and a mixture of DCM:MeOH (v:v = 50:1) as eluent (Scheme 9.2).

Dark red solid, m.p. > 350°C (decomp.); ^{1}H NMR (300 MHz, $CDCl_3$): δ = 1.62 (s, 30H); ^{13}C NMR (150 MHz, $CDCl_3$): δ = 94.3 (d, $J_{Rh\text{-}C}$ = 9 Hz), 9.6; IR (KBr): v = 2975, 2910, 1458, 1371, 1162, 1022, 733 cm^{-1}.

9.2.2.2 2-(2,6-Diiodophenyl)pyridine (8)

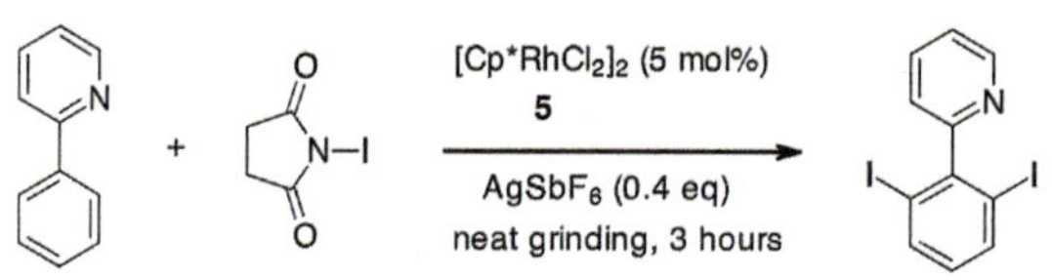

SCHEME 9.3

C—H functionalization of 2-phenylpyridine.

Procedure: A mixture of 2-phenylpyridine (**6**) (54.3 mg, 0.349 mmol), *N*-iodosuccinimide (**7**) (172.8 mg, 0.768 mmol), $AgSbF_6$ (48.07 mg, 0.139 mmol, 0.4 equiv.), and $[Cp^*RhCl_2]_2$ (**5**) (10.81 mg, 0.017 mmol, 5.0 mol%) was milled in a 10 mL stainless steel milling jar with one 10 mm stainless steel ball in a Retsch MM400 mixer mill at 30 Hz for 3 h. After the milling was completed, the product was isolated by column chromatography using silica and a mixture of pentane:EtOAc (v:v = 15:1) as eluent (Scheme 9.3).

Off-white solid, m.p. > 123–125°C; ^{1}H NMR (600 MHz, $CDCl_3$): δ=8.75 (ddd, 1H, J = 4.8, 1.1, 1.0 Hz), 7.92 (d, 2H, J = 8.1 Hz), 7.81 (dt, 1H, J = 7.7, 1.8 Hz), 7.34 (ddd, 1H, J = 7.7, 5.0, 1.1 Hz), 7.25 (dt, 1H, J = 7.8, 1.0 Hz), 6.74 (t, 1H, J = 8.0 Hz); ^{13}C NMR (150 MHz, $CDCl_3$): δ = 164.4, 149.5, 148.4, 139.2, 136.7, 131.3, 124.2, 123.4, 97.0; IR (KBr): v = 2923, 2328, 1748, 1560, 1405, 1265, 1179, 991, 757 cm^{-1}.

9.2.3 IMINE (SCHIFF BASE) SYNTHESIS AND METAL COMPLEXATION [7]

The manual grinding synthesis of imines, also known as the Schiff bases, has already found its way into the undergraduate curriculum, but the automated version using ball mills is not common in organic lab textbooks. By conducting the reactions outlined below, the students will gain practical knowledge on one-pot two-step mechanochemical synthesis consisting of the quantitative preparation of salen-type tetradentate ligand which is in the next step employed for complexation of a metal cation. Except for solvent-free or neat grinding, the students are introduced with the liquid-assisted grinding (LAG) as a method to accelerate the mechanochemical reactions and drive them to completion, leading to quantitative yields. As such, these reactions do not require workup, and the products are simply scraped off the walls of the grinding jars, which makes them particularly suitable for implementation into the undergraduate curriculum.

9.2.3.1 2,2′-[1,2-Ethanediylbis[(E)-nitrilomethylidyne]]bis-phenol (11)

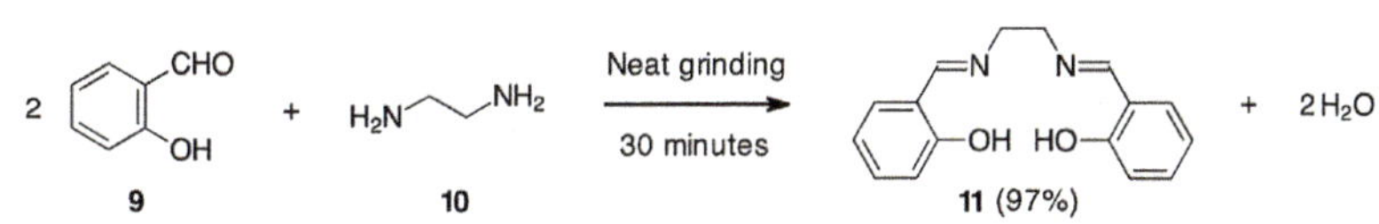

SCHEME 9.4

Mechanochemical synthesis of salen-type ligand.

Procedure: 2-Hydroxybenzaldehyde (**9**) (2.5 mL, 0.023 mmol) and ethylenediamine (**10**) (0.776 mL, 0.012 mmol) were added to a 25 mL screw-top grinding jar charged with single 15 mm stainless steel ball. The mixture was milled using a Retsch MM400 mill operating at 25 Hz for 30 min. The bright yellow product was scraped off the grinding jar walls and dried in the oven at 80°C for 36 h to obtain salenH_2 ligand **11** in 97% yield (2.99 g) (Scheme 9.4).

^{1}H NMR (300 MHz, $CDCl_3$): $\delta = 13.23$ (s, 2H), 8.36 (s, 2H), 7.32–7.22 (m, 5H), 6.96–6.93 (d, 2H, $J = 9$ Hz), 6.89–6.84 (t, 2H, $^2J = 7.5$ Hz, $^3J = 15$ Hz), 3.95 (s, 4H).

9.2.3.2 2,2′-[1,2-Ethanediylbis[(E)-nitrilomethylidyne]]-bis[phenolato]-κN,N′,O,O′-zinc(II) (12)

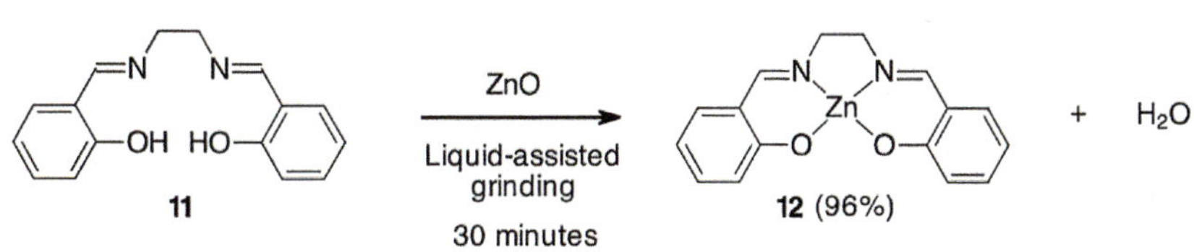

SCHEME 9.5

Mechanochemical complexation of zinc with salen-type ligand.

Procedure: 2-Hydroxybenzaldehyde (**9**) (2.5 mL, 0.023 mmol) and ethylenediamine (**10**) (0.776 mL, 0.012 mmol) were added to a 25 mL screw-top grinding jar charged with single 15 mm stainless steel ball and ground at 25 Hz for 30 min. Without isolation of the ligand **11**, zinc oxide (0.937 g, 0.012 mmol) and methanol (0.5 mL) were added and the reaction mixture was milled for a further 30 min. The pale yellow product was scraped off the grinding jar walls and dried in the oven at 80°C for 36 h to obtain Zn complex **12** in 96% yield (3.64 g) (Scheme 9.5).

^{1}H NMR (300 MHz, d_6-DMSO): $\delta = 8.43$ (s, 2H), 7.15–7.10 (m, 4 Hz), 6.62–6.60 (d, 2H, $J = 6$ Hz), 6.44–6.39 (t, 2H, $^{2}J = 6$ Hz, $^{3}J = 15$ Hz), 3.72 (s, 4H).

9.2.3.3 "All-at-Once Synthesis" of Zn Complex 12

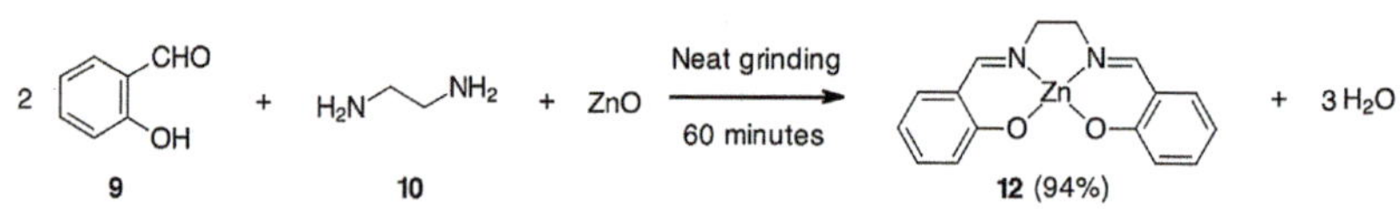

SCHEME 9.6

"All-at-once" synthesis of Zn–salen complex.

Procedure: 2-Hydroxybenzaldehyde (**9**) (2.5 mL, 0.023 mmol), ethylenediamine (**10**) (0.776 mL, 0.012 mmol), and zinc oxide (0.937 g, 0.012 mmol) were added to the grinding jar and set to shake at 25 Hz for 60 min. The pale yellow product was scraped off the grinding jar walls and dried in the oven at 80°C for 36 h to obtain Zn complex **12** in 94% yield (3.60 g) (Scheme 9.6).

^{1}H NMR (300 MHz, d_6-DMSO): $\delta = 8.43$ (s, 1H), 7.12 (dd, 2H, $J = 12.8$ Hz, $J = 4.7$ Hz), 6.61 (d, 1H, $J = 8.2$ Hz), 6.42 (t, 1H, $J = 7.3$ Hz), 3.72 (s, 2H).

9.2.4 AMIDE SYNTHESIS [8]

The rapid CDI-mediated synthesis of *N*-phenethylbenzamide provides students with an example of the mechanochemical activation of a carboxylic acid followed by amine addition, and coupled with simple aqueous workup. The ubiquity of amide functionality makes this type of chemical transformation an important teaching tool in preparative green organic chemistry, showcasing that amides can efficiently be accessed by routes other than solution-based ones. The students are expected to comprehend the concept of multistep milling and aqueous treatment of crude reaction mixtures and to develop knowledge on how to implement the green chemistry principles in practice through selection of the coupling reagent (CDI vs carbodiimides) and avoiding the use of organic solvents in the synthesis and purification steps.

9.2.4.1 N-phenethylbenzamide (15)

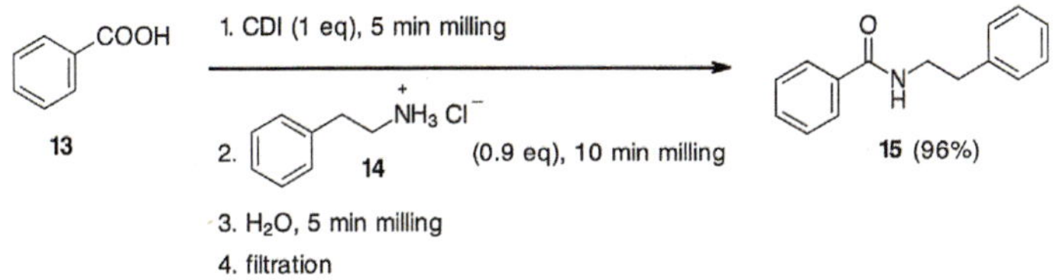

SCHEME 9.7

CDI-mediated mechanosynthesis of amide.

Procedure: Benzoic acid (**13**) (0.183 g, 1.5 mmol) and carbonyldiimidazole (CDI) (243 mg, 1.5 mmol) were introduced in a 12 mL stainless steel grinding bowl with 50 stainless steel balls (5 mm diameter). The bowl was closed and placed for 5 min within the Retsch PM100 planetary mill rotated at 500 rpm. Then, phenethylamine hydrochloride (**14**) (0.206 g, 1.31 mmol) was added and milling continued for 10 min at 500 rpm. After the amide-coupling reaction, 4 mL of deionized water was added, and the reaction mixture was milled at 500 rpm for additional 5 min. After this treatment, the fine suspension was filtered, washed with deionized water, and then dried under vacuo over P_2O_5 to obtain the amide product **15** in 96% yield (0.284 g) (Scheme 9.7).

Melting point 116–117°C; ^{1}H NMR (200 MHz, $CDCl_3$): δ=7.68–7.00 (m, 10H), 6.05 (s, 1H), 3.60–3.45 (m, 2H), 2.68 (t, 2H, J=6.5 Hz); ^{13}C NMR (300 MHz, $CDCl_3$): δ=167.7, 139.1, 134.8, 131.6, 129.0, 128.9, 128.8, 127.0, 126.8, 41.3, 35.9; MS (ESI): m/z 226.2 $[M+H]^+$.

9.2.5 DESYMMETRIZATION OF PHENYLENEDIAMINE: (THIO)UREA SYNTHESIS [9]

In the pool of different organic transformations amenable to mechanochemical treatment, the coupling reaction between amines and isothiocyanates leading to thioureas has been recognized as the example of an atom-efficient click reaction. Its simplicity, coupled with the inherently rigid structure of thioureas, their biological activity, and application as organocatalysts, anion sensors and transporters, makes the amine–isothiocyanate click coupling a suitable test reaction for exploring green chemistry principles of the solvent-free organic research laboratory at the undergraduate level. The proposed experiment demonstrates the utility of mechanochemistry to quantitatively desymmetrize small organic molecule by conducting a stoichiometric reaction (ie, avoiding reagents in excess). The mechanochemical desymmetrization will allow students to extend the scope of such click reactions by coupling the desymmetrized intermediate with an isocyanate reagent to afford a hybrid thiourea–urea framework. The set of learning objectives such as the efficiency of neat grinding versus LAG and the ability to conduct multistep milling synthesis will be further expanded by identifying mechanochemistry as a tool to increase the level of complexity in small organic molecules.

9.2.5.1 N[1]-(2-Aminophenyl)-N[2]-(4-chlorophenyl)thiourea (18)

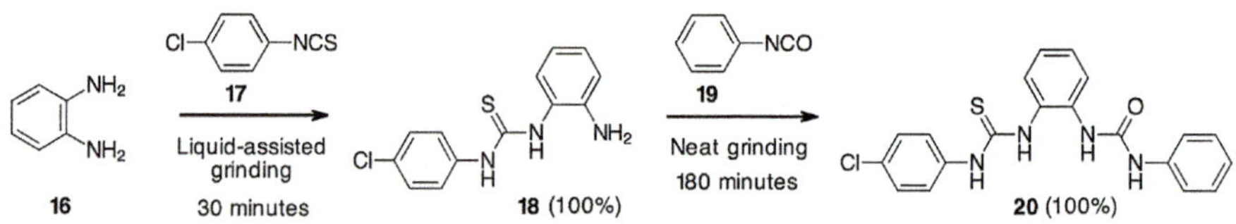

SCHEME 9.8

Mechanochemical desymmetrization of *o*-phenylenediamine.

Procedure (first step): An equimolar mixture of *o*-phenylenediamine (**16**) (0.45 mmol, 48.7 mg) and 4-chlorophenyl isothiocyanate (**17**) (0.45 mmol, 76.3 mg) was placed in a 10 mL grinding jar containing a stainless steel ball (12 mm diameter). The mixture was milled in the presence of 32 µL of dry methanol (η=0.25 µL mg^{-1}) at 30 Hz for 30 min using a Retsch MM400 mill. The product was scraped off the walls of the grinding jar affording thiourea **18** in quantitative yield (Scheme 9.8).

^{1}H NMR (300 MHz, d$_6$-DMSO): δ=4.90 (s, 2H, NH$_2$), 6.56 (t, 1H, J=7.5 Hz, Ar), 6.75 (d, 1H, J=8.0 Hz, Ar), 6.97 (t, 1H, J = 7.6 Hz, Ar), 7.07 (d, 1H, J=7.8 Hz, Ar), 7.35 (d, 2H, J=8.8 Hz, Ar), 7.55 (d, 2H, J = 8.8 Hz, Ar), 9.11 (s, 1H, NH), 9.61 (s, 1H, NH); ^{13}C NMR (75 MHz, d$_6$-DMSO): δ = 115.8, 116.2, 123.8, 125.1, 127.1, 127.9, 128.1 (2C overlapped), 138.7, 143.9, 180.1; HRMS-MALDI found: 278.0522; calc. for $C_{13}H_{13}ClN_3S$ (M+H$^+$): 278.0513.

9.2.5.2 N[1]-[N-(4-Chlorophenyl)thiocarbamoyl]-N[2]-[N-phenylcarbamoyl]-1,2-diaminobenzene (20)

Procedure (second step): An equimolar mixture of mono-thiourea **18** (0.45 mmol, 125.0 mg, prepared by milling) and phenyl isocyanate (**19**) (0.45 mmol, 53.6 mg, 49 µL) was ground neat in a 10 mL grinding jar with a 12 mm stainless steel ball at 30 Hz for 180 min. The product was scraped off the walls of the grinding jar affording mixed thiourea–urea **20** in quantitative yield.

^{1}H NMR (300 MHz, d$_6$-DMSO): δ=6.97 (t, 1H, J=7.3 Hz, Ar), 7.05 (t, 1H, J=7.5 Hz, Ar), 7.16–7.47 (m, 8H, overlapped Ar), 7.58 (d, 2H, J=8.6 Hz, Ar), 7.87 (d, 1H, J=8.0 Hz, Ar), 8.09 (s, 1H, NH), 9.26 (s, 2H, overlapped NH), 10.00 (s, 1H, NH); ^{13}C NMR (75 MHz, d$_6$-DMSO): δ=118.1, 121.7, 121.8, 122.6, 125.2, 126.7, 128.27, 128.33, 128.66, 128.72, 129.5, 135.1, 138.3, 139.6, 152.7, 180.9; HRMS-MALDI found: 397.0878; calc. for $C_{21}H_{20}ClN_4OS_2$ (M+H$^+$): 397.0884.

9.2.6 DIELS–ALDER REACTION [10]

In the following experiment, the Diels–Alder reaction, as one of the most important construction reactions in organic synthesis, is proposed to demonstrate the application of ball milling in the field of pericyclic reactions. The students will broaden their knowledge on the scope of mechanochemically assisted organic reactions by recognizing the ability of ball milling to stereospecifically yield only *endo*-isomer in short reaction time under solvent-free conditions. These benefits are complemented by the simple and straightforward isolation of the norbornene product.

9.2.6.1 Endo-4-anisyl-4-aza-tricyclo[5.2.1.0^{2,6}]-dec-8-ene-3,5-dione (23)

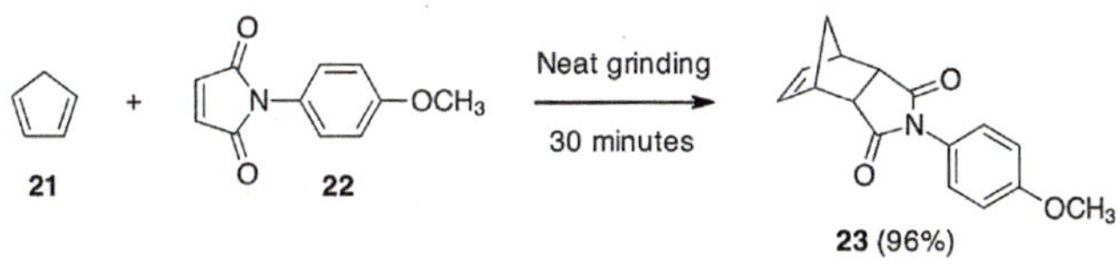

SCHEME 9.9

Solvnet-free mechanochemical Diels–Alder reaction.

Procedure: Freshly distilled cyclopentadiene (**21**) (5.2 mmol, 0.44 mL) and *N*-(4-methoxyphenyl)maleimide (**22**) (5.0 mmol), together with a stainless ball of 7.0 mm in diameter, were introduced into a 25 mL stainless jar. The grinding jar was then set to shake at 30 Hz for 30 min using a Retsch MM400 mill. TLC analysis on the resulting mixture demonstrated that the reaction proceeded completely and only the desired *endo*-product **23** was generated. Pure product was obtained by washing the jar with about 5 mL of hexane to remove the excess cyclopentadiene, the remaining solid was collected and dried in air (Scheme 9.9).

Melting point 170–172°C; IR (KBr): v = 2952, 2928, 1709, 1409, 1060, 934, 746 cm^{-1}; ^{1}H NMR (400 MHz, $CDCl_3$): δ = 1.59 (d, 1H, J = 8.8 Hz), 1.76 (d, 1H, J = 8.8 Hz), 3.40 (m, 2H), 3.48 (s, 2H), 3.78 (s, 3H), 6.24 (s, 2H), 6.93 (d, 2H, J = 8.9 Hz), 7.03 (d, 2H, J = 8.9 Hz); ^{13}C NMR (100 MHz, $CDCl_3$): δ = 45.5, 45.8, 52.2, 55.5, 114.4, 124.5, 127.9, 134.6, 159.5, 177.3; HRMS found: 269.1059; calc. for $C_{16}H_{15}NO_3$ (M^+): 269.1052.

REFERENCES

[1] Reed SM, Hutchison JE. Green chemistry in the organic teaching laboratory: an environmentally benign synthesis of adipic acid. J Chem Ed 2000;77:1627–9.

[2] McKenzie LC, Huffman LM, Hutchison JE. Greener solutions for the organic chemistry teaching lab: exploring the advantages of alternative reaction media. J Chem Ed 2009;86:488–93.

[3] Cave GWV, Raston CL. Green chemistry laboratory: benign synthesis of 4,6-diphenyl[2,2′]bipyridine via sequential solventless aldol and Michael addition reactions. J Chem Ed 2005;82:468–9.

[4] Leung SH, Angel SA. Solvent-free Wittig reaction: a green organic chemistry laboratory experiment. J Chem Ed 2004;81:1492–3.

[5] Klingensmith LM, Leadbeater NE. Ligand-free palladium catalysis of aryl coupling reactions facilitated by grinding. Tetrahedron Lett 2003;44:765–8.
[6] Hernández JG, Bolm C. $[Cp^{*}RhCl_2]_2$: mechanosynthesis and applications in C-H bond functionalisations under ball-milling conditions. Chem Commun 2015;51:12582–4.
[7] Ferguson M, Giri N, Huang X, Apperley D, James SL. One-pot two-step mechanochemical synthesis: ligand and complex preparation without isolating intermediates. Green Chem 2014;16:1374–82.
[8] Métro T-X, Bonnamour J, Reidon T, Sarpoulet J, Martinez J, Lamaty F. Mechanosynthesis of amides in the total absence of organic solvent from reaction to product recovery. Chem Commun 2012;48:11781–3.
[9] Štrukil V, Margetić D, Igrc MD, Eckert-Maksić M, Friščić T. Desymmetrisation of aromatic diamines and synthesis of non-symmetrical thiourea derivatives by click-mechanochemistry. Chem Commun 2012;48:9705–7.
[10] Zhang Z, Peng Z-W, Hao M-F, Gao J-G. Mechanochemical Diels–Alder cycloaddition reactions for straightforward synthesis of *endo*-norbornene derivatives. Synlett 2010;19:2895–8.

Author Index

N

O

P

X

Y

Z

Subject Index

'*Note:* Page numbers followed by "f" indicate figures and "t" indicate tables.'